The PWS
BookWare Companion Series ™

ADVANCED ENGINEERING
MATHEMATICS
USING MATLAB®V.4

The PWS
BookWare Companion Series ™

ADVANCED ENGINEERING MATHEMATICS
USING MATLAB® V.4

Thomas L. Harman

University of Houston/Clear Lake

James Dabney

Intermetrics

Norman Richert

University of Houston/Clear Lake

 PWS Publishing Company

I(T)P   An International Thomson Publishing Company

Boston • Albany • Bonn • Cincinnati • Detroit • London • Madrid • Melbourne • Mexico City
New York • Pacific Grove • Paris • San Francisco • Singapore • Tokyo • Toronto • Washington

PWS PUBLISHING COMPANY
20 Park Plaza, Boston, MA 02116-4324

MATLAB and PC MATLAB are trademarks of The Mathworks, Inc. The MathWorks, Inc. is the developer of MATLAB, the high-performance computational software introduced in this book. For further information on MATLAB and other MathWorks products—including SIMULINK™ and MATLAB Application Toolboxes for math and analysis, control system design, system identification, and other disciplines—contact The MathWorks at 24 Prim Park Way, Natick, MA 01760 (phone: 508-647-7000; fax: 508-647-7001; email: info@mathworks.com). You can also sign up to receive the MathWorks quarterly newsletter and register for the user group.Macintosh is a trademart of Apple Computer, Inc. MS-DOS is a trademark of Microsoft Corporation. BookWare Companion Series is a trademark of PWS Publishing Company.

I(T)P™
International Thomson Publishing
The trademark ITP is used under license

For more information, contact:

PWS Publishing Company
20 Park Plaza
Boston, MA 02116

International Thomson Publishing Europe
Berkshire House I68-I73
High Holborn
Loncon WC1V 7AA
England

Thomas Nelson Australia
102 Dodds Street
South Melbourne, 3205
Victoria, Australia

Nelson Canada
1120 Birchmount Road
Scarborough, Ontario
Canada M1K 5G4

International Thomson Editores
Campos Eliseos 385, Piso 7
Col. Polanco
11560 Mexico C.F., Mexico

International Thomson Publishing GmbH
Königswinterer Strasse 418
53227 Bonn, Germany

International Thomson Publishing Asia
221 Henderson Road
#05-10 Henderson Building
Singapore 0315

International Thomson Publishing Japan
Hirakawacho Kyyowa Building, 31
2-2-1 Hirakawacho
Chiyoda-ku, Tokyo 102
Japan

About the Cover : The BookWare Companion Series cover was created on a Macintosh Quadra 700, using Aldus FreeHand and Quark XPress. The surface plot on the cover, provided courtesy of The MathWorks, Inc., Natick, MA, was created with MATLAB® and was inserted on the cover mockup with a HP ScanJet IIP Scanner. It represents a surface created by assigning the values of different functions to specific matrix elements.

Editor: Bill Barter
Marketing Manager: Nathan Wilbur
Production Editor: Pamela Rockwell
Manufacturing Coordinator: Andrew Christensen
Cover Designer: Stuart Paterson, Image House, Inc.
Editorial Assistant: Monica Block
Cover Printer: Henry N. Sawyer, Inc.
Text Printer and Binder: Edwards Brothers, MI

Printed and bound in the United States of America.

99 00 — 10 9 8 7 6 5 4 3

Library of Congress Cataloging-in-Publication Data
Harman, Thomas L.
    Advanced engineering mathematics using MATLAB V.4 / Thomas L. Harman, James Dabney, Norman Richert.
        p.   cm.
    Includes bibliographical references and index.
    ISBN 0-534-94350-0
    1. Engineering mathematics—Data processing. 2. MATLAB. I. Dabney, James. II. Richert, Norman III. Title.
TA345.H32   1997
515 ' . 14 ' 028553—dc20   96-24466
                                                            CIP

A BC NOTE

Students learn in a number of ways and in a variety of settings. They learn through lectures, in informal study groups, or alone at their desks or in front of a computer terminal. Wherever the location, students learn most efficiently by solving problems, with frequent feedback from an instructor, following a worked-out problem as a model. Worked-out problems have a number of positive aspects. They can capture the essence of a key concept — often better than paragraphs of explanation. They provide methods for acquiring new knowledge and for evaluating its use. They provide a taste of real-life issues and demonstrate techniques for solving real problems. Most important, they encourage active participation in learning.

We created the BookWare Companion Series because we saw an unfulfilled need for computer-based learning tools that address the computational aspects of problem solving across the curriculum. The BC series concept was also shaped by other forces: a general agreement among instructors that students learn best when they are actively involved in their own learning, and the realization that textbooks have not kept up with or matched student learning needs. Educators and publishers are just beginning to understand that the amount of material crammed into most textbooks cannot be absorbed, let alone the knowledge to be mastered in four years of undergraduate study. Rather than attempting to teach students all the latest knowledge, colleges and universities are now striving to teach them to reason: to understand the relationships and connections between new information and existing knowledge; and to cultivate problem-solving skills, intuition, and critical thinking. The BookWare Companion Series was developed in response to this changing mission.

Specifically, the BookWare Companion Series was designed for educators who wish to integrate their curriculum with computer-based learning tools, and for students who find their current textbooks overwhelming. The former will find in the BookWare Companion Series the means by which to use powerful software tools to support their course activities, without having to customize the applications themselves. The latter will find relevant problems and examples quickly and easily and have instant electronic access to them.

We hope that the BC series will become a clearinghouse for the exchange of reliable teaching ideas and a baseline series for incorporating learning advances from emerging technologies. For example, we intend to reuse the kernel of each BC volume and add electronic scripts from other software programs as desired by customers. We are pursuing the addition of AI/Expert System technology to provide an intelligent tutoring capability for future iterations of BC volumes. We also anticipate a paperless environment in which BC content can flow freely over high-speed networks to support remote learning activities. In order for these and other goals to be realized, educators, students, software developers, network administrators, and publishers will need to communicate freely and actively with each other. We encourage you to participate in these exciting developments and become involved in the BC Series today. If you have an idea for improving the effectiveness of the BC concept, an example problem, a demonstration using software or multimedia, or an opportunity to explore, contact us.

Thank you one and all for your continuing support.

The PWS Electrical Engineering Team:

Bill_Barter@PWS.Com	Acquisitions Editor
Mary_ Thomas@PWS.Com	Developmental Editor
Nathan_Wilbur@PWS.Com	Marketing Manager
Pam_Rockwell@PWS.Com	Production Editor
Monica_Block@PWS.Com	Editorial Assistant

CONTENTS

*P*REFACE

This book may be used to supplement a traditional core text or as a non-traditional textbook as well as a resource for professionals. It presents mathematical techniques for applications in engineering, physics, and applied mathematics. The major topics are vector and matrix algebra, differential equations, Fourier analysis, advanced calculus, and partial differential equations. Our approach is to integrate analytical and computer solutions of problems that lead to understanding of advanced mathematical techniques. An important feature of the text is that there are many examples showing applications of mathematical principles using numerical and symbolic computer solutions.

Described in the following sections are the organization of the book and the files on the accompanying disk. Since the approach taken in this book represents a change from the traditional presentation of technical mathematics, two additional sections are included: "To Students and Professionals" explains the use of the text for self study; "To Instructors" is also provided to help instructors use the book successfully as a course text at a technical school or university.

As a programming language, we have chosen MATLAB because it is widely used both in industry and education. The language is especially popular in the engineering and applied mathematics communities. The basic MATLAB program performs a wide variety of numerical calculations and has an excellent graphics capability that is exploited in many examples in this text. Symbolic operations are possible with MATLAB's *Symbolic Math Toolbox*, which is based on Maple symbolic computation software. The disk includes MATLAB scripts for examples and solutions to problems as well as Maple scripts.

For the examples, we have created a number of MATLAB scripts (programs) that are contained on the disk accompanying this textbook. After each chapter, two sets of problems are included. One set of problems is called Reinforcement Exercises. Complete answers to these exercises are given in the text or on the disk if MATLAB scripts are necessary. The problems in the second set, called Exploration Problems, are generally intended to extend the knowledge presented in the chapters.

Prerequisites are a background in calculus and basic physics. Linear algebra would be useful, but it is not necessary since the important vector and matrix operations are developed in the book. Although expertise in programming is not required, some programming experience in a high-level language is assumed. In any case, MATLAB is straightforward to learn and use as a programming language.

BOOK ORGANIZATION

• Chapter 1 introduces MATLAB and describes many of its capabilities. Those readers familiar with MATLAB can cover this chapter very quickly.

• Chapter 2 discusses numbers and vectors including real and complex numbers as well as computer numbers. Vector spaces are also introduced. Chapter 3 treats matrices, matrix operations, and linear transformations. Eigenvalues and eigenvectors are the subjects of Chapter 4.

• Chapters 5 and 6 discuss differential equations. Chapter 5 is primarily concerned with linear differential equations with constant coefficients. Chapter 6 explains more advanced differential equations including those with variable coefficients, such as Bessel's and Legendre's equations.

• Chapter 7 presents various techniques to approximate functions and covers interpolation by polynomials and splines and least-squares curve fitting.

• Chapter 8 introduces Fourier series and Fourier transforms. Chapter 9 uses the discrete Fourier transform (DFT) and the fast Fourier transform (FFT) algorithm to compute Fourier transforms. The chapter also describes the errors involved in practical signal sampling.

• Chapter 10 begins the study of the calculus of multivariate functions and vector functions. The topics are typically included in an applied advanced calculus course.

• Chapter 11 presents the vector differential operators of gradient, divergence, curl, and Laplacian. Chapter 12 covers the integral calculus of multiple integrals, line integrals, and surface integrals and presents various applications of vector field theory.

• Chapter 13 presents an introduction to selected partial differential equations. Laplace's equation, the heat equation, and the wave equation are studied.

• The index includes a listing of the MATLAB commands used in the book.

CHAPTER ORGANIZATION

Each chapter begins with a preview, and, in addition to the text, contains numerous examples, exercises and problems, a bibliography, and answers to Reinforcement Exercises, and employs the following organizational elements:

Preview: An overview provides motivation for the material covered in the chapter.

Examples: Each chapter contains many examples to clarify the concepts and computational details. We consider the examples important reading that further explain or amplify the points being made in the textbook. For examples that use MATLAB, the script is included on the disk. Occasionally, a "What If?" discussion follows an example. These discussions are meant to stimulate thought and encourage a review of the pertinent material.

Reinforcement Exercises and Exploration Problems: The Reinforcement Exercises generally are straightforward problems that require application of the material in the chapter. In effect, these exercises are examples also, since complete answers are provided.

The Exploration Problems are intended to be more difficult than the Reinforcement Exercises. These problems were designed to extend the material covered in the chapter and to provide a greater challenge.

Annotated Bibliography: Sources of additional information are listed with a brief description of the reference.

Answers: The answers section includes results for all of the Reinforcement Exercises that do not require MATLAB scripts. Answers that require MATLAB scripts are on the disk.

SOFTWARE AND DISK FILES

———————————————— ■ ————————————————

MATLAB is an acronym for MATrix LABoratory. It is produced and supported by The MathWorks, Inc., of Natick, MA. The company provides a great deal of support, including technical assistance, technical notes, a periodic digest, and many MATLAB scripts at their FTP site (ftp.mathworks.com). More information is available at their World Wide Web (WWW) address (http://www.mathworks.com/).

The disk provided with this textbook contains MATLAB scripts for all the examples that use MATLAB and also scripts for the answers to Reinforcement Exercises that lend themselves to MATLAB solution. These scripts were developed using the *Student Edition of MATLAB*, Version 4, on a computer system using Windows.

Warning: Execution of the files with another version of MATLAB or another operating system may lead to error messages. Usually the problems can be resolved by modifying the scripts appropriately. Also, the text file (README.TXT) on the disk should be studied carefully before using the disk files.

MATLAB DISK FILES

The disk has a directory that contains two subdirectories for each chapter (CHn $n = 1, 2 \ldots, 13$) . These disk directories are as follows:

1. CLFLCHn ($n = 1, 2, \ldots, 13$) contain the MATLAB files that created MATLAB examples in the text. The prefix CLFL means *Clear Lake Files* with the name reflecting the fact that these files were created near Clear Lake in Houston where our university is located.

2. ANSCHn ($n = 1, 2, \ldots, 13$) contain MATLAB files that are answers to the Reinforcement Exercises that require MATLAB scripts or problems for which MATLAB scripts are helpful.

We attempted to design the examples in this book with the motto "The purpose of computing is insight, not numbers" in mind. This is the motto from R.W. Hamming's book *Numerical Methods for Scientists and Engineers*, McGraw-Hill, Inc. Also, the MATLAB scripts are written with clarity as the primary goal. Therefore, it is intended that the reader modify and improve the scripts in some cases, particularly with respect to the efficiency of the program in solving a specific problem.

MAPLE SUPPLEMENT

Maple is a programming language with numerical, graphical, and symbolic capabilities. As such, it is an excellent complement to the numerical capability of MATLAB. In fact, the MATLAB *Symbolic Math Toolbox* employs Maple for symbolic manipulation. A Maple supplement is included on the disk accompanying this book. The supplement contains Maple script files for most of the examples and Reinforcement Exercises in the textbook. For users of Maple, the scripts serve the same purpose as the MATLAB scripts included on the disk. The scripts were developed using Maple V. You may contact Waterloo Maple Software, Inc. of Waterloo, Ontario, Canada at their WWW address (http://www.maplesoft.com/) for more information about Maple.

ACKNOWLEDGEMENTS

—————————————◼—————————————

The authors would like to express their gratitude to all those who contributed to this textbook. The material and the scripts were developed over several semesters with many helpful suggestions from the students in the Engineering Applications class at the University of Houston-Clear Lake. In particular, Brenda Lee Moulton and Jim Watson gave a very complete review of the text. We also thank other reviewers whose suggestions improved the presentation including

>   Chris Frenzen, Naval Postgraduate School
>   Anastas Lazaridis, Widner University
>   Angrezej Manitius, George Mason University
>   Charles Neuman, Carnegie Mellon University

Our sincere appreciation goes to all the people who were involved in the production of the book. The editors and others at PWS Publishing Company were most helpful. Bill Barter and Mary T. Stone are especially to be thanked. Pam Rockwell did an excellent job of supervising production. We also wish to thank the technical staff at The MathWorks for their help. One other person deserving special mention is Jeanne Leslie. Of course, our families and friends are to be thanked for their support and patience.

We apologize to any other persons who helped in the endeavor but were not cited here. Please send any comments or criticisms to the authors at the University of Houston-Clear Lake, 2700 Bay Area Boulevard, Houston, Texas 77058.

TYPESETTING

The camera-ready copy for this book was set using the LaTeX typesetting macros developed for the PWS BookWare Companion Series by Norman Richert. Those interested in further information about LaTeX and TeX are encouraged to access the WWW site of the TeX Users Group (http://www.tug.org/).

T. L. Harman (harman@cl.uh.edu)
N. Richert (richert@cl.uh.edu)
J. B. Dabney (jdabney@owlnet.rice.edu)

To Students and Professionals

Our book is intended to present mathematical principles, concrete examples, and occasionally a MATLAB script (program) for plotting or calculation. It can be used to review various subjects in applied mathematics and this section explains the use of examples and MATLAB scripts assuming you are using the textbook for review purposes. The next section entitled *To Instructors* describes the use of the text as a course text or as a companion volume to another assigned textbook.

USE OF EXAMPLES AND WHAT IFS

We think that the most efficient learning comes from studying examples and then working problems with increasing levels of difficulty. Some examples are followed by "What If" questions that are intended to stimulate thinking about more intricate details of the concept being presented.

SPECIFIC
EXAMPLE

Consider Example 1.7 that requires the study of the function

$$y(t) = \frac{2\sqrt{3}}{9} e^{-4t} \sin\left(\omega t + \frac{\pi}{3}\right).$$

The text of the example discusses the mathematical properties of the function to determine the plotting parameters. Then, a MATLAB script is presented that plots the function. The "What If" question that follows the example asks you to determine the zero crossings of the function using the **zoom** command. This extends the basic concept and the script.

After an example, it is suggested that you complete some of the related Reinforcement Exercises. The answers to these problems are given at the end of each chapter. Exploration Problems are also included in each chapter. These generally require more thought and effort than the Reinforcement Exercises.

USE OF THE MATLAB DISK FILES

Although the book is structured so that basic mathematical principles can be studied without using MATLAB – many of the examples and the problems at the end of the chapters do not require MATLAB scripts– we have generally assumed that it will be used with the MATLAB (or Maple) software.

On the disk accompanying the text, a MATLAB script, also called an M-file, is provided for each of the examples and Reinforcement Exercises that uses MATLAB. This collection of MATLAB scripts can serve as a "library" of programs for mathematical applications. These scripts can be executed with the echo feature of MATLAB to see the effect of each instruction. Once the example in the text is reproduced and understood, you could execute the script with different input values or other changes to compare the results.

You are encouraged to use and modify the MATLAB scripts provided on the disk. We feel that computer work is a necessary part of the study of applied mathematics today, both as an aid to learning and as a technique useful in applications. The appropriate modifications might include adding additional comments that personalize the scripts. To increase efficiency, vectorized code could replace loops in some scripts. The ultimate goal should be to extend the application of the program to solve more sophisticated problems, perhaps those suggested in the Exploration Problems that require MATLAB. Before executing any programs, you are advised to read the readme file included on the disk.

To Instructors

In *Advanced Engineering Mathematics Using MATLAB*, we emphasize the explanation of mathematical principles and support those explanations by examples and simple MATLAB scripts. Since we omit most of the proofs to the theorems, and the examples are generally straightforward applications of the mathematics, the text is easy to read and substantially free of mathematical difficulties. The content, Examples, and Reinforcement Exercises are designed to present and reinforce the material we consider important for a general course in applied mathematics. On the other hand, the Exploration Problems in each chapter are included without answers to inspire the inquisitive students.

Described below are the conceptual structure of the textbook as it is divided into several sections; then, the use of MATLAB; and finally, suggestions for courses in which our textbook may be used as a primary textbook or as a companion volume to a more traditional book.

STRUCTURE OF THE TEXTBOOK

The first chapter is included for those students who are unfamiliar with MATLAB or those who wish to improve their use of the software. The design of the Examples, Reinforcement Exercises, and Exploration Problems involving MATLAB in other chapters is based on the assumption that the material in Chapter 1 is well understood. Portions of the MATLAB tutorials in *The Student Edition of MATLAB User's Guide* would be an appropriate supplement to Chapter 1.

LINEAR ALGEBRA AND DIFFERENTIAL EQUATIONS

Recognizing the increasing importance of applied linear algebra and computer techniques in mathematics instruction, we have included chapters on vectors, matrices, and eigenvalues and eigenvectors. These chapters serve to introduce important mathematical concepts as well as allow the student to take advantage of the computational ability of MATLAB. The emphasis that should be given to Chapter 2, Chapter 3, or Chapter 4 depends on the background of the students. However, the material in Chapter 4 covering eigenvalues and eigenvectors is useful for understanding parts of Chapters 5 and 6.

Differential equations are covered in the text in several chapters. Chapters 5 and 6 treat ordinary differential equations and Chapter 13 introduces partial differential equations. In Chapter 8, Fourier methods are applied to the solution of ordinary differential equations to determine the frequency response of systems.

APPROXIMATION OF FUNCTIONS

Chapter 7 presents several of the methods to approximate functions. Interpolation by both polynomial and spline fitting is discussed briefly. The equations of the least-squares approximation are presented in matrix form in the text. The vector form as a projection is treated in one of the Exploration Problems.

The discussion of orthogonal functions and polynomials begun in Chapter 2 and treated in Chapter 6 is continued in Chapter 7. The emphasis in Chapter 7 is on the properties of orthogonal functions useful for understanding Fourier series in Chapter 8.

FOURIER ANALYSIS

Chapter 8 presents the essentials of classical Fourier series and Fourier transform methods. MATLAB is used mainly to plot results calculated by hand. However, the errors involved in truncating the Fourier series and the Gibbs phenomenon are illustrated using MATLAB in the examples and in the problems at the end of the chapter.

The Fourier spectra for series and transforms are also emphasized in Chapter 8. Again, MATLAB plots are useful to help students visualize the results. As practical examples, both Fourier series and transforms are used to solve differential equations to determine the frequency response of a system modeled by a linear differential equation.

Chapter 9 presents the discrete Fourier transform (DFT) and the important fast Fourier transform (FFT) algorithm. The DFT is presented as an approximation to the Fourier transform and as a method of analyzing real signals. The errors involved in practical signal analysis, such as aliasing, are explored in some detail.

MULTIVARIATE AND VECTOR CALCULUS

Chapters 10, 11, and 12 describe multivariate and vector calculus and the ways in which MATLAB can be used to aid problem solving or plot useful results. For these chapters, the student is expected to be familiar with single-variable calculus.

Chapter 10 introduces differential calculus of functions of several variables. Various topics include limits, continuity, partial derivatives, the differential, the Jacobian, Taylor series, and finding extrema for functions of several variables. Vector functions and curvilinear coordinates are also presented. MATLAB scripts are used to plot multivariate functions to aid visualization as well as to perform other operations such as finding the minimum of a function.

Chapter 11 is concerned primarily with scalar and vector fields used to represent physical values that vary with position in space. The differential operators of gradient, divergence, curl, and Laplacian are studied in rectangular and curvilinear coordinate systems. Applications from mechanics, fluid flow, and electromagnetic theory are presented.

Chapter 12 begins with a brief review of integration and presents double and triple integrals in various coordinate systems. The Jacobian is used to explain change of variables in multiple integrals. Applications, numerical techniques, and MATLAB commands for integration are covered. Example applications of multiple integrals are two-dimensional Fourier transforms and joint probability functions.

Line and surface integrals are first defined in Chapter 12 as generalizations of ordinary integrals and then as line integrals and surface integrals of vector fields. The chapter next covers various theorems of vector integral calculus. The theorems of Green, Stokes, and Gauss are discussed and then applications are considered. Conservative forces, fluid flow, and Maxwell's equations of electromagnetic theory provide useful examples.

PARTIAL
DIFFERENTIAL
EQUATIONS

Chapter 13 introduces Laplace's partial differential equation, the heat equation, and the wave equation. The solution technique of separation of variables is emphasized since this chapter builds on the treatment of ordinary differential equations given in Chapter 6.

USE OF MATLAB

An important reason we chose MATLAB as the computer language for the book was its wide acceptance in the engineering community. The student edition is a relatively inexpensive investment for the student, particularly considering that the MATLAB program will be useful in many courses. Also, after graduation, students will probably encounter the professional version of MATLAB or a similar software package for calculation and simulation. The MathWorks provides Classroom Kits to instructors and it is possible to install MATLAB on the school's computer network for access by all students.

One advantage of MATLAB is that many of its functions (M-files) can be read as text files. For these functions, the algorithm is typically described in the script itself or in the *MATLAB User's Guide* so that interested students (or instructors) can study the computational methods in more detail. With a few exceptions, such as the Runge-Kutta algorithm in Chapter 6, we have not tried to analyze the algorithms in detail. We feel that such analysis properly belongs in a Numerical Methods course. Instead, we have tried to address the problems encountered when using the functions. Thus, the condition number of a matrix is considered in Chapter 2. As another example, our treatment of the fast Fourier transform (FFT) algorithm explains the errors made by improper sampling of a signal rather than numerical errors due to computation.

MATLAB TOOLBOXES

The *Student Edition of MATLAB* contains a *Symbolic Math Toolbox* and we have included examples to illustrate the symbolic capability of MATLAB in the textbook. We have found that students in a course will soon discover most of the symbolic commands and use them to advantage. The *Symbolic Math Toolbox* is really a MATLAB interface to a limited version of the Maple software. For Maple users, symbolic solutions to many examples and problems using Maple are included on the disk.

The *Student Edition of MATLAB* also contains the *Signals and Systems Toolbox*. These toolbox functions are useful for courses in Linear Systems or Digital Signal Processing.

One use of these toolboxes is to introduce them in a course that presents segments dealing with the mathematics of subjects such as communications, image processing, digital signal processing, or Fourier methods. This might be appropriate for an engineering mathematics course that provided the background for other courses in these applications areas.

MATLAB VERSUS MATHEMATICS

The fact that Chapter 1 introduces MATLAB might lead a reader to the (erroneous) conclusion that the textbook is really about MATLAB. This is not the case, and those instructors who wish to start immediately with the mathematics can begin with Chapter 2 and leave Chapter 1 for study by the students. If the students are familiar with MATLAB, they can try some of the Reinforcement Exercises or Exploration Problems at the end of Chapter 1 to test their skills.

It was our intention that the ratio of MATLAB to mathematics in the textbook should be such that mathematical principles rather than computations were emphasized. In a traditional mathematics course, the programming aspects could be downplayed and MATLAB could simply be used for calculations and plotting. Results derived by hand could be checked using commands from the *Symbolic Math Toolbox*.

However, more detailed programming assignments could be given in a course stressing MATLAB or in a computational laboratory course. Assigning some of the Exploration Problems involving MATLAB would be useful in such courses.

OTHER
LANGUAGES

Occasionally, in the authors' courses, students have used Mathematica or Maple to do the assignments from the textbook. These languages are not discussed in the textbook but the students have been successful in the course. Since these programs perform symbolic mathematical operations, some of the assignments using the numerical capabilities of MATLAB could be modified slightly to fit the situation.

SUGGESTIONS FOR COURSES

The material in the textbook has been used in one- and two-semester courses for students of engineering and physics. We include here suggestions for courses using the textbook as a primary text and as a companion volume.

1. PRIMARY
TEXTBOOK

If this textbook is used as a primary textbook, a one-semester introductory course in applied mathematics might cover the material in the first six or seven chapters. In two semesters, all of the book could be completed. Tables in this section list essential topics by section title in specific chapters. Suggestions for optional topics and topics that could be added are also included in the tables.

A more advanced one-semester course might start with Chapter 6 (Advanced Differential Equations) and cover most of the topics up through partial differential equations in Chapter 13. Assignments in such courses should include the Exploration Problems. For programming assignments, the MATLAB scripts included on the disk are a starting point. The examples or problems associated with them could be modified suitably and assigned as homework.

Students with a good background in linear algebra and calculus could start with Chapter 7 or Chapter 8 and finish most of the book in a one-semester course. For a more rigorous course, the instructor could, if desired, supply details of mathematical proofs and also cover additional topics not treated in our book. Additional topics are suggested in the chapter summaries to follow. Specific MATLAB commands are indicated by boldface type.

Chapter 1 Introduction (to MATLAB) The emphasis placed on this chapter depends on the background of the students and the purpose of the course. Table 0.1 lists the essential sections of Chapter 1 if MATLAB is to be discussed in detail. One essential point about MATLAB is that the array (matrix) is the basic data structure for MATLAB. For example, it is important to emphasize the difference in notation for element-by-element operations versus array operations. The discussion of the colon, semicolon, and period operators in Chapter 1 covers such notation. Many of the examples with scripts in Chapter 2 and Chapter 3 employ array manipulation. Another capability of MATLAB to emphasize is the extensive on-line help and the use of the **echo** command during execution of a script.

A number of the programming problems at the end of Chapter 1 employ MATLAB essentially as a calculator to give the students confidence in using the software. To highlight the programming language aspects of MATLAB, more complicated scripts such as those required by Problem 1.19 or Problem 1.20 could be assigned as homework.

If efficient programming is emphasized in the course, the difference between using loops and vectorizing is vital. Some of our MATLAB examples and the MATLAB solutions to Reinforcement Exercises can be improved in this regard. Improving the scripts on the disk gives the students additional motivation for studying the examples and exercises that use MATLAB.

TABLE 0.1 *Chapter 1 Topics*

Essential Sections	*Optional Topics*	*Topics To Add*
Introduction to MATLAB	**eval** (Example 1.3)	Programming project
MATLAB Commands for Display and Plotting	Problem Solving and Programming	Symbolic MATLAB
Creating MATLAB Programs		
MATLAB Programming Language		

As an optional topic, the MATLAB command **eval** is useful in some applications. The command is introduced in Example 1.3 and it should be compared to MATLAB functions later in the chapter after Example 1.8 is covered. The optional section "Problem Solving and Programming" might be discussed in a course involving a computational laboratory. A number of applications from physics or engineering might be combined to form a programming project for the semester. Also, the symbolic capability of the *Student Edition of MATLAB* could be presented and appropriate problems created and assigned.

In class, we try to make the point that MATLAB represents much more than just another programming language. Certainly MATLAB is a useful interactive environment for calculation. However, the capability for exploratory computing makes MATLAB the important program that it has become.

Chapters 2-4 Linear Algebra Topics   The motivation for including three chapters that cover topics from applied linear algebra is that modern applied mathematics combines calculus and algebra in many powerful algorithms that the computer has made feasible to apply. Thus, matrix methods are now used in applications to solve differential equations and in courses to explain algorithms such as the fast Fourier transform. Table 0.2 lists essential and optional topics if these chapters of the text are to be covered in detail in a course. Most of the topics in Chapter 2 and Chapter 3 are appropriate for such a course.

TABLE 0.2 *Linear Algebra Topics by Chapter*

Essential Topics	Optional Topics	Topics To Add
Chapter 2		
Most of the topics in this chapter	MATLAB Computer Numbers Vector Spaces of Functions	Expanded treatment of complex numbers
Chapter 3		
Most of the topics in this chapter	Homogeneous transformations	Discussion of numerical errors in matrix operations Symbolic matrix operations
Chapter 4		
General Discussion of Eigenvalues Eigenvalues and Eigenvectors Matrix Eigenvalue Theorems Complex Vectors and Matrices MATLAB Commands for Eigenvectors	Matrix Calculus Similar and Diagonalizable Matrices Special matrices Hermitian and unitary matrices Applications to Differential Equations	Motivational examples Computational methods

Chapter 2 briefly discusses MATLAB computer numbers and this section could be covered if the course emphasizes the numerical aspects and errors involved in computer computations. Vector spaces of functions is another optional topic; however, this section is useful background for the discussion of orthogonal functions in later chapters. A more complete discussion of complex numbers could be added. As now presented in Chapter 2, complex numbers are introduced and related to vectors.

Chapter 3 covers useful topics involving matrices. We consider the discussion of homogeneous transformations optional. Obviously, a great deal of material could be added since matrices play a role in many applications. A few suggestions are given in Table 0.2. In some courses, it would be appropriate to explore the numerical problems that arise when large matrices are treated. Such discussions could expand on the material concerning ill-conditioned matrices in the text. A starting point might be operations involving large Hilbert matrices or the Vandermonde matrix.

Chapter 4 can be presented in several ways. The introductory sections of the chapter are intended primarily to present the mathematics of eigenvalues and eigenvectors. In fact, the basic material about eigenvalues and eigenvectors was included mainly to serve as background material for differential equation problems in Chapter 5 and Chapter 6.

The resistor network used in Example 4.4 and Example 4.5 shows that the currents can be written as a linear combination of eigenvectors, as indicated in general by Equation 4.20. Later examples demonstrate the fact that the behavior of a vibrating system can be analyzed by knowing the frequencies and the mode shapes of oscillation, which are related to the eigenvalues and the eigenvectors of a dynamic system, respectively. To emphasize the analysis of mechanical systems, Example 4.18 and Example 5.18 with coupled masses or Example 6.12 showing the bending of a vertical column could be studied.

MATLAB has useful commands to compute eigenvalues and eigenvectors. Example 4.10 is worth some study. This example shows the use of iteration to compute the eigenvalues of a matrix using the QR decomposition method. The displayed results show how the matrix scheme converges and the relationship between the QR decomposition and the final eigenvalues. Such an iterative process could not reasonably be done by hand calculation. The students should be encouraged to execute the script (included on our disk) and observe each step as an aid to understanding the algorithm.

The sections on matrix calculus and the special matrices can be considered optional if matrix methods are not the main topics in a course. If matrix calculus is covered, the Cayley-Hamilton theorem could be studied in more detail with Example 4.17 illustrating the used of matrix calculus to solve a system of differential equations.

Other examples that could be added to this chapter include mechanical systems with a large number of degrees of freedom. Alternatively, such examples could be treated later after the chapters on differential equations. The generalized eigenvalue problem in the form

$$A\mathbf{x} = \lambda B\mathbf{x}$$

could also be presented and equations of the form $B\ddot{\mathbf{y}} + A\mathbf{y} = 0$ solved as additional examples in Chapter 5.

Since eigenvalues and eigenvectors are generally difficult to compute for large systems, numerical difficulties in computation could be discussed and suitable problems assigned.

Chapters 5-6 Differential Equation Topics   Chapter 5 is concerned with linear differential equations with constant coefficients. In class, we typically assign equations to be solved by hand and then encourage the use of MATLAB to study variations and plot comparative results. Chapter 6 considers more advanced differential equations and numerical solutions. Table 0.3 indicates essential and optional topics as well as possible additions for these chapters.

TABLE 0.3 *Differential Equation Topics by Chapter*

Essential Topics	*Optional Topics*	*Topics To Add*
Chapter 5		
Most of the topics in this chapter	State space representation	Various specific applications
Chapter 6		
Functions and Differential Equations	Numerical methods	Generalized inputs
Sequences and Series	Boundary Value Problems	Nonlinear equations
Taylor Series	Strum-Liouville equations	Stiff differential equations
Vector Equations		
Equations with Variable Coefficients		
Bessel and Legendre Equations		

Chapter 5 begins with a discussion of first-order equations; the MATLAB symbolic command **dsolve** is used to solve an equation with various initial conditions. Second-order differential equation solutions are presented next. Then, systems of differential equations are treated.

The section on transforming a differential equation into a system suitable for matrix solution should be treated with emphasis on Example 5.18 if students are unfamiliar with the technique. The MATLAB function **ode23** is introduced in this chapter but it is not discussed in detail until Chapter 6.

In Chapter 5, the discussion of state space could be skipped, particularly if the students are familiar with linear systems theory. Many topics could be added to the material including applications of specific interest to the students.

Chapter 6 treats differential equations with discontinuous inputs and equations with variable coefficients. A review of functions, sequences, and series may be necessary depending on the background of the students. Vector differential equations and equations with variable coefficients are the important topics to be covered. The Bessel and Legendre ordinary differential equations serve as examples. The solution functions are also used in Chapter 13 where partial differential equations are solved.

Optional topics in Chapter 6 include detailed discussion of numerical techniques to solve differential equations, boundary-value problems, and Sturm-Liouville equations.

Differential equations with generalized inputs such as impulse (delta) functions are not covered in the text. Neither are nonlinear or stiff differential equations. Such topics could be added to create a fairly complete course in ordinary differential equations.

Chapter 7 Approximation of Functions   The overall purpose of this chapter is to present several ways to represent a function in terms of other functions. The material covered would be selected according to the overall content of the course and particularly the emphasis placed on numerical methods. An application of curve fitting might be presented to motivate the students. Data values corrupted with noise could be used to simulate the measured values on a data acquisition channel. The MATLAB function **rand** can generate the simulated noise values.

If interpolation is to be covered, Example 7.1 should be discussed in detail. The example points out the errors that can arise using polynomials for interpolation over many points. Other interpolation schemes such as interpolation with arbitrarily spaced points (Lagrange interpolation) could be presented.

Spline fitting to data points eliminates the numerical instability associated with polynomial interpolation over a large number of points by using piecewise polynomial interpolation. We have omitted the derivation of the matrix equations for spline fitting but the equations are easily derived. One reason for the omission is that MATLAB has an efficient spline-fitting routine. Interested students can read the M-file (SPLINE.M) that is part of MATLAB. Various applications such as the use of splines in graphics could be presented in class or assigned as homework.

Chapter 7 also presents least-squares curve fitting and gives an example to show how MATLAB solves the problem. The Exploration Problems for Chapter 7 treat other aspects of the least-squares method. Since there are more equations than variables, the least-squares equations derived in the chapter are the normal equations of an overdetermined linear system. Problem 7.15 presents the vector (projection) solution to the least-squares equations.

The discussion of orthogonal functions should be reviewed before covering Fourier series in Chapter 8. Theorem 7.1 defines the Fourier coefficients as giving the best least-squares fit when a function is approximated by a sum of orthogonal functions. Example 7.4 and the discussion of orthogonal functions show that the famous Fourier trigonometric series is but one of many orthogonal series. Convergence of a series of orthogonal functions and Parseval's equality are other useful concepts to be employed also in Chapter 8.

Chapter 8-9 Fourier Topics   Chapter 8 presents Fourier series and Fourier transforms for continuous functions. Computer techniques for Fourier analysis are treated in Chapter 9. These two chapters are intended to be covered as a unit since Chapter 8 concentrates on theoretical results and Chapter 9 emphasizes the practical use of Fourier methods. Table 0.4 lists topics from the two chapters.

TABLE 0.4 *Fourier Topics by Chapter*

Essential Sections	Optional Topics	Topics To Add
Chapter 8		
Fourier Series	Power in a signal	Symbolic calculations
Properties of Fourier Series	Energy in a signal	Convolution
Fourier Transforms	Frequency response	Other transforms
		Complex analysis
Chapter 9		
Frequency Analysis of Signals	Practical Signal Sampling	FFT algorithm
Discrete and Fast Fourier	and DFT Errors	Difference equations
Transform		z-transforms
MATLAB Fourier Commands		
Practical Signal Analysis		

In class, we usually introduce these chapters with several motivational examples to explain the purpose of Fourier techniques. A key point in any example should be the mathematical and physical insights that are possible. Table 8.3 lists a number of applications for Fourier transforms.

Before covering Fourier series, the students should be familiar with the material on orthogonal functions in Chapters 2, 6 and 7. In our Fourier series examples, such as Example 8.2, MATLAB is used to plot analytical results only. However, MATLAB is ideal to sum the Fourier series for different numbers of terms and compare the results. In particular, the Gibbs phenomenon could be investigated in detail to show that the amplitude of the overshoot does not diminish with the number of terms in the series. Example 13.3, in which symbolic MATLAB commands are used, shows another approach to calculating Fourier coefficients. In fact, the *Symbolic Math Toolbox* has commands to compute Fourier, Laplace, and z transforms symbolically.

The treatment of the Fourier transform introduces the basic concepts and shows an application to a linear system by computing the frequency response. If linear systems theory is to be reviewed, the convolution theorem fits well here. Alternatively, a number of other examples could be presented after the transform is introduced.

Also, other transforms, such as the Laplace transform, could be studied as an extension of the material in Chapter 8. Various applications and other transforms were not included because courses such as Linear System Analysis, Digital Signal Processing, and other specific engineering courses typically cover Fourier and related methods in great detail.

For those interested in image processing or Fourier optics, the two dimensional transform is presented in Chapter 12. These students might also appreciate a discussion of wavelets, which are orthogonal functions that are a "hot topic" in certain applications areas. A good starting point for those not familiar with wavelets is the Internet. Also, the Math-Works offers a *Wavelet Toolbox* that works with the professional version of MATLAB.

Another topic to discuss might be functions, such as the unit step function, for which the ordinary Fourier transform does not exist. This is a useful function to compare Fourier and Laplace transforms. Transforms of generalized functions (impulse) deserve mention if linear systems applications are being stressed.

Chapter 9 is intended to present computer techniques to calculate the Fourier transform and, additionally, to introduce some aspects of practical signal analysis. For example, the limitations of the fast Fourier transform (FFT) algorithm in terms of the maximum frequency that can be analyzed should be stressed. Also, students should realize that the fact that the signal processed by the FFT and the resulting spectrum are effectively periodic can lead to misinterpretation of the results.

Understanding the phenomenon of aliasing in signal sampling and the consequences of the sampling theorem is necessary to properly apply the FFT to real signals. Example 9.6 demonstrates the error that can occur.

As an added topic, the FFT algorithm could be explained as a matrix with complex numbers as elements. The properties of complex numbers greatly reduces the computation involved. However, part of the efficiency of the algorithm also depends on clever indexing when the data points are being processed. We did not include these details because use of the FFT has become such a standard technique. The selection of the FFT parameters, such as the number of points to analyze, and the meaning of the results were considered the important features to present.

Discrete-time systems (difference equations) and z transforms could be presented as part of the discussion of discrete Fourier techniques. If signal processing is to be stressed in a course, programming projects could be assigned based on the *Signals and Systems Toolbox* of the MATLAB Student Edition. It is instructive to demonstrate the errors involved in signal sampling and to explore filtering or windowing techniques to reduce these errors.

Complex Analysis After deliberation, we decided not to include specific sections in the text that treat complex variables and complex integration. If the students need the material, complex integration could be presented and used to compute or invert Fourier and Laplace integrals. If only complex numbers are to be studied in detail in a course, the theory could be presented to help explain the fast Fourier Transform algorithm used in Chapter 9. Other suggestions for including complex analysis in a course are given in connection with the discussion of Chapter 12 and Chapter 13 below.

Chapters 10–12 Multivariate and Vector Calculus Topics In a course in applied mathematics beyond calculus, it is appropriate to cover almost all of the topics in Chapters 10, 11, and 12. The text contains the theory and a number of concrete examples for a one-semester course in the calculus of functions of several variables and vector analysis.

For successful completion of these chapters, it is assumed that the student is proficient in ordinary calculus of functions of a single variable. Students should know, or review, the basic rules of differentiation and integrations and techniques such as integration by parts. There is some review material in these chapters and elsewhere in the book. For example, Taylor series is presented in Chapter 6. Students should also be familiar with operations with vectors as treated in Chapter 2.

MATLAB is used extensively to plot functions in Chapters 10–12. For instance, MATLAB plotting commands are used to illustrate contour and surface plots in Example 10.1. Two-dimensional interpolation using MATLAB is demonstrated in Example 10.9. As shown in Example 10.14, plotting a function can help in determining the initial search point when finding the minimum. Plots of vector functions are shown in Chapter 11.

Numerical techniques for differentiation and integration are discussed here but not in great detail. This aspect of MATLAB could be emphasized, or alternatively the symbolic capability of the *Symbolic Math Toolbox*, which includes commands for symbolic differentiation and integration, computation of Jacobians, and Taylor series expansion, could be explored. Examples show differentiation (Example 10.11) and integration (Example 12.8) with MATLAB commands.

Chapter 10 Chapter 10 introduces functions of several variables. The main topics include partial derivatives, Jacobians for transformations, differentials and linear approximations, two-dimensional Taylor series, and extrema of multivariate functions. The section entitled "Vector Functions and Curvilinear Coordinates" can be stressed and expanded if desired. The electric dipole example (Example 10.16) may appeal only to electrical engineering students; other examples from mechanics or gravitation theory could be substituted to illustrate Taylor's expansion in coordinate systems other than rectangular.

Chapter 11 The chapter begins with a discussion of vector and scalar fields and introduces early on various MATLAB commands for vector calculus and plotting of fields. The gradient is derived after the directional derivative is explained. Students can use MATLAB commands to compute the numerical gradient (**gradient**) or the symbolic gradient (**jacobian**). If physical applications are to be stressed, we find it useful to present a thorough discussion of potential functions and Example 11.5.

Before vector field theory and its applications are presented, the chapter defines the divergence, curl, and Laplacian operators as well as Laplace's equation. A later section presents these operators in curvilinear coordinates.

A discussion of irrotational and solenoidal fields serves to introduce vector field theory. Helmholtz's theorem is presented to define a general vector field. As promised in the text of Chapter 11, vector fields are studied in more detail in Chapter 12, where the integral operations for vector fields are developed. The section on physical applications presents examples involving energy and work, fluid flow, and electromagnetic theory. Certainly many other examples could be presented but it is perhaps better to present the theorems of vector field theory in Chapter 12 before more complicated applications are studied. However, a simple example from electrostatics using Laplace's equation is presented in Example 11.13.

Chapter 12 Chapter 12 covers many applications of integrals. Its two primary purposes are to explore uses of single and multiple integrals and to illustrate the integral theorems of vector field theory. Summaries of these topics are presented in Table 12.1, Table 12.2, and Table 12.4.

Beginning with a discussion of integration and applications of single integrals such as calculations of arc length, the chapter then treats double and triple integrals, including change of variables formulas using the Jacobians of the transformations. Applications include standard problems such as finding volumes and mass as well as Fourier transforms and joint probability.

As indicated in Table 12.3, MATLAB provides commands for numerical and symbolic integration. Example 12.8 shows double integration using symbolic commands. Other MATLAB scripts for multidimensional integration are available on the Internet.

Line and surface integrals are explained as generalizations of ordinary integrals. These integrals are applied to vector fields to relate line integrals to work and surface integrals to flux. The chapter then relates vector differential calculus from Chapter 11 and vector integral calculus by means of the theorems of Green, Gauss, and Stokes. A variety of applications is presented including Maxwell's equations of electromagnetism.

A study of analytic functions, Cauchy-Riemann equations, and complex integration could be added after the material in Chapter 12 has been covered. Examples could be created by showing how to invert Fourier and Laplace integrals. Improper integrals might also be included.

Chapter 13 Partial Differential Equation Topics Applications of Laplace's equation, the heat equation, and the wave equation serve to introduce partial differential equations. Laplace's equation is solved in rectangular, cylindrical, and spherical coordinate systems. The Bessel functions for the cylindrical solution and the Legendre functions that result in spherical coordinates were discussed in some detail in Chapter 6.

The material in Chapter 13 also serves to tie together many of the topics presented earlier in the textbook, as Example 13.2, which uses the separation of variables technique to solve Laplace's equation, demonstrates. Complex analytic functions might be discussed as an addition to this chapter by showing such solutions to Laplace's equation. Conformal mapping techniques can be presented to solve the equation subject to various geometry of the boundary. For example, Laplace's equation in a 45° wedge is simplified by the conformal mapping that doubles angles. Poisson's equation solved by Green's function, an approach used in physics, could be presented. Engineers often refer to the result as the impulse response of the system.

2. COMPANION TEXTBOOK

According to the needs of the students and the interest of the instructor, the text can be used in a problem-solving course that incorporates MATLAB. With this approach, the book would be used as a companion volume with a more traditional textbook covering applied mathematics. Table 0.5 lists some of the many possibilities.

TABLE 0.5 *Table of Courses for Companion Volume*

Course	*Topics*
Applied Linear Algebra	Chapter 2, Chapter 3, Chapter 4 Parts of Chapter 7
Differential Equations	Chapter 5, Chapter 6 Chapter 8 (Fourier Solution) Chapter 13 (Partial DEs)
Fourier Analysis	Orthogonal functions (Chapters 2 and 7) Chapter 8, Chapter 9 2D Fourier (Chapter 10)
Advanced Calculus	Series (Chapter 6) Chapter 10, Chapter 11, Chapter 12
Applied Mathematics	Selected topics from Chapters 2-4 Series (Chapter 6), Fourier (Chapter 8) Chapter 10, Chapter 11, Chapter 12

PRACTICAL SUGGESTIONS

————————————————■————————————————

This section gives a few practical suggestions for using the textbook.

PREVIEWS

The Preview for each chapter is intended to provide physical or mathematical motivation for the material in the chapter. In some cases, we have emphasized the use of mathematical techniques to solve problems in electrical engineering. However, we also provide applications to mechanics and other areas of physics. Other suitable examples and discussion could be added if the students are better motivated by applications within a certain discipline.

USE OF
THEOREMS
AND PROOFS

The proofs of most mathematical theorems presented in the textbook have been omitted to make the text more readable to those who wish to emphasize applications and computations. In any case, the theorems are usually explained by examples or MATLAB programs.

Occasionally, a constructive proof is supplied, such as that for the theorem that relates linear transformations and matrices in Chapter 3. The proofs of other assertions can be supplied at the discretion of the instructor

REINFORCEMENT
EXERCISES AND
EXPLORATION
PROBLEMS

It was decided to include two classes of problems after each chapter: the Reinforcement Exercises contain generally "drill" type questions; the Exploration Problems extend the difficulty and range of questions.

The Reinforcement Exercises are intended to give students confidence in problem solving and in the use of MATLAB before attempting the more challenging Exploration Problems. Complete analytical answers or MATLAB programs are provided for the Reinforcement Exercises. We recommend that students be assigned these problems as nongraded homework except perhaps in cases where an answer as a MATLAB script on the disk is to be improved or modified.

The Exploration Problems are intended to cover more difficult applications or elaborate on theory not covered in detail in the text. These represent a starting point for exploring concepts that are the ultimate goal of applied mathematics. The emphasis placed on these problems depends on the way the textbook is used and the amount of rigor and difficulty suitable to the course.

Occasionally, problems require reference to other sources or literature searches. These problems can be assigned if appropriate. For specific applications, projects that combine various problems could be assigned. This approach can be particularly useful if the emphasis is on a particular discipline, such as electrical or mechanical engineering.

EXAMPLES AND DISK FILES　　The previous discussion *To Students and Professionals* explains the relationship of the Examples and What If questions as well as the use of the MATLAB disk files.

1   *INTRODUCTION*

MATLAB is an interactive computer program containing various commands and functions that aid mathematical analysis of problems in engineering and science. This first chapter is devoted to a discussion of the student version of MATLAB and introduces a selection of its many features. MATLAB examples will be presented throughout the textbook and applied to vector and matrix operations, ordinary differential equations, Fourier analysis, vector calculus, and partial differential equations.

Table 1.1 lists the general features of MATLAB and presents examples of the many operations that are possible. As indicated in the table, MATLAB is used for calculation, graphics, mathematical analysis, and programming.

For calculations, an important feature of MATLAB is that the program can perform valid mathematical operations using not only real and complex numbers but also vectors and matrices. MATLAB also has excellent graphics capability, enabling the user to visualize complicated functions in a variety of ways to aid analysis. The functions for mathematical analysis are useful for problem solving in areas such as calculus, differential equations, and linear algebra as well as more advanced studies. Finally, MATLAB is a complete programming language. Thus, programs can be created to solve many problems of practical interest.

TABLE 1.1 *MATLAB features*

Feature	Examples
Calculator:	
Arithmetic and trigonometric operations	$x \pm y$, xy, x/y, $\sqrt{x}$, e^x, $\log x$, x^y, $\sin x$, $\cos x$
Graphics:	
Two- and three-dimensional graphs	Plot $f(x)$, $f(x,y)$, and $f(x,y,z)$
Mathematical Analysis:	
Calculus	Differentiation and integration
Differential equations	Solve differential equations
General	Curve fitting and interpolation,
Linear algebra	Solve linear systems of equations
Special functions	Bessel and Legendre functions
Transforms	Fourier, Laplace, and z transforms
Statistics	Mean and standard deviation
Programming Language:	
Decision-making instructions	For, while, and if-then-else statements
I/O transfer	Load and save disk files
Symbolic Operations	*Symbolic Math Toolbox*
Signal Processing	*Signals and Systems Toolbox*

The *Student Edition of MATLAB* (Version 4) used for examples in this book also contains a set of commands (functions) for operations on variables and expressions defined symbolically. Collectively, these commands comprise the *Symbolic Math Toolbox*. There is also a *Signals and Systems Toolbox*. This toolbox is most useful in the study of signal processing or linear systems.

This chapter begins with descriptions of the basic mathematical and graphical capability of MATLAB. Then, several sections are devoted to MATLAB programming techniques. Many of the MATLAB features discussed in this introductory chapter are treated in more detail in later chapters. For example, the features and limitations of MATLAB for operations on numbers and vectors are discussed extensively in Chapter 2. Matrix operations are covered in Chapter 3.

This first chapter is intended to be somewhat interactive in nature. We suggest that the reader use the MATLAB program itself and the *MATLAB User's Guide* available from The Math Works, Inc., along with our text to clarify the points being discussed.

INTRODUCTION TO MATLAB

———————— ■ ————————

MATLAB is an interactive environment for numeric and symbolic computation. The program provides a mathematical library for matrix operations, the solution of differential equations, data analysis, and signal processing. The various operations are performed when MATLAB *commands* are recognized and executed. There are also a number of graphical commands that allow the display and visual analysis of data in two and three dimensions. In addition to the mathematical, scientific, and engineering functions that are the basic operations provided by MATLAB, the commands can be combined into programs to solve realistic problems.

The description in this chapter applies mainly to the student edition of MATLAB as used on personal computers (PCs) with the Windows operating system. Versions of the program for other types of computers may not have exactly the same features. The reader should refer to the MATLAB *User's Guide* for a specific computer to determine any differences. In the PC version, MATLAB commands can be entered through the keyboard or MATLAB can execute special programs called M-files to perform various operations, such as invert a matrix, evaluate an integral, or solve a system of differential equations and plot the results.

Since the student edition of MATLAB was introduced in 1991, it has been improved and a number of commands have been added to the original version. One important added capability is the ability to solve mathematical problems in symbolic form using the *Symbolic Math Toolbox*.

NOTATION

In the descriptions in this text, MATLAB commands are shown in boldface type, such as **help**, which displays the available MATLAB topics for which help is available. However, in examples that present executable MATLAB commands, the commands are shown exactly as typed, such as `help`. The names of files stored on disk are designated by capital letters.

The examples in this chapter illustrate the use of basic MATLAB commands. In later chapters, other commands will be defined as they are introduced. However, many of the commands have a number of variations that are not discussed in this text. These variations are explained in the MATLAB *User's Guide* or by the on-line help notes that are part of MATLAB itself. The inexperienced MATLAB user should study this chapter while experimenting with the commands on the computer.

MATLAB STRUCTURE

MATLAB executes commands entered from the keyboard or from files stored on disk. The variables used in a problem can be defined via keyboard input or read from disk files also. The structure of the MATLAB file system is shown in Figure 1.1.

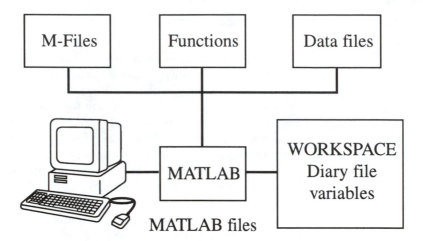

FIGURE 1.1 *MATLAB structure*

This figure shows the general relationship between the MATLAB program and various files on disk. The M-files and function files contain MATLAB commands to be executed when a file is called by entering the name of the file in response to the MATLAB prompt. A collection of MATLAB commands is usually called a MATLAB *script*.

Variables used during a MATLAB session are held in the workspace. The data files result when variables are saved with the **save** command. These variables can be loaded into the workspace with the **load** command. The diary file contains a copy of the commands executed and the results obtained during a MATLAB session.

A MATLAB SESSION

A MATLAB *session* is said to begin when the program prompts the user for a command with the >> prompt on the computer screen. This screen is MATLAB's *command window*. Then, the user may input variables or execute any one of the available commands to perform calculations, solve equations, and display or plot the results. The variables used in the session are held in the *workspace*, which is created whenever MATLAB is invoked. The computer directory in which variables and other data are saved in the current session is considered the *working directory*.

MATLAB responds to commands of the form:

>> <**command**>(CR)
>> <variable> = <**command**>(CR)
>> [<variables >] = <**command**>(CR)

The notation <**command**> means that any valid command can be substituted for **command** in the list just shown. This same notation is used for the variables. Since many commands require input values as

---

Chapter 1 ■ INTRODUCTION

arguments for the command, the reader should consult the MATLAB *User's Guide* for a complete discussion of any command.

After a command followed by a carriage return (CR) is entered at the keyboard, MATLAB will perform the specified action. Subsequent discussions will omit the CR from the text. In MATLAB examples, the CR is not printed since it is not displayed in an actual MATLAB session.

As an example, the command

```
>>demo
```

passes control to the demonstration program included with MATLAB. This is a good command to execute when learning MATLAB since the **demo** programs show many of the capabilities for calculation and plotting.

Other commands, such as **exp**, compute a value. The second form of a command assigns a value to a variable. Thus, the command

```
>>y = exp(x)
```

assigns the value of e^x to the variable y. Commands of this type require an argument. In the case of **exp**, the argument can be a scalar variable, a row of numbers corresponding to a vector, or an array of numbers entered row by row. The array corresponds to a matrix of numbers and it represents the fundamental variable type for MATLAB. For the vector and the array, the **exp** command computes the exponential value for each element.

Example 1.1 illustrates the forms of various MATLAB commands. The method of entering vector and array variables is shown in Example 1.2.

BUILT-IN FUNCTIONS AND M-FILES

The MATLAB program executes commands to perform specified operations. For example, executing the command

```
>>exp(x)
```

causes the function **exp** in MATLAB's directory of functions to calculate and display the value of e^x. Many functions, such as **exp**, are called *built-in functions* because they are a fundamental part of the MATLAB program. For these functions, the user cannot modify or even read the MATLAB computer code that performs the operation. The only information for built-in functions is obtained by reading the MATLAB *User's Guide* or using the **help** command during a session. Try the command

```
>>help exp
```

MATLAB can also read and execute programs called M-files that are stored on the computer's disk. The M-file is a text file called a script that contains sequences of MATLAB commands. Contents of the file can be displayed on the computer screen or printed for inspection. The commands in the script are executed when the name of the M-file is entered at the keyboard.

As far as the reader is concerned, the difference between the MATLAB built-in functions and M-files is not important at this point. Later in the chapter we will describe MATLAB programs and show how to create M-files.

□ EXAMPLE 1.1   *MATLAB Commands*

This example demonstrates the use of the three forms of MATLAB commands during a MATLAB session. The keystrokes to input commands and the MATLAB responses are reproduced in the accompanying MATLAB session. The typed commands and the responses were saved in a diary file on our computer's disk. The diary file was created by the command **diary**, as explained later. Only the example number was added before the record of the MATLAB session. In the text, statements preceded by >> were typed by the author. Lines that begin with a % are comments.

MATLAB Script _____

```
Example 1.1
>>% Sample MATLAB commands
>>format compact
>>% Assignment and Computation
>>x=1.0
x =
    1
>>exp(x)
ans =
    2.7183
% Assign the result to the variable y
>>y=exp(x);
>>y
y =
    2.7183
% Computer type and maximum size of arrays
>>[ctype,maxsize]=computer
ctype =
PCWIN
maxsize =
       8192
>>quit
```

The command **format compact** suppresses blank lines between MATLAB responses to commands. After x is assigned the value 1.0, the **exp** command computes e^x, with the result 2.7183, as expected. Since no variable was defined as the result, MATLAB assigns the result to the variable **ans**. Each calculation without a variable name assignment will be stored in **ans**. Typing

>>ans

will display the last value computed. The standard display of numbers shows four digits to the right of the decimal point.

Next a scalar variable y is defined as e^x. Notice the semicolon (;) following the assignment of y. This suppresses echoing the display of y on the screen when y is computed. The variable y is saved in the workspace and is available any time during the session. Typing y at the MATLAB prompt displays the value.

The command **computer** generates two results, indicating the type of computer (ctype) and the maximum size of an array allowed (maxsize). This result is for the student edition of MATLAB executed on a PC running Windows. The maximum size array could contain 8192 elements.

The variables x and y and the variables ctype and maxsize are saved during the current session only. After the **quit** command, the variables are removed from the workspace and the MATLAB session is finished.

□

WHAT IF?   If you are new to MATLAB, create a diary file for the MATLAB session of Example 1.1. For example, the command

```
>>diary examp1.dir
```

will open a diary file on the disk. Type the commands shown in the script and end the example with the command **diary off**. Then, display or print the diary file.

This simple exercise is to verify that MATLAB is working properly on your computer. If you encounter difficulties, consult the installation instructions in the MATLAB *User's Manual* for your computer.

USEFUL GENERAL COMMANDS

There are commands useful for general operations during a MATLAB session. Several of these are defined in Table 1.2.

TABLE 1.2 *MATLAB general commands*

Command	Action
demo	Executes the demonstration program
type	Displays the text of an M-file
help	Displays the help topics
what	Lists M-files in the working directory
dir	Displays the names of files in the working directory
diary	Creates a file with a copy of the current MATLAB session
quit	Ends the MATLAB session without saving any information

The M-file `filename.m` can be displayed with the command

```
>>type <filename>
```

The command

```
>>help <filename>
```

displays the comment lines (if any) at the beginning of the M-file. The command **what** lists only the M-files in the working directory. The directory command **dir** (directory) lists the names of all the files.

A diary file is created as a disk file with the **diary** command. The **diary** command was used to create the examples in this book that show MATLAB sessions. It is particularly helpful to specify a diary during debugging and testing of MATLAB programs. The diary contains all of the commands and data input via the keyboard as well as MATLAB's responses. The command **quit** terminates the current session.

Additional Help Several methods of getting help are available with MATLAB in addition to using the **help** command. Try typing **help help** and observe the result. The **lookfor** command searches through MATLAB files and returns the name and first line of a command containing a keyword. For example,

```
>> lookfor exponential
```

lists several files. Users of the Windows operating system can also get help on a variety of topics by using the Windows Help menu.

MATLAB VARIABLES

There are a number of ways to define the variables for MATLAB. The simplest way is to enter data from the keyboard. The variables can be scalars, vectors, arrays (matrices), or strings. Strings are variables containing text in the ASCII format. This is a standard format that is used to represent alphanumeric characters for input and display.

□ **EXAMPLE 1.2** *MATLAB Variables*

This example shows an interactive session to define a number of variables by typing in the values. The MATLAB session was recorded on a disk file using the **diary** command and the diary file that resulted is reproduced as the accompanying MATLAB session.

In the session, we define the values of a scalar named "scalar," a vector named "vector," and a matrix named "array." After the definition of a variable is entered, MATLAB echoes (displays) the values as shown, since the statement does not end with a semicolon. The command **quit** ends the session.

MATLAB Script _____

```
Example 1.2
>>% MATLAB Variables

>>format compact
>>scalar=3.5
scalar =
    3.5000
>>vector=[1 2 2 4 1]
vector =
    1    2    2    4    1
>>array=[1 1 3;3 4.0 2;1 5 1]
```

```
array =
     1     1     3
     3     4     2
     1     5     1
>>quit
```

□

WHAT IF?   Try the command **exp** with scalar, vector, and array variables, respectively, and compare the result. Does the command work as you expected?

Naming Variables   MATLAB variables can contain up to 19 characters, but they must begin with a letter. The remaining characters can be letters, numbers, or underscores. Thus, variables can have any names that are convenient to represent the variables being defined. However, when MATLAB executes, commands and variables are *case sensitive* to uppercase and lowercase letters. Thus, the variables `scalar` and `Scalar` are different variables.

Variables in the Workspace   The important commands that govern the manipulation of variables in the workspace are defined in Table 1.3.

TABLE 1.3 *MATLAB commands for variables*

Command	*Action*
clear	Clears the variables in the workspace
save	Saves variables in a file
load	Loads variables from a file
whos	Lists the variables in the workspace and their size

The **clear** command deletes any variables in the current workspace. Before exiting MATLAB, variables in the workspace may be saved in a file with the command

```
>>save <filename>
```

which allows the same variables to be used in a later MATLAB session. The **load** command in the form

```
>>load <filename>
```

reloads the variables during a subsequent session. The command **whos** lists the variables in the current workspace and defines their size. In Example 1.2, `vector` has size 1×5 and `array` has size 3×3.

String Variables A string variable contains readable text used for prompting for input data and labeling screen displays of data as well as creating special commands. Such variables are sometimes called *strings* or simply *text* for simplicity.

A string variable must be input as `'<string>'`, and must be enclosed by single apostrophes. Commands in Table 1.4 are useful for the display and evaluation of strings.

TABLE 1.4 *MATLAB commands for strings*

Command	Action
disp	Displays text or variables
eval	Evaluates the text in a string
fprintf	Displays a string on the screen

Text and the results of computations can be shown on the screen using the display command **disp**. The **eval** evaluates a text string and interprets it as a MATLAB command or a value. The **fprintf** and **disp** commands allow screen display of text and data. They are important for prompting the user or explaining the data displayed. The use of these commands is demonstrated later in the chapter.

□ EXAMPLE 1.3   *MATLAB* eval *Example (Optional)*

In some problems, it is convenient to define functions and values to be text strings, as shown in the accompanying MATLAB session. The use of the **eval** command allows these text strings to be executed as MATLAB commands. As an example, the MATLAB script using string variables defines the variable f1 to be the string `exp(x)`. The numerical values are computed if x is defined and the **eval** command is executed with f1 as argument.

In this session, the function e^x was first evaluated using **eval** at three points defined by the array variable x. The command y=exp(x) accomplishes the same result. However, if x is changed, **eval** can be used to compute the new values without retyping the function. The variable x can be a scalar, vector, or an array in subsequent definitions, as shown when x is redefined as a scalar in the script.

MATLAB Script ———————————————————
```
Example 1.3
>>% Use of Strings and eval
>>f1='exp(x)' % Define a string function
f1 =
exp(x)
```

```
>>x=[1 2 3]
x =
     1     2     3
>>eval(f1) % Compute the results
ans =
    2.7183    7.3891   20.0855
>> y=exp(x) % Compute the exponential without eval
y =
    2.7183    7.3891   20.0855
>>% ------------------
>>% Redefine x to be a scalar variable
>>x=0.5
x =
    0.5000
>>eval(f1)
ans =
    1.6487
```

□

The command **eval** is better employed when a complicated formula is to be repeatedly evaluated during a MATLAB session.

□ EXAMPLE 1.4   *Another Use of* eval *(Optional)*

The accompanying MATLAB session shows the definition of a polynomial as a string and evaluation of the polynomial at several points. The polynomial is

$$f(x) = x^5 - 15x^3. \qquad (1.1)$$

MATLAB Script _____

```
Example 1.4
>>% Define a polynomial function as a string
>>f2='x^5 - 15*x^3'
f2 =
x^5 - 15*x^3
>>x=3 % Compute a value
x =
     3
>>eval(f2)
ans =
  -162
>>x=-3 % Compute another value
x =
    -3
>>eval(f2)
ans =
   162
```

After the polynomial is defined as the MATLAB variable **f2**, the command **eval** computes the result that $f(3) = -162$ and $f(-3) = 162$.

Notice that once the function is defined as a string in the script, it does not have to be retyped during the session. In MATLAB, a text string is called a *macro* if it can be executed by the **eval** command. Macros are often useful to pass function names to function files, as will be demonstrated in several examples in later chapters.

□

INPUT AND CREATION OF DATA

In addition to keyboard input of data, there are other commands to generate data values that may be more convenient, as listed in Table 1.5.

TABLE 1.5 *MATLAB commands for input*

Command	Action
load	Loads variables from a file
input	Inputs the data to an M-file

The **load** command was discussed earlier. The **input** command can be used to input numerical data or text strings. It is usually used in an M-file to input data from the keyboard. This is particularly useful when the same M-file is to process various data values.

□ EXAMPLE 1.5   *MATLAB M-file Example*

The accompanying MATLAB script shows an M-file program to evaluate a function. This file was created with a text editor, as described later in the section on programming. The file is stored on the disk included with this textbook as file EX1_5.M. Before the M-file is executed, it is displayed with the **type** command. In the script, the function being evaluated is

$$y = a \times e^{-1.2t} - 3.0 \times e^{-2.0t}$$

where a and t are variables to be specified when the M-file executes. The purpose of writing this M-file is to rerun it with different values of a and t. As a specific example, $a = 2$ and $t = 1$ were input from the keyboard as the M-file was executed. The file is executed by the command

```
>>ex1_5
```

which causes line-by-line execution of the commands in the file, just as if they were typed at the keyboard.

The **fprintf** commands are used to explain the purpose of the M-file and to display the results. The \n in the **fprintf** command causes a line to be skipped in the display. Note the single apostrophes enclosing the text of the command; quotation marks are not legal MATLAB delimiters for text.

The first two input statements prompt the user to input the numerical values of a and t, respectively. The program then displays the complete function y using the numerical values for a and t. This is another use of the **fprintf** to display both text and numerical values. The term **%g** indicates that the values are to be displayed as either decimal values or in scientific notation. The most appropriate format is determined by MATLAB according to the size of the number, as discussed further in Chapter 2.

MATLAB Script _____

```
Example 1.5
>>format compact
>>type ex1_5

% M-file (EX1_5.M) to evaluate the function
%   y= a*exp(-1.2t) - 3.0*exp(-2t)
% INPUT:   Coefficient a and time t
% OUTPUT: y(t) displayed
fprintf('Compute y= a*exp(-1.2*t) - 3.0*exp(-2*t) \n')
a=input('Coefficient a, a= ');
t=input('Variable t, t= ');
%
% Display results
fprintf('\n Function: y=%g*exp(-1.2*%g)-3.0*exp(-2*%g) \n',a,t,t)
y=a*exp(-1.2*t) - 3.0*exp(-2*t)
% End M-file

>>% MATLAB session to execute EX1_5.M
>>ex1_5
Compute y= a*exp(-1.2*t) - 3.0*exp(-2*t)
Coefficient a, a= 2
Variable t, t= 1
 Function: y=2*exp(-1.2*1)-3.0*exp(-2*1)
y =
    0.1964
>>quit
 14 flops.
```

In the script, the parameter `flops` gives the approximate number of mathematical operations to compute the function.

□

W H A T I F ?   Suppose you wish to use the M-file in Example 1.5, perhaps with a modified function such as

$$y = a \times e^{-t} - 3.0 \times e^{-3.0t}.$$

The M-file is stored on the disk accompanying this textbook as file EX1_5.M. Simply transfer the M-file to a directory on your computer's disk and modify it with a text editor. It is recommended that you save the modified file with a new name to preserve the original. Then execute it with MATLAB.

MATLAB COMMANDS FOR DISPLAY AND PLOTTING

■

MATLAB commands are available for displaying and plotting data. The **disp** command can be used to annotate displayed results and thus improve the appearance and readability of the output. MATLAB also has a powerful graphics capability useful for engineering and scientific applications.

DISPLAY OF
DATA

Typing the name of a variable at the MATLAB prompt will cause the name of the variable and its value or values to be displayed. Other commands used for display of data are listed in Table 1.6.

TABLE 1.6 *MATLAB commands for display or clearing*

Command	Action
ans	Displays the results of a calculation
clc	Clears the command window
clear	Clears the workspace variables
disp	Displays text or a variable without displaying the name

The **ans** command was discussed in Example 1.1. This command is used to display the result of a calculation that is not associated with a named variable. The **clc** command is issued to remove clutter from the screen before beginning a new set of calculations. To free memory space, **clear** can be used to remove all the current variables from memory. Use of the **disp** command will be demonstrated in Example 1.6.

Default Values When MATLAB begins execution, the program uses *default* values for various parameters. These values can generally be changed using appropriate MATLAB commands. As an example, MATLAB uses the *format short* as its default format for display of numbers. This typically displays four decimal digits to the right of the decimal point unless the value is an exact integer. Other formats are set with the **format** command, as described in Chapter 2.

MATLAB Display Example

Consider the results of acceleration tests on a sports car, as shown in the accompanying table. The time in seconds represents the time from a stop to reach the final speed in miles per hour in the specified gear. The MATLAB script shows how to enter the data and display the values in columns with headings.

Gear	Time (s)	Final Speed (mi/h)
1	2.5	30
2	10.0	55
3	21.0	78
4	45.0	100
5	90.0	120

MATLAB Script _____

```
Example 1.6
>>type ex1_6
% EX1_6.M Display a table with Heading
test=[1 2.5 30
      2 10.0 55
      3 21.0 78
      4 45   100
      5 90   120];
disp('          Acceleration Tests  ')
disp('   Gear      Time(s)    Final Speed(mi/h)')
disp(test)
% End M-file
>>ex1_6
        Acceleration Tests
   Gear      Time(s)    Final Speed(mi/h)
   1.0000     2.5000    30.0000
   2.0000    10.0000    55.0000
   3.0000    21.0000    78.0000
   4.0000    45.0000   100.0000
   5.0000    90.0000   120.0000
>>quit
```

The script shows the diary file for the execution of the M-file named EX1_6.M. The array is first defined row by row by entering a CR after each row. Then, the **disp** commands display the table title, column headings, and data. The spacing of the column headings was determined with a bit of trial and error.

□

WHAT IF?   Suppose you wish to store the test data on disk. Execute the M-file and save the numbers in array **test** with the command

>>save test

which creates the disk file TEST.MAT. Then, modify EX1_6.M to use the **load** command and save the new version as EX1_6a.M. Execute the new script to verify the results. Also, you may wish to add a few more comments to describe the script.

GRAPHICS

MATLAB is capable of plotting both two-dimensional (2D) functions $f(x, y)$ and three-dimensional (3D) functions $f(x, y, z)$. In fact, for some applications the ability to plot complicated functions is one of the most powerful aspects of MATLAB. Executing the **demo** command allows the user to choose a number of graphics programs that illustrate MATLAB's graphics capability.

The **plot** command automatically scales the data for a 2D plot and draws x and y axes. Other commands are available to refine the plot and add axes, labels, and a title. Various commands useful for plotting results are listed in Table 1.7.

TABLE 1.7 *MATLAB commands for graphics*

Command	Action
Plotting:	
plot	Plots a linear x-y graph
subplot	Divides the graphics window for multiple plots
clf	Clears a figure
whitebg	Changes background to white
hold	Holds the plot on the screen
fplot	Plots a function defined by a string
ezplot	Plots a function defined symbolically
Labeling:	
axis	Allows manual scaling of axes
grid	Adds grid lines to a plot
title	Adds a title to a plot
xlabel	Adds a label to x-axis
ylabel	Adds a label to y-axis
Data:	
linspace	Creates linearly spaced values
logspace	Creates logarithmically spaced values
Printing:	
print	Prints the current figure window

The MATLAB command window previously discussed represents a *text screen* in which commands are entered and results are displayed. MATLAB also has a *graphics window* for the display of plotted data.

Figures are created in a graphics window by various plotting commands. It is also possible to create up to four independent plots in the same graphics screen with the **subplot** command. The command **clf** clears the current figure from the graph window. To reverse the background from its default black color, the command **whitebg** is used to create a white background. This is useful for printing the figure with black lines on a white background. The command actually "toggles" the background color, so that two consecutive executions of the **whitebg** command change the background back to black.

Multiple plots can be placed on the same screen with the **hold** command. However, the plot will use the already established scales for the axes.

The command **fplot** is useful to plot the graph of a function of one variable defined as a string variable. Thus, a function such as `'sin(x)'` can be plotted over a defined interval of **x**. The **fplot** command automatically determines the abscissa points in **x** to evaluate the function according to the variations of the function. Functions defined symbolically can be plotted with the command **ezplot**.

The commands **linspace** and **logspace** create vectors of points for function evaluation or plotting. Points generated by **linspace** are linearly spaced over a specified interval. For example, the command

```
>>x = linspace(0,10,11)
```

generates 11 points between 0 and 10 in the vector **x**. The command **logspace** creates logarithmically spaced points.

A figure can be printed using a printer attached to the computer system. Alternatively, the figure can be saved as a disk file for later printing or display. The **print** command has various options that define the output device and the format of the figure if saved on disk. The MATLAB figures in this text were saved in Encapsulated PostScript (EPS) form and then incorporated in the text during printing.

☐ EXAMPLE 1.7   *MATLAB Graphics Example*
Consider plotting the function

$$y(t) = \frac{2\sqrt{3}}{9} e^{-4t} \sin\left(\omega t + \frac{\pi}{3}\right) \tag{1.2}$$

for $t > 0$. Since the sin function varies between -1 and 1, we expect the function to vary at most between $\pm 2\sqrt{3}/9 = \pm 0.38$. Of course, the exponential term will drive the function to zero rapidly. When $t = 1$, the value is below 0.02. Thus, a plotting range in t between 0 and 1.5 seems reasonable with a vertical axis range between ± 0.5. The M-file EX1_7.M plots the function using the value

$$\omega = 4\sqrt{3}$$

for the radian frequency of the sine wave.

Using MATLAB's **plot** command, we have only to define a sequence of values of $y(t)$ versus t and not be concerned with the horizontal and vertical scales on the resulting graph unless the autoscaling ranges are not appropriate for the problem. However, even if autoscaling is used, the calculations of the range of data by hand are useful for checking the result approximately to ensure that there were no blunders in defining the equation for the MATLAB program.

MATLAB Script _____

```
Example 1.7
% EX1_7.M Define t values (151 points, 0.0, 0.01, ...) and
%   plot y(t)=(2*sqrt(3)/9)*exp(-4*t).*sin(w*t + pi/3)
t=[0:.01:1.5];
%  Define y(t)
w= 4*sqrt(3);              % Fixed frequency
y=(2*sqrt(3)/9)*exp(-4*t).*sin(w*t + pi/3);
% Plot the results with a grid
clf                        % Clear previous figures
plot(t,y)
grid
title('Plotting example')
xlabel('Time t')
ylabel('y(t)')
legend(['w=',num2str(w),' radians/sec'])
```

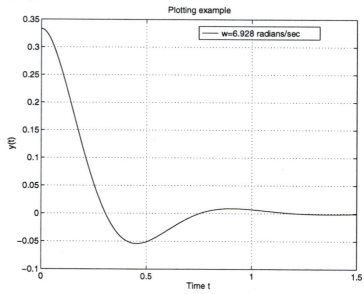

FIGURE 1.2 *MATLAB plot*

The vector of t values is first defined with an assignment statement of the form

$$t = [t_0 : \Delta t : t_f],$$

where t_0 is the starting point, Δt is the increment, and t_f is the final value. In the MATLAB workspace, t is actually a 1×151 row vector. The colon (:)

separating the arguments is called an *operator* when used in this manner to create a vector of values.

The function y in the script is defined using MATLAB's built in **sqrt**, **exp**, and **sin** functions. The definition of y illustrates several points about MATLAB's formats for equations. Multiplications are indicated by the $*$ between operands and the variable **pi** is the value π. The period (.) between the exponential and sine functions is an operator that is necessary because the intent is to perform *element-by-element* multiplication. More generally, if A and B are arrays with the same dimensions, then $A.*B$ yields the array whose elements are the products of the individual elements of A and B. This use of the period (.) operator is described in more detail later.

The **plot** command creates a graph of $y(t)$ versus t, and **grid** overlays a rectangular grid, as shown in Figure 1.2. The commands **xlabel** and **ylabel** are used in the script to annotate the axes. The command **legend** puts the frequency value on the graph by using the **num2str** command, which converts a number to an ASCII string for display.

$\square$

WHAT IF?   Suppose you wish to determine the zero crossings of the function plotted in Figure 1.2. The first value appears to be near $t = 0.3$. A closer estimate can be obtained by typing the command **zoom on**. Pressing the left mouse button changes the axes limits of a figure on the computer screen by a factor of two. By zooming in on the region near the zero crossing, an accuracy of several decimal places in defining y, where $y(t) \approx 0$, should be possible. Don't forget to turn the zoom feature off with **zoom off** after viewing the graph. Otherwise, zoom will be active during the entire MATLAB session.

The command [x,y]=**ginput** can be used to get the numerical values from a graph. Using the mouse, the command returns the x and y coordinates of the points selected on the graph shown on the computer screen. Refer to the MATLAB *User's Manual* for more details.

Suppose ω in Equation 1.2 is to be a variable parameter. Plot the function with various values of ω on the same figure using the **hold** command from Table 1.7.

Three-Dimensional Plotting   MATLAB also contains plotting commands to create surface and 3D plots. Figure 1.3 shows the plot and contour lines for the function

$$z = f(x, y) = \sqrt{x^2 + y^2 + 1},$$

which represents a surface in xyz-space. The contour lines shown in the xy-plane in this case are circles. They indicate the shape of the curves where the function is constant and show the locus of points

$$f(x, y) = c_1, \quad f(x, y) = c_2, \ \ldots$$

for various choices of the constants $c_1, c_2, \ldots$ defining each of the curves. Studying these level curves often provides a better understanding of the function than a plot of the surface itself.

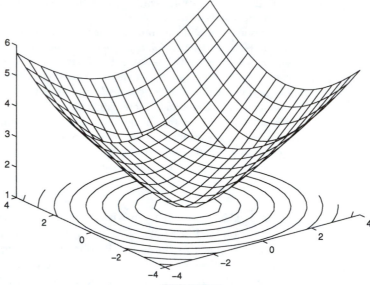

FIGURE 1.3 *Plot of surface $z = \sqrt{x^2 + y^2 + 1}$ and contour lines*

Only a few MATLAB commands are necessary to create Figure 1.3, as the accompanying MATLAB script shows.

MATLAB Script
```
[x,y]=meshgrid(-4:0.5:4);
z=sqrt(x.^2+y.^2+1);
meshc(x,y,z);
```

The **meshgrid** command defines the range of x and y on a square grid. The function $z = f(x, y)$ is defined for plotting using the **sqrt** command. Notice that the period operator (.) is used with the power operator (^) to square each element of the arrays. The command **meshc** plots the function and the contour lines. These commands will be used extensively to represent functions of interest in vector calculus, as presented in later chapters.

Colors   Unfortunately, a textbook with black and white figures cannot demonstrate the use of colors when creating MATLAB figures. The colors can be used for various purposes, such as differentiating between several graphs in one figure. However, there are more sophisticated uses of colors, particularly in plots of 3D functions. For example, different colors can be used to represent different heights in a contour plot. The MATLAB *User's Guide* describes such uses of colors.

CREATING MATLAB PROGRAMS

A MATLAB *program*, often called a *script*, consists of a number of MATLAB statements designed to solve an engineering, mathematical, or scientific problem. The simplest statements contain one or more MATLAB commands. As we have seen, the MATLAB commands refer to built-in functions, such as **exp**, for calculation and more general commands that are provided as M-files with MATLAB.

In addition to the commands provided as part of MATLAB, the user can define unique sequences of MATLAB commands and combine them into a script as in Example 1.5. The script can be stored on disk as an M-file. The complete program typically includes comments, statements for input and output of data, including graphical output, as well as the executable statements to perform calculations. This section explains the MATLAB M-files and functions.

MATLAB M-FILES AND FUNCTIONS

MATLAB is often used in a command-driven mode called the *immediate* mode. When single-line commands are entered during an interactive session, MATLAB processes the statement and displays the results immediately. The MATLAB examples in Example 1.1 through Example 1.4 are of this type. Once the MATLAB session is completed, the commands are no longer available, although they may be saved in a diary file for viewing, but not for execution.

If a sequence of commands is to be used repeatedly, perhaps with different data values, the MATLAB commands can be saved as a disk file called an M-file. An M-file may contain a complete MATLAB script for execution or a function. The previous MATLAB scripts, including the M-files of Example 1.5 and Example 1.6, have used only the built-in MATLAB functions and commands. A user-defined function must be created in a special way that is defined by the conventions of MATLAB.

MATLAB FUNCTIONS

A function file consists of sequences of MATLAB statements that serve to add a new function to the existing set provided by MATLAB. A function is stored on the disk as an M-file which typically provides a specific numerical result as the function is called.

A function file must begin with the word *function* as the first executable statement of the file. Values may be passed to the function for manipulation and output values can be returned. The format of the first statement for a function named **fname** is of the form

```
function [out1,out2,...]=fname(in1,in2,...)
```

where the input variables are placed within parentheses and outputs are enclosed in square brackets if there is more than one. This function executes when it is called by name as **fname** in a MATLAB session or M-file.

One important distinction between variables defined in a script and those in a function is that the function's variables are *local* variables. These local variables exist only during the execution of the function and do not appear in the MATLAB workspace.

☐ EXAMPLE 1.8   *MATLAB Function Example*

Example 1.7 contained the MATLAB script to compute values of the equation

$$y(t) = \frac{2\sqrt{3}}{9} e^{-4t} \sin\left(4\sqrt{3}\, t + \frac{\pi}{3}\right)$$

and to plot the result. If the equation is to be used in various MATLAB programs, it would be better to create a function that evaluates the expression. The accompanying MATLAB script shows the M-file of Example 1.7 rewritten to call the function **clxfunct**.

The function is also displayed in the second script. This function is contained in a disk file CLXFUNCT.M. It is necessary to create the M-file that calls the function separately from the function because each must be stored in a different disk file. The plotting statements from Example 1.7 could be added to the MATLAB script to produce the plot shown in Figure 1.2.

Notice that the variable names within the function are not related to the names used in the calling M-file. Because of this, a programmer writing an M-file that calls the function is not restricted to using specific variable names. The variables passed between the function and M-file must agree as to type, such as scalar or array.

MATLAB Script _____

```
Example 1.8
% EX1_8.M Plot x(t) defined by function clxfunct(t)
t=[0:.01:1.5];        % Define t values
%   Call function
x=clxfunct(t);
% Plot the results with a grid
plot(t,x)
grid
% End file
```

The function must be stored in the disk file CLXFUNCT.M. The contents of the file are shown in the next script. It is generally good practice to define the function call and the results in the prologue comments of the function. These comments appear if the **help** command is used with the function name. The statements in the function are executed when the function is called by another M-file such as EX1_8.M.

MATLAB Script _____

```
function xvalues=clxfunct(tvalues)
% Call:  xvalues=clxfunct(tvalues) to compute function
%   xvalues=(2*sqrt(3)/9)*exp(-4*tvalues) x
%        sin(4*sqrt(3)*tvalues+ pi/3)
xvalues=(2*sqrt(3)/9)*exp(-4*tvalues).*sin(4*sqrt(3)*tvalues + pi/3);
% End function
```

☐

CREATING M-FILES

MATLAB itself does not contain a text editor program, but any editor on the computer system can be specified to create M-files. An M-file can be created using any text editor or word processor that allows files to be saved in the ASCII format. The files written using a text editor are called ASCII files since they contain only standard ASCII characters, such as letters, numbers, and a few special characters, including Carriage Return to end a line and Line Feed to create a blank line.

Word processors normally allow the user to insert a number of formatting commands, such as a command to italicize a word. These special commands are not understood by MATLAB. Each statement in a MATLAB program must end with a Carriage Return, so the word processing capability to wrap text from line to line must not be used. Also, if a word processor is used to create a program, the file must be saved as an ASCII file. This is an option in most popular word processors.

MATLAB PROGRAMMING LANGUAGE

For mathematical programs, the primary characteristics of a high-level programming language, such as C, FORTRAN, or MATLAB, are defined by the allowed data types and the operations involving the data. Other features of a language include the control statements and the input/output (I/O) capability, as well as any special operations or commands.

MATLAB is particularly suited for vector and matrix operations. This makes the language convenient for solving problems in applied mathematics or physics. Moreover, MATLAB contains a number of special functions that can be used by a programmer to create powerful algorithms without programming many of the mathematical functions. Table 1.8 lists a few of the characteristics and examples of commands of the MATLAB language.[1] A selection of these commands is presented in this section. Example I/O commands were discussed previously.

TABLE 1.8 *MATLAB program characteristics*

Classification	Examples
Data types	Scalars, vectors, matrices, strings, and special values
Operators:	
Arithmetic	$+ - */ \setminus \hat{} =$
Logical	**& | ~ all, any, find**
Relational	$< <= > >= ~= ==$
Special	$\% \ ' \ . : ;$
Functions	Mathematics, signal processing, and symbolic math
Program statements:	
Control flow	**if, for, while, break**
Debugging	**echo on**, **pause**, **dbtype**
Evaluation	Timing commands such as **clock** and **etime**
I/O	Commands to input data, create graphics, and print

ARITHMETIC, LOGICAL, RELATIONAL, AND SPECIAL OPERATORS

MATLAB contains a number of operators for ordinary arithmetic, such as $+$ and $-$, raising to a power ($\hat{}$), and logical operations representing AND (**&**), OR (**|**), and NOT (**~**). However, these operators not only apply to single variables but such operators will also work on vectors and matrices when the operation is valid.

The relational operators are used to compare values or elements of arrays. If the relationship is true, the result is a logical variable whose value is one. Otherwise, the value is zero if the relationship is false. Thus, the statement

```
a == b
```

is testing the variables for equality. These operators are used frequently with other commands such as **all**, **any**, or **find** which test arrays for logical conditions. For example, assume the vector **V** contains numerical values that are all different. The commands

```
>> peakindx=find(V==max(V))
>> peakV=V(peakindx)
```

---

[1]These characteristics are those of the student edition of MATLAB, designated Version 4. The professional version has slightly different features, but the program characteristics are essentially the same for the student and professional versions.

find the location of the maximum value and assign that value to the variable `peakV`. In MATLAB, the expression `V(n)` indicates the nth item in the vector so that the value of `peakindx` selects the location of the maximum value in the vector. This variable is an integer called an *index* that locates a value in a vector. Notice that there is no need to specify the length of the vector as would be necessary in some high-level languages.

You may care to test the commands with a vector of values such as `V=[1.1,3.4,-2,7,1,8]`. What do the commands return if the vector is `V=[1,0,1,0,0,1]`? Hint: MATLAB creates a scalar variable, a vector, or an array automatically as needed in a script.

In contrast to the relational operator ($==$), the single equal sign in statements such as $x = y$ indicates an *assignment* statement that sets one variable equal to another.

Colon, Semicolon, and Period Operators

The colon (:), semicolon (;), and period (.) are called *operators* when used in MATLAB program statements because they cause operations to be performed by MATLAB commands. These versatile operators are not merely convenient but form an essential part of useful and efficient programs.

The MATLAB *User's Guide* calls the colon operator " one of the most useful operators in MATLAB." An important use of the *colon* operator was demonstrated in Example 1.8 by the statement

```
t=[0:.01:1.5]
```

which creates a vector of `t` values $0.0, 0.01, 0.02, \ldots, 1.5$. Another use of the colon is to designate sequential elements in a vector or an array. For `t`, the first element of the vector is `t(1)`. Thus, the statement `t(1:5)` designates the first through the fifth elements of the vector `t`. As shown in Chapter 3, this operator is vital for matrix operations. Problem 1.12 provides some practice using the colon operator.

The semicolon operator has been used in several examples in this chapter. Putting a semicolon at the end of a MATLAB statement prevents the display of the results as the statement executes. A semicolon is also used to separate several individual MATLAB statements typed on one line. Another application was presented in Example 1.2, where the semicolon was used to separate rows of the matrix `array` defined in that example.

Period operators are used primarily to define element-by-element operations for arrays. This is the operation desired when the array represents separate values of a function that are stored in an array for convenience. These operators are

```
.*    .^    and    ./
```

for element-by-element multiplication, raising to a power, and division, respectively.

In element-by-element multiplication and division, the arrays involved must have the same number of elements unless one is a scalar. Chapter 3 and 4 present the use of the mathematical operators when array operations rather than element-by-element operations are intended.

WHAT IF? Define two 3×3 arrays such as the array in Example 1.2. Call the arrays A and B and compare the results of the operations

$$A * B \quad \text{with} \quad A.*B \quad \text{and} \quad A\char`^B \quad \text{with} \quad A.\char`^B$$

using MATLAB.

STATEMENTS FOR CONTROL FLOW

The commands **if**, **for**, and **while** define decision making structures called control flow statements for the execution of parts of a script based on various conditions. Each of these structures is ended by an **end** command.

The **if** is a conditional command which causes following statements to execute if the given condition is true. For two conditions in a loop, the **else** command is used and the structure is **if-else-end**. There are many possible combinations as described in the MATLAB *User's Manual*. As an example, assume that **A** is an array of numbers with m rows and n columns. The script

MATLAB Script _____

```
[m n] = size(A)      % Determine size as [rows columns]
if A(:,1) == 0       % If all the elements of Column 1 are zero
  B=A(1:m,2:n)       %  set B to an m by n-1 array
else
  B=A                % B=A if first column of A is nonzero
end
```
_____

tests the entries in the first column of **A** for zero values. If all the values are zero, the columns of **A** are copied into the array **B** with the first column deleted. Otherwise, the arrays are set equal. Notice in the indexing of the array **A**, **1:m** refers to the first m rows and **2:n** selects the columns starting from the second column.

In a program, *loops* are used to execute a series of statements repetitively. Thus, loops serve to perform the iterative parts of an algorithm. Several MATLAB commands are available for looping a fixed number of times, as well as looping based on a condition. The **for** command is used to execute the series of MATLAB statements iteratively. For example, the loop

```
for I=2:2:6, v(I)=0, end
```

operates on the elements of v with index **I=2,4,6**. The commands set to zero the values in vector v with even index 2, 4, and 6.

The **while** command will repeat a series of statements while a given condition is true. The statements will be executed as long as a logical condition defined by a relational operator is true or a mathematical expression remains nonzero.

Consider a function that steadily increases from 0 to 1.0 (monotonically increasing) defined in the vector f. To find the first value that is greater than or equal to 0.5 and copy it to the variable f1, the statements

```
I=1; while f(I) < 0.5; I=I+1; end; f1=f(I)
```

could be used. In this program segment, semicolons are used to separate individual MATLAB statements and also to suppress display of partial results in the **while** loop.

In general, it is not necessary to use explicit looping commands when dealing with arrays. For example, the commands

```
>> k=find(f >=0.5)
>> f1=f(k(1))
```

will accomplish the same thing as the previously shown **while** loop. In this case, the variable k would specify the locations of *all* the values in f greater than or equal to 0.5.

A **break** command can terminate the execution of a **for** or **while** loop before completion of the entire looping process. This command usually follows an **if** conditional statement to terminate the loop when a specific condition occurs.

Suppose the vector f in the previously shown **while** loop or **find** command does not contain a value greater that 0.5. You should try this case to show that an error occurs for both methods. The simple script using the **break** command will stop the looping if all of the values in the vector are less than 0.5.

MATLAB Script _____

```
f1=0
I=1
while f(I) < 0.5
 I=I+1
  if I>length(f), break, end
end
if (I <= length(f)), f1=f(I), end
```
_____

The command **length** computes the number of elements (length) of a one-dimensional array (vector). Thus, the conditional (**if**) statement in the script tests the index I and causes the loop to end if the index exceeds the length of the vector f. When the script ends, the variable f1 either contains the first value equal or greater than 0.5 or zero. The zero value indicates that no value in the vector meets the criterion.

Vectorizing Usually there are several ways in MATLAB to accomplish the same programming task. An important point about the MATLAB programming language is that it is not similar to FORTRAN or C in terms of its treatment of arrays. In general, good MATLAB programmers avoid **for** loops if MATLAB will create an array with an alternative command or series of statements. The process of creating arrays without loops is called *vectorizing*. A simple example was shown in Example 1.7, where the statement

```
t=[0:.01:1.5]
```

was used to create an vector of time values. A **for** loop replacing this statement would be less efficient in terms of execution time.

Problem 1.21 requires comparison of the times of execution for vectorizing and looping program segments. Generally speaking, the programs in this textbook were written with clarity rather than efficiency as the primary goal. Therefore, improvements using vectorizing are certainly possible in some examples.

DEBUGGING COMMANDS

Normally, as a MATLAB script is executed, the commands within the file are not displayed on the screen. The command **echo on** causes the commands to be displayed for debugging or demonstrations. When it is desired to stop the display of commands, the command **echo off** is used.

The **pause** command causes the execution of a script to stop. This is useful for creating delays to view displays or partial results. Generally, a **pause** command would be used between graphic displays. Otherwise, the first display is only visible momentarily. When a key is pressed on the keyboard, execution continues.

MATLAB also has a set of debugging commands for more extensive debugging. For example, the command

```
dbtype <filename>
```

lists the statements in the named M-file with line numbers. The line numbers are often used to set *breakpoints* at a certain line in the script. At a breakpoint, control is temporarily returned to the terminal so the user can diagnose the program results that have been calculated up to the statement at the breakpoint. The command **dbstop** sets a breakpoint and **dbcont** resumes execution from a breakpoint. After the debugging session is complete, the command **dbclear** is used to clear any breakpoints.

EFFICIENCY OF THE PROGRAM

The MATLAB command **clock** returns a vector that contains the current time and date as indicated by the computer's time-of-day clock and calendar. The time between events in a program could be timed by comparing the values from **clock**. To calculate the elapsed time, command **etime** can be used. The sequence of MATLAB commands

```
tstart=clock;
    < MATLAB commands to be timed >
elapset=etime(clock,tstart)
```

sets the variable `elapset` to the difference in time between `tstart` and the time `elapset` is evaluated. Thus, `elapset` represents the time taken by the MATLAB commands executed between the two **clock** commands.

Exercises   You should now be able to do all the Reinforcement Exercises at the end of the chapter. Try using the debugging commands of MATLAB to find the errors if your program is not executing properly.

PROBLEM SOLVING AND PROGRAMMING (OPTIONAL)

It is perhaps presumptuous to discuss problem solving with readers studying advanced mathematics, since problem solving is a key part of all previous courses in engineering, mathematics, and physics. However, our present purpose is to integrate problem solving by traditional methods with MATLAB solutions. Therefore, problem solving and programming are two skills that must be combined. The basic approach is as follows:

1. Define the problem precisely in terms of the input and output variables and the algorithm to be used.

2. Design the MATLAB program and various test cases.

3. Create and debug the program.

4. Test the program thoroughly.

Each of these steps is discussed in detail.

PROBLEM DEFINITION   The first step in problem solving is to define the problem in a mathematical form that is amenable to computer solution. For example, an analysis of a physical problem system might lead to the conclusion that a quadratic equation of the form

$$ax^2 + bx + c = 0 \tag{1.3}$$

must be solved for the variable x with a, b, and c known constants. This might appear to define the problem adequately for computer solution; that is, find the roots of the equation and display these values of x. In fact, the physical problem may dictate that x can vary over only a certain interval of values. If this is the case, the specification of the problem must include any constraints on the solution. As an example, a constraint for the solution of the quadratic equation might be that only positive values of x would be allowed.

The next step is to define a solution method using the equations and constraints from the statement of the problem. This should include a definition of the input and output information. For the quadratic equation example, a, b, and c are input variables and x is the output variable. In more complicated situations, the output might be a graph including a display of the input variables themselves. For numerical values, the data formats should also be defined. For example, the output values could be represented as numbers in scientific notation with four decimal places of accuracy.

The purpose of specifying the inputs and outputs is to separate the data from the algorithm. The way the input values are manipulated to determine the output values constitutes the *algorithm* used to solve the problem. Selection of the algorithm can be one of the most difficult and involved aspects of the problem-solving process.

The accuracy and efficiency of a program depends on the algorithm chosen, which in turn may depend on the range of data values involved. The efficiency of a computer algorithm can be defined in terms of its speed of execution or its use of memory and disk storage. Thus, there can be a dependency on the programming language and even the computer used. Considering these factors, there is rarely one algorithm for solution of a given problem.

One approach to solve the quadratic Equation 1.3 is to apply the quadratic formula for x as

$$x = \frac{-b \pm \sqrt{b^2 - 4ac}}{2a}, \tag{1.4}$$

since we can write an explicit analytical solution. For computer solution, it is convenient to separate the input variables, the algorithm, and the output variables as shown in Figure 1.4.

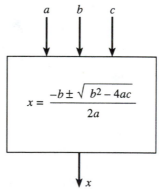

FIGURE 1.4 *Input variables (a, b, c), algorithm, and output variable (x) for the quadratic formula*

Without further information, such as a definition of the accuracy required for x, it is not possible to define the best algorithm to solve an equation even as simple as a quadratic one. In fact, a simple formulation of the solution, as in Equation 1.4, has a number of potential drawbacks. For example, what should the program do if $a = 0$? Any algorithm should have provisions for special data values that cause mathematical errors.

In conclusion, a certain amount of study and experimentation is required to define an acceptable method of solution because of the various factors that affect the correctness and efficiency of an algorithm.

WHAT IF? Consider the solution to Equation 1.3 for arbitrary values of a, b, and c. What error provisions would you incorporate?
Problem 1.17 explores the quadratic equation further.

PROGRAMMING CONSIDERA-TIONS
Once a problem to be solved by computer is defined in terms of data input and output, algorithm, and test cases, it is necessary to create a computer program that implements the algorithm. The aspects to be considered include

1. Program style,

2. Design,

3. Debugging,

4. Testing.

These general programming aspects can be discussed independently of the specific problem to be solved.

Notice that *program style* is first in the list of programming issues. Style refers mainly to the readability of the program and its general structure. Program design covers many factors, such as modularity, efficiency, and the provisions for debugging and testing. However, debugging and testing are so important in computer problem solving that they are considered separately here.

All these aspects of programming are affected by the intended use for a program. A commercial program, such as MATLAB itself, that will be widely used but not modified or even read in its C-language form by the user obviously should be as efficient as possible. On the other hand, efficiency in speed of execution may not be an important consideration for a scientific program used once to solve a specific problem. However, the program should be easy to read, debug, and test to save time for the programmer. Thus, programming always has subjective aspects, but there are certain guidelines developed by experienced programmers that should be followed in any programming project. Of course, the overriding concern is correctness of a program. Most of the guidelines have as their underlying purpose the creation of correct programs.

Program Style The style of a program is defined by the type of comments, variable names, indentation, and similar techniques to improve readability. For example, each program should begin with *preamble* comments. These describe what the program does and how to use it if any special instructions are necessary. Then, input and output data should be defined, as well as any other program or function called by the program. Any errors that are detected and the results of encountering an error should be described. Also, include the author's name and the date if the program is to be used by others.

Many *explanatory* comments should be included to explain parts of the program that do not perform obvious functions. Such comments help any reader of the program, but, most importantly, they aid the programmer in understanding and modifying the program. One caveat is that these comments should explain what the program does, *not* what the computer instructions do. The instructions are better explained in the User's Guide for the programming language rather than in the program.

Variable names should be chosen to reflect the programming, mathematical, or scientific quantities being represented. The programming variables include array indices and loop counters in repetitive segments of the program. In summary, liberal use of comments and careful selection of variable names can greatly enhance the readability of a program.

☐ EXAMPLE 1.9   *MATLAB Program Example*

The accompanying MATLAB script shows a possible preamble to a function that solves the quadratic equation $ax^2 + bx + c = 0$ using the quadratic formula solution of Equation 1.4.

MATLAB Script

```
function [x,Eflag] = quadrat(a,b,c)
%
% Function to solve the quadratic equation ax^2+bx+c=0
%
% CALL: [x,Eflag]=quadrat(a,b,c)
%
% INPUTS:  a, b, and c coefficients as real numbers
%
% OUTPUTS: x, a vector of the solutions as real or complex numbers
%      Eflag, error condition
%          if a=b=0, Eflag=1 since x is not defined
%          otherwise Eflag=0
%
% ALGORITHM: The quadratic formula is used in the form
% (1) x = -b/2a +/- sqrt(b^2-4ac)/2a, if a not = 0;
% (2) x = -c/b, if a =0 and b not = 0.
% -------------------------------
%
Eflag=0; % Assume result is correct
% (Program statements start here)
```

In the preamble, the calling method is specified and the input variables are named and their type defined. Here the variables are specified as real numbers, although MATLAB would allow complex numbers as variables. The solution to be output is defined as a vector x that would contain two entries in the general case. A variable usually called an error *flag* by programmers, is defined. This variable (Eflag) indicates the correctness of the computation based on the error criteria used in the function. In this function, the initial value of Eflag is zero. An error occurs if the coefficients a and b both are zero, in which case the error variable has the value 1 rather than zero. Another program using this M-file should check the value of the error flag after execution and take appropriate action if the value is not zero.

The description of the algorithm is simple in this example. According to the description of the algorithm, the equation

$$bx + c = 0$$

is also solved. This allows $a = 0$ as a legitimate value for the leading coefficient of the quadratic equation.

However, no mention is made of the accuracy of the solution that is achieved using the quadratic equation when $a \neq 0$. The results of Problem 1.17 indicate that this formulation may lose accuracy for certain values of the coefficients. A good numerical algorithm should specify the accuracy expected for the expected range of the input variables if the numerical value of the computed solution can be inaccurate in some cases.

□

WHAT IF?   Suppose you wish to have a MATLAB function to solve the quadratic equation. Create and test a function defined by the statement

```
function [x,Eflag] = quadrat(a,b,c)
```

where the input variables are the coefficients of the quadratic equation and the output variables are the solution and the error flag described in Example 1.9. Our disk file QUADRAT.M is an example function.

Program Design   There are several important issues in program design after the problem solution is well defined and ready for programming. One decision is whether the program will be general or very specific. Generality in a program refers to the capability of the program to solve a class of similar problems using variables, rather than a specific problem with fixed numerical values.

For example, a program to solve the equation $3x^2 + 2x + 4 = 0$ is obviously specific. A more general program solves $ax^2 + bx + c = 0$, where a, b, and c are variables that are input each time the program executes. Similarly, a program that performs 20 iterations to produce a result is less general than one that performs N iterations, where N is increased in the program until a certain accuracy in the solution is reached. If the program is to be used to solve a variety of similar problems, designing for generality will save time if modification of the program is to be avoided to solve each specific problem.

Complexity Another aspect to consider is the reduction in complexity of a program after a preliminary design has been created. Reading, debugging, and testing the program will be easier if the program is divided into independent segments that perform specific tasks. If possible, avoid programming constructs that "jump around" (transfer control) from segment to segment. Such transfers make the flow of the program difficult to follow for debugging.

Efficiency It is possible that following the programming guidelines just discussed can increase the size of the program and possibly increase the time of execution. However, experienced programmers also know that program efficiency can be improved *after* a correct program is produced. Thus, attempts to speed up the program should be made after the program is made correct, readable, general, and easy to debug and test. Generally speaking, increasing the efficiency means refining segments of the program to execute faster or to use less storage.

Since most high-level languages have commands to allow program segments to be timed, program segments that seem to be using excessive amounts of execution time can be found. Then, improving the program segment might involve changing the programming commands or even the algorithm.

Debugging Debugging is the art of locating and correcting errors in a program. These errors, often called *bugs*, may arise from a number of sources, including those listed in Table 1.9.

TABLE 1.9 *Program error sources*

Cause	Type of Error
Misunderstanding of the programming language	Syntax error or calculation error
Data Error	Incorrect inputs
Error in algorithm	Incorrect outputs

In many cases, MATLAB will discover and indicate a syntax error, such as the improper termination of a loop or an undefined variable used in a program. However, other errors resulting from a misunderstanding of a command or a sequence of instructions may not cause MATLAB to indicate the cause of an error. If a command is not completely understood, it is best to test the command with simple examples. This is particularly important with a program, such as MATLAB, that has a number of very sophisticated commands. The **eval** command discussed previously is one example.

Errors occurring because of incorrect data may not be detected by MATLAB. For example, using a value in degrees rather than radians in the **sin** command would cause a numerical error in the expected result but not a syntax error, since MATLAB would use the input value in radians. Adequate program testing should uncover such errors.

The algorithm or logic of the program may be in error if a program or program segment is free of the previously discussed bugs but produces incorrect answers. The *logic* of a program here means the paths taken as the program executes. A logic error could result from taking the wrong path after a decision or failure to consider one or more conditions, such as not testing for a negative value if a negative result would cause an error. Always check for every case, even if you are "sure" that a certain case could never occur. An input data error might cause an unexpected condition. One way to catch logic errors is to create a flowchart before starting the coded program and then to trace the proposed program logic with various conditions according to the flowchart.

An improper algorithm may solve the problem incorrectly, or it may solve certain special cases but not the entire problem. When the algorithm uses an iterative process, there is always a danger of numerical errors. In some cases, the method may produce a solution that diverges rather than converges to the expected solution. Some experimentation with different algorithms is often required to find a suitable method to solve a particular problem.

Testing Testing a program is typically considered an art, particularly by novice programmers. However, an experienced problem solver creating mathematical software will proceed with a well-prepared set of test data, a systematic test procedure, and a complete understanding of the expected results. Solutions to the test cases should be calculated by hand or the results should be obtained from a reliable source, such as a mathematical handbook. Then, the test cases should be compared with the computer output. Obviously, this approach may not work for some practical problems whose solutions are not known beforehand. However, it may be possible to formulate the problem so that a simpler version can be tested.

As an example, consider a method for solving the cubic equation

$$\alpha x^3 + ax^2 + bx + c = 0$$

using a numerical algorithm. One test case should consider $\alpha = 0$ so that the reduced equation becomes the quadratic equations. Another case could be the equation $(x - 1)^3 = 0$ with the three solutions $+1, +1, +1$. For a general case, the computed x values should be substituted in the cubic equation to see if the result is zero.

Modules Since solving a problem may involve a long and complicated sequence of operations, it is recommended that the program be implemented as a series of relatively short program segments, usually called *program modules.* Ideally, each module should perform only one task that is easily tested. Using MATLAB, each module would be an M-file consisting of a script or function.

Errors Any testing should be thorough and also include error conditions that might be expected. For example, if a problem is designed to deal with a certain range of input data, then values outside of the range should not be processed and an error should be indicated to the user.

REINFORCEMENT EXERCISES AND EXPLORATION PROBLEMS

You should now be able to create the MATLAB scripts required by the Reinforcement Exercises with little difficulty. The Exploration Problems are intended to provide more challenging opportunities to try your MATLAB programming skill.

REINFORCEMENT EXERCISES

P1.1. **Vector multiplication** Given the vectors

$$\mathbf{a} = [1, 2, 3, 4],$$
$$\mathbf{b} = [0.1, 0.2, 0.3, 0.4],$$

compute the following:

    a. $\mathbf{d} = 2\mathbf{a}$;

    b. $\mathbf{f}$ as the element-by-element product of $\mathbf{a}$ and $\mathbf{b}$, i.e.,
      $\mathbf{f}(i) = a_i \times b_i$ $i = 1, \ldots, 4$, using the period (.*) operator.

P1.2. **Function evaluation** Evaluate the following functions at the values indicated:

    a. $f(x) = x^2 + x - 3$ at $x = -2$;

    b. $f(x) = \dfrac{x(x+2)(x-1)}{(x+1)(x-2)}$ at $x = -2, 1, 4$;

    c. $f(x) = \left(3x^{3/2} + 20\sqrt{x}\right)^{1/3}$ at $x = 4$.

Comment: In Part b, vectorize x and evaluate at three points simultaneously. Be careful to do element-by-element multiplication.

P1.3. **MATLAB built-in functions** Let x be a real number, say, 2.0. How many MATLAB built-in functions can you find that compute $f(x)$? Compute the value of $f(x)$ for each function. For example, $\sqrt{2.0} \approx 1.414$. Then, try numbers that are not integers, such as π, and complex numbers, such as $3 + 4i$.

P1.4.   MATLAB table of powers   Create a table of squares, cubes, and fourth powers for $x = 0, 1, \ldots, 10$. Display the table in columns.

MATLAB converts a row of numbers **x** to a column using the transpose operator in the form **x'**.

Hint: This can be done with two MATLAB statements.

P1.5.   MATLAB conversion table   Create and print a conversion table between degrees and radians. The range of degrees is $0°$ to $360°$ in $10°$ increments. Label the table and each column.

P1.6.   Simple plot   Plot x versus y for the function created by the equations

$$x = 3\sin(t), \qquad y = 2\cos(t),$$

for $t = [0, 0.1, 0.2, \ldots, 2\pi]$. Also, put text at the center of the plot that says, " This is an ellipse." What values actually occur in vector t if the end point is designated 2π?

P1.7.   Use of eval   Using the MATLAB command **eval**, for each element a_i of the vector $\mathbf{a} = [0, 1, 2, 3, 4]$, compute

    a. $\mathbf{d}_i = e^{a_i}$;

    b. $\mathbf{f}_i = a_i^2$.

P1.8.   Plots   Given the function $y = 3\cos(7x) + 5\cos(13x)$ on the interval 0 to 10, plot:

    a. The function in rectangular coordinates;

    b. The function in polar coordinates.

Experiment by increasing the number of points to use for the plot until a smooth curve results.

P1.9.   MATLAB function   Write a MATLAB function that plots a function entered as a string using the variable x, in the interval from 1 to 10. An example of the usage of the function is

`>>pfun('funct(x)')`

where funct is a MATLAB function such as **sin**. Test the function using $x^2 - x^3$.

P1.10.   Program and plot   Write a MATLAB function that returns the ratio of the area of an isosceles triangle to the perimeter of the same triangle, given the base and height. Using that function, write a program that plots the ratio of area and perimeter versus the ratio of base to height. Use a fixed height of 10, and vary the base from 0 to 100.

EXPLORATION PROBLEMS

P1.11.   Fibonacci numbers   Write a program to compute and display the Fibonacci numbers

$$f(n+2) = f(n+1) + f(n), \qquad f(0) = f(1) = 1$$

for values of f less than 100.

Hint: Use a **while** conditional loop to stop the program.

P1.12. Matrix operations Given the matrix of numbers

$$A = \begin{bmatrix} 10 & 11 & 12 & 13 \\ 20 & 21 & 22 & 23 \\ 30 & 31 & 32 & 33 \\ 40 & 41 & 42 & 43 \end{bmatrix},$$

predict the outcome of the following MATLAB commands and then check the results on the computer.

    a. A(:,2)

    b. A(2,:)

    c. A(:,2:3)

    d. A(:)

    e. A(:,:)

    f. diag(A)

    g. diag(A,1)

P1.13. Plotting For $0 \leq t \leq 15$, plot the three functions

$$\begin{aligned} f_1(t) &= 3.0e^{-0.3t}\cos(2.5t + 150°) + 1.2, \\ f_2(t) &= 3.0e^{-0.3t} + 1.2, \\ f_3(t) &= 1.2. \end{aligned}$$

Plot these function on the same graph using different line styles. Include a title and axes labels as well as a legend defining the graphs.

P1.14. Circle plot Plot the unit circle $x^2 + y^2 = 1$ using the MATLAB **plot** command. Does it look like a circle on the screen or when printed? If not, explore the use of the **axis** command so that the circle looks symmetrical as it should be.

P1.15. Root finding Let $f(x) = e^x - x - 2$ and use MATLAB to find graphically the approximate roots of the equation $f(x) = 0$ on the real axis. Try to get the results to two or three decimal places and test your answers.

Hint: Use the **zoom** command to help find the roots.

P1.16. Vectorizing and timing Compute the sine of the numbers from 0 to 10 radians with a resolution of 0.01 in two ways:

    a. Use a **for** loop;

    b. Vectorize the loop.

Execute and time each approach on your computer. What is the improvement by vectorizing?

P1.17. Quadratic equation Suppose you wish to write a MATLAB program to solve the equation

$$f(x) = ax^2 + bx + c = 0.$$

Considering the discussion of program style, design, debugging, and testing in the text, create the program to solve for the roots of the equation. First, define the basic algorithm. Then, list all of the special cases and limitations of the program.

Document the program properly so it can be used with confidence by another programmer. Can your program fail for any values of a, b or c? Does the accuracy of the solution change with changes in the constants?

Test the program with the cases

$$
\begin{aligned}
f(x) &= x^2 + x + 1, \\
f(x) &= x^2 - 2, \\
f(x) &= 3x + 6, \\
f(x) &= x^2 - 10^5 x + 1, \\
f(x) &= x^2 - 1,000,000.000001x + 1.
\end{aligned}
$$

Comment: The textbook by Mathews listed in the Annotated Bibliography for Chapter 2 discusses various numerical errors, including difficulties that arise using the quadratic formula.

P1.18. Exponential Write a program to evaluate e^x approximately as

$$
e^x \approx \sum_{n=0}^{N} \frac{x^n}{n!} = 1 + x + \frac{x^2}{2!} + \cdots + \frac{x^N}{N!}.
$$

Compare the 10th and 20th partial sums for e with the numerical value returned by MATLAB's built-in **exp** function. Use the command **format long** to get the full decimal representation.

P1.19. Flyball program Write a program to compute and print the height versus time of a ball whose trajectory is governed by the equation

$$
y(t) = -16t^2 + 96t \quad \text{feet},
$$

where $y(t)$ is the ball's height above the ground at time t in seconds. Start at $t = 0$ and print the results at reasonable intervals of time, such as every 0.01 seconds.

Comment: Be sure to stop the program when the ball hits the ground $(y(t) = 0, t > 0)$.

P1.20. Rise time The *rise time* T_r of a function is a measure of the initial rate of increase of a function. Numerically, the rise time is defined as the time it takes the function to rise from 10% to 90% of its value.

   a. Write a program to calculate the rise time of a function.

   b. Plot the function and label the 10% and 90% points.

   c. Test the result with the function

$$
Y(t) = (1 - e^{-t/\tau})
$$

   using various values of τ such as $\tau = 0.5, 1.0, \ldots$.

Can you find an approximate relationship between the rise time T_r and the parameter τ?

P1.21. Program timing Compare the time needed to compute the function

$$y = \sqrt{\exp{(t/1000)}} \log I$$

by the following MATLAB methods:

a. ```
    for I=1:1000
        t=.001*I;
        y(I)=sqrt(exp(t)).*log(t);
    end
    ```

b.  ```
    T=[1:1000]; y=sqrt(exp(.001*T)).*log(T);
    ```

Discuss the reasons for the time difference. If you have a very fast computer, such as a Pentium-based system, put the program segments themselves in a loop and compute the average times. Also, are the element-by-element operations necessary in these scripts?

ANNOTATED BIBLIOGRAPHY

1. *The Student Edition of MATLAB—User's Guide.* Prentice Hall, Englewood Cliffs, NJ. *The manual is the most complete source of information about all the MATLAB commands in the student edition. Any reader using the professional version should refer to the appropriate manual since there are some differences between the student and professional versions.*

2. Garcia, Alejandro L., *Numerical Methods for Physics*, Prentice Hall, Englewood Cliffs, NJ, 1994. *The textbook applies MATLAB solutions to many numerical problems in physics. Many programming aspects of MATLAB are presented and MATLAB and FORTRAN programs are compared.*

3. Nakamura, Shoichiro, *Numerical Analysis and Graphic Visualization with MATLAB*, Prentice Hall, Upper Saddle River, NJ, 1996. *The text emphasizes visual analysis of engineering and scientific problems. An appendix describes the development of a graphical user interface (abbreviated GUI) that will be helpful to programmers creating menu-driven MATLAB applications.*

4. VanTassel, Dennie, *Program Style, Design, Efficiency, Debugging, and Testing*, Prentice Hall, Englewood Cliffs, NJ, 1978. *A very readable book for those who wish to improve their programming ability.*

ANSWERS

The answers to the Reinforcement Exercises in this chapter require MATLAB scripts. These scripts are stored as files on the disk included with this textbook. The file README.TXT on the disk explains the disk directories.

2   *VECTORS*

PREVIEW_____

Generally, the study of vectors in analytic geometry, calculus, and elementary physics takes the geometric point of view when vectors are spoken of as having magnitude and direction in two or three dimensions. However, an algebraic approach is necessary for computer representation and generalization beyond three dimensions. In many problems in physics and engineering, as well as in mathematics, the variables describing a physical situation are treated as functions of vectors. These vectors have components that may represent the values of the variable in different directions or at different times.

The chapter begins with a review of the properties of real and complex numbers, including their representation in MATLAB. Then, we present the basic properties of vectors in two and three dimensions. Many properties of three-dimensional vectors are easily generalized to a vector with $n > 3$ components. Also, collections of functions and other mathematical entities can be described as vectors in a vector space. That such generalizations are useful is an indication of the great power of mathematics when applied to physical problems. Just as the use of a three-dimensional vector to represent force is an abstraction from the physical world, these further abstractions will prove valuable in analyzing many physical systems.

The remainder of the chapter considers vectors and vector spaces for applications to problems in engineering and physics presented in later chapters. The important topics include linear independence for vectors, abstract vector spaces, and vector spaces of functions.

Table 2.1 lists various scalar and vector topics covered in this textbook. Vectors also play a role in other applications and chapters not listed, such as Chapter 8 dealing with Fourier series and Chapter 9 which introduces the Fast Fourier Transform (FFT).

TABLE 2.1 *Topics in vector analysis*

Chapter	Symbol	Topic
Chapter 2	$\alpha + \beta$	Sum of scalars
	$\alpha\beta$	Product of scalars
	$\alpha\mathbf{x}$	Scalar times vector
	$\mathbf{x} + \mathbf{y}$	Sum of vectors
	$\mathbf{x} \cdot \mathbf{y}$	Dot product
	$\mathbf{x} \times \mathbf{y}$	Cross product
	$\|\mathbf{x}\|$	Norm
	$\mathbf{R}^n$	Real vector space
	V	Abstract vector space
Chapter 3	$A\mathbf{x} = \mathbf{b}$	Linear equations
Chapter 4	$\mathcal{L}(\alpha\mathbf{x} + \beta\mathbf{y})$	Linearity
	$A\mathbf{x} = \lambda\mathbf{x}$	Eigenvectors
Chapter 5, 6	$\dfrac{d\mathbf{x}}{dt} = A\mathbf{x}$	Differential equations
Chapter 11	$\dfrac{df(t)}{dt}, \ \dfrac{\partial f(t)}{\partial t}$	Derivative of a function
	$\dfrac{d\mathbf{x}(t)}{dt}$	Derivative of a vector
	∇f	Gradient of a function
	$\nabla \cdot \mathbf{x}$	Divergence of a vector
	$\nabla \times \mathbf{x}$	Curl of a vector

In the table, α and β are real numbers (scalars); $\mathbf{x}$, $\mathbf{y}$, and $\mathbf{b}$ are vectors; and A is a matrix.

PROPERTIES OF REAL NUMBERS

Since real numbers are used extensively to describe physical phenomena, a review of the fundamental operations on real numbers is presented first in this chapter. Then, we consider numbers as represented by MATLAB.

REAL NUMBERS The set of real numbers consists of *integers*, *rational* numbers, and *irrational* numbers. The integers are $1, 2, 3, \ldots, -1, -2, -3, \ldots$, and 0. Fractions in the form p/q, where p and q are integers, are rational numbers. *Irrational* numbers can be represented by an infinite decimal expansion, such as $3.14159\ldots$ for π. However, irrational numbers cannot be represented exactly as a ratio of integers.

Real numbers that are not integers can be represented by *nonterminating* decimal expansions, as in the example of π just given. The sequence of decimal digits of π is both *nonterminating* and *nonrepeating*. In contrast, rational numbers have a repeating decimal expansion, such as $\frac{1}{4} = 0.25000000\ldots$, in which the repeated digit is zero. In most cases, the repeated digits will not be zero. For example, the fraction $\frac{7}{22} = 0.3181818\ldots$ has a nonterminating but repeating decimal expansion.

Since the representation of most real numbers requires an infinite series of digits, we expect that only certain numbers can be represented exactly for hand or computer calculations because the number of decimal places is limited in any such calculations. The importance of these observations will be explored later when MATLAB computer numbers are described.

Axioms for Real Numbers In mathematics, certain operations and basic properties involving the quantities being considered such as numbers, vectors or matrices are assumed to be true. These assumptions are called *axioms* or *postulates* and no proof is necessary. Most of the other properties are presented by *theorems*, which are proven using the postulates and perhaps other theorems. The theorems allow further generalizations and applications to be developed.

The basic axioms for real numbers define how numbers are combined by addition and multiplication. These axioms define the associative, commutative, and distributive laws as well as the identity and inverse elements for any real numbers $x, y,$ and z.

Associative laws. For real numbers, the grouping of the operations for addition or multiplication is not important, since

1. $(x + y) + z = x + (y + z)$,

2. $(x \times y) \times z = x \times (y \times z)$.

A number of basic operations are not associative. For example, subtraction is not associative, since

$$(x - y) - z \neq x - (y - z).$$

Thus, if $x = 5, y = 3, z = 1$, the left grouping yields 1, but 3 results from the operations on the right. For nonassociative operations, careful grouping is important.

Commutative laws. The order of addition or multiplication of real numbers does not matter:

1. $x + y = y + x$,

2. $x \times y = y \times x$.

Certain operations of importance in vector and matrix algebra that we will study later are not commutative with respect to multiplication.

Distributive laws. Addition and multiplication are connected by the distributive law:

$$x \times (y + z) = x \times y + x \times z.$$

Identity elements. The numbers 0 and 1 are called the *identity elements* for addition and multiplication, respectively:

1. $x + 0 = x$,

2. $1 \times x = x$.

Inverse elements. For each real number x there is an inverse element for addition, and if the real number is not zero there is an inverse for multiplication:

1. $x + (-x) = 0$,

2. if $x \neq 0$, $x \times (1/x) = 1$.

The inverse for addition is called the *negative* of the number and the inverse for multiplication is called the *reciprocal*. The negative can be used to define the operation of subtraction as addition of the negative, writing $x - y = x + (-y)$. The reciprocal can be used to define the operation of division as multiplication by the reciprocal, $x \div y = x \times (1/y)$ where $y \neq 0$.

It is interesting to note that the properties just listed do not uniquely define the real numbers. For example, the complex numbers and the binary numbers also satisfy these properties.

Generalization   Part of our mathematical education is the generalization of the concept of *number* from the positive integers (counting numbers) to fractions (rational numbers) and then to negative and irrational numbers. The generalization continues with the study of transcendental numbers such as π and finally the complex numbers. All these numbers satisfy the axioms just presented.

ABSOLUTE
VALUE

The real numbers can be represented as points on the *real number* line shown in Figure 2.1. For convenience, the entire real number line will be referred to as **R**. Thus, a statement about **R** applies to all the real numbers.

On **R**, the distance of any real number from the origin is the magnitude, or *absolute value*, of the number. The absolute value is defined as

$$|x| = \begin{cases} x, & \text{if } x \geq 0, \\ -x, & \text{if } x < 0. \end{cases} \tag{2.1}$$

Applying the formula, $+5$ has the absolute value 5, as expected. The number -5 also has the absolute value $-(-5) = 5$.

FIGURE 2.1 *The number line* **R**

The *distance* between the real numbers x and y is defined as $|x - y|$. This distance is the length of the line segment of **R** with the endpoints x and y. The distance on **R** between the numbers -5 and 5 is then $|-5 - (+5)| = 10$. Notice that the absolute value defines the positive distance of the number x from the origin where $x = 0$. These concepts will be generalized when vectors are studied.

MATLAB COMPUTER NUMBERS (OPTIONAL)

In modern computer systems, numbers are represented in the *binary* number system. The digits are 0 and 1 and the computer can store a finite number of binary digits to represent any number. For integers, the decimal value of a binary number is

$$N = \sum_{i=0}^{m-1} d_i \, 2^i = d_0 + d_1 \times 2 + d_2 \times 2^2 + \cdots + d_{m-1} \times 2^{m-1}, \tag{2.2}$$

where m is the number of digits and the d_i are binary digits.

The number range for a binary number with m digits is from

$$(000\ldots0000)_2 \quad \text{to} \quad (11111\ldots1111)_2,$$

where the notation $(N)_2$ means that N is a number in base 2.[1] The most positive value sums to the decimal value $2^m - 1$. An 8-digit positive binary

---

[1]In the text, numbers other than decimal numbers will have the base specified explicitly.

number thus has the maximum value of $2^8 - 1$, or 255. The number 1011_2 has the decimal value 11 using the expansion of Equation 2.2.

If a fraction is to be represented, a *binary point* is used to separate the integer from the fractional part of the number, just as the decimal point is used for decimal numbers. The binary number $(0.11111\ldots1111)_2$ has the decimal value

$$
\begin{aligned}
N &= \sum_{i=1}^{m} d_{-i}\, 2^{-i} = d_{-1} \times 2^{-1} + d_{-2} \times 2^{-2} + \cdots + d_{-m} \times 2^{-m} \\
&= \frac{1}{2} + \frac{1}{4} + \frac{1}{8} + \cdots + d_{-m} \times 2^{-m},
\end{aligned}
\tag{2.3}
$$

since $d_{-i} = 1$ for $i = 1, \ldots, m$. This is a geometric series whose sum converges to 1 as m goes to infinity, as you are asked to show in Problem 2.3. For a finite value of m digits, the sum is $1 - 2^{-m}$. With 16 binary digits, the value is

$$
1 - 2^{-16} = 0.99998
$$

to five places after the decimal.

FIXED-POINT NOTATION

The binary number representation just discussed assumes that the binary point was located in a fixed position, yielding either an integer or a fraction as the interpretation of the internal machine representation. Unfortunately, the binary point is not actually stored with the number, but its position must be remembered by the programmer (or the program) to display the results in the usual form. This method of representation is called *fixed point*.

It is a theorem that any real number x such that $0 < x \leq 1$ can be represented in a unique manner as a nonterminating binary fraction. In the computer, the binary series must be terminated after m digits to represent a decimal number. Except in special cases such as $1/2$, the binary representation will be an approximation. The error can be shown to be less than 2^{-m}.

In practice, the machine number also is limited to a finite range determined by the number of binary digits used in the representation. For a 32-digit binary number, the range of positive fixed-point integers is about $+2^{32}$ or $+10^{11}$, obtained by solving the equation $2^{32} = 10^y$ for the exponent $y = 32 \log_{10} 2$. This limited range of the fixed-point numbers is a drawback for certain applications.

FLOATING-POINT NOTATION

To overcome many of the limitations of fixed-point notation, a method that is the counterpart of scientific notation is used for number representation in computers. *Floating-point* notation represents a number as a fractional part times a selected base raised to a power. The choice for the base is usually 2, although base 16 is sometimes used. In the machine representation, only the fractional part and the value of the exponent are stored. The decimal equivalent is written as

$$N.n = f \times 2^e \qquad (2.4)$$

using base 2 in this case. Thus, 1.5 could be written as $(11)_2 \times 2^{-1}$. The fraction f is called the *mantissa* and e is the *exponent*.

In floating-point notation, the number of binary digits devoted to the fraction determines the precision and the length of the exponent in digits determines the range. The actual arithmetic operations may be carried out by a computer's Central Processing Unit (CPU) or by a separate hardware chip called a co-processor. In some cases, the arithmetic is performed by a software package containing routines for floating-point arithmetic.

MATLAB Floating-point Notation   MATLAB and many other programs for mathematics may employ the IEEE floating-point standard representation for machine numbers.[2] The results of a calculation for this format can contain as many as 16 significant decimal digits and have a range of from 10^{-308} to 10^{308}. However, the results may be displayed in a number of different ways.

Comment: Unless floating-point operations are implemented in software, the machine floating-point hardware actually determines the number format used. In fact, some machines do not conform to the IEEE standard. Try the MATLAB command **isieee** to determine if your computer uses IEEE arithmetic.

☐ EXAMPLE 2.1   *MATLAB Range of Numbers*
Although MATLAB's calculations are carried out to the maximum precision in floating point internally, the results may be displayed as fixed-point decimal numbers or as numbers in scientific notation. Table 2.2 lists the choices.

Although physical measurements rarely justify the precision yielded by the long display formats in the table, MATLAB computes values to the highest precision to reduce roundoff errors. Roundoff error is the difference between the perfectly accurate number used in mathematical analysis and the computed number with only a fixed number of significant digits. The hexadecimal format listed in the table is frequently used to read binary numbers in base 16 notation for convenience.

---

[2]IEEE, usually pronounced I-triple-E, is the Institute of Electrical and Electronics Engineers.

TABLE 2.2 *MATLAB formats for numbers*

Command	Format
format	Default setting (short)
format short	Fixed-point format with 5 digits
format long	Fixed-point format with 15 digits
format short e	Floating-point format with 5 digits
format long e	Floating-point format with 15 digits
format hex	Hexadecimal format

The accompanying MATLAB script is taken from a diary file to show various numerical values. The diary file was edited to display the results of each numerical input on the same line. The smallest fraction representable in the author's machine is shown in the MATLAB script as machine epsilon (**eps**). Using **eps** can sometimes avoid a divide by zero. For example, consider the limit

$$\lim_{x \to 0} \frac{\sin x}{x} = 1.$$

As computed by MATLAB, $\sin x/x$ with $x = 0$ would result in a divide-by-zero error. However, the statements

```
x=x+eps
y=sin(x)/x
```

would lead to the correct result. Adding **eps** to each element in an array should not affect the results of arithmetic operations unless the values are so small that **eps** is significant.

Various long and short formats are also shown in the MATLAB script. Each number was entered by typing the value at the MATLAB prompt, and the MATLAB representation is shown on the same line. Notice that the program warns the user if a division by zero occurs. Remember that machine zero is simply a number smaller than the smallest possible computer number. Mathematically, this number is not necessarily actually zero.

The answer **NaN** (Not a Number) is standard IEEE floating-point notation for an indeterminate result. Thus, if the IEEE standard is being used, MATLAB does not stop after calculations such as $0/0$ or ∞/∞, although the results are not meaningful. The purpose of introducing **NaN** in this way is so that a program can determine that a calculation was invalid and proceed accordingly.

Next in the script, the value **y** is entered in exponential form representing 10^{306}. When multiplied by 100, the number is still within MATLAB's range. If the new **y** is then multiplied by 5, the value overflows and MATLAB indicates that the product **z** is infinity. Other values, such as 10^{-307} and $\log_{10}(2)$, are also shown.

MATLAB Script

```
Example 2.1
>>% (Machine epsilon)
% EX1_2.M
>>eps
eps =  2.2204e-016
>>format short
1/3
ans =  0.3333
>>format long
>>1/3
ans =  0.33333333333333
>>1/0
Warning: Divide by zero
ans =  Inf
>>0/0
Warning: Divide by zero
ans =  NaN
>>y=1.0e306
y =    1.000000000000000e+306
>>y=100*y
y =    1.000000000000001e+308
>>z=5*y
z =   Inf
log10(2)
ans = 0.30102999566398
>>x=1e-307
  x = 9.999999999999999e-308
>>log10(2)
  ans =  0.30102999566398
>>quit
```

□

W H A T   I F ?   Suppose the MATLAB representation of a number is not accurate enough for an application. Investigate the variable precision arithmetic command **vpa** of the *Symbolic Math Toolbox*. Compare MATLAB's numerical value of π with the symbolic representation to more than 15 digits.

Problem 2.16 and Problem 2.22 treat MATLAB roundoff errors. Use the command **format long** to see the 15 digit fixed-point representation of the numbers or **format long e** for the floating-point representation.

COMPLEX NUMBERS

A *complex number* is written $z = x + iy$, where x and y are real numbers and i is the imaginary unit with the property $i^2 = -1$. The number x is called the *real part* of z and y is called the *imaginary part*. A complex number can be represented as a point $P(x, y)$ in the xy-plane as shown in Figure 2.2. Polar coordinates are associated with a complex number just as for points in the plane. From the figure, the radial distance r and the polar angle θ with respect to the positive real x-axis are

$$r = \sqrt{x^2 + y^2}$$
$$\tan \theta = \frac{y}{x} \tag{2.5}$$

when $x \neq 0$. If $x = 0$, $\theta = \pi/2$ when $y > 0$ and $\theta = -\pi/2$ when $y < 0$.

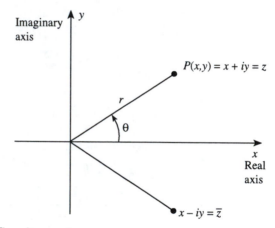

FIGURE 2.2 *Complex numbers*

The radial distance of z is written as $r = |z|$ and is called the *magnitude* of z. The magnitude is also called the *modulus* or length of z. The *complex conjugate* of z is the number $\bar{z} = x - iy$ in Figure 2.2 with polar angle $-\theta$. From the figure, it is evident that $|\bar{z}| = |z|$. The polar angle is called the *argument* of z, written arg z. We see that

$$\arg \bar{z} = -\arg z.$$

The complex numbers have operations of addition and multiplication that obey the properties of real numbers previously listed. The *sum* of two complex numbers is formed by adding the real and imaginary parts. Thus, if $z_1 = x_1 + iy_1$ and $z_2 = x_2 + iy_2$, the sum is

$$z_1 + z_2 = (x_1 + x_2) + i(y_2 + y_2).$$

The *product* of two complex numbers z_1 and z_2 is

$$z_1 z_2 = (x_1 x_2 - y_1 y_2) + i(x_1 y_2 + y_1 x_2).$$

From this formula, it follows that $z\bar{z} = |z|^2$.

POLAR FORM OF COMPLEX NUMBERS

Using the magnitude and argument of z in Figure 2.2, a complex number can be expressed in the form

$$z = r(\cos\theta + i\sin\theta). \tag{2.6}$$

In complex analysis, it is shown that the *complex exponential* function satisfies the relationship

$$e^{i\theta} = \cos\theta + i\sin\theta. \tag{2.7}$$

Thus, the polar form of a complex number can also be written as

$$z = re^{i\theta}. \tag{2.8}$$

This is a convenient form for certain applications, as discussed in Chapter 5. The formula of Equation 2.7 is called *Euler's formula*. Since the complex exponential function satisfies the same multiplication rule as the real exponential function e^x, the expression e^z has the representation

$$e^z = e^{x+iy} = e^x e^{iy} = e^x(\cos y + i\sin y). \tag{2.9}$$

MATLAB COMPLEX NUMBERS

MATLAB accepts complex numbers in both Cartesian and polar form. The Cartesian format for the complex number $z = 3 + 4i$ would be

```
>> z=3+4i
>> z=3+4j
```

since either i or j can be used to specify the imaginary number $\sqrt{-1}$. Because MATLAB would interpret i4 as a variable name, even though mathematically $4i = i4$, in MATLAB we must write i*4 in statements such as z=3+i*4.

The MATLAB polar form is

```
>> w=r*exp(i*theta)
```

where r and theta must be defined before w is used.

Comment: It is important not to use the variables i and j for loop variables or similar purposes in programs that also use MATLAB's complex numbers since the loop variables may override their definition as imaginary numbers.

VECTORS IN TWO DIMENSIONS AND THREE DIMENSIONS

The notion of *vectors* has proven invaluable in physics and mathematics. The vector in space is a combination of a positive number called the *magnitude* and a *direction*. One useful feature of vectors is that the equations of physics can be represented in terms of vectors independent of a particular coordinate system. Newton's second law

$$\mathbf{F} = m\mathbf{a}$$

provides an example. The vector form of the equation is not altered by a change in the coordinate axes. For example, the form shown for Newton's law holds whether $\mathbf{F}$ is defined in rectangular, polar, or spherical coordinate systems. Also, the vector equation is a convenient "shorthand" to represent complicated formulas. The use of vectors can lead to greater understanding of a problem as well as simplification of computations as will be shown in many examples in this text.

INTRODUCTION TO VECTORS

In some cases involving two-dimensional (2D) and three-dimensional (3D) space, visualization of a physical situation is aided by drawing the vectors involved as directed line segments to give the vectors a geometric interpretation. A vector can also be identified using coordinates that represent its origin and endpoint. This latter view is the algebraic view.

As discussed previously, the real numbers can be identified with the points of a straight line, as shown in Figure 2.1. Each number x is represented by a point on the line, so the value can be pictured geometrically with respect to other real numbers. Similarly, the origin and endpoint of vectors in the 2D plane can be represented as pairs of real numbers, and the components can be plotted in the xy-plane, as shown in Figure 2.3. When the origin of the vector is $[0, 0]$ in the xy-plane, the vector is typically defined only by the coordinates of its endpoint.

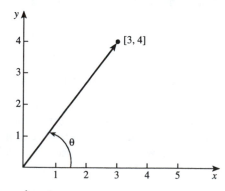

FIGURE 2.3 *Points and vectors*

For example, the 2D vector starting at $[0, 0]$ with endpoint $[3, 4]$ can be drawn as a line segment from the origin $[0, 0]$ to its endpoint 3 units in the x direction and 4 units in the y direction. These values of 3 and 4 represent the *components* of the vector in the x and y directions, respectively. In three dimensions, the vector's endpoints are represented by three real numbers so the vector can be designated as $[x_1, x_2, x_3]$.

A vector can thus be described geometrically by a line from its origin to its endpoint. The geometric view is used to visualize the problem at hand, as when the vectors represent distances or forces in various directions. The algebraic notation is useful when performing calculations such as addition with vectors. The legitimate operations that are allowed for vectors will be defined later in this chapter.

Notation In this text, individual vectors are usually designated by roman letters printed in boldface type ($\mathbf{a}, \mathbf{b}, \mathbf{A}, \mathbf{x} \ldots$). As an example, the equation $\mathbf{a} = [3, 4]$ defines $\mathbf{a}$ as the vector from the origin to the point $[3, 4]$. The same notation means that the vector $\mathbf{a}$ has *component* 3 in the x direction and *component* 4 in the y direction.

Numbers are often called *scalars* to emphasize the distinction between numbers and vectors. Thus, energy and power in physics are scalar quantities, whereas velocity and acceleration are vector quantities, since these later quantities have a direction associated with them.

The term *scalar multiple of a vector* means multiplication of the vector components by a number. The multiplication operation is denoted as $\alpha\mathbf{x}$ for the scalar α and the vector $\mathbf{x}$.

The notation $\mathbf{R}^2$ is used to denote the set of all 2D vectors, and $\mathbf{R}^3$ indicates all of the 3D vectors. The exponent defines the *dimension* of the vectors being discussed. Thus, $\mathbf{a} = [3, 4]$ is a 2D vector that is a member of $\mathbf{R}^2$.

Addition and Subtraction of Vectors Consider two arbitrary vectors $\mathbf{x} = [x_1, x_2]$ and $\mathbf{y} = [y_1, y_2]$. The sum and difference of the vectors are the new vectors

$$
\begin{aligned}
\mathbf{x} + \mathbf{y} &= [x_1 + y_1, x_2 + y_2], \\
\mathbf{x} - \mathbf{y} &= [x_1 - y_1, x_2 - y_2],
\end{aligned}
$$

in terms of the components of the original vectors. Notice that the resulting vectors would not generally have the origin as their beginning point. We can also draw the vectors and determine the sum and difference as shown in Figure 2.4.

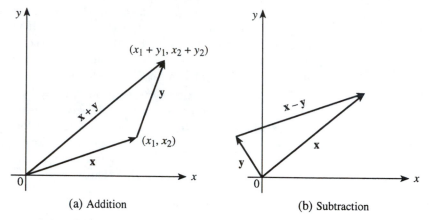

(a) Addition
(b) Subtraction

FIGURE 2.4 *Addition and subtraction of vectors*

In Figure 2.4(a), the vector $\mathbf{y}$ is shifted so that initial point of $\mathbf{y}$ is placed at the terminal point of $\mathbf{x}$. The vector $\mathbf{x} + \mathbf{y}$ is the vector from the origin to the end point of $\mathbf{y}$ in the figure. By drawing the diagram with $\mathbf{y}$ at the origin, you can show that $\mathbf{y} + \mathbf{x} = \mathbf{x} + \mathbf{y}$.

Figure 2.4(b) shows the subtraction of two vectors. As an aid to remembering the direction, note that $\mathbf{x} = \mathbf{x} - \mathbf{y} + \mathbf{y}$, so that $\mathbf{x} - \mathbf{y}$ is the vector that must be added to $\mathbf{y}$ to obtain $\mathbf{x}$. The vector $\mathbf{y} - \mathbf{x}$ has the same length as $\mathbf{x} - \mathbf{y}$ but points in the opposite direction.

Length of a Vector The *length* of a vector $\mathbf{x}$ is a nonnegative scalar quantity that measures the distance between its initial point and terminal point. In the Cartesian coordinate system, the components of a 3D vector represent the lengths along the three mutually perpendicular axes x, y, and z.

The length of a vector is also called its *magnitude*, especially when the vector represents a physical quantity. Thus, if $\mathbf{a}$ defines the acceleration vector for a body, the magnitude of the acceleration is written a in this textbook.

The length of a 2D vector with components $[x_1, x_2]$ in the xy-plane follows from the Pythagorean theorem as

$$\| \mathbf{x} \| = \sqrt{x_1{}^2 + x_2{}^2}. \tag{2.10}$$

The length of the 2D vector $\mathbf{a} = [3, 4]$ previously defined is thus

$$\| \mathbf{a} \| = \sqrt{3^2 + 4^2} = 5.$$

Generalization to three dimensions is straightforward.

The length of the vector $\mathbf{a}$ between points $[x_1, y_1]$ and $[x_2, y_2]$ can be found by forming $x = x_2 - x_1$ and $y = y_2 - y_1$ and then computing

$$\| \mathbf{a} \| = \sqrt{x^2 + y^2}.$$

Some authors use the notation | **a** | to indicate the length or magnitude of the vector **a**. This representation is quite common for vectors in $\mathbf{R}^2$ and $\mathbf{R}^3$. In a later section of this chapter, the length or magnitude of a vector will be called the *norm* of the vector. This is a generalization of the concept of length, which is useful for vectors with more than two or three components. The norm of **a** is written $|| \mathbf{a} ||$.

Unit Vectors   A vector **x** of length 1 ($|| \mathbf{x} || = 1$) will be called a *unit vector*. For any vector **x**, a unit vector **n** in the direction of **x** can be computed as

$$\mathbf{n} = \frac{\mathbf{x}}{|| \mathbf{x} ||}$$

assuming that $|| \mathbf{x} ||$ is not zero. This unit vector defines the direction of vector **x**.

A nonzero vector that is divided by its magnitude is said to be *normalized*. Such a vector, as for example the vector $\mathbf{n} = \mathbf{x}/||\mathbf{x}||$, has the property that $||\mathbf{n}|| = 1$.

DOT PRODUCT   An important operation for two vectors of the same dimension is the *scalar*, or *dot*, product. One definition of the dot product for two nonzero vectors with a common origin point is

$$\mathbf{x} \cdot \mathbf{y} \; = \; || \mathbf{x} || \; || \mathbf{y} || \; \cos \theta. \tag{2.11}$$

In the equation, θ is the angle between the vectors, usually measured in radians, taken from the interval $0 \leq \theta \leq \pi$.

The dot product of the 3D vectors $\mathbf{x} = [x_1, x_2, x_3]$ and $\mathbf{y} = [y_1, y_2, y_3]$ can also be shown to be

$$\mathbf{x} \cdot \mathbf{y} \; = \; x_1 y_1 + x_2 y_2 + x_3 y_3 \tag{2.12}$$

in terms of their components. Of course, for 2D vectors the terms in x_3 and y_3 are not included. Notice that the length of the vector **x** can be computed as $\sqrt{\mathbf{x} \cdot \mathbf{x}}$, as shown by letting $\mathbf{x} = \mathbf{y}$ in Equation 2.12, which defines the dot product in terms of the components of the vectors.

The two definitions just given for the dot product are equivalent. For example, if **x** and **y** are two nonzero vectors in three dimensions that originate from a common point, the cosine of the angle between them is

$$\cos \theta \;\; = \;\; \frac{(\mathbf{x} \cdot \mathbf{y})}{(|| \mathbf{x} || \; || \mathbf{y} ||)} \tag{2.13}$$

$$= \;\; \frac{x_1 y_1 + x_2 y_2 + x_3 y_3}{\sqrt{x_1^2 + x_2^2 + x_3^2} \; \sqrt{y_1^2 + y_2^2 + y_3^2}}. \tag{2.14}$$

This relationship can be proven, as indicated by Problem 2.25.

□ **EXAMPLE 2.2**   *Use of the Dot Product*

Let $\mathbf{x} = [2, 4, 0]$ and $\mathbf{y} = [6, -4, 2]$. Thus, the vectors have length

$$\begin{aligned} \| \mathbf{x} \| &= \sqrt{\mathbf{x} \cdot \mathbf{x}} &= \sqrt{4 + 16 + 0} &= 2\sqrt{5}, \\ \| \mathbf{y} \| &= \sqrt{\mathbf{y} \cdot \mathbf{y}} &= \sqrt{36 + 16 + 4} &= 2\sqrt{14}, \end{aligned}$$

and the dot product is

$$\mathbf{x} \cdot \mathbf{y} = (2)(6) + (4)(-4) + (0)(2) = -4.$$

By combining Equation 2.11 and Equation 2.10, the angle between the vectors is computed as

$$\begin{aligned} \theta &= \cos^{-1} \left[\frac{\mathbf{x} \cdot \mathbf{y}}{\| \mathbf{x} \| \ \| \mathbf{y} \|} \right] \\ &= \cos^{-1} \left[\frac{-4}{(2\sqrt{5})(2\sqrt{14})} \right] = \cos^{-1} \left[\frac{-1}{\sqrt{70}} \right] \approx 96.9°. \end{aligned}$$

If we define $\mathbf{d} = \mathbf{y} - \mathbf{x}$, then the distance between the endpoints of $\mathbf{y}$ and $\mathbf{x}$ is the length of $\mathbf{d}$. Thus, the distance is

$$\begin{aligned} \| \mathbf{d} \| &= \| \mathbf{y} - \mathbf{x} \| \\ &= \sqrt{(6-2)^2 + (-4-4)^2 + (2-0)^2} = \sqrt{84} \\ &= 2\sqrt{21}. \end{aligned}$$

□

Orthogonal Vectors   If two vectors have a nonzero length and

$$\mathbf{x} \cdot \mathbf{y} = 0,$$

the vectors are said to be perpendicular, or *orthogonal*, to each other. In 2D and 3D, the vectors thus meet at right angles, since $\theta = 90°$ when $\cos \theta = 0$. The fact that two vectors $\mathbf{x}$ and $\mathbf{y}$ are orthogonal can be represented by the notation $\mathbf{x} \perp \mathbf{y}$.

Standard Unit Vectors   The 3D *unit vectors*, defined as

$$\begin{aligned} \mathbf{i} &= [1, 0, 0], \\ \mathbf{j} &= [0, 1, 0], \\ \mathbf{k} &= [0, 0, 1], \end{aligned} \tag{2.15}$$

are called the *standard*, or *natural*, unit vectors in $\mathbf{R}^3$. The triplet $\mathbf{i}, \mathbf{j}, \mathbf{k}$ is also called a set of *basis vectors*, since any 3D vector can be represented as a sum of scalar multiples of these three unit vectors.

The standard unit vectors all have length 1 and are mutually orthogonal, as is easily shown by forming the dot product between any pair of these vectors. Such a set of vectors is said to form an *orthonormal* set.

□ **EXAMPLE 2.3**   *Use of Unit Vectors*

The unit vectors are frequently used to form an alternative notation for a vector in ordinary 3D space. Thus, the vector $\mathbf{F} = [f_x, f_y, f_z]$ can be written as

$$\mathbf{F} = f_x\mathbf{i} + f_y\mathbf{j} + f_z\mathbf{k}, \tag{2.16}$$

where f_x, f_y, and f_z represent the lengths of the vector along the x-, y-, and z-axes, respectively.

However, the unit vectors and the dot product have far greater application than notational convenience or geometrical interpretation. For example, assume that the force on an object is defined by the vector $\mathbf{F}$. Then, the component of the force in the x-direction is given by $\mathbf{F} \cdot \mathbf{i}$. The proof is straightforward, since

$$\mathbf{F} \cdot \mathbf{i} = f_x(\mathbf{i} \cdot \mathbf{i}) + f_y(\mathbf{j} \cdot \mathbf{i}) + f_z(\mathbf{k} \cdot \mathbf{i}) = f_x \tag{2.17}$$

after taking the dot product of the standard unit vectors.

□

Application of the Dot Product   The dot product is frequently used to find the component of one vector in the direction of another. For example, an important use of the dot product is to determine the work done by a constant force $\mathbf{F}$ acting on a body moving a distance and in a direction defined by the vector $\mathbf{s}$. The magnitude of $\mathbf{s}$ defines the distance moved, but only the component of $\mathbf{F}$ in the direction of $\mathbf{s}$ does work. The calculation $\mathbf{F} \cdot \mathbf{s}$ multiplies the component of $\mathbf{F}$ in the direction of motion times the distance traveled and hence defines the work done. In vector calculus, the dot product plays a role in computing certain line and surface integrals and in the definition of the divergence of a function ($\nabla \cdot \mathbf{F}$) to be studied in later chapters.

ROW AND COLUMN VECTORS

So far in this chapter, we have represented a vector as a row of numbers, called a *row vector*. The vector $\mathbf{x}$ can also be represented as a *column vector* in the form

$$\mathbf{x} = \begin{bmatrix} x_1 \\ x_2 \\ x_3 \end{bmatrix}.$$

These representations of vectors will be useful in Chapter 3 when matrix operations are considered. For two vectors $\mathbf{x}$ and $\mathbf{y}$ that have the same number of components, the dot product can be computed by writing one vector as a row vector and the other as a column vector and performing matrix multiplication to yield

$$\mathbf{x} \cdot \mathbf{y} = [x_1, x_2, x_3] \begin{bmatrix} y_1 \\ y_2 \\ y_3 \end{bmatrix} = x_1y_1 + x_2y_2 + x_3y_3. \tag{2.18}$$

In some contexts, the terms *scalar product* and *inner product* are also used to describe the product of Equation 2.18.

PROJECTIONS A very important application of the dot product of two vectors is in determining the *projection* of one nonzero vector along another. Figure 2.5 shows the geometry of the problem.

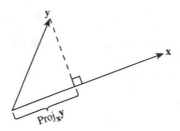

FIGURE 2.5 *Projections*

The figure shows the projection of **y** on **x** determined by dropping a line from the end of **y** that is perpendicular to **x** in the plane of **x** and **y**. Thus, the length of the projection measured from the origin of the vectors to the intersection of the line with **x** is $||\mathbf{y}|| \cos\theta$. We now wish to define the projection in terms of the dot product and also define the vector that specifies the projection. This is accomplished by first observing that

$$||\mathbf{y}|| \cos\theta = \frac{(||\mathbf{x}||)||\mathbf{y}|| \cos\theta}{||\mathbf{x}||}$$

and noting that $\mathbf{x}/||\mathbf{x}||$ is a unit vector in the direction of **x** and hence defines the direction of the projection. Using the notation $\text{proj}_\mathbf{x}\mathbf{y}$ to indicate the *projection* of **y** on **x**, we can write

$$\text{proj}_\mathbf{x}\mathbf{y} = \frac{(\mathbf{x} \cdot \mathbf{y})}{||\mathbf{x}||} \frac{\mathbf{x}}{||\mathbf{x}||} = \frac{(\mathbf{x} \cdot \mathbf{y})}{||\mathbf{x}||^2}\mathbf{x}. \qquad (2.19)$$

For the vectors in Figure 2.5, $\theta < \pi/2$ so $\mathbf{x} \cdot \mathbf{y} > 0$. If $\theta > \pi/2$, $\mathbf{x} \cdot \mathbf{y} < 0$.

□ EXAMPLE 2.4 *Projections*

a. The projection of $\mathbf{u} = 2\mathbf{i} + 3\mathbf{j}$ on the vector $\mathbf{v} = \mathbf{i} + \mathbf{j}$ is given by

$$\text{proj}_\mathbf{v}\mathbf{u} = \frac{5\mathbf{v}}{(\sqrt{2})^2} = \frac{5}{2}\mathbf{i} + \frac{5}{2}\mathbf{j}.$$

b. Let **x** be a nonzero vector. Then, for any other vector **y**, the vector

$$\mathbf{w} = \mathbf{y} - \frac{(\mathbf{y} \cdot \mathbf{x})}{||\mathbf{x}||^2}\mathbf{x} = \mathbf{y} - \text{proj}_\mathbf{x}\mathbf{y}$$

is orthogonal to **x**. To show this, form the dot product of **w** and **x** and expand as

$$\mathbf{w} \cdot \mathbf{x} = \left[\mathbf{y} - \frac{(\mathbf{y} \cdot \mathbf{x})}{||\mathbf{x}||^2}\mathbf{x}\right] \cdot \mathbf{x} = \mathbf{y} \cdot \mathbf{x} - \frac{(\mathbf{y} \cdot \mathbf{x})}{||\mathbf{x}||^2}\mathbf{x} \cdot \mathbf{x}$$

$$= \mathbf{y} \cdot \mathbf{x} - \frac{(\mathbf{y} \cdot \mathbf{x})||\mathbf{x}||^2}{||\mathbf{x}||^2} = \mathbf{y} \cdot \mathbf{x} - \mathbf{y} \cdot \mathbf{x} = 0.$$

Using the vectors of Part a, the component of $\mathbf{u}$ perpendicular to $\mathbf{v}$ is

$$\mathbf{u}_\perp = (-\mathbf{i} + \mathbf{j})/2.$$

$\square$

WHAT IF?   You should draw the vectors $\mathbf{u}$ and $\mathbf{v}$ in Example 2.4 to confirm that the projection is correct. Also, show that $\mathbf{u}$ can be written as the sum of the components parallel and perpendicular to $\mathbf{v}$.

CROSS PRODUCT

The *cross product* is an operation on vectors in $\mathbf{R}^3$ that has widespread use in physics. Whereas the scalar (dot) product of two vectors yields a scalar value, the cross product of two vectors is another vector. Thus, some authors use the term *vector product* for this operation.

The cross product is indicated as $\mathbf{c} = \mathbf{a} \times \mathbf{b}$ and satisfies the following relationship:

$$\mathbf{c} = \| \, \mathbf{a} \, \| \, \| \, \mathbf{b} \, \| \, \sin\theta \, \mathbf{n}, \tag{2.20}$$

where $\mathbf{n}$ is a unit vector that is perpendicular to both $\mathbf{a}$ and $\mathbf{b}$. Since the resulting vector is perpendicular to both $\mathbf{a}$ and $\mathbf{b}$, we can write

$$\mathbf{c} \perp \mathbf{a}, \ \mathbf{c} \perp \mathbf{b}. \tag{2.21}$$

The cross product is defined only in three dimensions, and it is natural to apply it to vectors whose components are the value of the vector along the x-, y-, and z-axes. For the vectors $\mathbf{a} = [a_x, a_y, a_z]$ and $\mathbf{b} = [b_x, b_y, b_z]$, the expression for $\mathbf{a} \times \mathbf{b}$ can be written in terms of the vector components as

$$\mathbf{a} \times \mathbf{b} = (a_y b_z - a_z b_y)\mathbf{i} + (a_z b_x - a_x b_z)\mathbf{j} + (a_x b_y - a_y b_x)\mathbf{k}. \tag{2.22}$$

A technique to remember the expansion of Equation 2.22 is to write the array

$$\mathbf{a} \times \mathbf{b} = \begin{vmatrix} \mathbf{i} & \mathbf{j} & \mathbf{k} \\ a_x & a_y & a_z \\ b_x & b_y & b_z \end{vmatrix} \tag{2.23}$$

and expand this as a determinant in the form

$$\mathbf{a} \times \mathbf{b} = \mathbf{i} \begin{vmatrix} a_y & a_z \\ b_y & b_z \end{vmatrix} - \mathbf{j} \begin{vmatrix} a_x & a_z \\ b_x & b_z \end{vmatrix} + \mathbf{k} \begin{vmatrix} a_x & a_y \\ b_x & b_y \end{vmatrix}. \tag{2.24}$$

This is not a numerical determinant in the ordinary sense, as treated in Chapter 3, but it is a useful mnemonic to expand the cross product.

Figure 2.6 shows the relationship of the vectors $\mathbf{a}$, $\mathbf{b}$, and $\mathbf{c}$ in the cross product $\mathbf{c} = \mathbf{a} \times \mathbf{b}$ for the positive direction of the cross product vector **c**. This orientation of the vectors defines a right-handed system of axes. Remember that a right-handed system is defined such that if the fingers of your right hand curl from **a** to **b**, your thumb points in the direction of the cross product vector.

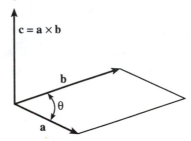

FIGURE 2.6 *Cross product of two vectors*

□ EXAMPLE 2.5   *Cross Product*
     A result of electromagnetic field theory is that the force **F** acting on a charge q moving with a velocity **v** in a magnetic field **B** is given by

$$\mathbf{F} = q\,\mathbf{v} \times \mathbf{B},$$

where q is measured in coulombs and **B** has units of teslas in the mks (meter-kilogram-second) system.[3]
     Let the physical values be

$$q = 1.6 \times 10^{-19} \text{ coulombs},$$
$$\mathbf{B} = 0.5\,\mathbf{i} - 0.8\,\mathbf{k} \text{ teslas},$$
$$\mathbf{v} = 10\mathbf{j} \text{ meters/second},$$

where the physical units apply to the magnitude of the vectors.
     Then, the force acting on the charge is

$$q\,\mathbf{v} \times \mathbf{B} = q \begin{vmatrix} \mathbf{i} & \mathbf{j} & \mathbf{k} \\ 0 & 10 & 0 \\ 0.5 & 0 & -0.8 \end{vmatrix}$$
$$= q\,[-8.0\mathbf{i} - 5.0\mathbf{k}].$$

The result is $\mathbf{F} = (-12.8\mathbf{i} - 8.0\mathbf{k}) \times 10^{-19}$ newtons. Although the particle is initially moving in the y-direction, the field gives the particle components of velocity in the x- and z-directions as well.

□

     Cross products are used extensively in mechanics to represent such quantities as angular momentum. In electromagnetic field theory, the cross product defines the direction of the force acting on a moving charged particle in a magnetic field. The cross product will also be used again in Chapter 11, where the expression $\nabla \times \mathbf{F}$ will be introduced as the *curl* of a vector function.

---

[3]The mks system is also called the SI (International System of Units) measurement system.

Examples of Applications   Table 2.3 lists a few of the applications of vectors commonly used in physics.

TABLE 2.3 *A few vector applications*

Meaning	Vector Form
Newton's second law	$\mathbf{F} = m\mathbf{a}$
Force on a charge	$\mathbf{F} = q\mathbf{E}$
Linear momentum	$\mathbf{p} = m\mathbf{v}$
Work	$W = \mathbf{F} \cdot \mathbf{s}$
Force on a charge	$\mathbf{F} = q\mathbf{v} \times \mathbf{B}$
Torque	$\mathbf{T} = \mathbf{r} \times \mathbf{F}$
Angular acceleration	$\mathbf{L} = \mathbf{r} \times \mathbf{p}$

VECTORS IN HIGHER DIMENSIONS

In elementary physics, the vectors of interest represent physical phenomena, such as the force or the electric field at a point in 2D or 3D space. However, the generalization of the vector concept to vectors having more than two or three components is useful in many areas of applied mathematics.

The notation $\mathbf{R}^n$ will denote the set of all vectors with n real components. In the previous section, $\mathbf{R}^2$ denoted the set containing all the 2D vectors with components of the form $[x, y]$. Similarly, $\mathbf{R}^3$ contains all the 3D vectors with three real components. The emphasis on the real components is made here because vectors may also have complex values for components. Unless otherwise noted in the text, the vectors discussed have real components.

When $\mathbf{x}$ is one of the vectors in the set $\mathbf{R}^n$, we say that $\mathbf{x}$ is in $\mathbf{R}^n$ or occasionally use the notation $\mathbf{x} \in \mathbf{R}^n$. Thus, the statement $\mathbf{a} \in \mathbf{R}^5$ means that $\mathbf{a}$ is a five-dimensional vector with five real components.

An n-dimensional vector $\mathbf{x}$ in $\mathbf{R}^n$ is defined as the n-tuple of real numbers written

$$\mathbf{x} = [x_1, x_2, \ldots, x_n], \qquad (2.25)$$

where each x_i is a real number. For convenience, such a vector is often referred to as an *n-vector* or simply as a *vector* if n is understood.

Basic operations are defined for vectors in $\mathbf{R}^n$ that are straightforward generalizations of the familiar operations of addition, multiplication by a scalar, and defining the length of vectors in $\mathbf{R}^2$ and $\mathbf{R}^3$.

The *addition* of two vectors $\mathbf{x}$ and $\mathbf{y}$ in $\mathbf{R}^n$ is defined as

$$
\begin{aligned}
\mathbf{x} + \mathbf{y} &= [x_1, x_2, \ldots, x_n] + [y_1, y_2, \ldots, y_n] \\
&= [x_1 + y_1, x_2 + y_2, \ldots, x_n + y_n].
\end{aligned} \tag{2.26}
$$

The *scalar multiple* of a vector $\alpha\mathbf{x}$ is formed by multiplication of each component of a vector in $\mathbf{R}^n$ by a real number α, with the result

$$
\alpha\mathbf{x} = [\alpha x_1, \alpha x_2, \ldots, \alpha x_n]. \tag{2.27}
$$

The term *norm* is sometimes used for the length of vectors in $\mathbf{R}^n$ when $n > 3$. The generalization of the vector length given in Equation 2.10 for a vector with n components is

$$
\|\mathbf{x}\| = \left(\sum_{i=1}^{n} x_i^{\,2} \right)^{1/2} = \sqrt{x_1^2 + x_2^2 + \cdots + x_n^2}. \tag{2.28}
$$

The norm is a scalar value that measures the size of the vector in n-space. However, there are different types of norms useful in some applications. Therefore, the norm in Equation 2.28 is properly referred to as the 2-norm to distinguish it from other norms. The 2 defines the power to which the components are raised. In physics, $\mathbf{R}^n$ with this norm is called the Euclidean n-space.

□ EXAMPLE 2.6   *Vectors of Higher Dimensions*

Many physical quantities can be represented by an n-vector. For example, suppose an analog signal sampled periodically yields n samples, denoted

$$
[s_1, s_2, \ldots, s_n].
$$

Let the values be represented by the vector $\mathbf{S}$, which could serve as a shorthand notation when referring to the samples.

However, since $\mathbf{S}$ is a vector, useful computations can be performed on $\mathbf{S}$. If the signal is input to a multiplier circuit that multiplies the input by α, then the sampled output is $\alpha\,\mathbf{S}$ if the input and output signals are sampled simultaneously. The norm of $\mathbf{S}$ also has physical meaning if the input signal is an electrical signal. In this case, the root-mean-square (RMS) value is the norm divided by $\sqrt{n}$.

Vector language and vectors of greater than three dimensions are used in solving problems in many areas of physics. Equations of a mechanical system are often described as having N *degrees of freedom* using N independent vectors to express the solutions. In statistical mechanics, an n-dimensional space is used. Relativity uses a four-dimensional space, and quantum mechanics even extends the space of n dimensions to one with an infinite number of components.

□

MATLAB VECTORS

One of MATLAB's fundamental data types is the n-dimensional vector. The vector **x** can be entered in the form

```
>> x=[x1 x2 ...xn]
```

where the `xi` are the components written in a row with spaces separating the values. This MATLAB form represents the vector $\mathbf{x} = [x_1, x_2, \ldots, x_n]$. The fundamental operations on vectors are defined in Table 2.4, in which the first column defines the operation, the second column shows the MATLAB format, and the third column presents the mathematical result.

TABLE 2.4 *MATLAB Vector Operations*

Operation	MATLAB Form	Mathematical Form
Vector	`[x1 x2 ...xn]`	$[x_1, x_2, \ldots, x_n]$
Addition	`x+y`	$[x_1 + y_1, x_2 + y_2, \ldots, x_n + y_n]$
Subtraction	`x-y`	$[x_1 - y_1, x_2 - y_2, \ldots, x_n - y_n]$
Transpose	`x'`	Column vector
Dot product	`x*y'` or **dot(x,y)**	$x_1 y_1 + x_2 y_2 + \cdots + x_n y_n$
Cross product	**cross(x,y)**	$\mathbf{x} \times \mathbf{y}$
Norm	**norm(x)**	$\|\mathbf{x}\|$
Special:		
$x_i \times y_i$	`x.*y`	$[x_1 y_1, x_2 y_2, \ldots, x_n y_n]$
x_i / y_i	`x./y`	$[x_1/y_1, x_2/y_2, \ldots, x_n/y_n]$
$x_i{}^m$	`x.^m`	$[x_1{}^m, x_2{}^m, \ldots, x_n{}^m]$
$f(x_i)$	`f(x)`	$[f(x_1), f(x_2), \ldots, f(x_n)]$
Augment	`[x y]`	$[x_1, x_2, \ldots, x_n, y_1, y_2, \ldots, y_n]$
Element x_i	`x(i)`	x_i
Row vector		
Integers m, n	`x=m:n`	$x = [m, m+1, \ldots, n]$
Real a, dx, b	`x=a:dx:b`	$x = [a, a+dx, a+2dx, \ldots, b]$

To compute the dot product of two vectors, the *transpose* of one of the vectors must be used in the product. The transpose of a vector **xt** is entered as

```
>> xt=[x1 x2 ...xn]'
```

or as

```
>> xt=[x1;x2;...;xn]
```

with semicolons separating the components. The result is the column vector

$$xt = \begin{bmatrix} x_1 \\ x_2 \\ \vdots \\ x_n \end{bmatrix}.$$

The MATLAB command

`>> xt=x'`

converts the row vector **x** into the column vector **xt**. The command **dot** computes the dot product of two vectors without the need to form the transpose of one of the vectors.

There are a number of special operations listed in the table that are not standard vector operations. MATLAB operators multiply (.*) or divide (./) two vectors element by element when the period operator follows the vector. Also, each element of a vector may be raised to a power. Most of the MATLAB functions, such as **exp** or **sqrt**, can be used to compute the function of each element of a vector.

The n components of a vector **x** can be augmented with the elements of a vector **y** having n components with the command

`>> z=[x y]`

after which **z** becomes a row vector with $2n$ elements. The ith element of a vector **x** is designated as **x(i)**. Thus, the third element of **x** is **x(3)**. Finally, a vector with fixed increments between the elements can be generated. For example, the command

`>> x=[1:0.5:3]`

creates the mathematical vector $\mathbf{x} = [1, 1.5, 2.0, 2.5, 3.0]$.

☐ EXAMPLE 2.7   *MATLAB Operations*

The accompanying MATLAB sessions illustrate various operations from previously presented discussions and examples. In the first script, the command **abs** is used to compute the absolute value of a number. Next, the function **clbindec** is called to convert a binary number to decimal using Equation 2.2. The binary digits are entered as elements in a vector. After the conversion, the result is displayed showing 111110_2 converted to 62. The function is also displayed for convenience. It is stored on disk as file CLBINDEC.M.

MATLAB Script ───────────────────────────────

```
Example 2.7
>>x=-5
x =
     -5
>>y=abs(x)
y =
      5
```

```
>>xbin=[1 1 1 1 1 0]; % Convert binary to decimal

>>Ndec=clbindec(xbin)

Ndec =
    62
%
function ndec=clbindec(xbin)
% CALL:  ndec=clbindec(xbin), Convert positive binary number xbin
%   to decimal value ndec. No error check if xbin is not binary
m=length(xbin);
ndec=0;
for I=1:m
 ndec=ndec+xbin(I)*2^(m-I);
end
```

The second session shows the vector operations of Example 2.2 performed with MATLAB commands. The commands compute the norms of the vectors, the dot product, and the angle between the vectors in radians and degrees.

MATLAB Script _____

```
Example 2.7
%  Norm, Dot Product and Angle
>>x1=[2 4 0]
x1 =
     2     4     0
>>y1=[6 -4 2]
y1 =
     6    -4     2
>>% Compute norms
>>nx1=norm(x1)
nx1 =
    4.4721
>>ny1=norm(y1)
ny1 =
    7.4833
>>% Dot product
>>xdoty=x1*y1'
xdoty =
    -4
>>% Angle between vectors
>>theta=acos(xdoty/(nx1*ny1))
theta =
    1.6906
>>thetadeg=theta*180/pi
thetadeg =
   96.8646
>>quit
```

□

Symbolic MATLAB The *Symbolic Math Toolbox* has commands to manipulate vectors defined in symbolic form. For example, the command **symop** performs symbolic operations on vectors and matrices. To compute the norm of a vector and the dot product of vectors, the symbolic command **transpose** is useful.

PROPERTIES OF VECTORS

Consider arbitrary vectors $\mathbf{x}$ and $\mathbf{y} \in \mathbf{R}^n$ and a scalar α. A basic property we assume to be true for any vectors in $\mathbf{R}^n$ is that the sum of the vectors is also a vector in $\mathbf{R}^n$. The scalar multiple of any vector in $\mathbf{R}^n$ is also a vector in $\mathbf{R}^n$. These statements are called *closure* properties and are stated mathematically as

$$\mathbf{x} + \mathbf{y} \in \mathbf{R}^n \quad \text{and} \quad \alpha\mathbf{x} \in \mathbf{R}^n$$

for vectors in $\mathbf{R}^n$.

Other properties of vectors in $\mathbf{R}^n$ define various addition and multiplication operations. For vectors $\mathbf{x}$, $\mathbf{y}$, and $\mathbf{z}$ and scalars α and β, these vectors satisfy the following relationships for addition and multiplication by a scalar.

The properties of *vector addition* are as follows:

1. $\mathbf{x} + \mathbf{y} = \mathbf{y} + \mathbf{x}$.   Commutative law

2. $(\mathbf{x} + \mathbf{y}) + \mathbf{z} = \mathbf{x} + (\mathbf{y} + \mathbf{z})$.   Associative law

3. $\mathbf{x} + \mathbf{0} = \mathbf{0} + \mathbf{x} = \mathbf{x}$, where $\mathbf{0} = [0, 0, ..., 0]$.   Zero element

4. $\mathbf{x} + (-\mathbf{x}) = \mathbf{0}$.

In Property 4, the negative of $\mathbf{x}$ is defined as $(-1)\mathbf{x} = -\mathbf{x}$ or in terms of the components,

$$-\mathbf{x} = [-x_1, -x_2, ..., -x_n].$$

Let α and β be scalars. Then, the scalar multiplication properties of vectors are

5. $\alpha(\mathbf{x} + \mathbf{y}) = \alpha\mathbf{x} + \alpha\mathbf{y}$.

6. $(\alpha + \beta)\mathbf{x} = \alpha\mathbf{x} + \beta\mathbf{x}$.

7. $(\alpha\beta)\mathbf{x} = \alpha(\beta\mathbf{x})$.

8. $\mathbf{x} = 1 \times \mathbf{x}$

□ **EXAMPLE 2.8**   *Commutative Law*

There are several ways to define the basic properties of vectors with real elements. Our approach is to postulate these properties without proof, as has just been done. For example, the commutativity property of vectors is a proposition that can be proven but was assumed. Thus, we state that addition of two vectors is *commutative* because

$$\mathbf{x} + \mathbf{y} = \mathbf{y} + \mathbf{x}. \tag{2.29}$$

Another approach to presenting the basic properties of vectors is to prove the relationships using the axioms that were presented earlier for real numbers. The properties of vectors listed in this section can all be proven from the definition of vector addition and multiplication of a vector by a scalar using the properties of real numbers. For example, given that $\alpha + \beta = \beta + \alpha$ for any real numbers α and β, proving the commutative law for vectors is straightforward. Since the vectors have real components, the definition of $\mathbf{x} + \mathbf{y}$ previously given by Equation 2.26 yields the jth component of the sum as

$$x_j + y_j \quad \text{for } j = 1, 2, \ldots, n. \tag{2.30}$$

Then, by the commutative law for real numbers, each component can also be written as

$$y_j + x_j \quad \text{for } j = 1, 2, \ldots, n. \tag{2.31}$$

This latter sum is $\mathbf{y} + \mathbf{x}$, so vector addition is commutative, as stated in Equation 2.29.

□

In the relationships that hold for vectors, the associative, commutative, and distributive laws of multiplication that apply to real numbers are not listed. Instead, the dot product of two vectors $\mathbf{x}$ and $\mathbf{y}$ is defined as the sum of the products of the respective components.

The dot product of two n-dimensional vectors $\mathbf{x}$ and $\mathbf{y}$ is defined as

$$\mathbf{x} \cdot \mathbf{y} = x_1 y_1 + x_2 y_2 + \cdots + x_n y_n = \sum_{i=1}^{n} x_i y_i. \tag{2.32}$$

From the definition, it is clear that the dot product is commutative, since $\mathbf{x} \cdot \mathbf{y} = \mathbf{y} \cdot \mathbf{x}$. A useful equation relates the norm of a vector in $\mathbf{R}^n$ and the dot product. Letting $\mathbf{x} = \mathbf{y}$ in Equation 2.32, the square of the norm is computed as

$$\| \mathbf{x} \|^2 = \mathbf{x} \cdot \mathbf{x}, \tag{2.33}$$

using the norm as previously defined in Equation 2.28.

Unfortunately, many of the geometric properties of vectors in $\mathbf{R}^2$ and $\mathbf{R}^3$ are lost when the dimension is greater than three. However, it is useful to define the concept of orthogonal vectors in $\mathbf{R}^n$.

In terms of the dot product, the nonzero vectors $\mathbf{x}$ and $\mathbf{y}$ are said to be *orthogonal* if $\mathbf{x} \cdot \mathbf{y} = 0$. Vectors in $\mathbf{R}^n$ that are mutually orthogonal and of unit length (norm is 1) are called *orthonormal*. Thus, a set of vectors $\mathbf{x}_1, \mathbf{x}_2, \ldots, \mathbf{x}_n$ is orthonormal if

$$\mathbf{x}_i \cdot \mathbf{x}_j = \begin{cases} 1, & \text{if } i = j, \\ 0, & \text{if } i \neq j. \end{cases}$$

The vector $\mathbf{0}$ is considered to be orthogonal to every vector in $\mathbf{R}^n$, since $\mathbf{x} \cdot \mathbf{0} = 0$ for every vector $\mathbf{x} \in \mathbf{R}^n$. In the next section, orthonormal sets of vectors will be used to represent vectors in $\mathbf{R}^n$.

When dealing with vectors in $\mathbf{R}^n$ and in some other applications, the dot product is often called the *inner product*. The operation is defined by the notation

$$\langle \mathbf{x}, \mathbf{y} \rangle = x_1 y_1 + x_2 y_2 + \cdots + x_n y_n \tag{2.34}$$

for two vectors in $\mathbf{R}^n$.

The notation for a set of orthonormal vectors can be simplified by using the inner product and introducing the *Kronecker delta* with the definition

$$\delta_{ij} = \begin{cases} 1, & \text{if } i = j, \\ 0, & \text{if } i \neq j. \end{cases}$$

Then, a set of vectors $\mathbf{x}_1, \mathbf{x}_2, \ldots, \mathbf{x}_n$ is an orthonormal set if and only if

$$\langle \mathbf{x}_i, \mathbf{x}_j \rangle = \delta_{ij} \qquad i, j = 1, 2, \ldots, n.$$

The set of all n-dimensional vectors with complex components will be designated as $\mathbf{C}^n$. The vectors in the set have the form

$$\mathbf{z} = [z_1, z_2, \ldots, z_n],$$

where each component is the complex number $z_i = x_i + iy_i$, $i = 1, 2, \ldots n$. The vector $\bar{\mathbf{z}}$ has the ith component $\bar{z}_i = x_i - iy_i$

It is possible to define inner product and length for complex vectors. As introduced earlier in the chapter, the length of the complex number $z = x + iy$ considered as a vector in xy-space is

$$||x + iy|| = (z\bar{z})^{1/2} = \sqrt{(x + iy)(x - iy)} = \sqrt{x^2 + y^2}.$$

Since x and y are real numbers, the length of the vector is a positive number. Thus, if z is not zero, $z\bar{z}$ is always real and positive.

With complex vectors, we wish to define an inner product such that the length of a vector $\mathbf{z}$ is a positive number. Consider the vectors

$$\mathbf{z} = [z_1, z_2, \ldots, z_n] \quad \text{and} \quad \mathbf{w} = [w_1, w_2, \ldots, w_n]$$

with complex components. The complex conjugate of such vectors is formed by conjugating each component of the vectors. The *inner product* of complex vectors $\mathbf{z}$ and $\mathbf{w}$ can be defined as

$$\langle \bar{\mathbf{z}}, \mathbf{w} \rangle = \bar{z}_1 w_1 + \cdots + \bar{z}_n w_n,$$

in which the conjugate of the first vector is used. For complex vectors, the inner product is thus formed by first conjugating $\mathbf{z}$ and then forming the sum of the products of the corresponding elements of $\bar{\mathbf{z}}$ and $\mathbf{w}$. Then, the norm or length of the vector $\mathbf{z}$ can be written as

$$\|\mathbf{z}\| = \langle \mathbf{z}, \bar{\mathbf{z}} \rangle^{1/2} = \langle \bar{\mathbf{z}}, \mathbf{z} \rangle^{1/2},$$

with the understanding that the complex conjugate must be used in the product if $\mathbf{z}$ has complex components. If $\mathbf{z} \neq \mathbf{0}$, it is simple to prove that $\langle \mathbf{z}, \bar{\mathbf{z}} \rangle > 0$.

For vectors with real components, the dot product previously defined and the inner product are identical since $z = \bar{z}$ if z is a real number and so it follows that $\mathbf{z} = \bar{\mathbf{z}}$ for real vectors.

VECTOR SPACES

■

Thus far we have moved from the easily visualized vectors in $\mathbf{R}^2$ and $\mathbf{R}^3$ to the more "abstract" vectors in $\mathbf{R}^n$. However, except for the cross product, the various vector properties and operations are equally valid independent of the dimension of the vectors. One further generalization dealing with vectors is of great importance in both theoretical studies and certain practical problems.

Our final generalization for vectors in $\mathbf{R}^n$ introduces the concept of a *vector space*. The vector space is a set of vectors together with the rules for vector addition and multiplication by a scalar. Vectors in $\mathbf{R}^n$ are said to form a vector space. These vectors satisfy the closure condition and the eight properties of vectors listed previously.

□ EXAMPLE 2.9   *Vector Space*

The introduction of n-dimensional vectors and vector spaces represents an important generalization from the 2D and 3D space of the physical world. Advantages of the extension will be shown in a number of ways in this text. The main principle is that once a fact about vector spaces in general is known to be true, we can apply that fact to every vector space, rather than having to prove the fact for each new vector space.

For example, the number zero (0) is itself a vector space that is sometimes called the *trivial* vector space. Thus, zero satisfies all the properties of a vector space such as $0 + 0 = 0$, $\alpha \times 0 = 0$, and $0 = 1 \times 0$. Note that the number 1 does not form a vector space since it violates the closure property; that is,

$1 + 1 = 2$ but 2 is not in the space. Extending this argument slightly shows that the vector $\mathbf{0}$ in $\mathbf{R}^n$ also forms a vector space. We will study much more interesting vector spaces in the sections to follow.

The set of points in $\mathbf{R}^2$ that lie on a line passing through the origin constitutes a vector space. The points $[x, y]$ are defined by the equation $y = mx$ where m is a scalar constant. Letting $\mathbf{x} = [x_1, y_1]$ and $\mathbf{y} = [x_2, y_2]$ with $y_1 = mx_1$ and $y_2 = mx_2$, it is easy to show that the closure property and the other eight properties for vector spaces are satisfied. However, the set of points on a line not passing through the origin ($y = mx + b$, $b \neq 0$) does not form a vector space, as you are asked to prove in Problem 2.10.

□

SUBSPACES

In some problems, a subset of the vectors in $\mathbf{R}^n$ is important, rather than the entire set. If the vectors in the subset satisfy the closure property of addition and scalar multiplication, the vectors also form a vector space called a subspace of $\mathbf{R}^n$. A set of vectors S that is a subset of the vectors in $\mathbf{R}^n$ forms a vector space if α is a scalar and for every vector $\mathbf{x}$ and $\mathbf{y}$ in S,

1. $\mathbf{x} + \mathbf{y} \in S$.   Closure under addition

2. $\alpha\, \mathbf{x} \in S$, α a scalar.   Closure under scalar multiplication

3. $\mathbf{0} \in S$.

4. $-\mathbf{x} \in S$.

A vector space with these properties is said to be a *subspace* of $\mathbf{R}^n$. The vector spaces $\mathbf{R}, \mathbf{R}^2, \ldots$, including $\mathbf{R}^n$ itself are subspaces of $\mathbf{R}^n$. Actually, Properties 3 and 4 follow from Property 2 and the fact that for a vector in $\mathbf{R}^n$, $0 \times \mathbf{x} = \mathbf{0}$. Similarly, Property 4 can be derived from Property 2 and the fact that $-\mathbf{x} = (-1)\mathbf{x}$ for vectors in $\mathbf{R}^n$.

There are two important properties of a subspace that are worth mentioning explicitly:

1. Every subspace of a vector space contains $\mathbf{0}$.

2. Any subspace of a vector space is a vector space, with the operations inherited from the original space.

If the vectors in the subset of $\mathbf{R}^n$ form a subspace, it is not necessary to verify that they satisfy the eight properties of vectors in a vector space because the properties are satisfied in the larger space and will be satisfied in every subspace. Conversely, if the subset does not form a subspace, some of the properties associated with a vector space will not apply.

LINEAR INDEPENDENCE AND BASES

Representing one mathematical quantity in terms of a combination of (presumably) simpler quantities is useful in many applications. As shown in an earlier section of this chapter, a 3D vector may be expressed as a sum of scalar multiples of three orthogonal unit vectors. In this section, this concept is extended to vectors in $\mathbf{R}^n$.

Extending the previous discussion of unit vectors in $\mathbf{R}^3$, it is natural to write a vector $\mathbf{x}$ in $\mathbf{R}^n$ in the form

$$\mathbf{x} = [x_1, x_2, \ldots, x_n] = x_1\mathbf{e}_1 + x_2\mathbf{e}_2 + \cdots + x_n\mathbf{e}_n, \qquad (2.35)$$

where

$$\begin{aligned} \mathbf{e}_1 &= [1, 0, \ldots, 0], \\ \mathbf{e}_2 &= [0, 1, \ldots, 0], \\ &\vdots \\ \mathbf{e}_n &= [0, 0, \ldots, 1]. \end{aligned} \qquad (2.36)$$

This set of unit vectors forms an orthonormal set, and any vector in $\mathbf{R}^n$ can be written as a sum of scalar multiples of these vectors. The importance of this fundamental set of vectors for $\mathbf{R}^n$ will be explored in more detail shortly.

□ EXAMPLE 2.10

Vector in terms of Unit Vectors

In $\mathbf{R}^3$, the standard unit vectors would be $\mathbf{e}_1$, $\mathbf{e}_2$, and $\mathbf{e}_3$. In most applications, they are written $\mathbf{i}$, $\mathbf{j}$, and $\mathbf{k}$, as used in Example 2.3. Thus, the vector

$$\mathbf{x} = [1, 1.5, 3, 7, 8]$$

can be written in terms of the unit vectors in $\mathbf{R}^5$ as follows:

$$\mathbf{x} = 1\mathbf{e}_1 + 1.5\mathbf{e}_2 + 3\mathbf{e}_3 + 7\mathbf{e}_4 + 8\mathbf{e}_5. \qquad (2.37)$$

A vector component in the direction of a unit vector is determined by taking the inner product of the vector and the unit vector. For the vector $\mathbf{x}$, the component in the direction of $\mathbf{e}_1$ is thus

$$\langle \mathbf{x}, \mathbf{e}_1 \rangle = 1\langle \mathbf{e}_1, \mathbf{e}_1 \rangle + 1.5\langle \mathbf{e}_1, \mathbf{e}_2 \rangle + \cdots + 8\langle \mathbf{e}_1, \mathbf{e}_5 \rangle = 1,$$

since the unit vectors form an orthonormal set. In terms of projections as defined in Equation 2.19, the component of $\mathbf{x}$ in the direction of $\mathbf{e}_1$ is

$$\text{proj}_{\mathbf{e}_1} \mathbf{x} = \frac{\langle \mathbf{e}_1, \mathbf{x} \rangle}{||\mathbf{e}_1||^2} \mathbf{e}_1 = 1\mathbf{e}_1.$$

□

Linear Independence of Vectors Many problems involving vector spaces are simplified if we find a set of vectors that can be used to generate any vector in the space. Such a set is called a *basis* for the space.

The properties of a basis set will be explored after the concept of linear independence is defined.

A vector $\mathbf{y}$ is called a *linear combination* of a set of vectors designated $\mathbf{x}_1, \mathbf{x}_2, \ldots, \mathbf{x}_n$ if $\mathbf{y}$ can be written as a sum of scalar multiples of these vectors in the form

$$\mathbf{y} = \alpha_1 \mathbf{x}_1 + \alpha_2 \mathbf{x}_2 + \cdots + \alpha_n \mathbf{x}_n, \qquad (2.38)$$

where each α_j is a scalar. The set of vectors

$$\mathbf{x}_1, \mathbf{x}_2, \ldots, \mathbf{x}_n \qquad (2.39)$$

is described as *linearly independent* if none of the vectors can be written as a linear combination of the others. An equivalent statement expresses the fact that a linear combination of independent vectors cannot be summed to yield the zero vector unless each coefficient is zero.

A set of vectors $\mathbf{x}_1, \mathbf{x}_2, \ldots, \mathbf{x}_n$ in $\mathbf{R}^n$ is said to be linearly independent if the equation

$$\alpha_1 \mathbf{x}_1 + \alpha_2 \mathbf{x}_2 + \cdots + \alpha_n \mathbf{x}_n = \mathbf{0} \qquad (2.40)$$

is true only when the scalars $\alpha_1 = \alpha_2 = \cdots = \alpha_n = 0$. Otherwise, the vectors are linearly *dependent*, and at least one of them is a linear combination of the others. Thus, if the relation in Equation 2.40 does hold with not all the α's equal to 0, the vectors are *linearly dependent*.

For two vectors, the definition means that the vectors are linearly dependent if and only if one is a scalar multiple of the other. When more than two vectors are involved, the test for independence results in a set of equations for the coefficients.

□ EXAMPLE 2.11   *Linearly Independent Vectors*

To test the vectors $\mathbf{i}, \mathbf{j}, \mathbf{k}$ for linear independence, we form the equation

$$a_1 \mathbf{i} + a_2 \mathbf{j} + a_2 \mathbf{k} = \mathbf{0},$$

which is equivalent to

$$[a_1, a_2, a_3] = [0, 0, 0].$$

By equating components, the vector equation leads to the conclusion

$$a_1 = 0, \quad a_2 = 0, \quad a_3 = 0.$$

Hence, the three unit vectors in $\mathbf{R}^3$ are linearly independent. In fact, it is easy to show that any set of mutually orthogonal nonzero vectors from $\mathbf{R}^n$ is linearly independent.

The vectors $[3, 2, 3], [1, 1, 0], [0, 1, -3]$ can be shown to be linearly dependent by solving the system of equations resulting from setting

$$a_1 [3, 2, 3] + a_2 [1, 1, 0] + a_3 [0, 1, -3] = [0, 0, 0]. \qquad (2.41)$$

The system

$$
\begin{aligned}
3a_1 &+ a_2 & &= 0 \\
2a_1 &+ a_2 &+ a_3 &= 0 \\
3a_1 & &- 3a_3 &= 0
\end{aligned}
\qquad (2.42)
$$

has a solution $[a_1, a_2, a_3] = [1, -3, 1]$, among others. Thus, the first vector is a linear combination of the other two since

$$[3, 2, 3] = 3\,[1, 1, 0] - 1\,[0, 1, -3].\qquad(2.43)$$

Although matrices and determinants are not treated until Chapter 3, a useful result from matrix theory can be applied to determine if n vectors in $\mathbf{R}^n$ are independent. First, form the matrix with the vectors as columns. Then, if the determinant of the matrix is nonzero, the vectors are independent. If the determinant is zero, the vectors are dependent. The determinant formed by the unit vectors is

$$\begin{vmatrix} 1 & 0 & 0 \\ 0 & 1 & 0 \\ 0 & 0 & 1 \end{vmatrix} = 1,\qquad(2.44)$$

so the vectors are independent. The determinant for the other set is

$$\begin{vmatrix} 3 & 1 & 0 \\ 2 & 1 & 1 \\ 3 & 0 & -3 \end{vmatrix} = 0,\qquad(2.45)$$

so the vectors are dependent.

$\square$

Basis Vectors   A *basis* for a vector space consists of a set of vectors that can be used to uniquely generate every vector in the space. Such a set of basis vectors is said to *span* the space. Important results from linear algebra allow us to determine the number of vectors that span the space and the characteristics of the vectors that can serve as the basis.

A set of vectors $\mathbf{x}_1, \mathbf{x}_2, \ldots, \mathbf{x}_n$ is said to form a *basis* for the vector space of $\mathbf{R}^n$ if the following two conditions hold:

1. The set of vectors $\mathbf{x}_1, \mathbf{x}_2, \ldots, \mathbf{x}_n$ is linearly independent.

2. Every vector in $\mathbf{R}^n$ can be written as a linear combination of the independent vectors.

Considering the vector space $\mathbf{R}^n$, it can be shown that every set of n linearly independent vectors in $\mathbf{R}^n$ is a basis for the space. Furthermore, although the basis set is not unique, every basis for $\mathbf{R}^n$ has exactly n vectors. Thus, if $\mathbf{x}$ is a vector in $\mathbf{R}^n$, then $\mathbf{x}$ can be written as

$$\mathbf{x} = \alpha_1 \mathbf{x}_1 + \alpha_2 \mathbf{x}_2 + \cdots + \alpha_n \mathbf{x}_n,\qquad(2.46)$$

where each α_j is a scalar if the vectors $\mathbf{x}_j$, for $j = 1, 2, \ldots, n$, are linearly independent and span the vector space.

The *dimension* of the vector space is the number of vectors in the basis, and this number is sometimes referred to as the number of *degrees of freedom* of the space. In a vector space of dimension n, such as $\mathbf{R}^n$, not more than n vectors can be linearly independent. Also, any set of basis vectors that span the space can contain no fewer than n vectors. Any subspace of $\mathbf{R}^n$ containing m vectors ($m \leq n$) has dimension m and requires m basis vectors. Thus, every basis of $\mathbf{R}^2$ must contain two independent vectors, and any three vectors in $\mathbf{R}^2$ must be linearly dependent.

□ **EXAMPLE 2.12**   *Orthonormal Basis for $\mathbf{R}^n$*

The orthonormal set of unit vectors

$$\mathbf{e}_1, \mathbf{e}_2, \ldots, \mathbf{e}_n \tag{2.47}$$

defined in Equation 2.36 forms a basis for $\mathbf{R}^n$. This set is often called the *standard basis*, or the *natural basis*, for the space. However, this set is not unique. Any set of n linearly independent vectors from $\mathbf{R}^n$ could serve as a basis. For example, the set of vectors formed by multiplying each vector in the natural basis by a nonzero constant would also serve as a basis for $\mathbf{R}^n$.

In $\mathbf{R}^1$, any nonzero vector can serve as a basis. This is evident since any two nonzero vectors $\mathbf{a}$ and $\mathbf{b}$ that lie along a line (collinear) can be written as $\mathbf{b} = k\mathbf{a}$, where k is a nonzero number. To write this in the form to test for linear dependence, let $k = -\alpha/\beta$ so that the relationship between the vectors becomes

$$\alpha\mathbf{a} + \beta\mathbf{b} = \mathbf{0},$$

where neither α or β is zero. Considering the definition of linear dependence, we conclude that two collinear vectors are always linearly dependent.

In two dimensions, two noncollinear vectors $\mathbf{a}$ and $\mathbf{b}$ are linearly independent. Thus, every vector in a plane can be represented as

$$\mathbf{c} = k_1\mathbf{a} + k_2\mathbf{b},$$

with the proper choice of the constants k_1 and k_2. The 2D vector space can be defined by the equation

$$\alpha\mathbf{a} + \beta\mathbf{b} + \gamma\mathbf{c} = \mathbf{0}.$$

By a similar argument, we can show that any vector in space can be represented as

$$\mathbf{d} = k_1\mathbf{a} + k_2\mathbf{b} + k_3\mathbf{c},$$

where $\mathbf{a}, \mathbf{b}, \mathbf{c}$ are three noncoplanar vectors in $\mathbf{R}^3$.

Generalizing these observations, we conclude that every n-dimensional vector can be represented as the linear combination of n linearly independent vectors. It is also clear that every set of more that n vectors must be linearly dependent.

□

A fundamental result of linear algebra is that n linearly independent vectors in $\mathbf{R}^n$ form a *basis* for $\mathbf{R}^n$. Thus, any vector in $\mathbf{R}^n$ can be expressed uniquely as a linear combination of the basis vectors. If $\mathbf{u}_1, \ldots, \mathbf{u}_n$ are linearly independent and $\mathbf{v}$ is any vector in $\mathbf{R}^n$, then

$$\mathbf{v} = \sum_{i=1}^{n} \alpha_i \mathbf{u}_i = \alpha_1\mathbf{u}_1 + \cdots + \alpha_n\mathbf{u}_n, \tag{2.48}$$

where the α_i are uniquely determined. To solve for the coefficients, each component can be written as

$$v_k = \alpha_1 u_{1k} + \cdots + \alpha_n u_{nk} = \sum_{i=1}^{n} \alpha_i u_{ik}, \qquad (2.49)$$

where the notation u_{ik} means the kth component $(k = 1, \ldots, n)$ of the vector $\mathbf{u}_i, i = 1, \ldots, n$. Thus, the ith vector in the linearly independent set has components

$$\mathbf{u}_i = (u_{i1}, u_{i2}, \ldots, u_{in}).$$

Equation 2.49 represents an $n \times n$ system of linear equations that can be solved for the coefficients in Equation 2.48, since the $\alpha_i, i = 1, \ldots, n$ are unique.

Now suppose that a set of vectors $\mathbf{u}_i, i = 1, \ldots, n$ in $\mathbf{R}^n$ are orthogonal. It can be shown that the vectors are linearly independent. The expansion of Equation 2.48 for any vector in $\mathbf{R}^n$ is again

$$\mathbf{v} = \sum_{i=1}^{n} \alpha_i \mathbf{u}_i, \qquad (2.50)$$

but in this case the coefficients are easy to find. Simply take the inner (dot) product of both sides of Equation 2.50 with each $\mathbf{u}_i$ and use the orthogonality property. Thus,

$$\langle \mathbf{v}, \mathbf{u}_i \rangle = \sum_{i=1}^{n} \alpha_i \langle \mathbf{u}_i, \mathbf{u}_i \rangle = \alpha_i \langle \mathbf{u}_i, \mathbf{u}_i \rangle. \qquad (2.51)$$

Solving for α_i, the result for the coefficients is

$$\alpha_i = \frac{\langle \mathbf{v}, \mathbf{u}_i \rangle}{\langle \mathbf{u}_i, \mathbf{u}_i \rangle} \qquad (2.52)$$

for $i = 1, \ldots, n$. Notice that if the vectors $\mathbf{u}_i$ are orthonormal, each coefficient α_i is simply the inner product of the vector $\mathbf{v}$ and $\mathbf{u}_i$ since $\langle \mathbf{u}_i, \mathbf{u}_i \rangle = 1$.

ORTHOGONAL SETS OF VECTORS

The importance of orthonormal sets is that they are equivalent to linearly independent sets in many aspects, but they also have an inner product associated with them. It is often more convenient to work with orthonormal sets because of the computational simplifications that arise. Two theorems relate linearly independent and orthonormal sets.

■ THEOREM 2.1　　　*Orthogonal sets*

An orthonormal set of vectors is linearly independent.

To show this result, let $\mathbf{u}_1, \ldots, \mathbf{u}_n$ be an orthonormal set and form the vector equation

$$c_1\mathbf{u}_1 + c_2\mathbf{u}_2 + \cdots + c_n\mathbf{u}_n = \mathbf{0}, \tag{2.53}$$

where the c_i's $(i = 1, 2, \ldots, n)$ are constants. The set of vectors will be linearly dependent if the only constants that satisfy Equation 2.53 are

$$c_1 = c_2 = \cdots = c_n = 0.$$

Taking the inner product of both sides of Equation 2.53 with $\mathbf{u}_1$ yields

$$\langle c_1\mathbf{u}_1 + c_2\mathbf{u}_2 + \cdots + c_n\mathbf{u}_n, \mathbf{u}_1 \rangle = \langle \mathbf{0}, \mathbf{u}_1 \rangle,$$

which can be written

$$c_1\langle \mathbf{u}_1, \mathbf{u}_1 \rangle + c_2\langle \mathbf{u}_2, \mathbf{u}_1 \rangle + \cdots + c_n\langle \mathbf{u}_n, \mathbf{u}_1 \rangle = 0.$$

Since $\langle \mathbf{u}_i, \mathbf{u}_1 \rangle = \delta_{i1}$, the conclusion is that $c_1 = 0$. Now taking the inner product of Equation 2.53 successively with $\mathbf{u}_2, \mathbf{u}_3, \ldots, \mathbf{u}_n$ shows that $c_2 = 0, \ldots, c_n = 0$. Summarizing, the results, we find that all of the constants are zero so the vectors are linearly independent.

Construction of an Orthonormal Set　　Once an orthonormal basis is found for a vector space, many computations are simplified since the vectors are orthogonal. For example, finding the coefficients in the expansion of an arbitrary vector in terms of the basis vectors is considerably simplified using the dot product, as previously shown.

Given a basis for a vector space, is there a way to construct an orthonormal basis? Fortunately, the answer is yes, as will be shown in the proof of the important theorem that states the existence of such an orthonormal set.

■ THEOREM 2.2　　　*Orthonormal vectors*

For every linearly independent set of vectors $\mathbf{x}_1, \ldots, \mathbf{x}_n$, *there exists an orthonormal set of vectors*

$$\mathbf{u}_1, \ldots, \mathbf{u}_n$$

such that each $\mathbf{u}_j, j = 1, 2, \ldots, n$, *is a linear combination of* $\mathbf{x}_1, \ldots, \mathbf{x}_j$.

To prove this theorem, an orthonormal set of vectors will be constructed using the Gram-Schmidt process.

Gram-Schmidt Process　　The *Gram-Schmidt process* is used to construct an orthonormal set from an independent set. Let $\mathbf{x}_1, \ldots, \mathbf{x}_n$ be a linearly independent set in a vector space. An orthonormal set $\mathbf{u}_1, \ldots, \mathbf{u}_n$ can be constructed by the following procedure:

1. Pick $\mathbf{x}_1$ and form a unit vector

$$\mathbf{u}_1 = \frac{\mathbf{x}_1}{||\mathbf{x}_1||}$$

so that $||\mathbf{u}_1|| = 1$.

2. Pick another element, say, $\mathbf{x}_2$, and form its projection on $\mathbf{u}_1$ by forming the vector $(\mathbf{x}_2 \cdot \mathbf{u}_1)\mathbf{u}_1$ and then letting

$$\mathbf{y}_2 = \mathbf{x}_2 - (\mathbf{x}_2 \cdot \mathbf{u}_1)\mathbf{u}_1.$$

Then, create the unit vector

$$\mathbf{u}_2 = \frac{\mathbf{y}_2}{||\mathbf{y}_2||}.$$

Notice that from the definition of $\mathbf{y}_2$ that it cannot be zero because that would imply that $\mathbf{x}_2$ and $\mathbf{u}_1$ are linearly dependent.

3. Continue in this way, successively computing $\mathbf{u}_1, \ldots, \mathbf{u}_j$, and form

$$\mathbf{y}_{j+1} = \mathbf{x}_{j+1} - (\mathbf{x}_{j+1} \cdot \mathbf{u}_1)\mathbf{u}_1 - \cdots - (\mathbf{x}_{j+1} \cdot \mathbf{u}_j)\mathbf{u}_j,$$

with

$$\mathbf{u}_{j+1} = \frac{\mathbf{y}_{j+1}}{||\mathbf{y}_{j+1}||}.$$

□ EXAMPLE 2.13   *Gram-Schmidt Process*

We first show that the vectors $\mathbf{x}_1 = (1, -1, 2)$ and $\mathbf{x}_2 = (1, 0, -1)$ are linearly independent. Then, we use the Gram-Schmidt process to produce an orthonormal set to be used as the basis for any vector in a plane in $\mathbf{R}^3$.

The vectors are linearly independent, since

$$c_1\mathbf{x}_1 + c_2\mathbf{x}_2 = 0$$

implies that $c_1 = c_2 = 0$, as you can easily prove by expanding the equation by components of the vectors. Applying the Gram-Schmidt process,

$$\mathbf{u}_1 = \frac{\mathbf{x}_1}{||\mathbf{x}_1||} = \frac{(1, -1, 2)}{\sqrt{6}}$$

and

$$\begin{aligned}\mathbf{y}_2 &= \mathbf{x}_2 - (\mathbf{x}_2 \cdot \mathbf{u}_1)\mathbf{u}_1 \\ &= (1, 0, -1) + (\frac{1}{6}, -\frac{1}{6}, \frac{1}{3}) = (\frac{7}{6}, -\frac{1}{6}, -\frac{2}{3}).\end{aligned}$$

Then

$$\mathbf{u}_2 = \frac{\mathbf{y}_2}{||\mathbf{y}_2||} = \frac{(\frac{7}{6}, -\frac{1}{6}, -\frac{2}{3})}{\sqrt{66/36}} = \frac{(7, -1, -4)}{\sqrt{66}}.$$

Since two *noncollinear* vectors can be used to define a plane, the points in the plane defined by the vectors $\mathbf{x}_1$ and $\mathbf{x}_2$ can be represented as

$$\alpha_1 \mathbf{x}_1 + \beta_1 \mathbf{x}_2,$$

where the coefficients are real numbers. Another representation is

$$\alpha_2 \mathbf{u}_1 + \beta_2 \mathbf{u}_2.$$

The representation in terms of $\mathbf{u}_1$ and $\mathbf{u}_2$ has the advantage that the vectors are perpendicular and form an orthonormal set.

$\square$

ABSTRACT VECTOR SPACES

In the earlier sections of this chapter, it was assumed that the vector space under discussion contained vectors with n real components. Although the concept of $\mathbf{R}^n$ as a vector space has great use in many applications, it is possible to define more general vector spaces. For example, a vector space can be defined in which the elements are complex numbers. Another vector space could be defined that consists of all continuous real-valued functions of a real variable. An example subspace of this vector space would be the set of all polynomials.

This section discusses the extension to vector spaces consisting of functions. Later chapters apply the technique to matrices, Fourier analysis, and the solution of differential equations. The approach frequently leads not only to simplified mathematical analysis and computation but also to useful physical insights in many problems of interest in physics and engineering.

Suppose V contains a collection of elements that may be vectors in $\mathbf{R}^n$, matrices, functions, or other elements. Further, assume that there are two operations given in the definition of the space, called addition and scalar multiplication. V is called an *abstract vector space* if the elements $\mathbf{a}, \mathbf{b}$, and $\mathbf{c} \in V$ and α and β are scalars with the closure properties

$$\mathbf{a} + \mathbf{b} \in V \quad \text{and} \quad \alpha \mathbf{a} \in V$$

and the additional properties:

1. $\mathbf{a} + \mathbf{b} = \mathbf{b} + \mathbf{a}.$ Commutative law for addition

2. $(\mathbf{a} + \mathbf{b}) + \mathbf{c} = \mathbf{a} + (\mathbf{b} + \mathbf{c}).$ Associative law for addition

3. There is zero element such that $\mathbf{a} + 0 = \mathbf{a}.$

4. There is an inverse element such that $\mathbf{a} + (-\mathbf{a}) = 0.$

5. $\alpha(\mathbf{a} + \mathbf{b}) = \alpha \mathbf{a} + \alpha \mathbf{b}.$

6. $(\alpha + \beta)\mathbf{a} = \alpha \mathbf{a} + \beta \mathbf{a}.$

7. $(\alpha \beta)\mathbf{a} = \alpha(\beta \mathbf{a}).$

8. There is an identity element such that $1 \times \mathbf{a} = \mathbf{a}$.

Comparing the properties of the abstract vector space with those of $\mathbf{R}^n$ previously defined shows that the properties are the same. The term *abstract* is used only to indicate that the vector space under discussion may consist of objects other than the vectors in $\mathbf{R}^n$. When no confusion could arise, the abstract vector spaces discussed are simply termed *vector spaces*. Also, when vector spaces of functions are considered, the boldface notation will be discontinued and functions will be represented in their ordinary mathematical form.

☐ EXAMPLE 2.14   *Polynomials*

The set of all polynomials

$$P(x) = a_0 + a_1 x + a_2 x^2 + \cdots + a_k x^k \qquad (2.54)$$

forms a vector space. In this case, vector addition is polynomial addition and scalar multiplication is multiplication of $P(x)$ by a constant.

Polynomials such as $1, x, x^2, \ldots, x^n$ are linearly independent, since one of these functions cannot be written as a linear combination of the others that is valid for all x. However, the vector space does not have a finite number of polynomials that span the space. In this case, the space is said to be an *infinite-dimensional* vector space. A number of vector spaces containing functions will have this property. This does not mean that the vector space does not have a basis but that the basis contains an infinite number of functions.

☐

VECTOR SPACES OF FUNCTIONS

We now come to perhaps the most important generalizations of vector theory in the chapter as they pertain to techniques in advanced mathematics. Reaching this point has involved extending the theory of vectors in $\mathbf{R}^2$ and $\mathbf{R}^3$ to those in $\mathbf{R}^n$ and then considering abstract vector spaces. The key points to be considered here include

1. Generalizing the dot product for vectors to the inner product for functions;

2. Extending the idea of length of a vector to define the norm of a function;

3. Introducing the concept of expressing a function in terms of a linear combination of orthogonal functions based on the expansion of a vector in terms of the basis vectors for the vector space.

The discussion will be rather short, but these subjects will be revisited in the chapters that treat differential equations, approximation of functions, and Fourier analysis, as well as elsewhere in the text.

When functions instead of vectors in $\mathbf{R}^n$ are the elements of a vector space, it is necessary to carefully specify the operations on the elements that are allowed. For example, it is possible to define a norm that measures the "length" of a function. Since a continuous function $f(x)$ is like a vector with a continuous range of components, adding the squares of the values obviously leads to an infinite result. However, a finite result is possible when the summation is replaced by integration. If the dot product for vectors is generalized to the *inner* product for functions, the norm can then be defined accordingly.

As for vectors, the basic operations for functions are addition and multiplication by a scalar. If $f(x)$ and $g(x)$ are continuous real-valued functions, the addition and scalar multiplication are defined as

1. $(f + g)(x) = f(x) + g(x)$,

2. $(\alpha f)(x) = \alpha f(x)$,

where α is a scalar and x is real.

The *inner product* of $f(x)$ and $g(x)$ is defined on the interval $[a, b]$ as the integral of the product

$$\langle f, g \rangle = \int_a^b f(x)\, g(x)\, dx. \tag{2.55}$$

Based on the definition of the inner product, the norm is written as

$$\|f\| = \langle f, f \rangle^{1/2} = \left[\int [f(x)]^2\, dx \right]^{1/2}. \tag{2.56}$$

If the inner product of two nonzero functions is zero, the functions are said to be *orthogonal*.

□ EXAMPLE 2.15

MATLAB Norm

The norm of $\sin x$ on the interval $[0, 2\pi]$ is computed as

$$
\begin{aligned}
\|\sin x\| &= \langle \sin x, \sin x \rangle^{1/2} \\
&= \left[\int_0^{2\pi} [\sin x]^2\, dx \right]^{1/2} = \left[\frac{1}{2} \int_0^{2\pi} [1 - \cos(2x)]\, dx \right]^{1/2} \\
&= \left[\frac{1}{2} \left(x - \frac{\sin(2x)}{2} \right) \Big|_0^{2\pi} \right]^{1/2} = \sqrt{\pi} \approx 1.7725.
\end{aligned}
$$

MATLAB Integration. The accompanying MATLAB script shows the computation of the norm of the sine both numerically and symbolically. The MATLAB command **quad** is used to compute the integral of a function. The format of the command is

```
>> quad('function',a,b)
```

where `function` is the name of a function file defining the function and `a,b` defines the interval of integration. The second script shows the function to return the values of $\sin^2(x)$. The norm is then the square root of the result of the integration performed by **quad**. MATLAB integration routines are treated in more detail in Chapter 12.

The symbolic command **int** will attempt to integrate a function defined symbolically. The limits $[a, b]$ can be defined as symbolic or numeric expressions. In the script, **int** is used to compute the symbolic indefinite and definite integral of the sine. Command **sympow** returns the square root of the symbolic value normsqn. Then, the command **numeric** converts the symbolic representation of the square root of π to a numerical value.

MATLAB Script ──────────────────────────────

```
Example 2.15
% EX2_15.M Compute the norm of sin(x) on the interval [0,2pi]
%   Compare symbolic and numerical result
%     (This script requires the Symbolic Math Toolbox)
% Numerical value; call function sinsq to compute (sin(x))^2
normsin1=sqrt(quad('sinsq',0,2*pi))    % Numerical value
%
% Symbolic
normsq=int('sin(x)^(2)');          % Perform symbolic integration
normsq=simple(normsq)              % Simplify the result
normsqn=int('sin(x)^2',0,2*pi)     % Definite integral
norm2=sympow(normsqn,1/2)          % Symbolic square root
normsin=numeric(norm2)             % Convert to a number


%
% Edited results from M-file EX2_15.M
%
>>ex2_15

normsin1 =  1.7725              % Numerical results
normsq   = -1/4*sin(2*x)+1/2*x  % Symbolic integration
normsqn  = pi                   % Symbolic definite integral
norm2    = pi^(1/2)             % Symbolic Square root
normsin  = 1.7725              % Numerical value of symbolic result
%
```
──────────────────────────────

The function `sinsq` returns the squared value of the sine of the argument x passed to the function when it is called by the M-file.

MATLAB Script ──────────────────────────────

```
Example 2.15
function yout=sinsq(x)
% CALL: yout=sinsq(x) returns the square of sine(x)
yout=sin(x).^2;
```
──────────────────────────────

$\square$

Consider a set of functions $\phi_n(x), n = 1, 2, \ldots$, each of which is continuous on the interval $[a, b]$. The set is *orthogonal* if the inner products of different nonzero functions in the set are zero. Thus, the set of functions is orthogonal if

$$\langle \phi_m(x), \phi_n(x) \rangle = \int_a^b [\phi_m(x)\phi_n(x)]dx = 0, \qquad m \neq n, \qquad \textbf{(2.57)}$$

and no $\phi_n(x)$ is identically zero except perhaps at a finite number of points. Furthermore, the system is *orthonormal* if the functions satisfy Equation 2.57 and

$$\langle \phi_n(x), \phi_n(x) \rangle = \int_a^b [\phi_n^2(x)]dx = 1, \qquad n = 1, 2, \ldots. \qquad \textbf{(2.58)}$$

Suppose a function $f(x)$ is continuous on the interval $[a, b]$. Then, following an approach similar to that of expanding a vector in terms of orthonormal basis vectors, we postulate that $f(x)$ can be expressed as

$$f(x) = \sum_{n=1}^{\infty} c_n \phi_n(x), \qquad \textbf{(2.59)}$$

where the coefficient c_m is determined as

$$c_m = \int_a^b f(x)\phi_m(x)\,dx. \qquad \textbf{(2.60)}$$

Of course, much is left to be said about the expansion of Equation 2.59. Questions about the accuracy of the approximation if a finite number of terms is used and other considerations to guarantee that the series actually represents $f(x)$ will be explored in other chapters. Here we will simply present a most famous set of orthogonal functions that have enormous utility in mathematics and science.

Trigonometric Functions   The set of functions

$$\cos x, \sin x, \cos 2x, \sin 2x, \ldots, \cos nx, \sin nx$$

defined for $-\pi \leq x \leq \pi$ span the vector space of trigonometric sums of the form

$$T(x) = \sum_{k=1}^{n} [a_k \cos kx + b_k \sin kx]. \qquad \textbf{(2.61)}$$

The set of trigonometric basis functions can be shown to be orthonormal with respect to the inner product,

$$\langle f, g \rangle = \frac{1}{\pi} \int_{-\pi}^{\pi} [f(x)\,g(x)]\,dx. \qquad \textbf{(2.62)}$$

Example 2.16 discusses these functions further.

A coefficient such as a_k is found by taking the inner product of each side of Equation 2.61 with $\cos kx$ and similarly for b_k using $\sin kx$ in the inner product. Thus,

$$a_k = \frac{1}{\pi} \int_{-\pi}^{\pi} [T(x) \cos kx]\, dx,$$

$$b_k = \frac{1}{\pi} \int_{-\pi}^{\pi} [T(x) \sin kx]\, dx \qquad k = 1, 2, \ldots. \qquad (2.63)$$

The numbers a_k and b_k are called the kth *Fourier coefficients*.

Fourier Techniques   The applications of Fourier techniques are discussed in detail in Chapter 8 and Chapter 9. The reader who is familiar with Fourier series to approximate functions should note that the trigonometric series in Equation 2.61 is not the complete series because it does not include a constant term. The issue here is the orthogonality of the trigonometric functions, not the approximation of arbitrary functions by a Fourier trigonometric series as treated in Chapter 8.

□ EXAMPLE 2.16   ***Orthogonal Functions***

The functions $\sin x$ and $\cos x$ are orthogonal over the interval $[-\pi, \pi]$, since

$$\langle \sin x, \cos x \rangle = \int_{-\pi}^{\pi} \sin x \cos x\, dx = 0 \qquad (2.64)$$

over this interval. This is easily seen if you recognize that the integrand is odd over a full period of the sinusoids.[4] Otherwise, make the substitution

$$\frac{1}{2} \sin 2x = \sin x \cos x$$

and integrate. For general integrals of this form, integrate by parts.

By integrating, we find that

$$\langle \cos nx, \cos mx \rangle = \int_{-\pi}^{\pi} \cos nx \cos mx\, dx = \begin{cases} 0, & n \neq m, \\ \pi, & n = m. \end{cases} \qquad (2.65)$$

These values for the inner product show that the cosine terms divided by $\sqrt{\pi}$ form an orthonormal set so that

$$\langle \frac{\cos nx}{\sqrt{\pi}}, \frac{\cos nx}{\sqrt{\pi}} \rangle = 1.$$

The result for $\langle \sin nx, \sin nx \rangle$ is the same. Thus, the factor $1/\pi$ is used to normalize the integral in Equation 2.62.

Polynomials can also be orthogonal over restricted intervals. For example, over the interval $-1 \leq x \leq 1$, the set of polynomials

$$P_0(x) = 1, \quad P_1(x) = x, \quad \text{and} \quad P_2(x) = \frac{3}{2}x^2 - \frac{1}{2}$$

are orthogonal, as you are asked to show in Problem 2.15. These are known as *Legendre polynomials*.

□

---

[4]Integrals of even and odd functions are discussed in Chapter 8.

REINFORCEMENT EXERCISES AND EXPLORATION PROBLEMS

REINFORCEMENT EXERCISES

In these problems, do the computations by hand unless otherwise indicated, and then check the solution with MATLAB for problems that have numerical or symbolic results.

P2.1. Binary system Show that the binary system with elements $(0,1)$ satisfies the properties of real numbers if the operations are defined as

+	0	1
0	0	1
1	1	0

×	0	1
0	0	0
1	0	1

P2.2. Associativity Using a simple example show that vector subtraction is not associative. Vector subtraction is defined as

$$\mathbf{a} - \mathbf{b} = \mathbf{a} + (-\mathbf{b}).$$

P2.3. Sum of series Sum the series

$$N = \sum_{i=0}^{\infty} d_i \, 2^{-i} = d_0 + d_1 \times 2^{-1} + d_2 \times 2^{-2} + \cdots$$

when each $d_i = 1$, and find the error when the series is truncated after m terms.

P2.4. Perpendicular vectors Determine the value α so that vectors

$$\mathbf{x} = 2\,\mathbf{i} + \alpha\,\mathbf{j} + \mathbf{k} \text{ and } \mathbf{y} = 4\,\mathbf{i} - 2\,\mathbf{j} - 2\,\mathbf{k}$$

are perpendicular. Compute $\mathbf{x} \cdot \mathbf{y}$ to verify the result.

P2.5. Unit vector Prove that if $\mathbf{x}$ is any nonzero vector, $\mathbf{u} = \mathbf{x}/\sqrt{\mathbf{x} \cdot \mathbf{x}}$ is a unit vector.

P2.6. Angle between vectors Find the angles in the triangle with the vertices:

$$[2, -1, 0], \quad [5, -4, 3], \quad \text{and} \quad [1, -3, 2].$$

P2.7. Cross product Show that the magnitude of the cross product $\mathbf{a} \times \mathbf{b}$ is the area of the parallelogram determined by $\mathbf{a}$ and $\mathbf{b}$.

P2.8. Basis vectors Given the following vectors, compute the sums and write them in terms of the standard basis vectors, and compute the dot product:

a. $\mathbf{x} = [0, 1, 4, -3]$, $\mathbf{y} = [2, 8, 6, -4]$;

b. $\mathbf{x} = [3, -5, 0, 8]$, $\mathbf{y} = [6, 1, -7, -2]$.

P2.9. Vectors and subspaces Show which of the following subsets of $\mathbf{R}^3$ are subspaces and which are not:

a. The set of vectors in $\mathbf{R}^3$ with first component 1;

b. All vectors with $x_3 = 0$;

c. The set of vectors with nonzero first component.

P2.10. Subspace   Using the properties of vector spaces, prove the following:

    a. The set of vectors in $\mathbf{R}^2$ that lie on a line passing through the origin $(y = mx)$ form a vector space.

    b. The set of vectors in $\mathbf{R}^2$ that lie on a line not passing through the origin $(y = mx + b,\ b \neq 0)$ does not form a vector space.

Why it is obvious that the set in (b) is not a subspace?

P2.11. Linear independent vectors   Determine if the following vectors are linearly independent:

    a. $[1, 2, -1, 2], [-2, -5, 3, 0], [1, 0, 1, 10]$;

    b. $[3, 0, 0, 2], [1, 0, 0, 4]$.

P2.12. Independence of polynomials   Are the following polynomials independent?

$$x^2 - 1, \ x^2 + x - 2, \ x^2 + 3x + 2$$

P2.13. Complex vectors   If we insist that the norm be a positive number, show that the ordinary definition of dot product for real vectors does not hold in the vector space containing complex numbers of the form $z = x + iy$, where x and y are real numbers and i is imaginary. What is the proper definition?

P2.14. Inner product   Compute the inner product of the following functions on the interval $[-\pi, \pi]$ with n and m distinct positive integers:

    a. $\langle \sin mx, \sin nx \rangle$;

    b. $\langle \cos mx, \cos nx \rangle$;

    c. $\langle \cos mx, \sin nx \rangle$;

    d. $\langle \cos nx, \cos nx \rangle$;

    e. $\langle \sin nx, \cos nx \rangle$.

Which functions are orthogonal?

P2.15. Orthogonal Legendre functions   Show that the Legendre polynomials in Example 2.16 are orthogonal over the interval in $\mathbf{R}^1$ such that $-1 \leq x \leq 1$.

P2.16. MATLAB roundoff error   Add the MATLAB value 10^{-6} in a loop N times, and compare the result with the value $N \times 10^{-6}$. If there is a difference, explain it. Try values such as $N = 100$ and $N = 1000$.

P2.17. MATLAB conversion   Write a MATLAB function to convert an N-digit octal (base 8) number to decimal. Test the function by converting 0.502_8.

P2.18. MATLAB vector operations   For the vectors

$$\mathbf{x} = [0, -1, 2, 3] \quad \text{and} \quad \mathbf{y} = [5, 1, 2, -3],$$

compute the following:

    a. $\mathbf{x} + \mathbf{y}$;

    b. $3\,\mathbf{x}$;

    c. $\mathbf{x} \cdot \mathbf{y}$;

    d. $\| \mathbf{x} \|$ and $\| \mathbf{y} \|$.

P2.19. MATLAB cross product Write a MATLAB function to compute the cross product of two 3D vectors. Test the function with the vectors in Example 2.5. Compare your version with the MATLAB command **cross**.

P2.20. MATLAB orthogonal vectors Are the vectors

$$\mathbf{x} = [\sqrt{2}/2, -1] \quad \text{and} \quad \mathbf{y} = [1, \sin(45°)]$$

orthogonal? Do the problem analytically and using MATLAB.

P2.21. MATLAB vector norm Write a MATLAB function to compute the norm and the corresponding unit vector for a vector in $\mathbf{R}^n$. Test the result with the vector $\mathbf{x} = [2, 3, 6]$.

EXPLORATION PROBLEMS

P2.22. MATLAB roundoff Write a MATLAB program using the command **floor** to input a real number r and an integer k and display r rounded to k decimal places. Test the program with numbers such as π.

Hint: Investigate the MATLAB **round** and **chop** commands.

P2.23. DeMoivre's theorem Given the complex numbers

$$z_1 = r_1(\cos\theta + i\sin\theta),$$
$$z_2 = r_2(\cos\phi + i\sin\phi)$$

and the result that

$$z_1 z_2 = r_1 r_2 [\cos(\theta + \phi) + i\sin(\theta + \phi)],$$

prove DeMoivre's theorem,

$$z^n = r^n(\cos n\theta + i\sin n\theta),$$

if $z = r(\cos\theta + i\sin\theta)$ and n is an integer.

Hint: Don't forget the cases $n \leq 0$.

P2.24. Complex functions Using the results of the previous problem, show that

$$\cos 3\theta = 4\cos^3\theta - 3\cos\theta.$$

P2.25. Dot product Prove the following theorem concerning the dot product.
If θ is the angle between two vectors $\mathbf{x}$ and $\mathbf{y}$ in $\mathbf{R}^3$, then

$$\cos\theta = \frac{\mathbf{x} \cdot \mathbf{y}}{||\mathbf{x}|| \, ||\mathbf{y}||}.$$

Hint: Apply the law of cosines to the triangle with sides $\mathbf{x}$, $\mathbf{y}$, and $\mathbf{x} - \mathbf{y}$.

P2.26. Area of a parallelogram Consider the parallelogram defined by two vectors **a** and **b** at an angle θ, as shown in Figure 2.7.

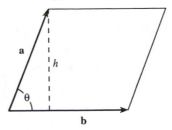

FIGURE 2.7 *Parallelogram*

Show that the area of the parallelogram is defined by

$$A = ||\mathbf{a}|| \, ||\mathbf{b}|| \, \sin \theta,$$

which is the cross product of the two vectors. Test the result with the vectors $\mathbf{a} = [3, 1, 4]$ and $\mathbf{b} = [-2, 5, 3]$.

Hint: The area is $17\sqrt{3}$.

P2.27. Distance in polar coordinates Determine the distance between the points $P_1(r_1, \theta_1)$ and $P_2(r_2, \theta_2)$ in the polar plane.

Hint: Write the points in rectangular coordinates and determine the distance.

Test the results with the points $P_1(1, 0)$ and $P_2(\sqrt{2}, \pi/4)$ with the angle measured in radians.

P2.28. Polynomials Consider the polynomial of degree n

$$P(x) = a_0 + a_1 x + a_2 x^2 + \cdots + a_n x^n.$$

Define the polynomial as the inner product of two vectors. Then, compare the results with the value computed by the MATLAB command **polyval** for the polynomial by testing both results with the polynomial

$$P(x) = x^5 + 5x^3 + 3x^2 + 4x + 2$$

evaluated at the points $1, 2, 3$ and -1.

P2.29. Projections Given two vectors **x** and **y** in $\mathbf{R}^n$, write the vector **x** as the sum of a vector $\mathbf{x}_{||}$ parallel to **y** and a vector $\mathbf{x}_\perp$ perpendicular to **y**. Verify the results.

Hint: Draw two typical vectors in $\mathbf{R}^2$ to define the equations.

Test the expressions for $\mathbf{x}_{||}$ and $\mathbf{x}_\perp$ for the vectors

$$\mathbf{x} = [4, -5, 3] \quad \text{and} \quad \mathbf{y} = [2, 1, -2].$$

P2.30. Orthogonal vectors Consider a linear combination of the vectors

$$
\begin{aligned}
\mathbf{u}_1 &= (1,1,1,1) \\
\mathbf{u}_2 &= (-1,-1,1,1) \\
\mathbf{u}_3 &= (-1,1,-1,1) \\
\mathbf{u}_4 &= (-1,1,1,-1)
\end{aligned}
$$

to represent the vector $\mathbf{v} = (2,3,-1,4)$. Show the vectors are linearly independent and find the expansion.

P2.31. MATLAB Gram-Schmidt process Write a MATLAB program to implement the Gram-Schmidt procedure. Test the program with the vectors from Example 2.13.

P2.32. Orthonormal set of vectors Use the Gram-Schmidt process to produce an orthonormal set from the complex vectors

$$
\mathbf{x}_1 = \begin{bmatrix} 1 \\ i \end{bmatrix} \quad \text{and} \quad \mathbf{x}_2 = \begin{bmatrix} 1-i \\ i \end{bmatrix}.
$$

P2.33. Test your computer Determine how fast your computer multiplies and what range of numbers is permitted by doing the following:

    a. Create a program to estimate the number of floating-point multiplications per second (flops) for your computer.

    b. Find the largest and smallest MATLAB numbers allowed by your computer.

P2.34. Summing-up questions

    a. Give an example of a set of orthogonal vectors that are not linearly independent.

    b. Give an example of a set of linearly independent vectors that are not mutually orthogonal.

    c. Suppose that $\mathbf{x}_1$, $\mathbf{x}_2$, $\mathbf{x}_3$, $\mathbf{x}_4$ are four vectors in $\mathbf{R}^3$. Can these vectors be linearly independent? Can these vectors be a basis for $\mathbf{R}^3$?

    d. Find all the vectors orthogonal to the vectors

$$
[1,1,1] \quad \text{and} \quad [1,-1,0].
$$

ANNOTATED BIBLIOGRAPHY

1. Cullen, Charles G., *Matrices and Linear Transformations*, Dover Publications, New York, 1972. *A brief but useful treatise.*

2. Grossman, Stanley I., *Elementary Linear Algebra*, Saunders College Publishing, Fort Worth, TX, 1991. *A good introduction to linear algebra with a number of MATLAB examples.*

3. Henrici, Peter. *Elements of Numerical Analysis*, John Wiley and Sons, New York, 1964. *A classic treatment of numerical methods, including errors that arise during numerical computation.*

4. Hill, David R., *Experiments in Computational Matrix Algebra*, Random House, New York, 1988. *The text shows the use of MATLAB to solve many problems in linear algebra and vector analysis.*

5. Mathews, John H., *Numerical Methods*, Prentice Hall, Englewood Cliffs, NJ, 1992. *The text provides a good discussion of machine arithmetic and errors.*

6. Strang, Gilbert, *Linear Algebra and Its Applications*, Saunders College Publishing, Fort Worth, TX, 1988. *A readable and interesting treatment of linear algebra.*

ANSWERS

——————————◼——————————

P2.1. **Binary system** In the binary system, letting x, y, and z be 0 or 1, the associative, commutative, and distributive laws are satisfied for both addition and multiplication. Identity elements obviously exist since $x + 0 = x$ and $1 \times x = x$. The equation $1 + 1 = 0$ shows that the system has an additive inverse.

P2.2. **Associativity** Notice that $(\mathbf{x} - \mathbf{y}) - \mathbf{z} = \mathbf{x} - \mathbf{y} - \mathbf{z} \neq \mathbf{x} - (\mathbf{y} - \mathbf{z}) = (\mathbf{x} - \mathbf{y}) + \mathbf{z}$. In even simpler terms, $(2 - 3) - 2 = -3 \neq 2 - (3 - 2) = 1$.

P2.3. **Sum of series** The series

$$S = \sum_{i=0}^{\infty} d_i \, 2^{-i} \tag{2.66}$$

is a *geometric series* in the form $S = \sum_{k=1}^{\infty} a \, r^{k-1}$, with $a = 1$ and $r = 1/2$. To sum the series, form the sum $S_N = a + ar + ar^2 + \cdots + ar^{N-1}$ and the difference $S_N - rS_N = a - ar^N$, and solve for S_N. The sum of the geometric series for N terms is thus $S_N = \frac{a(1-r^N)}{1-r}$. For the specific case of Equation 2.66,

$$S_N = \frac{1 - \left(\frac{1}{2}\right)^N}{1 - \frac{1}{2}},$$

which converges to 2 as N goes to infinity. The error for m terms is thus $S - S_m = 2^{-m+1}$.

P2.4. Perpendicular vectors Taking the dot product of the vectors yields $[2\,\mathbf{i} + \alpha\,\mathbf{j} + \mathbf{k}] \cdot [4\,\mathbf{i} - 2\,\mathbf{j} - 2\,\mathbf{k}] = [8 - 2\alpha - 2]$. For the vectors to be perpendicular, the dot product must be zero, so that $8 - 2\alpha - 2 = 0$, or $\alpha = 3$.

P2.5. Unit vector Assuming that $\mathbf{x}$ is a nonzero vector,

$$\mathbf{u} = \mathbf{x}/\sqrt{\mathbf{x} \cdot \mathbf{x}} = \frac{\mathbf{x}}{(||\mathbf{x}||^2)^{1/2}}.$$

Thus, forming $\mathbf{u} \cdot \mathbf{u}$, the result is

$$\frac{\mathbf{x}}{||\mathbf{x}||} \cdot \frac{\mathbf{x}}{||\mathbf{x}||} = \frac{1}{||\mathbf{x}||^2}(\mathbf{x} \cdot \mathbf{x}) = 1$$

so that $\mathbf{u}$ is a unit vector.

P2.6. Angle between vectors The vectors $\mathbf{A} = [2, -1, 0]$, $\mathbf{B} = [5, -4, 3]$, and $\mathbf{C} = [1, -3, 2]$ form a triangle, as shown in Figure 2.8.

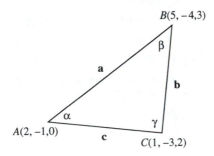

FIGURE 2.8 *Triangle in space*

Let the sides of the triangle be $\mathbf{a} = \mathbf{B} - \mathbf{A} = [3, -3, 3]$, $\mathbf{b} = \mathbf{C} - \mathbf{B} = [-4, 1, -1]$, and $\mathbf{c} = \mathbf{C} - \mathbf{A} = [-1, -2, 3]$.

The cosine of the angle between any two vectors is

$$\cos\theta = \frac{(\mathbf{x} \cdot \mathbf{y})}{(||\,\mathbf{x}\,||\,||\,\mathbf{y}\,||)}.$$

Computing the dot products and solving for the angles, we obtain $\alpha = 54.7356°$, $\beta = 35.2644°$, and $\gamma = 90.0000°$. The sum of the angles is $180°$, as expected.

P2.7. Cross product Consider the parallelogram with sides $\mathbf{a}$ and $\mathbf{b}$ that meet at an angle of θ degrees, as shown in Figure 2.9.

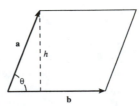

FIGURE 2.9 *Parallelogram formed by vectors $\mathbf{a}$ and $\mathbf{b}$*

The area of the parallelogram is the base time the height, or $h||\mathbf{b}||$. From the figure, the area is $h||\mathbf{b}|| = h||\mathbf{a}||\sin\theta||\mathbf{b}|| = ||\mathbf{a} \times \mathbf{b}||$.

P2.8. Basis vectors

    a. $\mathbf{x} + \mathbf{y} = [2, 9, 10, -7]$, $\mathbf{x} \cdot \mathbf{y} = 44$;

    b. $\mathbf{x} + \mathbf{y} = [9, -4, -7, 6]$, $\mathbf{x} \cdot \mathbf{y} = -3$

P2.9. Vectors and subspaces

    a. Considering the sum of two vectors in the space

$$[1, x_2, x_3] + [1, y_2, y_3] = [2, x_2 + y_2, x_3 + y_3],$$

it is clear that the sum of the vectors is not in the vector space since the sum has first component 2. Thus, the vectors in $\mathbf{R}^3$ with first component 1 do not form a subspace.

    b. All vectors with $x_3 = 0$ form a subspace.

    c. The set of vectors with nonzero first component form a subspace.

P2.10. Subspace To show that the vectors form a vector space, it is necessary to show that each property for a vector space holds.

    a. Suppose $\mathbf{x} = [x_1, y_1]$ and $\mathbf{y} = [x_2, y_2]$. To show the closure property, let $y_1 = mx_1$ and $y_2 = mx_2$. Then

$$
\begin{aligned}
\mathbf{x} + \mathbf{y} &= [x_1 + x_2, m(x_1 + x_2)] \in V, \\
\alpha \mathbf{x} &= [\alpha x_1, \alpha y_1] = [\alpha x_1, m(\alpha x_1)] \in V.
\end{aligned}
$$

Notice that $[0, 0] = \mathbf{0}$ is in V. Also, each vector in V is also in R^2, and R^2 forms a vector space.

    b. Points on the line not through the origin do not form a vector space because $\mathbf{0}$ is not in V. Also, closure under addition is violated.

P2.11. Linear independent vectors

    a. Since the vectors have the relationship $-5\mathbf{x}_1 - 2\mathbf{x}_2 + \mathbf{x}_3 = \mathbf{0}$, the vectors are dependent.

    b. The equations $\alpha[3, 0, 0, 2] + \beta[1, 0, 0, 4] = [0, 0, 0, 0]$ have the only solution $\alpha = \beta = 0$, so the vectors are independent.

P2.12. Independence of polynomials Letting

$$P_1 = x^2 - 1, \; P_2 = x^2 + x - 2, \; P_3 = x^2 + 3x + 2,$$

form the equation $c_1 P_1 + c_2 P_2 + c_3 P_3 = 0 + 0x + 0x^2$. Since the only solution is $[c_1, c_2, c_3] = [0, 0, 0]$, the polynomials are linearly independent.

P2.13. Complex vectors Consider two complex vectors with components

$$\mathbf{z}_1 = x_1 + iy_1, \quad \mathbf{z}_2 = x_2 + iy_2.$$

The ordinary dot product produces $\mathbf{z}_1 \cdot \mathbf{z}_2 = x_1 x_2 + (iy_1)(iy_1) = x_1 x_2 - y_1 y_1$, which could be a negative value. Forming the dot product with the conjugate vector yields the correct value

$$\mathbf{z}_1 \cdot \bar{\mathbf{z}}_2 = x_1 x_2 + (iy_1)(-iy_2) = x_1 x_2 + (y_1)(y_2).$$

P2.14. Inner product The inner product integrals are all zero except for $\langle \cos nx, \cos nx \rangle = \pi$.

P2.15. Orthogonal Legendre functions Integrating $P_0 P_1$ yields

$$\langle P_0, P_1 \rangle = \int_{-1}^{1} x \, dx = 0.$$

Similarly, $\langle P_0, P_2 \rangle = \langle P_1, P_2 \rangle = 0$ on the interval $[-1, 1]$.

P2.16. MATLAB roundoff error This problem is discussed in the December 1993 issue of the MathWorks *MATLAB Digest*, Volume 1, number 5. Since 10^{-6} is not exactly representable in the binary floating-point format, there is a possibility of round-off error at each step of a calculation. The difference between the product $N \times 10^{-6}$ and N sums of 10^{-6} ranges from about -1.7×10^{-19} for $N = 100$ to 7.935×10^{-14} for $N = 100,000$.

Comment: If the answers to problems for the Reinforcement Exercises in this chapter require MATLAB programs, these programs are stored as files on the disk included with this textbook.

3 *MATRICES*

Matrices are useful in almost every area of physics and engineering. Perhaps the most important use of matrices in problem solving is to represent systems of linear equations. In other applications, the matrices may represent linear transformations, as in graphics, where they are used for rotations and similar operations on points and lines in space. Matrix equations can also be used to represent certain systems of differential equations. The practical importance of matrices is that the methods for their manipulation are well known and algorithms for computer solutions of matrix equations are readily available.

Table 3.1 lists various matrix topics covered in this chapter and later in the book. In the table, α and β are scalars, $\mathbf{x}$, $\mathbf{y}$, and $\mathbf{b}$ are vectors, and A, B, and Q are matrices. Matrices also play a role in other applications treated in the text, including data analysis, fast Fourier transforms, and curvilinear coordinates in vector calculus.

This chapter begins with a discussion of matrices and their properties. Next, matrix methods are applied to the solution of linear systems of equations. Then, linear transformations are formulated as matrix operations. Much of the material in this chapter also serves as background material for the following chapters that treat eigenvectors and differential equations.

TABLE 3.1 *Topics in matrix analysis*

Chapter	Symbol	Topic
Chapter 3	$A + B$	Sum of matrices
	αA	Scalar times matrix
	AB	Product of matrices
	A^T	Matrix transpose
	A^{-1}	Matrix inverse
	$|A|$	Determinant
	$Q^{-1} = Q^T$	Orthogonal matrices
	$\text{rank}(A)$	Rank of A
	$A\mathbf{x} = \mathbf{b}$	Linear equations
	$\mathbf{y} = A\mathbf{x}$	Linear transformations
Chapter 4	$A\mathbf{x} = \lambda\mathbf{x}$	Eigenvalue equation
	$f(A)$	Matrix functions
Chapter 5, 6	$\dfrac{d\mathbf{x}}{dt} = A\mathbf{x}$	Differential equations
Chapter 7	$A^T A$	Least squares

BASIC PROPERTIES OF MATRICES

———————————————■———————————————

A *matrix* is a rectangular array of elements written as

$$A = \begin{bmatrix} a_{11} & a_{12} & \cdots & a_{1n} \\ a_{21} & a_{22} & \cdots & a_{2n} \\ \vdots & \vdots & \ddots & \vdots \\ a_{m1} & a_{m2} & \cdots & a_{mn} \end{bmatrix}. \tag{3.1}$$

This matrix A thus defined is spoken of as an $m \times n$ matrix, or alternatively as a matrix of *size* or *order* $m \times n$, referring to the fact that A has m rows and n columns. By convention, the row size is mentioned first. The general element in row i and column j is written a_{ij}, and the notation

$$A = (a_{ij})$$

means that the matrix A has the i, j element a_{ij}. Unless otherwise stated, it will be assumed that the elements are real numbers.

The matrices $A = (a_{ij})$ and $B = (b_{ij})$ are said to be *equal* if they are of the same size and the corresponding elements are equal. Thus, if $A = B$, for each i and j,

$$a_{ij} = b_{ij}. \tag{3.2}$$

Otherwise, the matrices A and B are not equal. Therefore, the matrix equality

$$\begin{bmatrix} 5x + 2y \\ x - 3y \end{bmatrix} = \begin{bmatrix} 7 \\ 1 \end{bmatrix}$$

states that $5x + 2y = 7$ and $x - 3y = 1$.

MATRIX OPERATIONS

Operations such as addition, multiplication by a scalar, multiplication of two matrices, and even a form of division can be defined for matrices if certain conditions are met by the two or more matrices involved in the operation. For the matrix operations to be defined, the sizes of the matrices involved must be compatible, as described when each operation is introduced in this chapter. When addition is defined, matrix addition is commutative. Matrix operations also have the distributive and associative properties when the appropriate multiplications are defined. Commutativity under multiplication $(AB = BA)$ is valid only in special cases, as defined in Chapter 4.

Matrix Addition   Two matrices of the same size can be added by performing element-by-element addition. The sum of matrices of different sizes is not defined. If A and B are both $m \times n$, the sum S is an $m \times n$ matrix with elements

$$s_{ij} = a_{ij} + b_{ij}. \tag{3.3}$$

Since the sum of two $m \times n$ matrices is another $m \times n$ matrix, the set of all $m \times n$ matrices is *closed under matrix addition*.

Multiplication by a Scalar   Multiplication of a matrix by a scalar is defined to mean that each element of the matrix is multiplied by the scalar. If A has elements a_{ij} and α is any scalar, the operation (αA) results in the elements (αa_{ij}).

Matrix Multiplication   If A is an $m \times n$ matrix and B is $n \times p$, the elements of the matrix *product AB* can be computed by writing

$$
\begin{aligned}
C &= AB \\
&= \begin{bmatrix} a_{11} & a_{12} & \cdots & a_{1n} \\ a_{21} & a_{22} & \cdots & a_{2n} \\ \vdots & \vdots & \ddots & \vdots \\ a_{m1} & a_{m2} & \cdots & a_{mn} \end{bmatrix} \begin{bmatrix} b_{11} & b_{12} & \cdots & b_{1p} \\ b_{21} & b_{22} & \cdots & b_{2p} \\ \vdots & \vdots & \ddots & \vdots \\ b_{n1} & b_{n2} & \cdots & b_{np} \end{bmatrix},
\end{aligned}
$$

multiplying the successive elements of a row of A by elements of the corresponding column of B, and summing to find an element of C. This yields the matrix C with elements

$$c_{ij} = a_{i1}b_{1j} + a_{i2}b_{2j} + \cdots + a_{in}b_{nj} \tag{3.4}$$

for $i = 1, 2, ..., m$ and $j = 1, 2, ..., p$. Matrix C has size $m \times p$. The result C is thus

$$C = \begin{bmatrix} c_{11} & c_{12} & \cdots & c_{1p} \\ c_{21} & c_{22} & \cdots & c_{2p} \\ \vdots & \vdots & \ddots & \vdots \\ c_{m1} & c_{m2} & \cdots & c_{mp} \end{bmatrix}.$$

Let the size of A be $m \times n$ and that of B be $s \times p$, then the following conclusions apply:

1. The product AB is defined if and only if
   (number of columns of A) = (the number of rows of B),
   or $n = s$.

2. The product BA is defined if and only if
   (number of columns of B) = (the number of rows of A)
   or $p = m$.

3. Even if both AB and BA are defined, they are not necessarily equal.

Thus, the existence of AB is independent of whether BA is defined and vice versa. In general, $AB \neq BA$. If $AB = BA$, we say that the matrices *commute*. This will be the exception, not the rule, for matrix multiplication.

☐ EXAMPLE 3.1   *Matrix Multiplication*

Let A be a 2×3 matrix and let B be a 3×2 matrix as follows:

$$A = \begin{bmatrix} 2 & -1 & 0 \\ 4 & 3 & -1 \end{bmatrix}; \qquad B = \begin{bmatrix} -4 & 0 \\ 6 & -4 \\ 1 & 6 \end{bmatrix}.$$

Using Equation 3.4 to compute the elements,

$$\begin{aligned} AB &= \begin{bmatrix} (2)(-4) + (-1)(6) + (0)(1) & (2)(0) + (-1)(-4) + (0)(6) \\ (4)(-4) + (3)(6) + (-1)(1) & (4)(0) + (3)(-4) + (-1)(6) \end{bmatrix} \\ &= \begin{bmatrix} -14 & 4 \\ 1 & -18 \end{bmatrix}. \end{aligned}$$

Thus, the product is a 2×2 matrix, since $(2 \times 3)(3 \times 2) \rightarrow (2 \times 2)$.

The product BA is a 3×3 matrix, $(3 \times 2)(2 \times 3) \rightarrow (3 \times 3)$, so the result cannot be equal to AB. The product BA is computed as

$$BA = \begin{bmatrix} (-4)(2)+(0)(4) & (-4)(-1)+(0)(3) & (-4)(0)+(0)(-1) \\ (6)(2)+(-4)(4) & (6)(-1)+(-4)(3) & (6)(0)+(-4)(-1) \\ (1)(2)+(6)(4) & (1)(-1)+(6)(3) & (1)(0)+(6)(-1) \end{bmatrix},$$

with the results

$$BA = \begin{bmatrix} -8 & 4 & 0 \\ -4 & -18 & 4 \\ 26 & 17 & -6 \end{bmatrix}.$$

□

□ EXAMPLE 3.2   *MATLAB Matrix Multiplication*

The following MATLAB session shows the data entry and multiplication of the matrices in Example 3.1. Each row of a matrix ends with a semicolon (;). Alternatively, a CR (Carriage Return) can be entered after each row. Notice that there is no need to worry about the size (dimension) of the matrices. MATLAB creates a matrix of the proper size when data entry is complete.

MATLAB has an extensive set of matrix operators as well as functions that perform special operations. Many more of these features of MATLAB for matrix manipulation will be introduced in later sections of this chapter.

MATLAB Script _____

```
Example 3.2
>>A=[2 -1 0;4 3 -1]
A =
      2     -1      0
      4      3     -1

>>B=[-4 0;6 -4;1 6]
B =
     -4      0
      6     -4
      1      6

>>C=A*B
C =
    -14      4
      1    -18

>>C1=B*A
C1 =
     -8      4      0
     -4    -18      4
     26     17     -6

>>quit
```
_____

□

Properties of Matrix Operations   The operations on matrices A, B, and C have the following properties when they have the sizes listed and α is a scalar:

1. $A + B = B + A$ when A and B are both $m \times n$.

2. $(A + B) + C = A + (B + C)$ if $A, B,$ and C are $m \times n$.

3. $A(B + C) = AB + AC$ if A is $m \times k$ and B and C are $k \times n$.

4. $\alpha(A + B) = \alpha A + \alpha B$;

5. $(A + B)C = AC + BC$ if A and B are $m \times k$ and C is $k \times n$.

6. $A(BC) = (AB)C$ if A is $m \times k$, B is $k \times r$, and C is $r \times n$.

These properties will be assumed for matrices, but they can each be proven using the previous definitions and the fact that the matrix elements are real numbers.

The set of all $m \times n$ matrices forms a *vector space* in which the "vectors" are $m \times n$ matrices, vector addition is matrix addition, and multiplying a vector by a scalar becomes scalar multiplication of a matrix. Properties of vector spaces were defined in Chapter 2.

Transpose of a Matrix If A is an $m \times n$ matrix with elements a_{ij}, the $n \times m$ matrix obtained from A by interchanging the rows and columns is called the *transpose* of A and is written A^T. The element in the ith row and jth column of A^T is thus a_{ji}.

For example, the transpose of the matrix

$$C = \begin{bmatrix} -14 & 4 & x \\ 1 & -18 & y \end{bmatrix}$$

is the matrix

$$C^T = \begin{bmatrix} -14 & 1 \\ 4 & -18 \\ x & y \end{bmatrix}.$$

In the matrix, x and y are any real numbers. The transpose is formed by making the rows of C the columns of C^T.

Properties of the Transpose Several properties of the transpose are as follows:

1. The transposition operation is *reflective*; that is, $\left(A^T\right)^T = A$.

2. The transpose of the product of two matrices is equal to the product of their transposes in the reverse order; that is,

$$(AB)^T = B^T A^T.$$

The vector $\mathbf{y} = [y_1, y_2, \ldots, y_m]$ can be considered to be a $1 \times m$ matrix or a *row* vector. An $m \times 1$ matrix can be considered a *column* vector. The length of the vector is usually referred to as the *dimension* of the vector, rather than its size, as for matrices. Thus,

$$\begin{bmatrix} 1 \\ 3 \\ 0 \end{bmatrix}$$

is an example of a 3D column vector, and $[x, 5, -\sqrt{x}, 0]$ is a 4D row vector.

By the definition of matrix multiplication, the product of an $m \times n$ matrix and the n-dimensional row vector $\mathbf{y}$ should be written $A\mathbf{y}^T$, since $\mathbf{y}^T$ is an $n \times 1$ matrix that is an n-dimensional column vector. Where no confusion would result, the product of a matrix A and a vector $\mathbf{x}$ will be written as $A\mathbf{x}$. However, it is understood that the vector $\mathbf{x}$ must have size $n \times 1$ if A is $m \times n$.

MATLAB MATRIX OPERATIONS

The strength of MATLAB, indeed its purpose, is matrix manipulation. Table 3.2 lists a number of the fundamental operations. Addition, subtraction, multiplication, multiplication by a scalar, and matrix transpose follow the rules previously defined for matrices. Multiplication by a scalar applies to individual elements of matrices, and reference to a_{ij} in the table means that the operation applies to every element of the matrix.

TABLE 3.2 *MATLAB Basic Matrix Operations*

Operation	MATLAB Form	Mathematical Form
Matrix	`[a11 ... a1n; ... amn]`	A
Addition	`A+B`	$A + B$
Subtraction	`A-B`	$A - B$
Transpose	`A'`	A^T
Multiply	`A * B`	$A * B$
Multiply by scalar	`s*A`	$s \times a_{ij}$
Special:		
Multiply	`A.*B`	$a_{ij} \times b_{ij}$
Power	`A.^m`	a_{ij}^m
Function	`f(A)`	$f(a_{ij})$
Add scalar	`A+s`	$a_{ij} + s$

☐ EXAMPLE 3.3   *MATLAB Matrix Operations*

The MATLAB representation of the transpose of the vector **x** is **x**'. Similarly, the transpose of *A* is written *A*'. The accompanying MATLAB session illustrates the properties of the transpose. The matrix C is first defined and its transpose Ct is computed by MATLAB. The transpose of Ct is clearly C. Then, a matrix A is defined and the product C*A is found. Finally, the transpose of the product is shown to be equal to the product of the individual transposes taken in the opposite order.

MATLAB Script _____

```
Example 3.3
>>format compact
>>C=[-14 4;1 -18]
C =
   -14    4
     1  -18
>>Ct=C'   % C transpose
Ct =
   -14    1
     4  -18
>>Ct'     % Transpose of C transpose = C
ans =
   -14    4
     1  -18
>>A=[2 1;3 0]
A =
     2    1
     3    0
>>AxB=C*A
AxB =
   -16  -14
   -52    1
>>CAt=(C*A)'     % (C*A) transpose
CAt =
   -16  -52
   -14    1
>>A'*C'
>>% (A transpose * C transpose) = (C*A) transpose
ans =
   -16  -52
   -14    1
>>quit
```

_____

☐

The *Symbolic Math Toolbox* contains a number of commands that manipulate matrices whose elements are symbolic expressions. The symbolic command **sym** allows the creation of a symbolic matrix. The commands **symadd**, **symmul**, and **symsub** are used for matrix addition, multiplication, and subtraction, respectively.

SQUARE AND SYMMETRIC MATRICES

A number of special matrices are important in both applications and theoretical studies. These include square matrices, identity matrices, diagonal matrices, and symmetric matrices.

SQUARE MATRICES

A *square* matrix has the same number of rows and columns. Thus, A is a square matrix if A is $n \times n$. If $A = (a_{ij})$ are the elements of A, the elements with $i = j$ are the diagonal elements of A and consist of the set

$$a_{11}, a_{22}, \ldots, a_{nn}.$$

The notation $\text{diag}(a_{11}, a_{22}, \ldots, a_{nn})$ is also used to specify the diagonal elements of a matrix with elements $a_{ii}, i = 1, \ldots n$. The other elements in the matrix are called the *off-diagonal* elements.

Diagonal Matrices   An important type of square matrix is the *diagonal* matrix. A diagonal matrix D is defined as a matrix in which all the off-diagonal elements are zero.

☐ **EXAMPLE 3.4**   ***Matrix Operations***

Let matrices be defined as

$$A = \begin{bmatrix} 1 & -1 & 1 \\ 0 & 1 & 0 \\ 2 & 0 & 3 \end{bmatrix}, \quad B = \begin{bmatrix} 3 & 0 & 0 \\ 0 & 1 & 0 \\ 0 & 1 & 1 \end{bmatrix}, \quad D = \begin{bmatrix} 3 & 0 & 0 \\ 0 & 2 & 0 \\ 0 & 0 & 2 \end{bmatrix}.$$

Since $A + B = B + A$, these matrices meet the commutative property of addition. They also satisfy the distributive properties previously defined because the sum $A + B$ as well as the products AB, AC, and BC are defined. However,

$$AB = \begin{bmatrix} 1 & -1 & 1 \\ 0 & 1 & 0 \\ 2 & 0 & 3 \end{bmatrix} \begin{bmatrix} 3 & 0 & 0 \\ 0 & 1 & 0 \\ 0 & 1 & 1 \end{bmatrix} = \begin{bmatrix} 3 & 0 & 1 \\ 0 & 1 & 0 \\ 6 & 3 & 3 \end{bmatrix},$$

$$BA = \begin{bmatrix} 3 & 0 & 0 \\ 0 & 1 & 0 \\ 0 & 1 & 1 \end{bmatrix} \begin{bmatrix} 1 & -1 & 1 \\ 0 & 1 & 0 \\ 2 & 0 & 3 \end{bmatrix} = \begin{bmatrix} 3 & -3 & 3 \\ 0 & 1 & 0 \\ 2 & 1 & 3 \end{bmatrix},$$

so the matrices A and B do not commute.

The diagonal elements are

$$\text{diag}\,(A) = (1,1,3),$$
$$\text{diag}\,(B) = (3,1,1),$$
$$\text{diag}\,(D) = (3,2,2),$$

but only D is a diagonal matrix.

It is easily shown that the product DA multiplies the ith row of A by d_{ii}. In this case, D is said to *premultiply* A. Similarly, the product AD multiplies the ith column of A by d_{ii} when A is *postmultiplied* by D. The products are

$$DA = \begin{bmatrix} 3 & -3 & 3 \\ 0 & 2 & 0 \\ 4 & 0 & 6 \end{bmatrix} \quad \text{and} \quad AD = \begin{bmatrix} 3 & -2 & 2 \\ 0 & 2 & 0 \\ 6 & 0 & 6 \end{bmatrix}. \tag{3.5}$$

$\square$

Identity Matrices   The $n \times n$ diagonal matrix such that $(a_{ii}) = 1$ for $i = 1, 2, ...n$ is called the *identity* matrix. The nth-order identity matrix is

$$I = \begin{bmatrix} 1 & 0 & 0 & 0 & 0 & \cdots & 0 \\ 0 & 1 & 0 & 0 & 0 & \cdots & 0 \\ 0 & 0 & 1 & 0 & 0 & \cdots & 0 \\ 0 & 0 & 0 & 1 & 0 & \cdots & 0 \\ 0 & 0 & 0 & 0 & 1 & \cdots & 0 \\ \vdots & \vdots & \vdots & \vdots & \vdots & \ddots & \vdots \\ 0 & 0 & 0 & 0 & 0 & \cdots & 1 \end{bmatrix}.$$

If A and I are square matrices of the same size, $AI = IA$. The identity matrix will play an important role in many of the problems to be presented later.

MATLAB SPECIAL MATRICES

Table 3.3 lists a number of MATLAB commands to create special matrices.

TABLE 3.3 *MATLAB special matrices*

Command	Matrix
A=[ ]	Empty matrix
diag(A)	Main diagonal of A
eye(n)	$n \times n$ identity matrix $(a_{ii} = 1)$
eye(size(A))	Identity matrix $[\text{size(I)} = \text{size(A)}]$
ones(m,n)	$m \times n$ matrix $(a_{ij} = 1)$
rand(m,n)	$m \times n$ matrix of random numbers
zeros(m,n)	$m \times n$ matrix of zeros

MATLAB Special Matrices

The empty matrix is useful to define matrices that are needed as inputs for certain commands but without specifying the size. The command **diag**(A) returns a column vector with elements from the main diagonal of A.

The other commands create matrices with special elements. The command **eye** creates the matrix with elements:

$$a_{ij} = \begin{cases} 1, & i = j, \\ 0, & i \neq j. \end{cases}$$

The name eye is used to distinguish this identity matrix name from the variable I, which is used frequently in MATLAB programs as an index. The command **ones** creates a matrix with 1 as each element.

The command **zeros** can be used to create a large *sparse* matrix of size $n \times n$ whose elements are mostly zero when the command

```
>>A=zeros(n)
```

is followed by commands to define the nonzero elements. For example, if A(1,2) is to be 1, the command

```
>>A(1,2)=1
```

will set the element to 1 and leave the other elements zero.

The **rand** command creates vectors or matrices whose elements are random numbers. This can be useful to create a number of test matrices if random numbers are suitable for an algorithm.

In the commands with row and column arguments, (1,n) designates a row vector with n columns and (m,1) designates a column vector. When the argument has the form size(A), the resulting matrix has the same dimensions as A.

□

W H A T   I F ?   In Chapter 1, the colon (:) and period (.) operators were described. Using the matrices in Example 3.4, predict the outcome of the following operations and then determine the MATLAB results:

```
A(:,1);  A(3,:);  A*B;  A.*B;  A^2;  A.^2;  A(1:2,B)
```

SYMMETRIC MATRICES

The square matrix A with elements $a_{ij} = a_{ji}$ is called a *symmetric* matrix. The interesting fact is that

$$A = A^T$$

for a symmetric matrix A. Symmetric matrices arise in many physical problems.

Symmetric Matrices

Consider the electrical circuit shown in Figure 3.1, which is composed of three loops of resistors with values in ohms.

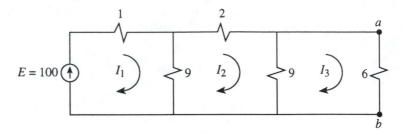

FIGURE 3.1 *A circuit composed of linear, bilateral elements*

Kirchhoff's voltage law states that the sum of the voltages around a closed loop is zero. The electromotive force is $E = 100$ volts. The voltage in each resistor R is given by Ohm's law, $E = I \times R$, where I is the current in amperes flowing through the resistor. In loop 1, the voltage across the 1-ohm resistor is $1 \times I_1$. Note that the voltage for the 9-ohm resistor between loop 1 and loop 2 is $9 \times (I_1 - I_2)$. The usual electrical problem is to solve for $\mathbf{I} = [I_1, I_2, I_3]$. However, in this example, the purpose is to study the form of the equations that result from applying Kirchhoff's law to the three loops shown. The result is

$$
\begin{array}{rcrcrcr}
10I_1 & - & 9I_2 & + & 0I_3 & = & 100, \\
-9I_1 & + & 20I_2 & + & -9I_3 & = & 0, \\
0I_1 & - & 9I_2 & + & 15I_3 & = & 0,
\end{array}
$$

or in matrix form

$$
\begin{bmatrix} 10 & -9 & 0 \\ -9 & 20 & -9 \\ 0 & -9 & 15 \end{bmatrix} \begin{bmatrix} I_1 \\ I_2 \\ I_3 \end{bmatrix} = \begin{bmatrix} 100 \\ 0 \\ 0 \end{bmatrix}.
$$

This system of equations is easily solved for the current $\mathbf{I}$ using one of the methods to be discussed in later sections of this chapter.

The matrix relating the currents and the voltage is symmetric so that $a_{ij} = a_{ji}$ for $i \neq j$. This is because the coupling between the loops is *bilateral*. In this case, this means that if current in one loop produces a voltage in another, the same current in the second circuit would produce the same voltage in the first. Ordinary electrical and mechanical elements are symmetrically bilateral, at least in their ideal models.

Furthermore, the resistors are *linear* elements since the resistance is assumed constant, regardless of current or voltage. Thus, circuits with linear, bilateral elements lead to linear systems of equations with symmetric matrices. This will also be found to be true if the equations describing the system result in differential equations. If such systems are analyzed as in this circuit example and the matrix equation is not symmetric, it is an indication of a mistake.

□

□ EXAMPLE 3.7   *Equations with Matrices*

The form of an equation relating physical variables yields information about the mathematical model being used. Many models lead to *scalar* equations. For example, Ohm's law, $E = IR$ (introduced in Example 3.6) relates only the magnitude of current and voltage. There is no indication of the spatial properties of the resistor, which in reality may be a 3D cylinder of carbon or other material.

Many of the electrical and mechanical properties of materials are defined by relations between vector forces and flows or displacements. Newton's second law written in the form $\mathbf{a} = \mathbf{F}/m$ indicates that the acceleration of an object with constant mass is in a direction parallel to the applied force. In effect, the resistor in Ohm's law and the object in Newton's law are treated as point objects. Further, $\mathbf{a} = \mathbf{F}/m$ can be treated as a shorthand notation for the equations

$$a_x = \frac{F_x}{m}, \quad a_y = \frac{F_y}{m}, \quad a_z = \frac{F_z}{m}.$$

Thus, the acceleration in the x direction is a function only of F_x and the constant m, and so is independent of the forces in the other directions.

Rewriting Ohm's law in a slightly more sophisticated form results in the equation

$$\mathbf{J} = \sigma \mathbf{E},$$

where $\mathbf{J}$ is the current density in amperes per square meter in two dimensions or amperes per cubic meter in three dimensions. The quantity σ is conductivity. This is the reciprocal of resistance measured in mhos. $\mathbf{E}$ is the electric field applied to the body with conductivity σ. In one dimension, $\mathbf{J}$ becomes the current I, $\sigma = 1/R$, and the field $\mathbf{E}$ is replaced by the voltage E.

The equation $\mathbf{J} = \sigma \mathbf{E}$ is a vector equation and implies that the conductivity of the material is *isotropic* if it is a constant, no matter what the direction of the applied field. This is true for materials constructed from cubic crystals in which the x, y, and z directions are equivalent. In matrix form, this relationship could be written

$$\begin{bmatrix} J_x \\ J_y \\ J_z \end{bmatrix} = \begin{bmatrix} \sigma & 0 & 0 \\ 0 & \sigma & 0 \\ 0 & 0 & \sigma \end{bmatrix} \begin{bmatrix} E_x \\ E_y \\ E_z \end{bmatrix}.$$

In many materials, the resultant current is not in the same direction as the applied field. This fact leads to many interesting electrical and optical properties of materials. In fact, crystals of noncubic structure may be expected to have *anisotropic* properties. The more general relationship between current density and electric field can be written as

$$\begin{aligned} J_x &= \sigma_{11}E_x + \sigma_{12}E_y + \sigma_{13}E_z, \\ J_y &= \sigma_{21}E_x + \sigma_{22}E_y + \sigma_{23}E_z, \\ J_z &= \sigma_{31}E_x + \sigma_{32}E_y + \sigma_{33}E_z. \end{aligned}$$

These equations show that an electric field in the x direction, for example, creates a current in all three directions, assuming the σ_{i1} ($i = 1, 2, 3$) are not zero. Thus, a matrix can be used to define the relationship between an applied electric field and the directions and magnitude of the resulting current in a

---

material. Similarly, stress and strain relations for a deformable body can be related by a matrix equation.

□

DETERMINANTS AND MATRIX INVERSES

―――――――――――――■―――――――――

Two important concepts that are useful from the study of square matrices are the determinant and the inverse matrix. First, we present approaches to compute the determinant and inverse of a matrix A that are used for theoretical studies and calculations by hand when small matrices are involved. Then, more practical computational methods and MATLAB solutions are presented.

DETERMINANT OF A MATRIX

The *determinant* of a square matrix is a number written symbolically as $\det(A)$ or $|A|$.[1] A common technique to compute the determinant is by *Laplace expansion*. However, a few definitions are necessary first to specify the elements in the expansion.

Minors and Cofactors The minors and cofactors of a determinant are derived as follows:

a. For an $n \times n$ matrix A, the *minor* of the element a_{ij} designated M_{ij} is the $n - 1 \times n - 1$ submatrix of A obtained by deleting row i and column j of A;

b. The *cofactor* of a_{ji}, designated A_{ij}, is the value $(-1)^{i+j} \times \det(M_{ij})$.

We note that some authors designate the minor as the determinant of the submatrix in Item a rather than the submatrix itself. In that case, the cofactor expression in Item b is modified in that M_{ij} rather than $|M_{ij}|$ is used.

The determinant of A can be determined by the Laplace expansion using any element of a row or a column as follows:

a. The expansion of $|A|$ by row k is the sum

$$\sum_{j=1}^{n} (-1)^{k+j} a_{kj} \det(M_{kj}).$$

―――――――――――――――――――――――

[1] The notation for the determinant $|A|$ should not be confused with the notation for the absolute value of a number, which is a nonnegative quantity. The numerical value of a determinant with real entries can be any real number.

b. The expansion of $|A|$ by column k is the sum

$$\sum_{i=1}^{n}(-1)^{k+i}a_{ik}\det(M_{ik}).$$

The expansion of $|A|$ by any row or column yields the same value for the determinant. Although the Laplace expansion is not practical for computations with large matrices, it is useful to explore the properties of the determinant. One conclusion is that because the value of $\det(A)$ is the same if expanded about any row or column, the determinant of a matrix is zero if the matrix has a row or column of zeros.

☐ EXAMPLE 3.8   *Determinant Expansion*

A number of examples will be given to illustrate the use of Laplace's expansion to find the determinant.

a. For an $n \times n$ matrix written as

$$A = \begin{bmatrix} a_{11} & a_{12} & \cdots & a_{1n} \\ a_{21} & a_{22} & \cdots & a_{2n} \\ \vdots & \vdots & \ddots & \vdots \\ a_{n1} & a_{n2} & \cdots & a_{nn} \end{bmatrix},$$

the expansion in terms of the cofactors using row 1 is

$$\begin{aligned} \det(A) &= \sum_{j=1}^{n}(-1)^{1+j}a_{1j}\det(M_{1j}) \\ &= a_{11}A_{11} - a_{12}A_{12} + \cdots + (-1)^{n+1}a_{1n}A_{1n}. \end{aligned}$$

b. For the 2×2 matrix

$$A = \begin{bmatrix} a_{11} & a_{12} \\ a_{21} & a_{22} \end{bmatrix} = \begin{bmatrix} 4 & 2 \\ 0 & 3 \end{bmatrix},$$

the determinant of A can be computed by expansion using row 1, row 2, column 1, or column 2. Using row 1, the value is

$$|A| = 4 \times 3 - 2 \times 0 = 12.$$

Notice that expanding with row 2 or column 1 requires only one multiplication since $a_{21} = 0$. The result is

$$|A| = 3 \times 4 \text{ (using row 2)} = 4 \times 3 \text{ (using column 1)} = 12.$$

☐

Determinants and Matrix Inverses

We now consider the question of the existence of a multiplicative inverse of a square matrix A. Given two $n \times n$ matrices A and B, B is called the inverse of A if

$$AB = BA = I, \tag{3.6}$$

where I is the $n \times n$ identity matrix. The matrix B is written as A^{-1}, and this matrix A^{-1} is called the *inverse* of A. From Equation 3.6, if the inverse exists, it has the property that A and A^{-1} commute, so that

$$A^{-1}A = AA^{-1} = I,$$

where I is the identity matrix of order n if A is $n \times n$.

If A has an inverse, it is said to be *invertible*. An invertible matrix is also called a *nonsingular* matrix. If A does not possess an inverse, it is said to be *noninvertible*, or more commonly *singular*. Note that inverses are defined only for square matrices.

If A is invertible, the elements of the inverse of a $n \times n$ matrix A can be computed using its determinant and the transpose of the matrix of its cofactors. First, we form the matrix composed of the cofactors of A with the elements

$$b_{ij} = (-1)^{i+j} \det(M_{ij}). \tag{3.7}$$

Then, the *adjoint* of A is defined as the transpose of the matrix B. Thus, adj(A) is the matrix B^T. The inverse is then

$$A^{-1} = \frac{1}{|A|} \operatorname{adj}(A). \tag{3.8}$$

Remember that the *transpose* of the matrix of cofactors is used to form the adjoint. Since it was assumed that A is invertible, $|A| \neq 0$.

☐ EXAMPLE 3.9　　*Matrix Inverse*

Let

$$A = \begin{bmatrix} 1 & 4 & 8 \\ 1 & 0 & 0 \\ 1 & -3 & -7 \end{bmatrix}.$$

The determinant is $+4$, since

$$\det(A) = -1 \begin{vmatrix} 4 & 8 \\ -3 & -7 \end{vmatrix} = -1(-28 + 24) = 4$$

using row 2 to expand.

The matrix of cofactors (B) is computed to be

$$B = \begin{bmatrix} 0 & 7 & -3 \\ 4 & -15 & 7 \\ 0 & 8 & -4 \end{bmatrix}.$$

Equation 3.8 for the inverse in terms of the adjoint yields

$$A^{-1} = \frac{1}{4}B^T = \begin{bmatrix} 0 & 1 & 0 \\ 7/4 & -15/4 & 2 \\ -3/4 & 7/4 & -1 \end{bmatrix}.$$

Forming the product $AA^{-1} = I$ verifies that we have found the inverse.

□

SUMMARY OF DETERMINANT AND MATRIX PROPERTIES

The following is a list of the properties of the determinants of $n \times n$ matrices A and B:

1. If A has a row or column of zeros, $\det(A) = 0$.

2. If any two rows (or columns) of A are interchanged, the sign of $\det(A)$ is changed.

3. If all the elements of one row or of one column of a matrix are multiplied by the same number α, the determinant is multiplied by α.

4. The determinant of a diagonal matrix is the product of the diagonal elements.

5. $\det(A^T) = \det(A)$.

6. $\det(AB) = \det(A) \times \det(B)$.

Determinants can be used to check if a matrix is singular or not. If the determinant is not zero, A is nonsingular and thus has an inverse. The determinant can also be used in the computation of the inverse as in Equation 3.8. However, these direct methods using the determinant are not generally used in computer solutions or even for numerical solutions by hand, because of the amount of computation necessary. The *reduction* methods presented in later sections of this chapter are usually more efficient for numerical calculations.

Properties of Square Matrices The determinant and inverse are defined only for square matrices. The following statements summarize some of the important properties of square matrices:

a. If both A and B are $n \times n$, the products AB and BA are defined, but it is not generally true that the product $AB = BA$.

b. The determinant of a matrix A is only defined for a square matrix.

c. Only a square matrix can have an inverse defined as A^{-1} in the equation $A^{-1}A = AA^{-1} = I$.

d. If $\det A \neq 0$, then A is invertible.

Note that the identity matrices are square, as are diagonal and symmetric matrices.

By hand, the calculation of the determinant and the inverse for a given matrix is generally tedious so hand calculations are used mainly for theoretical studies. Fortunately, MATLAB provides commands to compute the numerical value of the determinant and the inverse of a matrix. The command **det** computes the determinant, and **inv** returns the inverse for matrices with numerical entries.

Symbolic commands **determ** and **inverse** are the equivalent commands for matrices defined symbolically. The command **numeric** converts a matrix with elements defined as symbolic expressions to a matrix with numerical entries.

☐ EXAMPLE 3.10

MATLAB Matrix Inverse

The MATLAB commands **det** and **inv** compute the determinant and inverse of a matrix, respectively. The accompanying MATLAB session illustrates the use of these commands and several of the properties of determinants and inverses taken from Example 3.9. The command **rats** (rational) converts the decimal values of the elements in the matrix to a rational fraction approximation. In this case, the entries are rational numbers so the equivalence is exact.

MATLAB Script _____

```
Example 3.10
>>A=[1 4 8;1 0 0;1 -3 -7]
A =
       1      4      8
       1      0      0
       1     -3     -7

>>detA=det(A)
detA =
       4

>>Ainv=inv(A)
Ainv =
            0     1.0000          0
       1.7500    -3.7500     2.0000
      -0.7500     1.7500    -1.0000

>>rats(Ainv)
ans =
            0          1          0
          7/4      -15/4          2
         -3/4        7/4         -1

>>Ainv*A
ans =
       1.0000          0          0
            0     1.0000     0.0000
            0          0     1.0000
```

```
>>At=A'
At =
        1      1      1
        4      0     -3
        8      0     -7

>>det(At)
ans =
        4
>>quit
```

□

ORTHOGONAL AND TRIANGULAR MATRICES

Orthogonal and triangular matrices have special properties that are particularly useful in applications. These types of matrices are introduced in this section and used in other sections of this chapter and later in the book.

ORTHOGONAL MATRICES

An *orthogonal matrix* Q is an $n \times n$ matrix that is invertible with the property that its transpose is also its inverse, so that

$$Q^{-1} = Q^T. \tag{3.9}$$

This relationship is equivalent to $Q^T Q = Q Q^T = I$, where I is the identity matrix. An important example in two dimensions is the matrix

$$Qr = \begin{bmatrix} \cos\theta & -\sin\theta \\ \sin\theta & \cos\theta \end{bmatrix}.$$

This matrix is associated with rotations of vectors in a plane as described later in this chapter. Clearly, $QrQr^T = I$ for this matrix since

$$QrQr^T = \begin{bmatrix} \cos\theta & -\sin\theta \\ \sin\theta & \cos\theta \end{bmatrix} \begin{bmatrix} \cos\theta & \sin\theta \\ -\sin\theta & \cos\theta \end{bmatrix} = \begin{bmatrix} 1 & 0 \\ 0 & 1 \end{bmatrix}.$$

The following theorem states the relationship between orthogonal matrices and sets of orthonormal vectors. The theorem also shows how to construct orthogonal matrices.

■ THEOREM 3.1   *Orthogonal Matrix*

The $n \times n$ matrix Q is orthogonal if and only if the columns (and rows) of Q form an orthonormal system.

———————————————————————■

Let $\mathbf{q}_1, \mathbf{q}_2, \ldots, \mathbf{q}_n$ be the orthonormal vectors in $\mathbf{R}^n$ that form the columns of matrix Q. By the definition of orthonormal vectors,

$$\langle \mathbf{q}_i, \mathbf{q}_j \rangle = \mathbf{q}_i^T \mathbf{q}_j = \delta_{ij}, \tag{3.10}$$

where δ_{ij} is the Kronecker delta: $\delta_{ij} = 0$ if $i \neq j$ and $\delta_{ij} = 1$ if $i = j$. Since Q is orthogonal, $Q^{-1}Q = Q^T Q = I$, which in terms of the column vectors is

$$
Q^T Q =
\begin{bmatrix}
q_{11} & q_{21} & \cdots & q_{n1} \\
q_{12} & q_{22} & \cdots & q_{n2} \\
\vdots & \vdots & \ddots & \vdots \\
q_{1n} & q_{2n} & \cdots & q_{nn}
\end{bmatrix}
\begin{bmatrix}
q_{11} & q_{12} & \cdots & q_{1n} \\
q_{21} & q_{22} & \cdots & q_{2n} \\
\vdots & \vdots & \ddots & \vdots \\
q_{n1} & q_{n2} & \cdots & q_{nn}
\end{bmatrix}
$$

$$
=
\begin{bmatrix}
\mathbf{q}_1^T \mathbf{q}_1 & \mathbf{q}_1^T \mathbf{q}_2 & \cdots & \mathbf{q}_1^T \mathbf{q}_n \\
\mathbf{q}_2^T \mathbf{q}_1 & \mathbf{q}_2^T \mathbf{q}_2 & \cdots & \mathbf{q}_2^T \mathbf{q}_n \\
\vdots & \vdots & \ddots & \vdots \\
\mathbf{q}_n^T \mathbf{q}_1 & \mathbf{q}_n^T \mathbf{q}_2 & \cdots & \mathbf{q}_n^T \mathbf{q}_n
\end{bmatrix}
= I. \tag{3.11}
$$

The last equality shows that if the matrix Q is orthogonal, the columns are orthonormal.

Conversely, suppose the column vectors of a matrix Q are orthonormal. Then, the off-diagonal entries of the matrix in Equation 3.11 are zero, considering the definition of Equation 3.10. Hence, $Q^T Q = I$, as Equation 3.11 shows. Similarly, $QQ^T = I$, so that $Q^T = Q^{-1}$ is the unique inverse and Q^T is orthogonal.

Orthogonal Transformations   When applied to arbitrary vectors in $\mathbf{R}^n$, multiplication by an orthogonal matrix preserves the length of a vector and the value of the inner product. Recall from the discussion in Chapter 2 that the inner product of two column vectors $\mathbf{x}$ and $\mathbf{y}$ can be written

$$\mathbf{x} \cdot \mathbf{y} = \langle \mathbf{x}, \mathbf{y} \rangle = \mathbf{x}^T \mathbf{y},$$

and the length or norm of a vector is

$$||\mathbf{x}|| = \sqrt{\mathbf{x} \cdot \mathbf{x}} = \sqrt{\mathbf{x}^T \mathbf{x}}.$$

Applying an orthogonal matrix Q to the vectors $\mathbf{x}$ and $\mathbf{y}$, the properties of the inner product and the norm can be defined as follows:

a. $\langle Q\mathbf{x}, Q\mathbf{y} \rangle = \langle \mathbf{x}, \mathbf{y} \rangle$.

b. $||Q\mathbf{x}|| = ||\mathbf{x}||$.

Let Q be an orthogonal matrix and consider the products

$$\mathbf{u} = Q\mathbf{x} \quad \text{and} \quad \mathbf{v} = Q\mathbf{y},$$

which are called orthogonal transformations. That the value of the inner product is preserved by an orthogonal transformation is shown by forming the products

$$
\begin{aligned}
\langle \mathbf{u}, \mathbf{v} \rangle &= \mathbf{u}^T\mathbf{v} = (Q\mathbf{x})^T Q\mathbf{y} = \mathbf{x}^T Q^T Q\mathbf{y} \\
&= \mathbf{x}^T I \mathbf{y} = \langle \mathbf{x}, \mathbf{y} \rangle
\end{aligned}
$$

using the fact that $Q^T Q = Q^{-1} Q = I$. Applying this result with the substitution $\mathbf{y} = \mathbf{x}$ leads to the conclusion

$$\langle Q\mathbf{x}, Q\mathbf{x} \rangle = \langle \mathbf{x}, \mathbf{x} \rangle$$

or

$$\| Q\mathbf{x} \| = \| \mathbf{x} \|,$$

showing that lengths are preserved after an orthogonal transformation.

Orthogonal matrices and transformations will be considered in more detail later in the chapter after linear transformations are introduced.

UPPER AND LOWER TRIANGULAR MATRICES

Square matrices whose nonzero elements form a triangular pattern play an important role in applications and in computational linear algebra. If A has elements (a_{ij}), then the following definitions apply:

1. A matrix is *upper triangular* if all its elements below the diagonal are zero ($a_{ij} = 0$ for $i > j$).

2. A matrix is *lower triangular* if all its elements above the diagonal are zero ($a_{ij} = 0$ for $i < j$).

3. A matrix is *diagonal* if all its elements not on the diagonal are zero ($a_{ij} = 0$ for $i \neq j$).

Note that a diagonal matrix is both upper and lower triangular. All the triangular matrices have special properties that make them useful. For example, the determinant and inverse of a triangular matrix are particularly easy to compute.

Determinant of a Triangular Matrix   Consider the lower triangular matrix

$$
L = \begin{bmatrix}
a_{11} & 0 & 0 & 0 \\
a_{21} & a_{22} & 0 & 0 \\
a_{31} & a_{32} & a_{33} & 0 \\
a_{41} & a_{42} & a_{43} & a_{44}
\end{bmatrix}
$$

The determinant can be computed by the Laplace expansion using cofactors as previously discussed to yield

$$
\begin{aligned}
\det(A) &= a_{11}A_{11} + 0A_{12} + 0A_{13} + 0A_{14} = a_{11}A_{11} \\
&= a_{11}\begin{vmatrix} a_{22} & 0 & 0 \\ a_{32} & a_{33} & 0 \\ a_{42} & a_{43} & a_{44} \end{vmatrix} \\
&= a_{11}a_{22}\begin{vmatrix} a_{33} & 0 \\ a_{43} & a_{44} \end{vmatrix} \\
&= a_{11}a_{22}a_{33}a_{44},
\end{aligned}
$$

which shows that $\det(A)$ is the product of the diagonal elements of A. We generalize this result and state the following theorem for an $n \times n$ matrix.

■ THEOREM 3.2   *Determinant of triangular matrix*
Let $A = (a_{ij})$ be an upper or lower triangular $n \times n$ matrix. Then

$$det(A) = a_{11}a_{22} \cdots a_{nn}.$$

———————————————————————— ■

The theorem is proven by induction, as you should show in Problem 3.32.

The modern approach and the approach used by MATLAB to compute the determinant of a matrix is to factor the matrix into a product of triangular matrices and apply the theorem that defines the determinant of a product of two matrices.

■ THEOREM 3.3   *Determinant of matrix product*
Let A and B be $n \times n$ matrices, then

$$det(AB) = det(A)det(B).$$

———————————————————————— ■

WHAT IF?   The proof of the product rule for determinants is rather involved, but the result is easily demonstrated for two arbitrary matrices. However, by testing several matrices it will become evident that the determinant of a sum of matrices is not always equal to the sum of the determinants.

You can create arbitrary numerical matrices with the MATLAB command **rand** to test the conjectures for determinants. The symbolic command **determ** creates matrices with symbolic entries. For example, the command **sym**

```
determ(sym('a,b;c,d'))
```

yields the result a*d-b*c, as expected.

———

Chapter 3 ■ MATRICES

Suppose the matrix A can be *factored* into the product of a lower triangular matrix L and an upper triangular matrix U so that $A = LU$. From Theorem 3.3 concerning the product of determinants,

$$\det(A) = \det(LU) = \det(L)\det(U). \qquad (3.12)$$

Assuming that A and B are invertible $n \times n$ matrices, the inverse of the product is the product of the inverses in the opposite order, so that

$$(AB)^{-1} = B^{-1}A^{-1}.$$

Thus, if $A = LU$, the inverse can be computed as

$$(A)^{-1} = U^{-1}L^{-1}. \qquad (3.13)$$

Before discussing the details of the procedure to determine $A = LU$, called *LU factorization*, we present an example to demonstrate that such factorization is possible.

☐ EXAMPLE 3.11   *LU Factorization*
Consider the matrix equation

$$A = \begin{bmatrix} a_{11} & a_{12} & a_{13} \\ a_{21} & a_{22} & a_{23} \\ a_{31} & a_{32} & a_{33} \end{bmatrix} = \begin{bmatrix} 1 & 0 & 0 \\ l_{21} & 1 & 0 \\ l_{31} & l_{32} & 1 \end{bmatrix} \begin{bmatrix} u_{11} & u_{12} & u_{13} \\ 0 & u_{22} & u_{23} \\ 0 & 0 & u_{33} \end{bmatrix}, \qquad (3.14)$$

where $L = (l_{ij})$ is lower triangular with diagonal elements 1 and $U = (u_{ij})$ is upper triangular.

The system can be solved for the elements of L and U assuming that any elements that act as divisors are nonzero. Thus, equating the elements shows that

$$\begin{array}{lll} a_{11} = u_{11}, & a_{12} = u_{12}, & a_{13} = u_{13}, \\ a_{21} = l_{21}u_{11}, & a_{22} = l_{21}u_{12} + u_{22}, & a_{23} = l_{21}u_{13} + u_{23}, \\ a_{31} = l_{31}u_{11}, & a_{32} = l_{31}u_{12} + l_{32}u_{22}, & a_{33} = l_{31}u_{13} + l_{32}u_{23} + u_{33}. \end{array} \qquad (3.15)$$

We can solve for the unknown elements of L and U from this series of equations. There are several methods of solution that are adequate for small matrices. The *Doolittle method* will be used here.

In the equations relating the elements of A and those of L and U, notice that the first row of U is the same as the first row of A. Then, we solve the equations for the first column of L (l_{21}, l_{31}), the second row of U (u_{22}, u_{23}), and so on.

As an example matrix consider

$$A = \begin{bmatrix} 4 & 1 & 1 \\ 2 & -1 & 1 \\ 2 & 1 & 1 \end{bmatrix}.$$

Direct solution of Equations 3.15 leads to the results

$$u_{11} = 4, \; u_{12} = 1, \; u_{13} = 1,$$

$$l_{21} = 2/4, \; l_{31} = 2/4,$$
$$u_{22} = -1 - (1/2)(1) = -1.5, \; u_{23} = 1 - (1/2)(1) = 0.5,$$
$$l_{32} = \frac{1}{-1.5}(1 - (1/2)(1)) = -1/3,$$
$$u_{33} = 1 - (1/2)(1) - (-1/3)(1/2) = 2/3.$$

Thus, the LU factorization of (A) is

$$A = \begin{bmatrix} 4 & 1 & 1 \\ 2 & -1 & 1 \\ 2 & 1 & 1 \end{bmatrix} = \begin{bmatrix} 1 & 0 & 0 \\ 0.5 & 1 & 0 \\ 0.5 & -1/3 & 1 \end{bmatrix} \begin{bmatrix} 4 & 1 & 1 \\ 0 & -3/2 & 0.5 \\ 0 & 0 & 2/3 \end{bmatrix}$$

It is also easily verified that

$$\det(A) = \det(L)\det(U) = (1)(-4) = -4.$$

The MATLAB command **lu** performs the LU factorization, and for the case of the matrix A just presented, the MATLAB result is identical. Although the direct solution method can be generalized to large matrices, another method for LU factorization is used by the MATLAB algorithm. The algorithm will be discussed after Gaussian elimination for systems of equations is presented.

$\square$

SYSTEMS OF LINEAR EQUATIONS

Systems of linear equations arise in many fields of mathematics and engineering. The solution of n equations for n unknowns is a well-known problem studied in linear algebra. There are a great number of techniques to solve the equations both analytically and numerically. Before exploring various methods of solution for a set of linear equations, we define a slightly more general situation where there are m equations but n variables. The system of m linear equations in n variables $x_i, i = 1, 2, \ldots, n$ can be written as

$$
\begin{aligned}
a_{11}x_1 &+ a_{12}x_2 + \ldots + a_{1n}x_n = b_1, \\
a_{21}x_1 &+ a_{22}x_2 + \ldots + a_{2n}x_n = b_2, \\
&\qquad\qquad\qquad \vdots \\
a_{m1}x_1 &+ a_{m2}x_2 + \ldots + a_{mn}x_n = b_m.
\end{aligned}
\tag{3.16}
$$

These equations are called *linear* because each variable x_i appears to the first power only. The variables are called *unknowns* when the purpose is to find the values of the variables that satisfy the equations.

In matrix form, the system of Equations 3.16 can be written as

$$A\mathbf{x} = \mathbf{b}, \tag{3.17}$$

in terms of a matrix A, vector $\mathbf{x}$, and a constant vector $\mathbf{b}$. Considering Equation 3.16, the matrix A in Equation 3.17 is the $m \times n$ *matrix of coefficients*

$$A = \begin{bmatrix} a_{11} & a_{12} & \cdots & a_{1n} \\ a_{21} & a_{22} & \cdots & a_{2n} \\ \vdots & \vdots & \ddots & \vdots \\ a_{m1} & a_{m2} & \cdots & a_{mn} \end{bmatrix}. \tag{3.18}$$

The $n \times 1$ matrix (vector) $\mathbf{x}$ of unknowns in Equation 3.17 is

$$\mathbf{x} = \begin{bmatrix} x_1 \\ x_2 \\ \vdots \\ x_n \end{bmatrix}, \tag{3.19}$$

and the vector of constants $\mathbf{b}$ is the $m \times 1$ matrix

$$\mathbf{b} = \begin{bmatrix} b_1 \\ b_2 \\ \vdots \\ b_m \end{bmatrix}. \tag{3.20}$$

Techniques to solve the system of equations $A\mathbf{x} = \mathbf{b}$ for the unknowns in $\mathbf{x}$ have been studied for well over a century. Several methods of solution are treated in this chapter, including MATLAB techniques for numerical solution.

TERMINOLOGY A *solution* to the system of linear equations in Equation 3.16 is a set of n scalars $x_1, x_2, \ldots, x_n$ that satisfy the equations. These scalars are the elements of the vector $\mathbf{x}$ that satisfy Equation 3.17. Before discussing the solution methods, we will review some of the important terminology used to describe a system of equations.

A linear system of equations is said to be *consistent* if it has a solution. If there is no solution, the system is called *inconsistent*. For the linear system $A\mathbf{x} = \mathbf{b}$, the system is either

1. Consistent, with a unique (one) solution $\mathbf{x}$;

2. Consistent, with infinitely many possible solutions;

3. Inconsistent, with no solutions.

If the constant vector $\mathbf{b}$ in Equation 3.17 is zero, the linear system $A\mathbf{x} = \mathbf{0}$ is called *homogeneous*. This system always has the zero vector $\mathbf{x} = \mathbf{0}$ ($\mathbf{0} \in \mathbf{R}^n$) as a solution, which is termed the *trivial* solution. The system with $\mathbf{b} \neq \mathbf{0}$ is *nonhomogeneous*. In this chapter, solutions of the nonhomogeneous system will be sought. Homogeneous systems of equations will play a role in later chapters.

If $n > m$ in Equation 3.16, the system has more unknowns than equations and is called an *underdetermined system*. If the system is consistent, some of the variables can be chosen arbitrarily and the remaining variables are defined in terms of the arbitrary variables.

If $n < m$, the system has more equations than unknowns and is called *overdetermined*. Such equations can arise from measurements that yield a large number of data values but involve only a few variables in the mathematical model. In many problems, no attempt is made to solve underdetermined or overdetermined systems exactly, but methods such as least squares approximation are applied. Several of the references in the Annotated Bibliography at the end of the chapter cover such solution techniques in detail. Least squares solutions are treated in Chapter 7 of this text.

Of course, if $n = m$ the system has the same number of unknowns as equations. When the system is consistent, there is only one solution vector **x** that satisfies the equations. This is the most important case treated in this chapter and later chapters of the text.

☐ EXAMPLE 3.12   *Linear Systems*

A system of equations may possess no solutions, only one solution, or many solutions. For example, the 2×2 system

$$
\begin{aligned}
x_1 + x_2 &= 1 \\
x_1 + x_2 &= 2
\end{aligned}
$$

is said to be *inconsistent*, since it has no solutions. In 2D space ($\mathbf{R}^2$), these equations represent two parallel lines so there is no intersection.

The consistent 2×2 system

$$
\begin{aligned}
x_1 + x_2 &= 1 \\
x_1 - x_2 &= 0
\end{aligned}
$$

has the unique solution $x_1 = x_2 = \frac{1}{2}$. In $\mathbf{R}^2$, the two lines intersect at one point.

The 2×2 system

$$
\begin{aligned}
x_1 + x_2 &= 1 \\
2x_1 + 2x_2 &= 2
\end{aligned}
$$

really represents one equation in two unknowns since the second equation is a scalar multiple of the first. The result is an infinite number of solutions since $x_1 = 1 - x_2$ is a solution for each x_2 value. This is an underdetermined system in which x_2 (or x_1) can be chosen arbitrarily. Notice that only the determinant of the consistent system with a unique solution is nonzero. For an $n \times n$ system, this guarantees that a unique solution exists.

☐

This section treats the general linear system

$$A\mathbf{x} = \mathbf{b},$$ (3.21)

which consists of m equations in n unknowns. Thus, A is an $m \times n$ matrix. The basic questions concerning the solutions are as follows:

1. Do solutions exist?

2. If a solution exists, is it unique?

3. How are solutions to be found?

If A is an $n \times n$ square matrix, the questions about solutions are usually easily answered when the size of matrix A is relatively small, such as 3×3. However, some systems of interest are very large, perhaps containing thousands of equations. In this latter case, the equations can often be solved using computer algorithms. Various methods of solution of the equation $A\mathbf{x} = \mathbf{b}$ are presented first in this section without concern for numerical considerations and errors introduced by computer solution.

SOLUTION BY
MATRIX
INVERSE

The question as to how to find a solution to $A\mathbf{x} = \mathbf{b}$ is first answered for an $n \times n$ system by formally writing the solution for the unknown vector $\mathbf{x}$ as

$$\mathbf{x} = A^{-1}\mathbf{b},$$

where A^{-1} is the inverse of A, as previously defined. Unfortunately, this "solution" $\mathbf{x} = A^{-1}\mathbf{b}$ is valid only if the inverse of A exists and can be computed. A powerful theorem from linear algebra defines when this solution is valid and unique.

■ THEOREM 3.4   *Unique solution*

Let A be an $n \times n$ matrix. Then, the system $A\mathbf{x} = \mathbf{b}$ has a unique solution if and only if A is invertible (nonsingular). In this case, the unique solution is

$$\mathbf{x} = A^{-1}\mathbf{b}.$$

Thus, the solution to the $n \times n$ system of equations can be computed if the inverse of A exists and it can be found. From previous discussions, we know that an $n \times n$ matrix A has an inverse if and only if the determinant of A is nonzero. However, solving the system of equations by computing the inverse is only one way to find the unknowns in $\mathbf{x}$. In many problems, finding the inverse directly is less efficient than applying other methods of solution.

Most of the modern computer methods for solving systems of linear equations, finding the inverse of a matrix, or computing the determinant do not utilize the formal methods previously discussed. A more efficient method for computation is to reduce the system of equations or the matrix being inverted to a simpler form. This may be accomplished by a series of *elementary row operations* that can be applied to any matrix. MATLAB uses a variation of this method, as described later.

The three types of elementary operations performed on the rows of a matrix are as follows:

1. Replace a row of the matrix by a nonzero multiple of the row.

2. Replace a row by the sum of that row and a multiple of another row.

3. Interchange two rows of the matrix.

The importance of these operations to the solution of linear equations is expressed in a theorem defining the equivalence of two linear systems.

■ THEOREM 3.5

Equivalence of systems

Suppose the system $A\mathbf{x} = \mathbf{b}$ is converted to the system $A_r\mathbf{x} = \mathbf{b}_1$ by applying a sequence of elementary row operations to A to yield A_r and exactly the same sequence of operations to $\mathbf{b}$ to get $\mathbf{b}_1$. Then, the two systems are equivalent in the sense that they have the same solutions. The matrix A_r is called the reduced *matrix of A.*

───■

□ EXAMPLE 3.13

Elementary Matrix Operations

Consider the equations

$$\begin{bmatrix} 1 & 1 & 1 \\ -1 & 2 & -4 \\ 1 & 3 & 9 \end{bmatrix} \begin{bmatrix} x \\ y \\ z \end{bmatrix} = \begin{bmatrix} 2 \\ 2 \\ 0 \end{bmatrix}$$

to be solved for x, y, z. Let the notation R_j denote the jth row. To perform elementary row operations on each side of the equation, we form the *augmented* matrix with the constant vector as the fourth column and reduce the system using elementary operations. The augmented matrix then becomes

$$\left[\begin{array}{ccc|c} 1 & 1 & 1 & 2 \\ -1 & 2 & -4 & 2 \\ 1 & 3 & 9 & 0 \end{array}\right].$$

The object of the reduction is to add multiples of one row to other rows or to exchange rows, so that the first three columns form the identity matrix if a unique solution exists; if not, a row of zeros will appear. The following operations will solve this example:

$$\left[\begin{array}{ccc|c} 1 & 1 & 1 & 2 \\ -1 & 2 & -4 & 2 \\ 1 & 3 & 9 & 0 \end{array}\right] \longrightarrow \left[\begin{array}{ccc|c} 1 & 1 & 1 & 2 \\ 0 & 3 & -3 & 4 \\ 1 & 3 & 9 & 0 \end{array}\right] \quad R_2 \leftarrow R_2 + R_1,$$

$$\longrightarrow \begin{bmatrix} 1 & 1 & 1 & 2 \\ 0 & 1 & -1 & 4/3 \\ 0 & 2 & 8 & -2 \end{bmatrix} \begin{matrix} \\ R_2 \leftarrow \frac{1}{3} R_2 \\ R_3 \leftarrow R_3 - R_1, \end{matrix}$$

$$\longrightarrow \begin{bmatrix} 1 & 1 & 1 & 2 \\ 0 & 1 & -1 & 4/3 \\ 0 & 0 & 1 & -7/15 \end{bmatrix} \begin{matrix} \\ R_3 \leftarrow R_3 - 2R_2 \\ R_3 \leftarrow \frac{1}{10} R_3, \end{matrix}$$

$$\longrightarrow \begin{bmatrix} 1 & 1 & 0 & 37/15 \\ 0 & 1 & 0 & 13/15 \\ 0 & 0 & 1 & -7/15 \end{bmatrix} \begin{matrix} R_1 \leftarrow R_1 - R_3 \\ R_2 \leftarrow R_2 + R_3, \end{matrix}$$

$$\longrightarrow \begin{bmatrix} 1 & 0 & 0 & 24/15 \\ 0 & 1 & 0 & 13/15 \\ 0 & 0 & 1 & -7/15 \end{bmatrix} \begin{matrix} R_1 \leftarrow R_1 - R_2. \end{matrix}$$

In these calculations, the order of the operations does not matter. The result is that $[x, y, z] = [24/15, 13/15, -7/15]$.

$\square$

The use of elementary operations, as in Example 3.13, reduced the matrix of coefficients A to the identity matrix because A was square and nonsingular. Thus, the number of equations was equal to the number of unknowns and a *unique* solution was guaranteed since A was not singular. However, if A is not invertible or if A is not square, other approaches must be used to determine if a system of equations is consistent. The problem can be studied in terms of the rank of the matrices involved. We first define the rank of a square matrix to clarify the idea.

Rank of a Square Matrix The *rank* of a matrix A, written rank(A), can be defined as the number of linearly independent rows (or columns). An important result from linear algebra is that the rows (considered as vectors in $\mathbf{R}^n$) of a nonsingular $n \times n$ matrix A are linearly independent. The n columns, considered as vectors in $\mathbf{R}^n$, are also linearly independent. This result was used in Chapter 2 when the determinant was used to test vectors for linear independence. If n vectors in $\mathbf{R}^n$ form the columns (or rows) of an $n \times n$ matrix A, the vectors are linearly independent if and only if $\det(A) \neq 0$.

If the $n \times n$ matrix A is invertible, the matrix has a nonzero determinant, and the number of independent rows (or columns) is n and rank(A) $= n$ also. Thus, one method to determine if the rank is n is to compute $\det(A)$. If $\det(A) \neq 0$, rank(A) $= n$. Otherwise, the rank of A is less than n, but the rank is not determined. The following theorem states that a nonsingular matrix has rank n.

■ THEOREM 3.6　　***Rank of nonsingular matrix***

The $n \times n$ matrix A is nonsingular if and only if the rank of A is n.

─────────────────────────────────────── ■

The usual method for determining the rank of a matrix by hand is to reduce the matrix, by elementary operations, to a matrix whose rank is obvious. Thus, to determine the rank, the matrix A is first reduced by elementary operations, as described in Example 3.13. The resulting matrix is called the *reduced* matrix and will be designated as A_r. Elementary row operations do not change the rank of A_r, so the rank of A_r is equal to the rank of A.

A square matrix can always be reduced to upper triangular form by a sequence of elementary operations. As previously discussed, a matrix with elements a_{ij} is said to be *upper triangular* if each $a_{ij} = 0$ for $i > j$, so that all the elements below the diagonal are zero. It is possible that the upper triangular matrix could have a row of zeros. The rank of A is the number of nonzero rows in the reduced matrix. Since the reduced matrix in Example 3.13 was the identity matrix, $\text{rank}(A) = n$.

Rank of a Nonsquare Matrix For a nonsquare matrix, the procedure to determine the rank is the same as for a square matrix, as just described. The statement that applies to any matrix is

$$\text{rank}(A) = \text{ number of nonzero rows of the reduced matrix.}$$

Even for a nonsquare matrix, the rank is equal to the number of linearly independent rows. The number of linearly independent columns is the same.

Rank and the Solution of Linear Equations Consider the system of linear equations $A\mathbf{x} = \mathbf{b}$ and the augmented matrix for the system $[A \mid \mathbf{b}]$. Then, the equations have a solution if the rank of A is equal to the rank of the augmented matrix.

■ THEOREM 3.7   ***Solution and rank of augmented matrix***

The nonhomogeneous system of equations $A\mathbf{x} = \mathbf{b}$ has a solution if and only if $\text{rank}(A) = \text{rank}([A \mid \mathbf{b}])$.

————————————————————————————————■

The actual computation involves the comparison of the rank of the reduced matrix of A with the entire reduced augmented matrix. The reduced form of A is the first n columns of the reduced augmented matrix if A is an $m \times n$ matrix.

Summary Assume the reduced augmented matrix for the system of equations $A\mathbf{x} = \mathbf{b}$ is computed to be $[\,A_r \mid \mathbf{b}_1\,]$. The following observations are valid for solutions of the system of equations:

1. No solution exists if the reduced augmented matrix $([\,A_r \mid \mathbf{b}_1\,])$ has more nonzero rows than A_r; that is, there is no solution when the rank of the augmented matrix is greater than the rank of the matrix of coefficients A.

2. If A_r is $m \times n$ implying m equations in n unknowns, solutions are possible if the rank of the augmented matrix $([\,A_r \mid \mathbf{b}_1\,])$ is the same as the rank of A_r.

3. For a square matrix, if $A_r = I$, then $\mathbf{x} = \mathbf{b}_1$ is the unique solution of $A\mathbf{x} = \mathbf{b}$.

INVERSE   It has already been shown that a nonsingular square matrix can be reduced to the identity matrix by a sequence of elementary operations. By similar methods, the inverse of A can be found.

■ THEOREM 3.8   *Inverse matrix*

If an $n \times n$ matrix A can be converted to the $n \times n$ identity matrix I by a sequence of elementary operations, then the inverse A^{-1} is equal to the result of applying the same sequence of elementary operations to I.

———————————————————————————■

MATLAB MATRIX FUNCTIONS
———————————————■———————————————

If possible, we prefer to use MATLAB or Maple to perform the operations necessary to determine a matrix inverse or to solve a system of equations. MATLAB algorithms are very efficient for such computations. A summary of MATLAB matrix operations is listed in Table 3.4.

TABLE 3.4 *MATLAB matrix operations*

Operation	MATLAB	Mathematical Form
Determinant	**det**	$\det(A)$
Inverse	**inv**	A^{-1}
Rank	**rank**	$\text{rank}(A)$
Reduced matrix	**rref, rrefmovie**	A_r
Solution of linear systems:		
A is $n \times n$	A\b	Solve $A\mathbf{x} = \mathbf{b}$
A is $m \times n$	A\b	Solve system with least squares
Symbolic Commands:		
Determinant	**determ**	
Inverse	**inverse**	
Linear system	**linsolve**	

The MATLAB numerical commands **det** and **inv** are used to compute the determinant and the inverse of a square matrix when the inverse exists. MATLAB also has numerical commands to determine rank (**rank**), reduce a matrix by elementary operations, and to solve systems of linear equations.

The command **rref** produces a reduced form of a matrix. An interesting variation is **rrefmovie** that shows a "movie" on the screen of the computer terminal as the command executes.

If A is an $n \times n$ matrix and **b** is a column vector with n components, the *backslash* command **x= A\b** returns the solution vector if A is not singular. In case A is an $m \times n$ matrix, the backslash command creates a solution for an underdetermined or overdetermined using least squares approximation, as explained in Chapter 7.

Commands in the *Symbolic Math Toolbox* are available to find the determinant (**determ**) and inverse (**inverse**) of a symbolic matrix. The command **linsolve** solves a system of linear equations.

☐ EXAMPLE 3.14   *MATLAB Solution of Linear System*

Consider the following four linear systems:

a. An inconsistent system,

$$
\begin{aligned}
2x &- y + z = 1, \\
x &+ y - z = 2, \\
3x &- y + z = 0;
\end{aligned}
$$

b. An underdetermined system,

$$
\begin{aligned}
-x_1 + x_2 + 3x_3 &= -2, \\
x_2 + 2x_3 &= 4;
\end{aligned}
$$

c. A consistent system with a unique solution,

$$
\begin{aligned}
x - 2y &= -1, \\
2x + 3y &= 7;
\end{aligned}
$$

d. An inconsistent system,

$$
\begin{aligned}
x - 2y &= -1, \\
-x + 2y &= 7.
\end{aligned}
$$

The accompanying MATLAB session shows the results of MATLAB analysis. Many of the MATLAB partial results from the command **rrefmovie** are not included. It is fun to watch the reduction, but the details are not important to us here.

Attempting to reduce the system in Case a leads to an inconsistent system, since the rank of the reduced coefficient matrix A is 2, but the rank of the reduced augmented matrix is 3. The two independent rows of A are row 1 and row 2 because the reduced matrix has leading 1's in those columns.[2] The third row is dependent and can be written as

$$
\text{row } 3 = \frac{4}{3}\text{row } 1 + \frac{1}{3}\text{row } 2.
$$

---

[2]See the reference by David R. Hill listed in the Annotated Bibliography.

In Case b, the ranks of the reduced matrix and the reduced augmented matrix are the same. Thus, there are solutions to this system in terms of an arbitrary variable. Let $x_3 = \alpha$, where α is a number. Then, $x_2 = 4 - 2\alpha$ and $x_1 = 6 + \alpha$.

In Case c and Case d, the MATLAB backslash operator is used to attempt to solve the systems of equations. Case c yields a solution. Notice that MATLAB warns that the system in Case d is singular to working precision. Although the determinant is actually zero, numerical computation may not yield precisely zero for the result. However, in this case, the determinant is so small that MATLAB considers the matrix singular.

MATLAB Script _____

```
Example 3.14
%Case (a)
>>AUG=[2 -1 1 1;1 1 -1 2;3 -1 1 0]
AUG =
     2    -1     1     1
     1     1    -1     2
     3    -1     1     0

>>rrefmovie(AUG)
swap rows 1 and 3
A =
     3    -1     1     0
     1     1    -1     2
     2    -1     1     1
pivot = A(1,1)
A =
    1.0000   -0.3333    0.3333         0
    1.0000    1.0000   -1.0000    2.0000
    2.0000   -1.0000    1.0000    1.0000
   .
   .
   .

ans =
     1     0     0     0
     0     1    -1     0
     0     0     0     1

%Note: rank of AR is less that [AR | B ] so there is no solution
-----------------------------------------------------------------
%Case (b)
>> AUG=[-1 1 3 -2;0 1 2 4]
AUG =
    -1     1     3    -2
     0     1     2     4

>> rrefmovie(AUG)
pivot = A(1,1)
   .
   .
```

```
A =
     1    -1    -3     2
     0     1     2     4


ans =
     1     0    -1     6
     0     1     2     4

%Rank is the same, there are many solutions
----------------------------------------
%Case (c)
>>A=[1 -2;2 3]
A =
     1    -2
     2     3

>>b=[-1 7]
b =
    -1     7
>>b=b'
b =
    -1
     7
>>x=A\b

x =
    1.5714
    1.2857
% A unique solution
-------------------------------------------
%Case (d)
>>A=[1 -2;-1 2]
A =
     1    -2
    -1     2

>>b=[-11 13]'
b =
   -11
    13
>> x=A\b
Warning: Matrix is singular to working precision.
```

□

□ EXAMPLE 3.15   *MATLAB Matrix Inverse*
This example uses the MATLAB command **rrefmovie** to determine the
inverse of a 3×3 matrix A by elementary operations on the augmented matrix

$[A \mid I]$. Only part of the sequence of reductions is shown. The result is given by the 3×6 matrix in which the first three columns considered as a matrix have been reduced to I and the last three columns are the columns of A^{-1}. If only the inverse of A is required, use of the command **inv** is more appropriate.

MATLAB Script _____

```
Example 3.15
>> AUG=[2   4   8 1 0 0
        1   0   0 0 1 0
        1  -3  -7 0 0 1]

AUG =

     2     4     8     1     0     0
     1     0     0     0     1     0
     1    -3    -7     0     0     1

>>rrefmovie(AUG)
 .
 .
 .
ans =
    1.0000         0         0         0    1.0000         0
         0    1.0000         0    1.7500   -5.5000    2.0000
         0         0    1.0000   -0.7500    2.5000   -1.0000
```

_____

□

MATLAB SOLUTION OF SYSTEMS OF LINEAR EQUATIONS

The MATLAB algorithms to compute the determinant (**det**) and inverse (**inv**) of a matrix and solve equations $A\mathbf{x} = \mathbf{b}$ use a method called *Gaussian elimination with partial pivoting*. Assuming that A is an $n \times n$ matrix that is nonsingular, the basic approach is to express this square matrix as the product of two triangular matrices before computing the determinant, or inverse or solving a system of equations.[3]

From the previous discussion leading to Equations 3.12 and 3.13, the determinant and inverse are computed from the LU factorization of A as

$$\det(A) = \det(L)\det(U),$$
$$(A)^{-1} = U^{-1}L^{-1}.$$

Systems of Equations Let A be an $n \times n$ invertible (nonsingular) matrix and $\mathbf{b}$ be a vector in $\mathbf{R}^n$. Further, assume that A can be factored into the product of a lower triangular matrix L and an upper triangular matrix U. Then, the system of equations $A\mathbf{x} = \mathbf{b}$ can be written

$$A\mathbf{x} = LU\mathbf{x} = \mathbf{b}. \tag{3.22}$$

_____

[3]For matrices with special structure, other algorithms may be used. See the *MATLAB User's Guide* for details.

_____

To solve the system for $\mathbf{x}$, we define a new vector such that $\mathbf{z} = U\mathbf{x}$, so that the system equation becomes

$$A\mathbf{x} = L(U\mathbf{x}) = L\mathbf{z} = \mathbf{b}.$$

Once $\mathbf{z}$ is found, the solution $\mathbf{x}$ of the system $A\mathbf{x} = \mathbf{b}$ is obtained by solving

$$U\mathbf{x} = \mathbf{z}.$$

As Example 3.16 shows, solving the triangular systems of equations is particularly simple. The factoring approach is an alternative to the method of reducing the augmented matrix $[\,A \mid \mathbf{b}\,]$, as shown in Example 3.13.

☐ EXAMPLE 3.16   *LU Solution of* $A\mathbf{x} = \mathbf{b}$

To demonstrate the LU factorization of Equation 3.22, consider the system of equations defined by

$$A\mathbf{x} = \begin{bmatrix} 4 & 1 & 1 \\ 2 & -1 & 1 \\ 2 & 1 & 1 \end{bmatrix} \begin{bmatrix} x_1 \\ x_2 \\ x_3 \end{bmatrix} = \begin{bmatrix} 9 \\ 3 \\ 7 \end{bmatrix}.$$

The matrix was presented in Example 3.11, and it has the LU factored form

$$A = \begin{bmatrix} 4 & 1 & 1 \\ 2 & -1 & 1 \\ 2 & 1 & 1 \end{bmatrix} = \begin{bmatrix} 1.0 & 0 & 0 \\ 0.5 & 1 & 0 \\ 0.5 & -1/3 & 1 \end{bmatrix} \begin{bmatrix} 4 & 1 & 1.0 \\ 0 & -3/2 & 0.5 \\ 0 & 0 & 2/3 \end{bmatrix}.$$

Considering the equation $L\mathbf{z} = \mathbf{b}$

$$\begin{bmatrix} 1.0 & 0 & 0 \\ 0.5 & 1 & 0 \\ 0.5 & -1/3 & 1 \end{bmatrix} \begin{bmatrix} z_1 \\ z_2 \\ z_3 \end{bmatrix} = \begin{bmatrix} 9 \\ 3 \\ 7 \end{bmatrix},$$

we solve for $\mathbf{z}$ by *forward substitution*, with the result

$$
\begin{aligned}
z_1 &= 9 \\
z_2 &= 3 - 0.5z_1 = -1.5 \\
z_3 &= 7 - 0.5z_1 + \frac{1}{3}z_2 = 2
\end{aligned}
$$

so that $\mathbf{z} = [9, -1.5, 2]^T$.

Next, we solve $U\mathbf{x} = \mathbf{z}$ in the form

$$\begin{bmatrix} 4 & 1 & 1.0 \\ 0 & -3/2 & 0.5 \\ 0 & 0 & 2/3 \end{bmatrix} \begin{bmatrix} x_1 \\ x_2 \\ x_3 \end{bmatrix} = \begin{bmatrix} 9 \\ -1.5 \\ 2 \end{bmatrix}$$

by *back substitution* to obtain

$$
\begin{aligned}
x_3 &= 3 \\
x_2 &= -\frac{2}{3}[-1.5 - 0.5x_3] = 2 \\
x_1 &= \frac{1}{4}[9 - x_2 - x_3] = 1
\end{aligned}
$$

so that $\mathbf{x} = [1, 2, 3]^T$.

□

WHAT IF?   Factoring the matrix A in Example 3.16 yielded matrices L and U, as shown in the example. However, if the rows of A are interchanged to form a *row permutation* of matrix A, the MATLAB command **lu** yields a different lower triangular matrix. In fact, the matrix is rearranged to reduce numerical errors. Experiment a bit with permutations of A from the example and perhaps with matrices with random entries to determine what command **lu** is doing.

GAUSS ELIMINATION AND NUMERICAL ERRORS

The method of solving systems of linear equations using elementary row operations is often known as the *Gaussian elimination method*. MATLAB algorithms use a variation of this method to solve

$$A\mathbf{x} = \mathbf{b}$$

when A is a general square matrix. The algorithm is described in the *MATLAB User's Guide* and analyzed in some detail in the textbook by Forsythe, Malcolm, and Moler listed in the Annotated Bibliography for this chapter. Their text also explains the *condition number of a matrix* that is used as an indication of the numerical accuracy of the results of matrix inversion or linear equation solution.

Numerical analysts say a problem is *ill conditioned* if a small relative error in data causes a large relative error in the computed solution. In the case of $A\mathbf{x} = \mathbf{b}$, they would seek a measure of the sensitivity of $\mathbf{x}$ to changes in A and $\mathbf{b}$. If the system is ill conditioned, the solution may be inaccurate. An ill-conditioned matrix is close to singular, and computation of the inverse may be impossible due to numerical errors.

The numerical quantity associated with the condition of the matrix A is called the *condition number*. Datta's textbook listed in the Annotated Bibliography for this chapter presents a detailed analysis of the linear system problem. In general, the larger the condition number, the more ill conditioned are the matrix and the associated linear system. In effect, the condition number defines how close the matrix is to singular.

MATLAB's **cond** command computes the condition number of a matrix. A matrix with a condition number of 1 is considered well conditioned. An experimental result is that roundoff errors in solving a system of linear equations by Gaussian elimination can result in the loss of

$$\log_{10}(\text{cond}(A))$$

decimal places if cond(A) is the condition number of the system matrix.

Numerical Errors

Consider the linear system

$$
A = \begin{bmatrix}
1 & 1/2 & 1/3 & 1/4 & 1/5 \\
1/2 & 1/3 & 1/4 & 1/5 & 1/6 \\
1/3 & 1/4 & 1/5 & 1/6 & 1/7 \\
1/4 & 1/5 & 1/6 & 1/7 & 1/8 \\
1/5 & 1/6 & 1/7 & 1/8 & 1/9
\end{bmatrix}
\begin{bmatrix}
x_1 \\ x_2 \\ x_3 \\ x_4 \\ x_5
\end{bmatrix}
=
\begin{bmatrix}
137/60 \\ 87/60 \\ 459/420 \\ 743/840 \\ 1879/2520
\end{bmatrix},
$$

which has been designed so that the exact solution is $\mathbf{x} = [1, 1, 1, 1, 1]$. The matrix A is an example of a *Hilbert matrix* of order five. The Hilbert matrices are known to be ill conditioned in the sense that numerical errors may destroy the accuracy of a solution to the system $A\mathbf{x} = \mathbf{b}$.

The accompanying MATLAB script computes the condition number of the matrix A and the also displays the error compared to the known solution.

MATLAB Script _____

```
Example 3.17
% CLCONDNO.M Solve A*x=b for A a 5x5 Hilbert matrix
%   x=[1 1 1 1 1] is answer; xcomp is computed vector
% OUTPUT: Condition number (condA) and xerror=xcomp-x
%
format rat          % Show rational values
format compact      %   and suppress blank lines
A=hilb(5);          % Hilbert matrix
b=[137/60 87/60 459/420 743/840 1879/2520]';
x=[1 1 1 1 1];      % Exact answer
% Check condition number and solve
fprintf('Condition number and error in xcomp:\n')
condA=cond(A)       % Display condition number
format long         % Format decimal values
xcomp=A\b;          % Computed value of solution
fprintf('Error in calculated value xcomp(I), I=1,5\n')
xerror=xcomp - x'
%-----------------------------------------------------
%  Results
Condition number and error in xcomp:
condA =
 476607
Error in calculated value xcomp(I), I=1,5
xerror =
  1.0e-011 *
 -0.00681676937120
  0.12230216839271
 -0.51394444255948
  0.76161299489286
 -0.36720626539477
```

The condition number for the Hilbert matrix is about 4.8×10^5, so it is expected that numerical problems may arise. The error here is on the order of 10^{-11}, which is accurate enough for most applications. However, with such a large condition number, we expect that relatively small changes in A or $\mathbf{b}$ would cause rather large changes in the solution vector $\mathbf{x}$. You are asked to investigate these cases in Problem 3.26.

□

WHAT IF? The error of the results in Example 3.17 does not seem too large. However, suppose that we increase the size of the matrix to 20×20? Redo the script and find the error that MATLAB makes if the Hilbert matrix is that large. Our file EX317WIF.M on the disk solves the problem. Try it without peeking first.

LINEAR TRANSFORMATIONS

Many operations discussed in this chapter can be described as performing operations or transformations on vectors in an abstract vector space. A function or operation denoted $\mathcal{L}$ will be called a *linear function* or *linear operation* if the following relationship holds:

$$\mathcal{L}[\alpha\mathbf{a} + \beta\mathbf{b}] = \alpha\mathcal{L}(\mathbf{a}) + \beta\mathcal{L}(\mathbf{b}), \qquad (3.23)$$

where $\alpha, \beta \in \mathbf{R}$, and $\mathbf{a}, \mathbf{b}$ are vectors in the vector space.

LINEARITY FOR FUNCTIONS

For real-valued continuous functions and α and β scalars, the linear relationship can be written

$$\mathcal{L}[\alpha f(t) + \beta g(t)] = \alpha\mathcal{L}[f(t)] + \beta\mathcal{L}[g(t)], \qquad (3.24)$$

where α and $\beta \in \mathbf{R}$ are constants. Thus, the derivative operator is linear, since

$$\frac{d}{dt}[\alpha f(t) + \beta g(t)] = \alpha\frac{df(t)}{dt} + \beta\frac{dg(t)}{dt}. \qquad (3.25)$$

Operations such as differentiation and integration are linear. Also, many important transformations, such as the Laplace and Fourier transforms, are linear.

Linearity Example

Let α, β, and m be scalars. The function $y = f(x) = mx$ is shown to be linear using Equation 3.24 by forming

$$
\begin{aligned}
f(\alpha x_1 + \beta x_2) &= \alpha m x_1 + \beta m x_2 \\
&= \alpha f(x_1) + \beta f(x_2)
\end{aligned}
$$

and noting that the second equation is equivalent to the first by substituting $f(x_1) = mx_1$ and $f(x_2) = mx_2$.

The function $y = mx + b$ is not linear if $b \neq 0$, according to Equation 3.24. Letting $\alpha = \beta = 1$ for simplicity, $f(x) = mx + b$ so that

$$f(x_1 + x_2) = m(x_1 + x_2) + b,$$

but

$$f(x_1) + f(x_2) = (mx_1 + b) + (mx_2 + b) = m(x_1 + x_2) + 2b,$$

which is not the same unless $b = 0$. Remember from the discussion of vector spaces in Chapter 2 that the set of points that satisfy the equation $y = mx$ form a vector space, but the solutions to $y = mx + b$ do not if $b \neq 0$.

It is perhaps a curious fact that the equation for a "line" in $\mathbf{R}^2$ of the form $y = mx + b$, $b \neq 0$ is not linear according to the definition given by Equation 3.24. Thus, the terms *linear* and *line* must be used appropriately.

□

MATRICES AS LINEAR TRANSFORMS

Considering matrix multiplication of a linear combination of vectors

$$
\begin{bmatrix}
a_{11} & a_{12} & \cdots & a_{1n} \\
a_{21} & a_{22} & \cdots & a_{2n} \\
\vdots & \vdots & \ddots & \vdots \\
a_{m1} & a_{m2} & \cdots & a_{mn}
\end{bmatrix}
\left(
\alpha
\begin{bmatrix}
x_1 \\
x_2 \\
\vdots \\
x_n
\end{bmatrix}
+ \beta
\begin{bmatrix}
y_1 \\
y_2 \\
\vdots \\
y_n
\end{bmatrix}
\right),
$$

it is clear that the operation is linear, since performing the multiplications would be equivalent to the linear equation

$$A(\alpha \mathbf{x} + \beta \mathbf{y}) = \alpha A\mathbf{x} + \beta A\mathbf{y},$$

where α and β are scalars.

From the properties of matrices, multiplication of a $n \times 1$ vector $\mathbf{x}$ in $\mathbf{R}^n$ by an $m \times n$ matrix A represents a linear operation, since

$$A(\alpha \mathbf{x} + \beta \mathbf{y}) = \alpha A\mathbf{x} + \beta A\mathbf{y}, \tag{3.26}$$

where the vectors on the right-hand side $A\mathbf{x}$ and $A\mathbf{y}$ are $m \times 1$; that is, the vectors are in $\mathbf{R}^m$. In fact, it is a general result that an operation or function that transforms a vector in $\mathbf{R}^n$ to a vector in $\mathbf{R}^m$ is linear if and only if it coincides with multiplication by some $m \times n$ matrix.

The function that converts a vector in $\mathbf{R}^n$ to a vector in $\mathbf{R}^m$

$$\mathbf{R}^n \xrightarrow{\ f\ } \mathbf{R}^m$$

is said to have *domain* in $\mathbf{R}^n$ and *range* in $\mathbf{R}^m$. The terms transformation and mapping are also used to describe f rather than function. When the function is linear, there exists a unique $m \times n$ matrix A_T such that

$$f(\mathbf{x}) = A_T \mathbf{x}$$

for every $\mathbf{x}$ in $\mathbf{R}^n$. The proof of this assertion is important since the proof actually shows how to construct the matrix corresponding to the linear function.

First, consider that any vector $\mathbf{x} \in \mathbf{R}^n$ can be written in terms of the natural basis vectors as

$$\mathbf{x} = x_1 \mathbf{e}_1 + x_2 \mathbf{e}_2 + \cdots + x_n \mathbf{e}_n,$$

where $\mathbf{e}_j$ is the unit vector with 1 for the jth entry and zero in all other entries. If a function f is linear, we know that

$$
\begin{aligned}
f(\mathbf{x}) &= f(x_1 \mathbf{e}_1 + x_2 \mathbf{e}_2 + \cdots + x_n \mathbf{e}_n) \\
&= x_1 f(\mathbf{e}_1) + x_2 f(\mathbf{e}_2) + \cdots + x_n f(\mathbf{e}_n)
\end{aligned}
\tag{3.27}
$$

using the definition of a linear function.

Now, construct the matrix equivalent to the linear function by defining the matrix A as the matrix whose jth column is the m-dimensional vector $f(\mathbf{e}_j)$ for $j = 1, 2, \ldots, n$. Thus, A has m rows and n columns. Using Equation 3.27, the function $f(\mathbf{x})$ can be written

$$
f(\mathbf{x}) = x_1 \begin{bmatrix} a_{11} \\ a_{21} \\ \vdots \\ a_{m1} \end{bmatrix} + \cdots + x_n \begin{bmatrix} a_{1n} \\ a_{2n} \\ \vdots \\ a_{mn} \end{bmatrix}
$$

$$
= \begin{bmatrix} a_{11}x_1 + \cdots + a_{1n}x_n \\ a_{21}x_1 + \cdots + a_{2n}x_n \\ \vdots \\ a_{m1}x_1 + \cdots + a_{mn}x_n \end{bmatrix},
$$

which by the definition of matrix multiplication is

$$f(\mathbf{x}) = A\mathbf{x}.$$

We can summarize these result in a theorem that relates the linear function f and the matrix A as follows:

■ THEOREM 3.9   *Linear function*

Let $\mathbf{R}^n \xrightarrow{f} \mathbf{R}^m$ be a linear function, and let A be the matrix whose jth column is $f(\mathbf{e}_j)$. Then $f(\mathbf{x}) = A\mathbf{x}$ for every $\mathbf{x} \in \mathbf{R}^n$.

The relationship between vector spaces and linearity for the points in $\mathbf{R}^2$ is that if a set of points represented by vectors $\mathbf{x}_1, \mathbf{x}_2, \ldots, \mathbf{x}_n$ all lie on one line through the origin, any linear combination of them lies in the same line since they are all multiples of the same vector. Also, the zero vector is in the set. We say that a subset S of a vector space V is a *linear subspace* if every linear combination of elements of S is also in S. Usually, the linear subspace is simply called a *subspace*, as was the case in our discussions in Chapter 2.

TRANSFORMATIONS IN THE PLANE AND THREE-DIMENSIONAL SPACE

In computer graphics, a typical problem is to display a view of a 2D or 3D object on a video screen. Using matrix algebra, new views of the object can be generated by rotation, translation, and scaling. Describing the motion (kinematics) of a robot manipulator is an important problem in robotics. Matrices can be used to define the position and orientation of the manipulator at any time with respect to the coordinates of the robot's reference frame.

In general, matrices can be used to allow points and vectors to be rotated about coordinate axes, translated in space, and referenced relative to other reference frames. As shown in a later chapter, we may also use matrices to map the coordinates of spherical or cylindrical reference frames into xyz coordinates (Cartesian space), or vice versa.

Rotations in the Plane In Figure 3.2, a vector $\mathbf{x} = [x, y]$ is rotated through the angle θ to become the vector $\mathbf{x}' = [x', y']$.

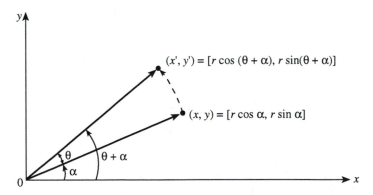

FIGURE 3.2 *Rotation of a two-dimensional vector*

The length r of the vector is not changed by rotation. From the geometry, the coordinates of $\mathbf{x}$ in terms of the angles involved are

$$x = r \cos \alpha \quad \text{and} \quad y = r \sin \alpha.$$

It is desired to obtain the coordinates of the rotated vector in terms of x, y, and the angle θ. Thus, the rotated vector $\mathbf{x}'$ in Figure 3.2 has coordinates

$$
\begin{aligned}
x' &= r\cos(\theta + \alpha) = r\cos\theta\cos\alpha - r\sin\theta\sin\alpha, \\
y' &= r\sin(\theta + \alpha) = r\sin\theta\cos\alpha + r\cos\theta\sin\alpha.
\end{aligned}
$$

Thus, the relationship between the endpoint coordinates is

$$
\begin{aligned}
x' &= x\cos\theta - y\sin\theta, \\
y' &= x\sin\theta + y\cos\theta.
\end{aligned}
$$

Defining the 2D *rotation* matrix as

$$
R_\theta = \begin{bmatrix} \cos\theta & -\sin\theta \\ \sin\theta & \cos\theta \end{bmatrix}, \tag{3.28}
$$

we see that

$$
R_\theta \begin{bmatrix} x \\ y \end{bmatrix} = \begin{bmatrix} x' \\ y' \end{bmatrix}
$$

by multiplying $\mathbf{x}^T$ by R_θ. If the angle θ is changed to $-\theta$, the sign of the off-diagonal terms are changed. Of course, the rotation matrix should become the identity matrix if $\theta = 0°$.

Notice that in these operations, the vectors are considered 2×1 column vectors. Also, the series of operations $R_{\theta_1} R_{\theta_2} \mathbf{x}^T$ results in a rotation by angle $\theta_2 + \theta_1$. If the order of matrix multiplications is reversed, the rotation angle remains the same. This is always the case when a series of rotations of a vector from the origin are made in the plane.

Three-dimensional Rotations   Figure 3.3 illustrates the 3D reference frame we will use for rotations.

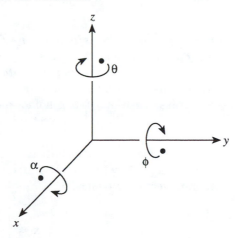

FIGURE 3.3 *Cartesian reference frame in three dimensions*

In three dimensions, the axis of rotation must be specified. The subscripts for a rotation matrix will indicate the axis of rotation and the angle. The rotation matrices for the three axes are as follows:

1. Rotation by an angle α about the x axis:

$$R_{x,\alpha} = \begin{bmatrix} 1 & 0 & 0 \\ 0 & \cos\alpha & -\sin\alpha \\ 0 & \sin\alpha & \cos\alpha \end{bmatrix}; \qquad (3.29)$$

2. Rotation by an angle ϕ about the y axis:

$$R_{y,\phi} = \begin{bmatrix} \cos\phi & 0 & \sin\phi \\ 0 & 1 & 0 \\ -\sin\phi & 0 & \cos\phi \end{bmatrix}; \qquad (3.30)$$

3. Rotation by an angle θ about the z axis:

$$R_{z,\theta} = \begin{bmatrix} \cos\theta & -\sin\theta & 0 \\ \sin\theta & \cos\theta & 0 \\ 0 & 0 & 1 \end{bmatrix}. \qquad (3.31)$$

When performing a series of 3D rotations about several axes by matrix multiplications, the order in which the rotations are performed is important, since the rotation matrices do not commute in general. Except for special cases, applying the rotation matrices to a vector in different order will generate a different result. As a check on the results, the rotation matrices should become identity matrices when the angles involved are set to zero. Also, the rotation matrices are orthogonal matrices, so the columns (or rows) must be orthonormal.

Homogeneous Transformations (Optional)   The 3×3 rotation matrix does not provide for translation or scaling of a vector. The concept of *homogeneous coordinate representation* is introduced to develop matrix transformations that include rotation, translation, scaling, and perspective transformation. Using homogeneous transformations, the transformation of an n-dimensional vector is performed in an $(n+1)$-dimensional space.

In homogeneous transformations, the true vectors in $\mathbf{R}^3$ are written as vectors in $\mathbf{R}^4$ with a scaling factor as the last component. Let the 3D column vector be defined as

$$\mathbf{x} = \begin{bmatrix} x \\ y \\ z \end{bmatrix}.$$

Then, the homogeneous representation is

$$\mathbf{x}_h = \begin{bmatrix} sx \\ sy \\ sz \\ s \end{bmatrix},$$

where s is a numerical scaling value. We will assume that this scaling factor is 1 for our present purposes. The rotation matrices that multiply this vector must be 4×4. This is accomplished by adding another row and column to the rotation matrices previously defined. For example, the homogeneous rotation matrix about the z axis is

$$Rh_{z,\theta} = \begin{bmatrix} \cos\theta & -\sin\theta & 0 & 0 \\ \sin\theta & \cos\theta & 0 & 0 \\ 0 & 0 & 1 & 0 \\ 0 & 0 & 0 & 1 \end{bmatrix}.$$

A homogeneous translation matrix in effect adds a translation vector to the vector being transformed. Assume that the translation values are $[t_1, t_2, t_3]$. Then, a homogeneous translation matrix has the form

$$T = \begin{bmatrix} 1 & 0 & 0 & t_1 \\ 0 & 1 & 0 & t_2 \\ 0 & 0 & 1 & t_3 \\ 0 & 0 & 0 & 1 \end{bmatrix}.$$

☐ **EXAMPLE 3.19**   *MATLAB Homogeneous Transformations*
      Linear transformations are easily accomplished using MATLAB. The accompanying scripts show the M-file (CLXROTZ.M) as well as the function (clxrotzf) that perform a homogeneous transformation to rotate a 3D vector by an arbitrary angle in degrees around the z axis. A test case is shown from the diary file created when the M-file was executed. The test vector is input as [1 2 3] and the rotation angle is $60°$. The diary file showing the results is included with the M-file script.

MATLAB Script _____

```
Example 3.19
% CLXROTZ.M  Rotate a vector around the z axis
%  Input the vector [x y z] and the angle in degrees.
%  Function clxrotzf is called to perform rotation
v1=input(' Vector [x y z]= ')
theta=input(' Input rotation angle (degrees)= ')
v11=[v1 1]';               % Form homogeneous vector
vrotz=clxrotzf(v11,theta); % Rotate
fprintf('Rotated vector\n')
vrotz                      % Display result
%
% -------------------------------------------------
```

```
% Results (Rotation matrix and rotated vector}
%
>>clxrotz
Vector [x y z]= [1 2 3]

v1 =
     1     2     3

 Input rotation angle (degrees)= 60

theta =
    60
Rotated vector
vrotz =
   -1.2321
    1.8660
    3.0000
    1.0000
>>quit
```

The function clxrotzf rotates the vector passed to it by the specified angle in degrees about the z axis. The homogeneous transformation yields a 4×1 vector as the result. In space, the vector $[1, 2, 3]^T$ is rotated to a new vector with the coordinates

$$\begin{bmatrix} -1.2321 \\ 1.8660 \\ 3.0000 \end{bmatrix}.$$

MATLAB Script _____

```
Example 3.19
function yh3rotz=clxrotzf(xto_rot,theta_rot)
% CALL: yh3rotz=clxrotzf(xto_rot,theta_rot)
%  Rotate the vector xto_rot by the angle theta_rot
%     around the z axis
%  xto_rot must be a 4 x 1 column vector, theta_rot in degrees.
theta_rot=theta_rot*pi/180;      % Convert to radians
Arotz=[cos(theta_rot) -sin(theta_rot) 0 0;sin(theta_rot)...
 cos(theta_rot) 0 0
0 0 1 0;0 0 0 1];
yh3rotz=Arotz*xto_rot;
```

☐

Chapter 3 ∎ MATRICES

REINFORCEMENT EXERCISES AND EXPLORATION PROBLEMS

REINFORCEMENT EXERCISES

In these problems, do the computations by hand, unless otherwise indicated, and then check those that yield numerical or symbolic results with MATLAB.

P3.1. **Matrix operations** Define the matrices A and B as

$$A = \begin{bmatrix} 3 & 4 & 0 \\ -1 & 0 & 2 \end{bmatrix} \quad \text{and} \quad B = \begin{bmatrix} 6 & -1 & 2 \\ 0 & 1 & 5 \\ -1 & 3 & 4 \end{bmatrix}.$$

Compute the following:

    a. AB;

    b. A^T.

P3.2. **Matrix operations** Define the matrices

$$A = \begin{bmatrix} 1 & 2 & 3 \end{bmatrix} \quad \text{and} \quad B = \begin{bmatrix} -2 \\ 4 \\ 1 \end{bmatrix}.$$

Compute the following:

    a. AB;

    b. BA.

P3.3. **Matrix operations** Define the matrices

$$A = \begin{bmatrix} 1 & 2 \\ 3 & 0 \end{bmatrix}, \quad B = \begin{bmatrix} 2 & -1 \\ 3 & 4 \end{bmatrix}, \quad \text{and} \quad C = \begin{bmatrix} 2 & -2 \\ 1 & 3 \\ 4 & -1 \end{bmatrix}.$$

    a. Compute $C(A + B)$.

    b. Show that the multiplication is distributive with respect to addition; that is

$$C(A + B) = CA + CB.$$

P3.4. **Transpose** Show that $(AB)^T = B^T A^T$ for the matrices

$$A = \begin{bmatrix} 2 & 3 \\ 0 & -1 \end{bmatrix} \quad \text{and} \quad B = \begin{bmatrix} 1 & 5 \\ 2 & 4 \end{bmatrix}.$$

P3.5. **Matrix size** Let

$$A = \begin{bmatrix} 1 & -1 & 1 \\ 0 & 1 & 0 \\ 2 & 0 & 3 \end{bmatrix} \quad \text{and} \quad B = \begin{bmatrix} 3 & 3 & 1 \\ 0 & 1 & 0 \\ -2 & -2 & 1 \end{bmatrix}.$$

    a. What are the sizes and types of A and B?

b. Compute AB.

c. Compute BA.

P3.6. **Matrix operations.** Let

$$A = \begin{bmatrix} 3 & 2 & 0 \\ 2 & 2 & -2 \\ 0 & -2 & 4 \end{bmatrix} \quad \text{and} \quad B = \begin{bmatrix} 1 & 0 & 1 \\ 0 & 2 & 0 \\ 1 & 0 & 3 \end{bmatrix}.$$

a. What are the sizes and types of A and B?

b. Compute AB.

c. Compute BA.

d. Is AB or BA symmetric?

P3.7. **Symmetric matrices** Using the properties of the transpose of a symmetric matrix, prove the following theorem.

If A and B are symmetric matrices, then AB is a symmetric matrix if and only if $AB = BA$.

P3.8. **Planes in 3D space** The linear equation in three variables x, y and z represents a plane in 3D space.

a. Write the system of equations for two planes and define the possible solutions;

b. What are the possible solutions for three planes?

P3.9. **Intersection of lines** Find the point of intersection (if any) for the following lines:

a.
$$\begin{aligned} x - 2y + 1 &= 0, \\ 2x + 3y - 7 &= 0. \end{aligned}$$

b.
$$\begin{aligned} x - 2y + 11 &= 0 \\ -x + 2y - 13 &= 0 \end{aligned}$$

P3.10. **Determinants** Let

$$A = \begin{bmatrix} -5 & -6 & 7 \\ 8 & -9 & 0 \\ -3 & 4 & 2 \end{bmatrix}.$$

Find $\det(A)$.

P3.11. **Inverse** Let

$$A = \begin{bmatrix} 2 & 3 & 4 \\ 5 & 6 & 7 \\ 8 & 9 & 0 \end{bmatrix}.$$

Compute the inverse of A in the following way:

a. Compute the determinant matrix B such that $b_{ij} = \det(A_{ij})$ and use the cofactors.

b. Transpose and divide by $\det(A)$.

P3.12. **System of linear equations** Solve the system of equations

$$\begin{aligned} x + y + z &= 2, \\ -x + 2y - 4z &= 2, \\ x + 3y + 9z &= 0 \end{aligned}$$

by computing $\mathbf{x} = A^{-1}\mathbf{b}$. Check that $A\mathbf{x} = \mathbf{b}$.

P3.13. Elementary matrix operations Solve the systems using elementary row operations:

a.
$$\begin{bmatrix} 1 & 2 & 3 \\ 1 & 0 & 0 \\ -4 & 1 & 2 \end{bmatrix} \begin{bmatrix} x \\ y \\ z \end{bmatrix} = \begin{bmatrix} 2 \\ -5 \\ 0 \end{bmatrix};$$

b.
$$\begin{aligned} x + 3y + z &= 6, \\ 3x - 2y - 8z &= 7, \\ 4x + 5y - 3z &= 17; \end{aligned}$$

c.
$$\begin{aligned} 3x_1 - 3x_2 + 6x_3 &= -3, \\ 2x_3 &= 0, \\ 3x_1 - 3x_2 + 7x_3 &= 1, \\ 10x_1 - 10x_2 + 24x_3 &= -2. \end{aligned}$$

P3.14. Matrix row operations Find the inverse by applying elementary row operations for the matrix:

$$A = \begin{bmatrix} 4 & -1 & 0 & 0 \\ 0 & 5 & -2 & 0 \\ 0 & 0 & 6 & -3 \\ 0 & 0 & 0 & 7 \end{bmatrix}.$$

P3.15. Matrix operations First, solve the system

$$\begin{aligned} x + 2y + 3z &= 0, \\ -x + y &= 0, \\ 5y + 5z &= 0 \end{aligned}$$

in terms of z. Then, let $z = 1$ and solve the system.

P3.16. Invertible matrix properties Assume that A is an $n \times n$ invertible matrix. Which statements are true?

  a. The system $Ax = b$ has a unique solution for every vector b in $\mathbf{R}^n$.

  b. The rows (and columns) of A are linearly independent.

  c. $\det(A) = 0$.

  d. A can be reduced (by elementary operations) to the identity matrix.

  e. The rank of A is n.

  f. The rows of A span $\mathbf{R}^n$.

P3.17. Homogeneous transformations Write the homogeneous transformation matrix that represents a rotation of α degrees about the x axis, followed by a translation of a units along the x axis. The axes are defined in Figure 3.3.

P3.18. Linear independence Consider the equations of combustion in which a mixture of CO, H_2, and CH_4 are burned with O_2 to form CO, CO_2, and H_2O:

$$\begin{aligned} CO + \frac{1}{2}O_2 &= CO_2, \\ H_2 + \frac{1}{2}O_2 &= H_2O, \\ CH_4 + 2O_2 &= CO_2 + 2H_2O, \\ CH_4 + \frac{3}{2}O_2 &= CO + 2H_2O. \end{aligned}$$

Treating the compounds as real variables, determine if the equations are independent. If not, write the dependent equation(s) in terms of the independent ones.

P3.19. MATLAB matrices Without typing in all the elements, use MATLAB to create a 6×6 matrix with 4's on the main diagonal and -1 on the first upper diagonal and first lower diagonal. This matrix is called a *band matrix* with the form

$$I = \begin{bmatrix} 4 & -1 & 0 & 0 & 0 & 0 \\ -1 & 4 & -1 & 0 & 0 & 0 \\ 0 & -1 & 4 & -1 & 0 & 0 \\ 0 & 0 & -1 & 4 & -1 & 0 \\ 0 & 0 & 0 & -1 & 4 & -1 \\ 0 & 0 & 0 & 0 & -1 & 4 \end{bmatrix}.$$

P3.20. MATLAB matrix operations Define the matrices

$$\mathbf{a} = \begin{bmatrix} 3 \\ 1 \\ 4 \end{bmatrix}, \quad \mathbf{B} = \begin{bmatrix} 0 & 1 \\ 0 & -2 \\ 2 & 3 \end{bmatrix}, \quad \mathbf{C} = \begin{bmatrix} 1 & 0 & -1 \\ 2 & 3 & 0 \\ 0 & 3 & 4 \end{bmatrix}, \quad \mathbf{d} = \begin{bmatrix} 1 & 0 & 2 \end{bmatrix}.$$

If the result exists, compute the following:

  a. $\mathbf{C} * \mathbf{B}$, $\mathbf{B}^T * \mathbf{C}^T$, $\mathbf{B} * \mathbf{C}^T$.

  b. $\mathbf{C}^2$, $\mathbf{C} * \mathbf{C}^T$, $\mathbf{C}^T * \mathbf{C}$.

  c. $\mathbf{B}^T * \mathbf{a}$, $\mathbf{B} * \mathbf{d}$, $\mathbf{a} * \mathbf{d}$.

  d. $|B|$, $|C|$, $\mathbf{B}^{-1}$, $\mathbf{C}^{-1}$.

P3.21. MATLAB matrix solution Solve the problem of Example 3.6 for the current vector **I** and verify Kirchhoff's law for each loop.

P3.22. MATLAB symmetric matrix program Write a MATLAB program to determine if an arbitrary $n \times n$ matrix is symmetric and display the answer. Test the program on the matrices of Problem 3.6 and a matrix of random numbers.

P3.23. MATLAB matrix rank Using the rank of the matrix whose rows are the vectors

$$[3\ 0\ 2\ 2], \quad [-6\ 42\ 24\ 54], \quad [21\ -21\ 0\ -15],$$

do the following:

  a. Determine the number of linearly independent vectors.

  b. If the set of vectors is dependent, write the dependent vector in terms of the independent ones.

P3.24. MATLAB matrix rank Given the matrix

$$A = \begin{pmatrix} 1 & 0 & 7 \\ 0 & 1 & 5 \\ 2 & -1 & 9 \end{pmatrix},$$

write the MATLAB instructions to reduce the matrix by elementary operations. Then, determine the rank. Check the result with the **rref** and **rank** commands.

P3.25. MATLAB matrix rank Determine the rank of A and A^T for the matrix

$$A = \begin{pmatrix} 1 & 2 & -3 \\ 2 & 1 & 0 \\ -2 & -1 & 3 \\ -1 & 4 & 2 \end{pmatrix}.$$

P3.26. MATLAB matrix solution. Consider the following matrix equation:

$$
\begin{bmatrix}
1 & 1/2 & 1/3 \\
1/2 & 1/3 & 1/4 \\
1/3 & 1/4 & 1/5
\end{bmatrix}
\mathbf{x} =
\begin{bmatrix}
11/6 \\
13/12 \\
47/60
\end{bmatrix}.
$$

This equation is slightly ill conditioned, but the exact solution was chosen so that $\mathbf{x} = [\,1\ 1\ 1\,]$. Compute

    a. The MATLAB solution;

    b. The number of decimal places of accuracy using MATLAB;

    c. The condition number of the matrix.

Optional: Determine the condition numbers of Hilbert matrices of various orders from 3 to 15.

P3.27. MATLAB transformations Write an M-function that rotates the 2D vector $\mathbf{x}$ by θ degrees. Test the function with vectors $(1, 0)$ and $(0, 1)$ using $\theta = 30°$.

P3.28. MATLAB transformation Write an M-function to rotate a triangle in the xy plane, and do the following:

    a. Test the function with the triangle having endpoints $(1, 1)$, $(3, 1)$, and $(3, 2)$ rotated by $\theta = 45°$.

    b. Plot the original triangle and the rotated one on a square grid 4×4.

EXPLORATION PROBLEMS

P3.29. Image matching Consider the following two matrices:

$$
Atmpl =
\begin{bmatrix}
0 & 0 & 0 & 0 & 0 & 1 & 1 \\
0 & 1 & 1 & 1 & 1 & 1 & 1 \\
0 & 1 & 1 & 1 & 1 & 1 & 1 \\
0 & 0 & 0 & 1 & 1 & 1 & 1 \\
0 & 1 & 1 & 1 & 1 & 1 & 1 \\
0 & 1 & 1 & 1 & 1 & 1 & 1 \\
0 & 1 & 1 & 1 & 1 & 1 & 1
\end{bmatrix},
\quad
Atest =
\begin{bmatrix}
1 & 1 & 1 & 1 & 1 & 1 & 0 \\
1 & 1 & 1 & 1 & 0 & 1 & 0 \\
1 & 1 & 1 & 1 & 1 & 1 & 0 \\
1 & 1 & 1 & 1 & 1 & 0 & 0 \\
1 & 1 & 1 & 1 & 1 & 1 & 0 \\
1 & 1 & 1 & 1 & 1 & 1 & 0 \\
1 & 1 & 0 & 0 & 0 & 0 & 0
\end{bmatrix}.
$$

The array *Atmpl* represents a template that is a binary image of a known character. The array *Atest* resulted from a scanned image. It is desired to know if the scanned image is equivalent to the known image but perhaps rotated by $0°$, $90°$, $180°$, or $270°$. One way to test if the images are equivalent is to compare the sum of the squares of the differences between the elements in *Atmpl* and *Atest*. (Using MATLAB you must sum the square of the column differences).

    a. Compute the best alignment of *Atest* with *Atmpl* by comparing the test image at the four rotation angles.

    b. In the MATLAB program, print the best alignment between the images in degrees.

P3.30. Area of parallelogram Let the vectors $\mathbf{a}$ and $\mathbf{b}$ in $\mathbf{R}^2$ define a parallelogram in a plane. Show that the area of the parallelogram is equal the absolute value of the determinant formed using the vectors as columns.

Hint: Start with the relationship area $= \|\mathbf{a}\|\,\|\mathbf{b}\|\,|\sin \theta|$ (see problems for Chapter 2) and compute the area of the parallelogram in terms of the components of the vectors.

P3.31. Matrix theorems Prove the following theorems for matrices A and B:

    a. $(A^{-1})^{-1} = A$.

    b. $(AB)^{-1} = B^{-1}A^{-1}$.

    c. $(A^T)^T = A$.

P3.32. Determinant of triangular matrix. Let $A = (a_{ij})$ be an upper or lower triangular $n \times n$ matrix. Then, the determinant is

$$A = a_{11}a_{22}\cdots a_{nn}.$$

Demonstrate this for a 4×4 lower triangular matrix and prove the theorem for an upper triangular matrix.

Hint: Prove by induction.

P3.33. Matrix multiplication Give examples to demonstrate the following properties of matrices:

    a. If $AB = 0$, we cannot conclude that $A = 0$ or $B = 0$.

    b. If $A^2 = 0$, we cannot conclude that $A = 0$.

    c. For which matrices does $(A + B)(A - B) = A^2 - B^2$?

P3.34. Determinant by reduction Given the matrix

$$A = \begin{bmatrix} 3 & 1 & -1 & 2 & 1 \\ 0 & 3 & 1 & 4 & 2 \\ 1 & 4 & 2 & 3 & 1 \\ 5 & -1 & -3 & 2 & 5 \\ -1 & 1 & 2 & 3 & 2 \end{bmatrix},$$

find the determinant by elementary reduction and then use MATLAB to check the result.

P3.35. Matrix inverse Find the inverse and verify the result for the matrix

$$Q = \begin{bmatrix} \sin\phi\cos\theta & \sin\phi\sin\theta & \cos\phi \\ -\sin\theta & \cos\theta & 0 \\ \cos\phi\cos\theta & \cos\phi\sin\theta & -\sin\phi \end{bmatrix}.$$

Hint: This is the matrix that relates spherical and rectangular coordinates.

P3.36. Matrix as differentiation operator Given the vector space of polynomials of third degree with basis vectors

$$P_1(x) = 1, \quad P_2(x) = x, \quad P_3(x) = x^2, \quad P_4(x) = x^3,$$

find a matrix that corresponds to differentiation of a polynomial.

Hint: Find the matrix such that $AP_n(x) = (n-1)P_{n-1}(x)$.

P3.37. MATLAB triangular matrix program Write MATLAB statements to determine if a matrix A is upper triangular or lower triangular.

Hint: Investigate the commands **any**, **all**, **tril**, and **triu**.

P3.38. Partitioned matrix Consider an $n \times n$ matrix E *partitioned* into four submatrices A, B, C, E, where A and D are square matrices of order p and q, respectively, and $p + q = n$. Letting

$$E = \left[\begin{array}{c|c} A & B \\ \hline C & D \end{array} \right],$$

find the inverse matrix as a partitioned matrix with the same partitioning.

Hint: Form the product $EE^{-1} = Ip$, where this partitioned matrix Ip has identity matrices on its diagonal.

Test the approach with the matrix

$$E = \left[\begin{array}{cc|c} 2.5 & 3.8 & 9.6 \\ -2.1 & 1.3 & 4.5 \\ \hline 1.0 & 3.2 & 6.7 \end{array} \right].$$

P3.39. Summing-up questions Answering these questions should require little or no calculation.

    a. For an $n \times n$ matrix A, how are $\det(2A)$ and $\det(-A)$ related to $\det(A)$?

    b. Suppose A, B and D are 2×2 matrices. Find the determinant of the 4×4 matrix formed as

$$\left[\begin{array}{cc} A & B \\ 0 & D \end{array} \right]$$

    c. If $B = M^{-1}AM$, show that $\det(B) = \det(A)$.

P3.40. MATLAB homogeneous transformations Write a MATLAB function to rotate a 3D vector by an arbitrary angle around each of the three axes. Given $\mathbf{u} = [1, 2, 3]^T$ as a 3D vector, compare the resulting vector in the following cases:

    a. Rotation of $60°$ about the z axis followed by a rotation of $-90°$ about the y axis;

    b. Rotation of $-90°$ about the y axis followed by a rotation of $60°$ about the z axis

ANNOTATED BIBLIOGRAPHY

1. Datta, Biswa Nath, *Numerical Linear Algebra and Applications*, Brooks Cole Publishing Company, Pacific Grove, CA, 1995. *The text analyzes numerical algorithms for the solution of systems of linear equations and presents potential errors that may arise.*

2. Forsythe, George E., M. A. Malcolm, and C. B. Moler, *Computer Methods for Mathematical Computations*, Prentice Hall, Englewood Cliffs, NJ, 1977. *A classic text, which describes a number of algorithms that were adapted for MATLAB.*

3. Garcia, Alejandro L., *Numerical Methods For Physics*, Prentice Hall, Englewood Cliffs, NJ, 1994. *The text covers MATLAB applications in areas of physics and includes discussions of roundoff errors and the use of matrices with random entries.*

4. Hill, David R., *Experiments in Computational Matrix Algebra*, Random House, New York, 1988. *The text discusses many of the topics treated in this chapter. It contains especially good discussions of the computational errors that arise using MATLAB commands.*

5. Klafter, Richard D., T. A. Chmielewski, and M. Negin, *Robotics Engineering*, Prentice Hall, Englewood Cliffs, NJ, 1989. *A practical text for robotics with a number of examples using homogeneous transformations.*

6. Williamson, Richard E., R. H. Crowell, and H. F. Trotter, *Calculus of Vector Functions*, Prentice Hall, Englewood Cliffs, NJ, 1972. *A good treatment of vectors, matrices, and linear transformations, as well as vector calculus.*

ANSWERS

P3.1. Matrix operations

a. The product matrix AB is $AB = \begin{bmatrix} 18 & 1 & 26 \\ -8 & 7 & -10 \end{bmatrix}$;

b. The transpose of A is $A^T = \begin{bmatrix} 3 & -1 \\ 4 & 0 \\ 0 & 2 \end{bmatrix}$.

P3.2. Matrix operations

a. $AB = 9$;

b. $BA = \begin{bmatrix} -2 & -4 & -6 \\ 4 & 8 & 12 \\ 1 & 2 & 3 \end{bmatrix}$.

The matrices do not commute and AB and BA are not even the same size.

P3.3. **Matrix operations** $C(A + B) = CA + CB$.

a. $C(A + B) = \begin{bmatrix} -6 & -6 \\ 21 & 13 \\ 6 & 0 \end{bmatrix}$;

b. $(CA + CB) = \begin{bmatrix} -4 & 4 \\ 10 & 2 \\ 1 & 8 \end{bmatrix} + \begin{bmatrix} -2 & -10 \\ 11 & 11 \\ 5 & -8 \end{bmatrix} = \begin{bmatrix} -6 & -6 \\ 21 & 13 \\ 6 & 0 \end{bmatrix}$.

P3.4. **Transpose**

$$(AB)^T = \begin{bmatrix} 8 & -2 \\ 22 & -4 \end{bmatrix} = B^T A^T.$$

P3.5. **Matrix size**

a. Using the MATLAB command (or by inspection), the sizes are 3×3, and both matrices are square.

b. $AB = \begin{bmatrix} 1 & 0 & 2 \\ 0 & 1 & 0 \\ 0 & 0 & 5 \end{bmatrix}$;

c. $BA = \begin{bmatrix} 5 & 0 & 6 \\ 0 & 1 & 0 \\ 0 & 0 & 1 \end{bmatrix}$.

P3.6. **Matrix operations**

a. Using the MATLAB command **size** (or by inspection), the sizes are 3×3 and both are symmetric;

b. $AB = \begin{bmatrix} 3 & 4 & 3 \\ 0 & 4 & -4 \\ 4 & -4 & 12 \end{bmatrix}$;

c. $BA = \begin{bmatrix} 3 & 0 & 4 \\ 4 & 4 & -4 \\ 3 & -4 & 12 \end{bmatrix}$;

d. The product of the matrices is not symmetric.

P3.7. **Symmetric matrices** Assume that AB is symmetric, then

$$(AB) = (AB)^T = (B)^T (A)^T = BA$$

since A and B are symmetric. Thus, the matrices commute.
If $(AB) = (BA)$, the product of the matrices commutes, and

$$(AB)^T = (BA)^T = (A)^T (B)^T = AB.$$

Thus, AB is symmetric.

P3.8. **Planes in 3D space** The equations for two planes would be

$$a_{11}x + a_{12}y + a_{13}z = b_1,$$
$$a_{21}x + a_{22}y + a_{23}z = b_2.$$

The planes could intersect in a straight line. Otherwise, the planes could be parallel or they could be the same plane.

P3.9. Intersection of lines

a. $[x, y] = [11/7, 9/7]$;

b. The lines are parallel, and MATLAB indicates the system is singular.

P3.10. Determinants $\det(A) = 221$.

P3.11. Inverse

$$A^{-1} = \frac{1}{|A|}\begin{bmatrix} -63 & 56 & -3 \\ 36 & -32 & 6 \\ -3 & 6 & -3 \end{bmatrix}^T = \frac{1}{30}\begin{bmatrix} -63 & 36 & -3 \\ 56 & -32 & 6 \\ -3 & 6 & -3 \end{bmatrix}$$

$$= \begin{bmatrix} -2.1000 & 1.2000 & -0.1000 \\ 1.8667 & -1.0667 & 0.2000 \\ -0.1000 & 0.2000 & -0.1000 \end{bmatrix}.$$

P3.12. System of linear equations

$$\mathbf{x} = \begin{bmatrix} 24/15 & 13/15 & -7/15 \end{bmatrix}^T = \begin{bmatrix} 1.6000 \\ 0.8667 \\ -0.4667 \end{bmatrix}.$$

P3.13. Elementary matrix operations   Use the MATLAB command **rrefmovie** to see the reductions. Notice how MATLAB swaps rows. This is to use the largest pivot element for reduction, as explained in the *MATLAB User's Guide*.

a. $\mathbf{x} = [-5, 74, -47]^T$;

b. Reduction leads to a row of zeros so a possible solution is to choose z and use $y = 1 - z$,$x = 3 + 2z$.

c. This is an overdetermined system with no solution.

P3.14. Matrix row operations

$$A^{-1} = \begin{bmatrix} 0.2500 & 0.0500 & 0.0167 & 0.0071 \\ 0 & 0.2000 & 0.0667 & 0.0286 \\ 0 & 0 & 0.1667 & 0.0714 \\ 0 & 0 & 0 & 0.1429 \end{bmatrix} = \begin{bmatrix} 1/4 & 1/20 & 1/60 & 1/140 \\ 0 & 1/5 & 1/15 & 1/35 \\ 0 & 0 & 1/6 & 1/14 \\ 0 & 0 & 0 & 1/7 \end{bmatrix}.$$

P3.15. Matrix operations   Trying to solve the system leads to a singular matrix. If z is chosen so that $z = 1$, $x = -1$, and $y = -1$.

P3.16. Invertible matrix properties   a. True; b. True; c. False; d. True; e. True; f. True.

P3.17. Homogeneous transformations   The transformation is

$$RT = \begin{bmatrix} 1 & 0 & 0 & a \\ 0 & 1 & 0 & 0 \\ 0 & 0 & 1 & 0 \\ 0 & 0 & 0 & 1 \end{bmatrix}\begin{bmatrix} 1 & 0 & 0 & 0 \\ 0 & \cos\alpha & -\sin\alpha & 0 \\ 0 & \sin\alpha & \cos\alpha & 0 \\ 0 & 0 & 0 & 1 \end{bmatrix} = \begin{bmatrix} 1 & 0 & 0 & a \\ 0 & \cos\alpha & -\sin\alpha & 0 \\ 0 & \sin\alpha & \cos\alpha & 0 \\ 0 & 0 & 0 & 1 \end{bmatrix}.$$

P3.18. Linear independence   The 4×4 matrix from the system of equations has rank 3, so only three independent reactions are necessary to describe the process. Equation 4 is the same as Equation 3 minus Equation 1.

Comment: MATLAB programs are included on the disk accompanying this textbook.

4

EIGENVALUES AND EIGENVECTORS

PREVIEW————————————————————————————

After the solution of systems of linear equations, the next most frequently
encountered matrix problem is that of calculation of the *eigenvalues* and
eigenvectors of a square matrix. Chapter 3 considered the problem of
solving the equation $A\mathbf{x} = \mathbf{b}$ and applied the result to the solution of
systems of linear equations. This chapter deals with the eigenvalue equation

$$f(\mathbf{x}) = \lambda\mathbf{x},$$

where f is a linear function that transforms the vector $\mathbf{x}$ into a scalar
multiple of itself. Our goal is to find the eigenvalues and eigenvectors that
belong to various linear operators, including matrices. The matrix equation
to be solved is $A\mathbf{x} = \lambda\mathbf{x}$.

   The eigenvalues and eigenvectors are useful in many problems,
including the solution of systems of linear differential equations as treated in
Chapter 5. Furthermore, the eigenvalues can have important physical
significance in many problems. For example, the eigenvalues may represent
the natural frequencies of vibration of a mechanical system or the
fundamental frequencies of oscillation in certain electrical systems. The
eigenvectors serve as a basis for the solution set in such problems.

This chapter begins with the formal definition of eigenvalues and eigenvectors. Since one method of determining the eigenvalues involves solving a polynomial equation that may have complex numbers as roots, a section is devoted to a brief review of the solution of such equations. Then, various properties and applications of eigenvalues and eigenvectors are presented.

Matrix calculus is introduced in this chapter to define the differentiation, integration, raising to a power, and exponentiation of a matrix. Also, the properties of eigenvalues for special matrices, such as symmetric, orthogonal, and Hermitian matrices, are presented. Most importantly, the material in this chapter will be used extensively in the study of differential equations. Example 4.18 shows a typical system to be analyzed using eigenvalues and eigenvectors. Chapter 5 presents the solutions for such systems in detail.

GENERAL DISCUSSION OF EIGENVALUES

If $\mathbf{x}$ is a vector in a vector space, certain linear operations may transform $\mathbf{x}$ into a scalar multiple of itself. Suppose that $\mathbf{x}$ is a nonzero vector in a vector space and f is a linear function that transforms $\mathbf{x}$ according to the relationship

$$f(\mathbf{x}) = \lambda \mathbf{x}, \tag{4.1}$$

where λ is a constant. Then $\mathbf{x}$ is said to be an *eigenvector* of the linear function f and λ is its associated *eigenvalue*. Although $\mathbf{x} = \mathbf{0}$ is always a solution to this equation since f is linear, we will rule this out and say that the zero vector cannot be an eigenvector. However, an eigenvalue may be zero. Any scalar multiple $\alpha \mathbf{x}$ of $\mathbf{x}$ is also an eigenvector for any number $\alpha \neq 0$ since if $f(\mathbf{x}) = \lambda \mathbf{x}$, then it follows from the linearity of f that

$$f(\alpha \mathbf{x}) = \alpha f(\mathbf{x}) = \lambda \alpha \mathbf{x}.$$

The word *eigen* is the German word for "own" or "proper," so eigenvalues are also called *proper* values in some applications. In engineering problems, the term *characteristic vector* is sometimes used for eigenvector, and the corresponding eigenvalue is termed the *characteristic value*.

Eigenvalue Example
The exponential function $e^{\lambda t}$ is an *eigenfunction* of various operations, such as differentiation, integration, and time shifting. Thus, when these operations are performed on $e^{\lambda t}$, the result is a constant times the exponential function.

Since for any constant λ,

$$\frac{d}{dt}(e^{\lambda t}) = \lambda(e^{\lambda t}), \tag{4.2}$$

the exponential is an eigenfunction of the differentiation operator.

If the time shift operator is defined as $T_s[f(t)] = f(t - t_0)$, then

$$T_s[e^{\alpha t}] = e^{\alpha(t-t_0)} = e^{-\alpha t_0}\, e^{\alpha t}. \tag{4.3}$$

Since the first term in the product is constant, setting $\lambda = e^{-\alpha t_0}$ results in the eigenvalue equation

$$T_s[e^{\alpha t}] = \lambda e^{\alpha t}. \tag{4.4}$$

This relationship will become very important in our study of differential equations in the next chapter.

☐

In terms of matrices, Equation 4.1 becomes

$$A\mathbf{x} = \lambda\mathbf{x}, \tag{4.5}$$

with λ a scalar and $\mathbf{x}$ in $\mathbf{R}^n$. Since the size of $\lambda\mathbf{x}$ is the same as $\mathbf{x}$, the matrix A must be square.

From Equation 4.5, it is evident that the eigenvector is not unique, since the vector $\alpha\mathbf{x}$ also satisfies the equation for any nonzero scalar α.

Equation 4.5 involves the unknowns λ and $\mathbf{x}$. The formal solution involves determinants and the eigenvalues will be shown to be the roots of the polynomial that results from forming $\det(A - \lambda I)$.

☐ EXAMPLE 4.2   *Matrix Eigenvalue Example*
An example of an eigenvalue equation with matrix multiplication is

$$\begin{bmatrix} 3 & 1 \\ 1 & 3 \end{bmatrix}\begin{bmatrix} 1 \\ -1 \end{bmatrix} = 2\begin{bmatrix} 1 \\ -1 \end{bmatrix}, \tag{4.6}$$

so 2 is the eigenvalue in this equation. The vector $[1, -1]^T$ is an eigenvector with eigenvalue 2, as is any vector of the form $[x, -x]^T$, where x is a nonzero constant.

☐

REVIEW OF POLYNOMIALS

Before embarking on a detailed study of eigenvalues and eigenvectors, we present a review of the properties of polynomials. The theoretical approach to finding the eigenvalues of a matrix involves finding the roots of a polynomial that results from the equation $A\mathbf{x} = \lambda\mathbf{x}$, as shown later.

Polynomials   Let $a_0, a_1, \ldots, a_n$ be $n+1$ arbitrary numbers with $a_n \neq 0$. Then, the function

$$P(z) = a_n z^n + a_{n-1} z^{n-1} + \cdots + a_0 \tag{4.7}$$

is a *polynomial* of degree n. The $n + 1$ constants $a_0, a_1, \ldots, a_n$ are the *coefficients* of the polynomial. A polynomial is a *real polynomial* if all its coefficients are real numbers. This text considers only polynomials with real coefficients unless otherwise stated, because these are associated with mathematical models of physical systems.

The numbers z that are solutions to the equation

$$P(z) = 0 \tag{4.8}$$

are called the *roots* or sometimes the *zeros* of the polynomial. The values of the roots are not necessarily real numbers. Thus, a root z may have the form $z = x + iy$, where i is the imaginary number $\sqrt{-1}$. In electrical engineering problems, this is often written j so that no confusion would result with the current if it is designated by i. As described in Chapter 2, the number $\bar{z} = x - iy$ is the complex *conjugate* of z. The notation z^* is also used to designate the complex conjugate of z.

Important properties of *real* polynomials and their roots are as follows:

1. A polynomial of degree $n \geq 1$ has n roots.

2. A polynomial of odd degree has at least one real root.

3. If z is a complex root of a real polynomial, then the complex conjugate $\bar{z}$ is a root also.

The polynomial $P(z) = a_n z^n + a_{n-1} z^{n-1} + \cdots + a_0$ can always be written in the form

$$P(z) = (z - z_1)(z - z_2) \cdots (z - z_n) a_n \tag{4.9}$$

as the product of linear factors using the roots $z_i, i = 1, 2, \ldots, n$ of

$$P(z) = 0.$$

☐ **EXAMPLE 4.3**   ***Polynomials***

The polynomial

$$P(z) = z^4 - 4z^3 + 6z^2 - 4z + 1$$

can be written as

$$P(z) = (z - 1)^4 = (z - 1)(z - 1)(z - 1)(z - 1),$$

which has the root $+1$ with multiplicity 4. In this case, the roots are all real.

The equation $z^2 + 1 = 0$ has solutions $z = \pm\sqrt{-1}$, and the polynomial can be written

$$P(z) = z^2 + 1 = (z + i)(z - i).$$

Since the polynomial has real coefficients, the roots occur in complex conjugate pairs, as stated before.

☐

There are formulas for the solution of polynomial equations of degrees $1, 2, 3$ and 4. For example, the first-degree equation and its solution are

$$ax + b = 0, \qquad x = -\frac{b}{a}$$

if $a \neq 0$. For the quadratic equation

$$ax^2 + bx + c = 0,$$

the solutions are

$$x = \frac{-b \pm \sqrt{b^2 - 4ac}}{2a}.$$

For the general third- and fourth-degree equations, the formulas are complicated but well defined. The roots of a general polynomial equation of degree higher than the fourth cannot be expressed by a formula that gives the roots exactly.[1] Techniques to determine approximate roots for a polynomial equation are discussed in Chapter 10.

MATLAB Polynomials The MATLAB command **roots** produces a column vector whose elements are the roots of a polynomial. The argument of **roots** is a row vector containing the coefficients ordered by descending powers. It is interesting that the algorithm to find the roots, as described in the *MATLAB User's Guide*, actually finds the eigenvalues of an equivalent matrix equation. This reverses the classical approach of finding eigenvalues from the roots of a polynomial equation, as will be presented later in this chapter.

For a polynomial defined symbolically, the command **factor** will factor the polynomial. The command **solve** will solve for the symbolic roots of the polynomial or return a numerical value if a symbolic solution cannot be found.

WHAT IF? Use MATLAB to find the roots of the polynomials in Example 4.3. Also, experiment with the symbolic MATLAB command **solve** to find the roots of a general third-degree polynomial. The command will also solve for the roots if the polynomial coefficients are numeric.

---

[1] The text by Uspensky listed in the Annotated Bibliography for this chapter describes the formulas for the cubic and fourth degree equation.

EIGENVALUES AND EIGENVECTORS

■

Since only square matrices can have eigenvalues and eigenvectors, the matrices treated in the remainder of this chapter are square matrices. The matrix form of the problem for the square matrix A is stated as

$$A\mathbf{x} = \lambda\mathbf{x}, \tag{4.10}$$

where λ is a scalar and $\mathbf{x}$ is a nonzero vector. This equation can be rearranged to read

$$A\mathbf{x} - \lambda\mathbf{x} = A\mathbf{x} - \lambda I\mathbf{x} = \mathbf{0}. \tag{4.11}$$

In this equation, $\lambda\mathbf{x}$ is written $\lambda I\mathbf{x}$. The identity matrix I is used in the equation to satisfy the rules of matrix multiplication.

Finally, Equation 4.11 becomes

$$(A - \lambda I)\,\mathbf{x} = \mathbf{0}. \tag{4.12}$$

This is a homogeneous equation that has a solution other than the zero vector ($\mathbf{0}$) if and only if the determinant is zero. Thus, to solve Equation 4.12 the equation for the determinant is formed as

$$|A - \lambda I| = 0. \tag{4.13}$$

The expanded determinant form of Equation 4.13 is

$$|A - \lambda I| = \begin{vmatrix} a_{11} - \lambda & a_{12} & \cdots & a_{1n} \\ a_{21} & a_{22} - \lambda & \cdots & a_{2n} \\ \vdots & \vdots & \ddots & \vdots \\ a_{n1} & a_{n2} & \cdots & a_{nn} - \lambda \end{vmatrix} = 0, \tag{4.14}$$

which is a polynomial in λ that can be written

$$C(\lambda) = (-\lambda)^n + c_{n-1}(-\lambda)^{n-1} + \cdots + c_1(-\lambda) + c_0 = 0. \tag{4.15}$$

This nth-order polynomial $C(\lambda)$ is called the *characteristic polynomial* of the matrix A.

The result is written as

$$C(\lambda) = |A - \lambda I|, \tag{4.16}$$

which can be factored with n roots as

$$C(\lambda) = (\lambda_1 - \lambda)(\lambda_2 - \lambda) \cdots (\lambda_n - \lambda) = \prod_{i=1}^{n}(\lambda_i - \lambda) \tag{4.17}$$

using the product operator $\prod$ to define the result.

A characteristic polynomial has n roots, which may include real roots, complex roots, and repeated roots, according to the previous discussion of polynomials. When the matrix equations represent a mathematical model of a physical system, the roots might define particular characteristics of the system, such as the frequency of oscillation or the rate of decay.

Using the formulation of Equation 4.14 or Equation 4.17, many important properties of the eigenvalues can be proven. Before listing these properties and proving some of them, we present a simple example using eigenvalues and eigenvectors.

Practical considerations. We emphasize that the method of determining the eigenvalues of a matrix introduced first in this chapter is used for small-size matrices and theoretical studies. The numerical method actually used by MATLAB to compute the eigenvalues will be presented later in the chapter.

□ **EXAMPLE 4.4** *Eigenvectors I*

Consider the resistive network of Figure 4.1. The problem to be discussed involves solving for the currents in the loops given the values of the voltages V_1 and V_2. Our solution method is to formulate the problem as a matrix equation and solve it using eigenvectors.

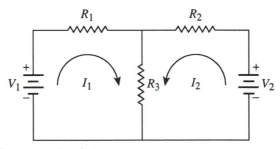

FIGURE 4.1 *Resistor network*

Applying Kirchhoff's voltage law, the sum of the voltages in the direction of the currents yields the system of equations:

$$\begin{aligned} V_1 &= R_1 I_1 + R_3(I_1 + I_2), \\ V_2 &= R_2 I_2 + R_3(I_1 + I_2), \end{aligned}$$

which leads to the matrix equation

$$\begin{bmatrix} V_1 \\ V_2 \end{bmatrix} = \begin{bmatrix} R_1 + R_3 & R_3 \\ R_3 & R_2 + R_3 \end{bmatrix} \begin{bmatrix} I_1 \\ I_2 \end{bmatrix}, \tag{4.18}$$

or $\mathbf{v} = R\,\mathbf{i}$ in matrix notation.

For simplicity, choose $R_1 = R_2 = R_3 = 1$. Notice that the matrix R would be symmetrical no matter what the choice of resistor values, since R_3 is the only common element. Then, the solution for $\mathbf{i}$ can be found by inverting

Equation 4.18 to yield

$$\begin{bmatrix} I_1 \\ I_2 \end{bmatrix} = \frac{1}{3} \begin{bmatrix} 2 & -1 \\ -1 & 2 \end{bmatrix} \begin{bmatrix} V_1 \\ V_2 \end{bmatrix},$$

which we write as $\mathbf{i} = G\mathbf{v}$, defining the matrix $G = R^{-1}$. Thus, given values for the two voltages, the current in each loop is found easily. By convention, a negative value of current flows in the opposite direction from that shown in Figure 4.1.

Solving the eigenvector problem for this network yields useful information about the currents and voltages. The equation
$|G - \lambda I|$ becomes

$$\begin{vmatrix} 2/3 - \lambda & -1/3 \\ -1/3 & 2/3 - \lambda \end{vmatrix} = 0$$

and yields the polynomial $(2/3 - \lambda)^2 - 1/9 = 0$, for which the eigenvalues are

$$\lambda_1 = \frac{1}{3} \quad \text{and} \quad \lambda_2 = 1.$$

The eigenvectors will be designated $\mathbf{v}_a$ and $\mathbf{v}_b$. Forming the eigenvector equation $G\mathbf{v}_a = \lambda_1 \mathbf{v}_a$ leads to the equation

$$\frac{1}{3} \begin{bmatrix} 2 & -1 \\ -1 & 2 \end{bmatrix} \begin{bmatrix} V_1 \\ V_2 \end{bmatrix} = \frac{1}{3} \begin{bmatrix} V_1 \\ V_2 \end{bmatrix},$$

which yields the result $V_1 = V_2$. Solving $G\mathbf{v}_b = \lambda_2 \mathbf{v}_b$ yields $V_2 = -V_1$. Thus, the eigenvectors can be written as

$$\mathbf{v}_a = V_1 \begin{bmatrix} 1 \\ 1 \end{bmatrix} \quad \text{and} \quad \mathbf{v}_b = V_1 \begin{bmatrix} 1 \\ -1 \end{bmatrix}.$$

If V_1 is set to 1, the dot product of $\mathbf{v}_a$ and $\mathbf{v}_b$ is zero, and the vectors are orthogonal. Thus, these two vectors are linearly independent, and they form a basis for $\mathbf{R}^2$, which is the dimension of the vector space for the voltages. Any other voltage vector can be written as a linear combination of these vectors.

The two *modes* for the system are shown in Figure 4.2.

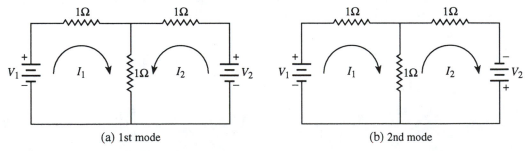

(a) 1st mode   (b) 2nd mode

FIGURE 4.2 *Resistor network modes*

In the first mode, corresponding to $\mathbf{v}_a$, the voltages cause the currents to flow in the same direction through the common resistor. The other mode, indicated by $\mathbf{v}_b$, shows the currents opposing. We shall return to this problem shortly to show how easily any solution can be formulated in terms of the eigenvectors.

□

LINEAR FUNCTIONS AND EIGENVECTORS

Vector spaces and basis vectors were introduced in Chapter 2. If the basis vectors are also the eigenvectors of a linear function f that operates on the vector space, a particularly useful result can be derived.

Assume that f is a linear function defined on a vector space V with a basis set of vectors $\{\mathbf{x}_1, \mathbf{x}_2, \ldots, \mathbf{x}_n\}$ consisting of eigenvectors of f. Then,

$$f(\mathbf{x}_k) = \lambda_k \mathbf{x}_k, \qquad k = 1, 2, \ldots, n,$$

because the action of f on any eigenvector is to multiply it by the associated eigenvalue. Since the eigenvectors are assumed to form a basis for V, any vector $\mathbf{x}$ in V can be written as a linear combination of the eigenvectors as

$$\mathbf{x} = c_1 \mathbf{x}_1 + c_2 \mathbf{x}_2 + \cdots + c_n \mathbf{x}_n,$$

where the c_i are constants. Using the fact that f is linear and the $\mathbf{x}$'s are eigenvectors and a basis, we conclude

$$
\begin{aligned}
f(\mathbf{x}) &= c_1 f(\mathbf{x}_1) + c_2 f(\mathbf{x}_2) + \cdots + c_n f(\mathbf{x}_n) \\
&= c_1 \lambda_1 \mathbf{x}_1 + c_2 \lambda_2 \mathbf{x}_2 + \cdots + c_n \lambda_n \mathbf{x}_n.
\end{aligned}
\tag{4.19}
$$

Thus, the result of any linear function f acting on a vector $\mathbf{x}$ that can be written as a linear combination of eigenvectors of f is simply to multiply each coefficient by the corresponding eigenvalue. In summary,

$$f\left(\sum_{k=1}^{n} c_k \mathbf{x}_k\right) = \sum_{k=1}^{n} \lambda_k c_k \mathbf{x}_k, \tag{4.20}$$

where each $\mathbf{x}_k$ is an eigenvector of f corresponding to the eigenvalue λ_k. So the action of f can be interpreted as a succession of multiplications.

From the discussions of vector spaces in Chapter 2, we know that every set of n linearly independent vectors in $\mathbf{R}^n$ is a basis for $\mathbf{R}^n$. In Chapter 3, linear transformation of a vector was seen to be equivalent to multiplication by a matrix. Thus, if the eigenvectors of the $n \times n$ matrix A are linearly independent, they can serve as a basis for $\mathbf{R}^n$. Furthermore, the result of Equation 4.20 can be applied to matrix multiplication of any vector in $\mathbf{R}^n$ that is written as a linear combination of the eigenvectors of A.

As discussed later, the $n \times n$ matrix A has n linearly independent eigenvectors if all the eigenvalues are distinct. Even though every $n \times n$ matrix has n eigenvalues, it is possible that the eigenvalues are not all different, so the matrix may not have n linearly independent eigenvectors.

Eigenvectors II

Returning to the resistive network of Example 4.4, we can show the application of Equation 4.20, since the eigenvectors form a basis for any vector of voltages. Consider two arbitrary voltages, V_1 and V_2, applied to the circuit of Figure 4.1. In terms of the eigenvectors for the circuit matrix G,

$$\begin{bmatrix} V_1 \\ V_2 \end{bmatrix} = c_1 \begin{bmatrix} 1 \\ 1 \end{bmatrix} + c_2 \begin{bmatrix} 1 \\ -1 \end{bmatrix},$$

because the eigenvectors form a basis for the space of voltage vectors. This linear system of equations is easily solved for the constants c_1 and c_2, yielding

$$c_1 = \frac{V_1 + V_2}{2} \quad \text{and} \quad c_2 = \frac{V_1 - V_2}{2}.$$

Thus, according to the result of Equation 4.20, the matrix problem $\mathbf{i} = G\mathbf{v}$ to solve for the currents in the circuit becomes

$$\begin{bmatrix} i_1 \\ i_2 \end{bmatrix} = \frac{1}{3} \left(\frac{V_1 + V_2}{2} \right) \begin{bmatrix} 1 \\ 1 \end{bmatrix} + 1 \left(\frac{V_1 - V_2}{2} \right) \begin{bmatrix} 1 \\ -1 \end{bmatrix},$$

since the eigenvalues are $\frac{1}{3}$ and 1.

Two numerical results are shown in Figure 4.3.

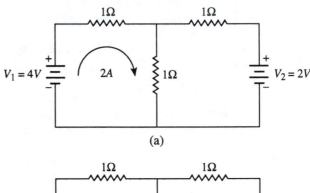

(a)

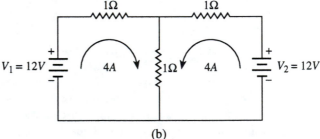

(b)

FIGURE 4.3 *Resistor network currents*

In the first circuit of Figure 4.3(a), $V_1 = 4$ volts and $V_2 = 2$ volts. The currents are

$$i_i = \frac{1}{3}(3)(1) + (1)(1)(1) = 2 \text{ amperes},$$

$$i_2 = \frac{1}{3}(3)(1) + (1)(1)(-1) = 0 \text{ amperes}.$$

When the voltage vector corresponds to one of the eigenvectors, the solution is simply a multiple of that eigenvector. For example, if $V_1 = 12$ volts and $V_2 = 12$ volts, the solution is

$$i_i = \frac{1}{3}(12)(1) = 4 \text{ amperes},$$

$$i_2 = \frac{1}{3}(12)(1) = 4 \text{ amperes},$$

as shown by the second circuit in Figure 4.3(b).

Power. The power dissipated in a resistor of resistance R ohms is $i^2 R$ watts if i is the current in amperes flowing through the resistor. For a battery, the power supplied is Vi, where V is the battery voltage and i is the current delivered to a circuit. Thus, for circuits such as those shown in Figure 4.3, the power delivered by the batteries is equal to the power dissipated in the resistors. The power from the batteries can be calculated as the inner product

$$P = \langle \mathbf{v}, \mathbf{i} \rangle.$$

For the circuit of Figure 4.3(a), the vectors are

$$\mathbf{v} = \begin{bmatrix} 4 \\ 2 \end{bmatrix}, \qquad \mathbf{i} = \begin{bmatrix} 2 \\ 0 \end{bmatrix}$$

so the power in the circuit is $\mathbf{v}^T \mathbf{i} = 8$ watts. This is easily verified by adding the powers in the resistors.

□

Electrical engineering students in particular may comment that the resistor circuit example uses a rather involved technique to solve a simple problem. However, the example shows the use of Equation 4.20 for a passive network of resistors.

The approach to solving problems involving matrices by using the eigenvectors has far-reaching applications. More typical examples involving mechanical systems are treated in Example 4.18 and in Chapter 5 after systems of differential equations are studied.

NUMBER OF INDEPENDENT EIGENVECTORS

If we are asked how many eigenvectors a matrix can have, the answer is clearly an infinite number, since any scalar multiple of an eigenvector is also an eigenvector. The more useful question is, How many independent eigenvectors can a matrix have? First, we state that an $n \times n$ matrix can have *at most* n linearly independent eigenvectors. The following theorem relates the linear independence of the eigenvectors to the values of the eigenvalues.

■ THEOREM 4.1

Distinct eigenvalues

Eigenvectors corresponding to the distinct eigenvalues of a matrix are linearly independent.

■

The theorem guarantees that if an $n \times n$ matrix has n distinct eigenvalues, all the eigenvectors are linearly independent. Hence, these eigenvectors form a basis for $\mathbf{R}^n$. However, the question as to number of linearly independent eigenvectors has not been answered if the number of distinct eigenvalues is less than n. The theorem states only that if an $n \times n$ matrix has m different (distinct) eigenvalues ($m \leq n$), then the matrix has *at least* m linearly independent eigenvectors.

When the eigenvalues of a matrix are repeated, it is required to compute the rank of the matrix $A - \lambda I$ to determine the number of independent eigenvectors associated with the repeated eigenvalue λ. This problem is treated in the next section.

REPEATED EIGENVALUES

When some eigenvalues are repeated for the $n \times n$ matrix A, the number of independent eigenvectors may range from 1 to n. The procedure to determine the number of linearly independent eigenvectors that share the repeated eigenvalue is to first find the eigenvalues for the matrix and then compute the rank of the matrix in the eigenvalue equation $A - \lambda I$ for the repeated eigenvalues. The following theorem defines how to relate the rank of the eigenvalue equation to the number of linearly independent eigenvectors.

■ **THEOREM 4.2**

Repeated eigenvalues

If λ is an eigenvalue of multiplicity k of an $n \times n$ matrix A, then the number of independent eigenvectors N of A associated with λ is given by

$$N = n - \text{rank}(A - \lambda I), \qquad (4.21)$$

and, furthermore, $1 \leq N \leq k$.

———————————————————————■

The techniques to determine the rank of the matrix in the eigenvalue equation were discussed in Chapter 3.

□ **EXAMPLE 4.6**

Matrix Eigenvalue Example

Consider the matrix

$$\begin{bmatrix} 1 & -1 & 0 \\ 0 & 1 & 1 \\ 0 & 0 & -2 \end{bmatrix}. \qquad (4.22)$$

Writing $|A - \lambda I| = 0$ leads to the determinant

$$\begin{bmatrix} 1-\lambda & -1 & 0 \\ 0 & 1-\lambda & 1 \\ 0 & 0 & -2-\lambda \end{bmatrix} = -(\lambda-1)^2(\lambda+2) = 0. \qquad (4.23)$$

Thus, the eigenvalues are $+1$, $+1$, and -2.

The repeated root $\lambda = 1$ may yield one or two linearly independent eigenvectors. One eigenvector is found by forming the equation

$$A\mathbf{x} = +1\mathbf{x} \quad \text{or} \quad (A - 1I)\mathbf{x} = 0, \qquad (4.24)$$

which becomes

$$\begin{bmatrix} 0 & -1 & 0 \\ 0 & 0 & 1 \\ 0 & 0 & -3 \end{bmatrix} \begin{bmatrix} x_1 \\ x_2 \\ x_3 \end{bmatrix} = \begin{bmatrix} 0 \\ 0 \\ 0 \end{bmatrix}. \tag{4.25}$$

The result is that $x_2 = x_3 = 0$, but x_1 can be chosen arbitrarily.

Choosing $x_1 = \alpha$, $\alpha \neq 0$, leads to the eigenvector

$$\mathbf{x}_1 = 1 \times \begin{bmatrix} \alpha \\ 0 \\ 0 \end{bmatrix} = \begin{bmatrix} \alpha \\ 0 \\ 0 \end{bmatrix} \tag{4.26}$$

corresponding to the eigenvalue $+1$.

Reducing $A - \lambda I$ with $\lambda = 1$ by elementary row operations, as discussed in Chapter 3, shows that the rank is 2 for this matrix. Using Equation 4.21 with $n = 3$ and $k = 2$, we expect only $3 - 2 = 1$ independent eigenvector.

The eigenvalue of -2 yields another linearly independent eigenvector. The eigenvector equation $(A + 2I)\mathbf{x} = 0$ corresponding to the eigenvalue -2 is

$$\begin{bmatrix} 3 & -1 & 0 \\ 0 & 3 & 1 \\ 0 & 0 & 0 \end{bmatrix} \begin{bmatrix} x_1 \\ x_2 \\ x_3 \end{bmatrix} = \begin{bmatrix} 0 \\ 0 \\ 0 \end{bmatrix}, \tag{4.27}$$

leading to the solution in terms of x_1 as system of equations

$$\begin{aligned} x_2 &= 3x_1, \\ x_3 &= -3x_2, \end{aligned} \tag{4.28}$$

with solutions if one of the variables is given an arbitrary nonzero value. Assuming that $x_1 = \beta$, an eigenvector corresponding to the eigenvalue -2 is

$$\mathbf{x}_2 = \begin{bmatrix} \beta \\ 3\beta \\ -9\beta \end{bmatrix}. \tag{4.29}$$

As an example, the vector

$$\mathbf{x} = \begin{bmatrix} 1 \\ 3 \\ -9 \end{bmatrix}$$

would serve as an eigenvector.

For this matrix, there are only two independent eigenvectors. Notice that the matrix A is nonsingular, since its determinant is -2. Thus, the rows (and columns) are linearly independent and there is a unique solution to the equation $A\mathbf{x} = \mathbf{b}$.

□

MATRIX EIGENVALUE THEOREMS

Important properties of an $n \times n$ matrix A with eigenvalues

$$\lambda_1, \lambda_2, \ldots, \lambda_n$$

are given by the following theorems:

1. λ *is an eigenvalue of A if and only if $|A - \lambda I| = 0$.*

2. *If λ is an eigenvalue of A, any nontrivial solution to $(A - \lambda I)\mathbf{x} = \mathbf{0}$ is an eigenvector of A corresponding to eigenvalue λ.*

3. *The determinant of A is the product of its eigenvalues so that*

$$det(A) = \lambda_1 \lambda_2 \cdots \lambda_n.$$

4. A *is singular if and only if it has an eigenvalue of zero.*

5. *The sum of the diagonal elements of A (called the trace) of A is equal to the sum of its eigenvalues, so that*

$$a_{11} + a_{22} + \cdots + a_{nn} = \lambda_1 + \lambda_2 + \cdots + \lambda_n.$$

6. *The eigenvalues of a triangular matrix are its diagonal entries.*

Applying these theorems can help hand computation of the eigenvalues.

Many of these theorems can be proven by considering the formula for computing the eigenvalues as given by Equation 4.14,

$$|A - \lambda I| = \begin{vmatrix} a_{11} - \lambda & a_{12} & \cdots & a_{1n} \\ a_{21} & a_{22} - \lambda & \cdots & a_{2n} \\ \vdots & \vdots & \ddots & \vdots \\ a_{n1} & a_{n2} & \cdots & a_{nn} - \lambda \end{vmatrix} = 0,$$

and the characteristic polynomial of Equation 4.17,

$$C(\lambda) = (\lambda_1 - \lambda)(\lambda_2 - \lambda) \cdots (\lambda_n - \lambda) = \prod_{i=1}^{n} (\lambda_i - \lambda).$$

To prove the assertion of Property 3 that the determinant of A is the product of its eigenvalues, form the characteristic polynomial

$$C(\lambda) = \det(A - \lambda I) = (\lambda_1 - \lambda)(\lambda_2 - \lambda) \cdots (\lambda_n - \lambda)$$

and notice that $C(0) = \det(A) = c_0$, a constant. Thus, we conclude that

$$C(0) = \det(A) = \lambda_1 \lambda_2 \cdots \lambda_n. \tag{4.30}$$

This leads immediately to the statement of Property 4, according to the discussion of singular matrices in Chapter 3. Since $\det(A) = 0$ if and only if A is singular, Equation 4.30 leads to the result that A is singular if and only if it has a zero eigenvalue.

To show that the eigenvalues of a triangular matrix are the elements of the diagonal, we use the properties of determinants explained in Chapter 3. If $A = (a_{ij})$ is an $n \times n$ triangular matrix, then the matrix $(A - \lambda I)$ is also triangular with diagonal elements

$$a_{11} - \lambda, a_{22} - \lambda, \ldots, a_{nn} - \lambda.$$

The determinant of a triangular matrix is the product of its diagonal entries, so that

$$|A - \lambda I| = \prod_{i=1}^{n}(a_{ii} - \lambda).$$

Thus, the eigenvalues of A are $a_{11}, a_{22}, \ldots, a_{nn}$ when the matrix is triangular.

□ EXAMPLE 4.7   *Examples of Theorems*

It was found in Example 4.6 that the eigenvalues of

$$A = \begin{bmatrix} 1 & -1 & 0 \\ 0 & 1 & 1 \\ 0 & 0 & -2 \end{bmatrix} \tag{4.31}$$

are $\lambda_1 = 1, \lambda_2 = 1, \lambda_3 = -2$. These are the diagonal entries, as expected from Property 6. Verifying Property 3 yields

$$\det(A) = -2 = \lambda_1 \lambda_2 \lambda_3 = (1)(1)(-2). \tag{4.32}$$

Forming trace(A) shows that

$$\text{trace}(A) = 1 + 1 - 2 = 0 = \lambda_1 + \lambda_2 + \lambda_3, \tag{4.33}$$

which verifies Property 5. This is obvious for a triangular matrix, but the trace of a general matrix is also equal to the sum of its eigenvalues.

□

COMPLEX VECTORS AND MATRICES

The eigenvalues and eigenvectors of a matrix can be complex, even though the matrix has real elements and hence real coefficients in its characteristic polynomial. To cover these cases, we introduce the vector space $\mathbf{C}^n$ of vectors with n complex components. For example, a complex vector $\mathbf{z}$ in $\mathbf{C}^3$ can be written as

$$\mathbf{z} = \begin{bmatrix} z_1 \\ z_2 \\ z_3 \end{bmatrix},$$

with $z_j = x_j + iy_j$, where x_j and y_j are real numbers and $j = 1, 2, 3$.

The notation $\bar{\mathbf{z}}$ means the complex conjugate of the vector. As an example, if $\mathbf{z}$ has elements $z_j = x_j + iy_j$, then

$$\bar{\mathbf{z}} = \begin{bmatrix} \bar{z}_1 \\ \bar{z}_2 \\ \bar{z}_3 \end{bmatrix},$$

where $\bar{z}_j = x_j - iy_j$ for $j = 1, 2, 3$. For the complex vectors, all the rules of addition and matrix multiplication are valid. However, the *length* of the vector has to be redefined, as described in Chapter 2. The inner product of two complex vectors $\mathbf{z}_1$ and $\mathbf{z}_2$ is formed as

$$\langle \bar{\mathbf{z}}_1, \mathbf{z}_2 \rangle. \tag{4.34}$$

The length of a complex vector $||\mathbf{z}||$ is thus $\langle \bar{\mathbf{z}}, \mathbf{z} \rangle^{1/2}$. Notice that these definitions also are valid for real vectors $\mathbf{x}$ in $\mathbf{R}^n$, since $\bar{\mathbf{x}} = \mathbf{x}$ when the elements are real.

If the vectors are column vectors, then the inner product of $\mathbf{z}_1$ and $\mathbf{z}_2$ is

$$\langle \bar{\mathbf{z}}_1, \mathbf{z}_2 \rangle = \bar{\mathbf{z}}_1^T \mathbf{z}_2. \tag{4.35}$$

The notation for the *conjugate transpose* is often designated by the superscript H as

$$\bar{\mathbf{z}}^T = \mathbf{z}^H, \tag{4.36}$$

which is called the *Hermitian transpose*. Using the inner product as defined in Equation 4.35, the orthogonality condition for two column vectors with complex elements becomes

$$\langle \bar{\mathbf{z}}_1, \mathbf{z}_2 \rangle = \bar{\mathbf{z}}_1^T \mathbf{z}_2 = 0.$$

If the matrix A has elements a_{ij}, the conjugate will be designated as $\bar{A}$ with the elements $\bar{a}_{ij}$. The conjugate transpose of A will be designated

$$\bar{A}^T = A^H, \tag{4.37}$$

with elements $\bar{a}_{ji}$. The notation A^* is also used to designate the Hermitian transpose. Although the importance of the Hermitian transpose will be explored in a later section, the definition is given here because it is very seldom that we compute only the transpose of a matrix with complex entries without taking the conjugate of the elements.

☐ EXAMPLE 4.8   *Complex Matrices and Eigenvalues*

Several examples will show the manipulation of complex vectors and matrices. The first example shows the conjugate transpose of a complex matrix. In Part b, the rotation matrix in $\mathbf{R}^2$ is shown to have complex eigenvalues.

a. Using the definition of Equation 4.37,

$$A^H = \begin{bmatrix} 5+i & 12 \\ t-i & 5i \\ 0 & t^2 \end{bmatrix}^H = \begin{bmatrix} 5-i & t+i & 0 \\ 12 & -5i & t^2 \end{bmatrix},$$

assuming that the variable t is real.

b. Consider the rotation matrix defined in Chapter 3 as

$$R_\theta = \begin{bmatrix} \cos\theta & -\sin\theta \\ \sin\theta & \cos\theta \end{bmatrix}.$$

Forming the determinant equation $\det(R_\theta - \lambda I) = 0$ yields the characteristic equation

$$\lambda^2 - 2\lambda\cos\theta + 1 = 0$$

with solutions

$$\cos\theta \pm i\sin\theta = e^{\pm i\theta}$$

using the Euler relationship defined in Chapter 2. The eigenvectors satisfy the equation

$$R_\theta \mathbf{x} = e^{\pm i\theta}\mathbf{x},$$

and we conclude that there are no eigenvectors in $\mathbf{R}^2$ unless $\theta = 0$ or $\theta = \pi$. In general, for other angles, the eigenvectors are

$$\begin{bmatrix} 1 \\ -i \end{bmatrix} \quad \text{and} \quad \begin{bmatrix} 1 \\ i \end{bmatrix},$$

as you are asked to show in Problem 4.11.

☐

MATLAB COMMANDS FOR EIGENVECTORS

━━━━━━━━━━━━■━━━━━━━━━━━━

MATLAB has several commands that compute numerical values of eigen-values and eigenvectors. The basic command forms are summarized in Table 4.1.

TABLE 4.1 *MATLAB eigenvector functions*

Operation	MATLAB	Result
Eigenvalues and eigenvectors	**eig**(A)	Eigenvalues of A
	[V,D] = **eig**(A)	Solves for eigenvectors and eigenvalues of A V is a matrix whose columns are eigenvectors with norm equal to 1 D is a diagonal matrix whose elements are the eigenvalues $(AV = VD)$
Characteristic polynomial	**poly**	Characteristic polynomial for $|\lambda I - A|$
Roots	**roots**	Roots of polynomial $p(x)$
Symbolic operation	**eigensys**	Eigenvectors and eigenvalues of a symbolic matrix
	charpoly	Symbolic characteristic polynomial

□ **EXAMPLE 4.9**   ***MATLAB Eigenvalue Example***

The accompanying MATLAB session uses MATLAB commands to find the eigenvalues and eigenvectors for the matrix of Example 4.6. The command **eig** computes the eigenvalues of A. Next, the command **poly** is used to determine the coefficients of the characteristic polynomial $x^3 - 3x + 2$. The command

```
>>[V,D]=eig(A)
```

returns the eigenvectors as columns of the vector V, and the corresponding eigenvalues are the values on the diagonal of the matrix D. The eigenvalues are $1, 1, -2$, as before, and the eigenvectors are normalized to have a unit norm. There are only two independent eigenvectors, as determined in Example 4.6. Thus, we notice that the columns of V are not mutually orthogonal. Since MATLAB computes eigenvectors with norm of 1, we convert the third eigen-vector to one with integer entries to agree with the results of Example 4.6.

```
Example 4.9
% Eigenvalues
>>A=[1 -1 0;0 1 1;0 0 -2]
A =
     1    -1     0
     0     1     1
     0     0    -2

>>EIG=eig(A)          % Determine eigenvalues
EIG =
     1
     1
    -2

>>POLY=poly(A)        %  and characteristic polynomial
POLY =
     1     0    -3     2

>>[V,D]=eig(A)        % V contains eigenvectors
%                         diag(D) is eigenvalues
V =
    1.0000    1.0000   -0.1048
         0    0.0000   -0.3145
         0         0    0.9435
D =
     1     0     0
     0     1     0
     0     0    -2
>>% Columns of V are eigenvectors
>>V =
    1.0000    1.0000   -0.1048
         0    0.0000   -0.3145
         0         0    0.9435

>>V3=V(:,3) % Take third column and set V(1,3)=1
V3 =
   -0.1048
   -0.3145
    0.9435

>>v33=V3/V(1,3)
v33 =
     1
     3
    -9
>>quit
```

Chapter 3 presented the numerical technique used by MATLAB to solve a system of linear equations $A\mathbf{x} = \mathbf{b}$. Using a modified Gaussian elimination technique, the system was transformed into another system with the same set of solutions $\mathbf{x}$. Thus, the transformed or reduced system was *equivalent* to the original system in this sense, but the solutions were easier to find by machine computation, so the solutions were probably more accurate.

For the matrix eigenvalue problem, the approach is to convert the matrix into an equivalent representation that has the same set of eigenvalues. To be useful, the new representation must have better properties for numerical computation than the original matrix.

MATLAB has various algorithms to compute the eigenvalues and eigenvectors of a matrix. Here we briefly discuss one particular algorithm called the QR *decomposition* algorithm. The purpose is to factor an $n \times n$ matrix A in the form

$$A = QR, \tag{4.38}$$

where Q is an orthogonal matrix and R is upper triangular. The QR method can be viewed as a modification of the Gram-Schmidt process for creating an orthonormal set of vectors as presented in Chapter 2. The claim for the QR algorithm is that it is more reliable (stable) and less susceptible to errors caused by roundoff than the Gram-Schmidt method.

QR **Method**   The existence and uniqueness of the QR matrices is guaranteed by the following theorem, which is proven using a constructive proof in the book by Dahlquist and Björck listed in the Annotated Bibliography for this chapter.

■ THEOREM 4.3   *Uniqueness of QR decomposition*
 Let A be an $m \times n$ matrix with $m \geq n$ and linearly independent columns. Then, there exists a unique $m \times n$ matrix Q such that

$$Q^T Q = D, \quad D = diag(d_1, \ldots, d_n), \quad d_k > 0,$$

for $k = 1, \ldots, n$, and a unique upper-triangular matrix R, with

$$r_{kk} = 1, \quad k = 1, \ldots, n,$$

such that

$$A = QR.$$

As stated in the review of polynomials at the beginning of the chapter, there is no formula for the roots of a polynomial of degree higher than four. Since the eigenvalues of a matrix are the n roots of the characteristic polynomial, any numerical eigenvalue method for an arbitrary matrix must be iterative in nature. Many of the details of the algorithm and various numerical problems that may arise are not presented here, but the MATLAB commands **qr** and **eig** are discussed.[2]

The basic QR algorithm starts with the matrix $A_0 = A$ and computes a sequence of matrices

$$
\begin{aligned}
A_0 &= Q_0 R_0, \\
A_1 &= R_0 Q_0 = Q_1 R_1, \\
A_2 &= R_1 Q_1 = Q_2 R_2,
\end{aligned}
$$

with the kth result

$$ A_k = Q_k R_k = R_{k-1} Q_{k-1}, $$

where k is the number of the iteration. The algorithm starts with matrix A_0 and factors it using the Gram-Schmidt process or a modification of it and then reverses the factors Q and R. As the algorithm executes, A_k approaches a triangular form, and thus its diagonal entries approach its eigenvalues. These are the eigenvalues of the original matrix, since each new matrix is *similar* to the original A matrix. As described later in this chapter, a similarity transform on a matrix A preserves eigenvalues.

☐ EXAMPLE 4.10   *MATLAB* QR *Decomposition and Eigenvalues*

To show the relationship between the MATLAB commands **qr** and **eig**, we test the matrix

$$ A = \begin{bmatrix} 1 & -1 & 0 & 0 \\ 3 & 5 & 0 & 0 \\ 0 & 0 & 1 & 5 \\ 0 & 0 & -1 & 1 \end{bmatrix} \tag{4.39} $$

with eigenvalues $2, 4, 1 \pm i\sqrt{5}$.

The MATLAB script first uses **eig** to find the eigenvalues and eigenvectors of A. Then, the QR factorization command **qr** is used in a loop to reduce A to QR form. After the MATLAB results for the 1st, 5th, and 25th loop are shown, the results will be analyzed.

MATLAB Script ————————————————————————————————
```
Example 4.10
% EX4_10.M Compare qr and eig commands to find eigenvalues
%  The script pauses after each qr iteration
A = [1 -1 0 0;3 5 0 0;0 0 1 5;0 0 -1 1]
[V,D]=eig(A)
pause
% Compute QR sequence; R*Q converges to quasi-diagonal
N=input('Number of Iterations, N= ')
for I=1:N,[Q,R]=qr(A);,,I,A=R*Q,pause,end
```
——

The following results of the MATLAB **qr** command for 1, 5, and 25 loops are taken from the diary file created for the M-file EX4_10.M.

---

[2]The textbook by Hill listed in the Annotated Bibliography for this chapter presents more details on the MATLAB techniques for finding eigenvalues, including potential problems leading to inaccurate results.

Example 4.10
%
>>% Eigenvectors and eigenvalues of A
>> ex4_10   % Execute the M-file

```
A =
     1    -1     0     0
     3     5     0     0
     0     0     1     5
     0     0    -1     1
V =
  -0.7071   -0.3162         0               0
   0.7071    0.9487         0               0
        0         0    0.9129          0.9129
        0         0    0 + 0.4082i     0 - 0.4082i
D =
   2.0000         0         0               0
        0    4.0000         0               0
        0         0    1.0000 + 2.2361i     0
        0         0         0          1.0000 - 2.2361i
```

% MATLAB results for 1, 5 and 25 loops of qr.
Number of Iterations, N= 25

```
I =
     1
A =
     5.2000   -1.6000        0         0
     2.4000    0.8000        0         0
          0         0   -1.0000   -3.0000
          0         0    3.0000    3.0000
```
% Fifth loop
```
I =
     5
A =
     4.0771   -3.9595        0         0
     0.0405    1.9229        0         0
          0         0    1.9845   -4.7409
          0         0    1.2591    0.0155
```
% 25th Loop
```
I =
    25
A =
     4.0000   -4.0000        0         0
     0.0000    2.0000        0         0
          0         0    0.1123   -4.7922
          0         0    1.2078    1.8877
```
% Use eig to find eigenvalues (Compare to diag(D))
>>An=A

```
An =
    4.0000   -4.0000        0        0
    0.0000    2.0000        0        0
         0         0   0.1123  -4.7922
         0         0   1.2078   1.8877
>>eig(An)

ans =
    4.0000
    2.0000
    1.0000 + 2.2361i
    1.0000 - 2.2361i
>>quit

% Computations
 656090 flops.
```

Analysis. Consider the matrix computed in Example 4.10 after 25 iterations of **qr**:

$$An = \begin{bmatrix} 4.0000 & -4.0000 & 0 & 0 \\ 0.0000 & 2.0000 & 0 & 0 \\ 0 & 0 & 0.1123 & -4.7922 \\ 0 & 0 & 1.2078 & 1.8877 \end{bmatrix}.$$

Since A of Equation 4.39 is real, the QR method uses real arithmetic to calculate the factors. In this case, two of the eigenvalues are complex, so the matrix

$$An = RQ$$

converges to a "quasi-triangular" form. As shown in Problem 4.19, finding the determinant and eigenvalues of matrix A can be simplified by considering A to be a *partitioned matrix*. In the example, the MATLAB command **eig** finds the correct eigenvalues of A.

Using MATLAB command **eig** will generally produce a more accurate result when computing the eigenvalues and eigenvectors of a matrix. This is because the **eig** algorithm is more sophisticated for computing these values than the approach using **qr** iteratively. The algorithm is described in some detail in the *MATLAB User's Guide*.

$\square$

Numerical Errors   Matrices without a full set of linearly independently eigenvalues are sometimes referred to as *defective* matrices. An $n \times n$ matrix cannot be defective if the matrix has n distinct eigenvalues. If eigenvalues are repeated, the matrix may or may not have n independent eigenvectors. Using any numerical algorithm to compute the eigenvalues of a matrix could lead to errors in determining if a matrix has repeated eigenvalues.

Warning. Numerical solutions to eigenproblems for some matrices often lead to difficulties due to roundoff errors in the large number of calculations involved. Textbooks or technical journals dealing with numerical analysis should be consulted before attempting practical problems. The books by Hill or Datta listed in the Annotated Bibliography for this chapter are good beginning references.

MATRIX CALCULUS

This section defines various operations on a matrix, such as differentiation, integration, raising to a power, and exponentiation. The derivative or integral of a matrix is defined as differentiation or integration of each element of the matrix. Other matrix functions, such as powers, polynomials and exponentials, are conveniently computed in terms of the eigenvalues of the matrix.

INTEGRATION AND DIFFEREN- TIATION

Consider the $n \times n$ matrix with elements that are functions of time. The notation $A(t) = [a_{ij}(t)]$, emphasizing the dependence of the elements on a variable t, will be used to designate such a matrix.

Derivative If each of the elements $a_{ij}(t)$ is a *differentiable* function of time, the derivative of $A(t)$ is

$$\frac{dA(t)}{dt} = \left[\frac{da_{ij}(t)}{dt} \right].$$ **(4.40)**

The notion $\dot{A}(t)$ is commonly used to indicate the time derivative of $A(t)$.

Integral If each of the elements $a_{ij}(t)$ is a *integrable* function of time, the integral of $A(t)$ is

$$\int A(t)\, dt = \left[\int a_{ij}(t)\, dt \right].$$ **(4.41)**

POLYNOMIALS AND POWERS

A polynomial function of a matrix A is defined to be a function of the form

$$P_n(A) = a_n A^n + a_{n-1} A^{n-1} + \cdots + a_0 I,$$ **(4.42)**

where the a_i are scalars and $A^2 = AA$, $A^3 = A^2 A$, ..., $A^n = A^{n-1}A$.

Direct hand calculation of the terms in Equation 4.42 is possible. For 2×2 matrices raised to low powers of n and for matrices with special structure, such as diagonal matrices, multiplication of the matrices to compute the polynomial may be reasonable. In the next section, the Cayley-Hamilton theorem will be used to aid our calculation.

EXPONENTIAL The function e^A, where A is an $n \times n$ matrix, is defined in terms of the Maclaurin series for $\exp(x)$.[3] In terms of the series, the exponential is

$$e^A = \sum_{k=0}^{\infty} \frac{A^k}{k!} = I + \frac{A}{1!} + \frac{A^2}{2!} + \cdots. \qquad \textbf{(4.43)}$$

This expression for e^A in terms of an infinite number of sums of matrix products looks even more formidable than the expression for a matrix polynomial in Equation 4.42. Fortunately, we can also compute e^A, at least in principle, using the Cayley-Hamilton theorem. Before presenting that powerful theorem, we give a few examples of matrix functions using MATLAB.

MATLAB
MATRIX
FUNCTIONS

Table 4.2 lists the MATLAB commands that compute matrix functions. Only the symbolic command **int** operates element by element. The others use various algorithms to determine functions of a matrix.

TABLE 4.2 *MATLAB matrix functions*

Operation	MATLAB	Result
Numerical	**expm**	Matrix exponential of A
functions	**funm**	Functions of a matrix
	logm	Inverse function of **expm**
	sqrtm	Square root of a matrix
Symbolic	**int**	Element-by-element integration of a symbolic matrix
	sympower	Computes A^n

☐ EXAMPLE 4.11 *Matrix Calculus*

Consider the diagonal matrix

$$A = \begin{bmatrix} 1 & 0 & 0 \\ 0 & 2 & 0 \\ 0 & 0 & 3 \end{bmatrix}.$$

It is desired to compute various functions such as A^k and e^A. The accompanying MATLAB script computes the result for A^2, A^5. Then, the **expm** command is used to compute e^A. The script also shows the use of the symbolic command **sympower** to compute the square of the symbolic matrix

$$A\text{sym} = \begin{bmatrix} a & b \\ c & d \end{bmatrix}.$$

---

[3] Taylor and Maclaurin series for functions are treated in Chapter 6.

Example 4.11

```
%EX4_11.M Example to show matrix operations (A^2, A^5, exp(A))
A=[1 0 0;0 2 0;0 0 3]
Asq=A^2         % Matrix power A*A
Afifth=A^5      % A*A*A*A*A
expA=expm(A) % Exponential expm(A)
%
% Symbolic calculation of A*A
Asym=sym('a, b ;c, d')
k=2;
Asymsq = sympow(Asym,k)
%----------------
% The MATLAB results are as follows:
%
A =
     1     0     0
     0     2     0
     0     0     3
Asq =
     1     0     0
     0     4     0
     0     0     9
Afifth =
     1     0     0
     0    32     0
     0     0   243
expA =
    2.7183         0         0
         0    7.3891         0
         0         0   20.0855
Asym =
[a, b ]
[c, d ]

Asymsq =
[a^2+b*c, a*b+b*d]
[c*a+d*c, b*c+d^2]
```

□

From these examples, it is reasonable to conclude that if Λ is an $n \times n$ diagonal matrix,

$$\Lambda = \begin{bmatrix} \lambda_1 & 0 & \cdots & 0 \\ 0 & \lambda_2 & \cdots & 0 \\ \vdots & \vdots & \ddots & \vdots \\ 0 & 0 & \cdots & \lambda_n \end{bmatrix},$$

then

$$\Lambda^m = \begin{bmatrix} \lambda_1^m & 0 & \cdots & 0 \\ 0 & \lambda_2^m & \cdots & 0 \\ \vdots & \vdots & \ddots & \vdots \\ 0 & 0 & \cdots & \lambda_n^m \end{bmatrix}. \tag{4.44}$$

A specific example of this result can be demonstrated by letting $b = c = 0$ in the 2×2 symbolic matrix of Example 4.11. Also,

$$e^\Lambda = \begin{bmatrix} e^{\lambda_1} & 0 & \cdots & 0 \\ 0 & e^{\lambda_2} & \cdots & 0 \\ \vdots & \vdots & \ddots & \vdots \\ 0 & 0 & \cdots & e^{\lambda_n} \end{bmatrix}. \tag{4.45}$$

We shall see that diagonalizing a matrix is very useful for certain computations.

WHAT IF? First, define a 3×3 identity matrix and compare the results of the MATLAB commands **exp** and **expm**. Also, it is instructive to read the MATLAB M-files that can be used to compute the matrix exponential. Compare the methods by reading the *MATLAB User's Guide* and the M-files that apply.

CAYLEY-
HAMILTON
THEOREM

The amazing Cayley-Hamilton theorem states that a matrix satisfies its own characteristic equation. This powerful theorem forms the basis for a method to compute functions of matrices. We present examples that show use of the theorem for computation of polynomials and exponentials of a matrix.

■ **THEOREM 4.4**   *Cayley-Hamilton theorem*

     Every square matrix A satisfies its own characteristic equation $|A - \lambda I| = 0$, so that if

$$\lambda^n + a_{n-1}\lambda^{n-1} + \cdots + a_1\lambda + a_0 = 0,$$

then the matrix equation is

$$A^n + a_{n-1}A^{n-1} + \cdots + a_1 A + a_0 = 0.$$

                                                                 ■

When the positive form of the eigenvalues is used in the characteristic equation rather than the form shown in Equation 4.15, the product of the eigenvalues is $(-1)^n a_0$, and

$$\det(A) = (-1)^n a_0.$$

□ EXAMPLE 4.12   *Cayley-Hamilton Theorem*

We can explore many properties of matrices using a general 2×2 matrix

$$A = \begin{bmatrix} a & b \\ c & d \end{bmatrix}. \tag{4.46}$$

Although we show the Cayley-Hamilton result for a 2×2 matrix, the result can be generalized to larger matrices.

According to the Cayley-Hamilton theorem, the 2×2 matrix of Equation 4.46 satisfies its characteristic equation. Thus, forming $|A - \lambda I| = 0$ yields

$$\lambda^2 - (a + d)\lambda + (ad - bc) = 0.$$

Substituting A for λ results in the matrix equation

$$A^2 - (a + d)A + (ad - bc)I,$$

which becomes

$$\begin{bmatrix} a & b \\ c & d \end{bmatrix}^2 - \begin{bmatrix} a^2 + ad & ba + bd \\ ca + cd & ad + d^2 \end{bmatrix} + \begin{bmatrix} ad - bc & 0 \\ 0 & ad - bc \end{bmatrix}. \tag{4.47}$$

Computing A^2 and adding the matrices in Equation 4.47 shows that the sum of the matrices is the 2×2 zero matrix, so A satisfies its own characteristic equation with A substituted for λ.

□

There is another result that is very useful to compute functions $f(A)$ of a matrix. It is based on a theorem of algebra that states if $f(\lambda)$ and $d(\lambda)$ are polynomials, there exist polynomials $q(\lambda)$ and $r(\lambda)$ such that

$$f(\lambda) = d(\lambda)q(\lambda) + r(\lambda),$$

where $r(\lambda)$ is called the remainder and is of degree $n - 1$. In this equation, either $r(\lambda) = 0$ or the degree of $r(\lambda)$ is less than $r(\lambda) = 0$. If $r(\lambda) = 0$, the polynomials $d(\lambda)$ and $q(\lambda)$ are the factors of $f(\lambda)$. Extending the argument to polynomials of a matrix, it is the case that

$$f(A) = d(A)q(A) + r(A), \tag{4.48}$$

which follows because A commutes with itself and two polynomials in A can be multiplied in the same manner as two polynomials in the real variable x.

Suppose that $d(A)$ in Equation 4.48 is the characteristic equation $\det(A - \lambda I)$. Then, $f(A) = r(A)$ by the Cayley-Hamilton theorem. The important conclusion is that *any* polynomial in A can be written as a polynomial of degree $n - 1$.

In fact, the results for polynomials can be generalized and used to compute arbitrary functions of a matrix. In the case A has *distinct* eigenvalues, the function of a matrix $f(A)$ can be expressed in a series up to A^{n-1}, since

$$f(A) = r(A) = \beta_0 I + \beta_1 A + \cdots + \beta_{n-1}A^{n-1}.$$

Thus, for an $n \times n$ matrix,

$$f(A) = \sum_{m=0}^{n-1} \beta_m A^m, \tag{4.49}$$

where the $\beta_m, m = 1, \ldots, n-1$ are constants. To compute the constants, the formula is

$$f(\lambda) = \sum_{m=0}^{n-1} \beta_m \lambda^m \tag{4.50}$$

for each of the n distinct eigenvalues of the matrix.

Problem 4.15 outlines the procedure when the eigenvalues are repeated. The text by Bronson listed in the Annotated Bibliography for this chapter treats cases in which the eigenvalues of the matrix are not distinct.

Polynomials of Matrices From Equation 4.49, we conclude that for a 2×2 matrix,

$$A^k = \beta_1 A + \beta_0 I,$$

no matter what the value of the exponent k. For the $n \times n$ matrix, the equations can be written as an $n \times n$ system of equations in the form

$$
\begin{aligned}
f(\lambda_1) &= \beta_{n-1}\lambda_1^{n-1} + \cdots + \beta_1 \lambda_1 + \beta_0, \\
f(\lambda_2) &= \beta_{n-1}\lambda_2^{n-1} + \cdots + \beta_1 \lambda_1 + \beta_0, \\
&\vdots \\
f(\lambda_n) &= \beta_{n-1}\lambda_n^{n-1} + \cdots + \beta_1 \lambda_1 + \beta_0.
\end{aligned} \tag{4.51}
$$

☐ EXAMPLE 4.13   ***Matrix Polynomial***

Consider the matrix of Example 4.11,

$$A = \begin{bmatrix} 1 & 0 & 0 \\ 0 & 2 & 0 \\ 0 & 0 & 3 \end{bmatrix},$$

for which we wish to compute A^5 using the approach of Equation 4.49 and Equation 4.51. Since A is 3×3, the matrix formulation that corresponds to Equation 4.49 is

$$A^5 = \beta_2 A^2 + \beta_1 A + \beta_0 I,$$

and the linear system of Equation 4.51 defining the constants becomes

$$
\begin{aligned}
1^5 &= \beta_2(1)^2 + \beta_1(1) + \beta_0, \\
2^5 &= \beta_2(2)^2 + \beta_1(2) + \beta_0, \\
3^5 &= \beta_2(3)^2 + \beta_1(3) + \beta_0.
\end{aligned}
$$

Solving for $\beta = [\beta_2, \beta_1, \beta_0]^T$ (using MATLAB), we find that

$$\beta = \begin{bmatrix} 90 \\ -239 \\ 150 \end{bmatrix}$$

so the result is

$$
\begin{aligned}
A^5 &= 90A^2 - 239A + 150 \\
&= 90 \begin{bmatrix} 1 & 0 & 0 \\ 0 & 4 & 0 \\ 0 & 0 & 9 \end{bmatrix} - 239 \begin{bmatrix} 1 & 0 & 0 \\ 0 & 2 & 0 \\ 0 & 0 & 3 \end{bmatrix} + 150 \begin{bmatrix} 1 & 0 & 0 \\ 0 & 1 & 0 \\ 0 & 0 & 1 \end{bmatrix} \\
&= \begin{bmatrix} 1 & 0 & 0 \\ 0 & 32 & 0 \\ 0 & 0 & 243 \end{bmatrix}.
\end{aligned}
$$

The MATLAB M-file EX4_13.M on our disk contains the script to compute these results.

$\square$

Exponential   Other functions of a matrix can be calculated using Equation 4.49, including the exponential of a matrix. The function e^A has a number of useful properties, such as

$$e^0 = I,$$

where 0 is the zero matrix. Also, the inverse is

$$(e^A)^{-1} = e^{-A}.$$

The matrix e^{At} is very important in the study of certain differential equations. This matrix has the property

$$\frac{de^{At}}{dt} = Ae^{At} = e^{At}A \tag{4.52}$$

if A is a constant matrix. This property and the other properties of e^{At} presented are easily proven by substituting the series form of the exponential from Equation 4.43 for e^{At} in the equations defining the properties. We treat the exponential of a matrix further in this chapter after similar and diagonalizable matrices are discussed.

SIMILAR AND DIAGONALIZABLE MATRICES

Matrices that are *diagonalizable* are of particular interest, since matrix functions of diagonal matrices are easily computed. The approach is to find a transformation that converts a matrix to diagonal form and preserves the values of the eigenvalues of the matrix.

The important results presented in this section are the definition of a diagonalizable matrix in terms of similar matrices and an explanation of the method to find the transformation that diagonalizes a matrix. For a matrix that has a full set of independent eigenvectors, the transformation is easily defined. The *similarity transformation* accomplishes the desired result when such a transformation exists.

SIMILAR MATRICES

Two $n \times n$ matrices A and B are said to be *similar* if there exists an invertible $n \times n$ matrix P such that

$$A = P^{-1}BP. \tag{4.53}$$

It is straightforward to prove that similar matrices have the same characteristic equation and, therefore, the same eigenvalues, as you are asked to show in Problem 4.18. The function defined in Equation 4.53 that takes the matrix B into the matrix A is called a *similarity transformation*.

DIAGONALIZABLE MATRICES

A matrix is *diagonalizable* if it is similar to a diagonal matrix. This requires that the matrix have a full set of linearly independent eigenvectors but the eigenvalues are not necessarily distinct.

■ **THEOREM 4.5**   *Independent eigenvectors*

An $n \times n$ matrix A is diagonalizable if and only if it possesses n linearly independent eigenvectors.

―――――――――――――――――――――――■

When A has n linearly independent eigenvectors $\mathbf{x}_1, \mathbf{x}_2, \ldots \mathbf{x}_n$, we can form the *modal* matrix

$$M = [\mathbf{x}_1 \, \mathbf{x}_2 \, \ldots \, \mathbf{x}_n]$$

in which the columns of M are the eigenvectors of A. We also define the diagonal matrix for which the eigenvalue λ_j in position $D(j,j)$ corresponds to the eigenvector $\mathbf{x}_j$ in M, with the form

$$D = \begin{bmatrix} \lambda_1 & 0 & \cdots & 0 \\ 0 & \lambda_2 & \cdots & 0 \\ \vdots & \vdots & \ddots & \vdots \\ 0 & 0 & \cdots & \lambda_n \end{bmatrix}.$$

This matrix is sometimes called the *spectral* matrix for A.

Multiplying A and M written in terms of the columns yields

$$
\begin{aligned}
AM &= A\left[\mathbf{x}_1\,\mathbf{x}_2\,\ldots\,\mathbf{x}_n\right]\\
&= \left[A\mathbf{x}_1\,A\mathbf{x}_2\,\ldots\,A\mathbf{x}_n\right]\\
&= \left[\lambda_1\mathbf{x}_1\,\lambda_2\mathbf{x}_2\,\ldots\,\lambda_n\mathbf{x}_n\right]\\
&= \left[\mathbf{x}_1\,\mathbf{x}_2\,\ldots\,\mathbf{x}_n\right]D\\
&= MD. \tag{4.54}
\end{aligned}
$$

Since the columns of M are linearly independent, the rank of M is n, the determinant of M is nonzero, and M^{-1} exists. The interested reader should work out the steps of Equation 4.54 in detail.

Premultiplying Equation 4.54 by M^{-1} yields the result

$$
D = M^{-1}AM,
$$

which shows that D is similar to A. Letting $P = M^{-1}$, it follows that

$$
A = P^{-1}DP = MDM^{-1}
$$

and A is similar to D. Thus, we conclude that the modal matrix and its inverse are used to diagonalize A when A has n linearly independent eigenvectors.[4]

Functions of a Diagonalizable Matrices   Computation of matrix functions is simplified when the matrix is diagonalizable. For example, assume that

$$
A = MDM^{-1}.
$$

Then, powers of A are easily computed, since

$$
\begin{aligned}
A^2 = AA &= (MDM^{-1})(MDM^{-1})\\
&= (MD)(M^{-1}M)(DM^{-1})\\
&= MD(I)DM^{-1} = MD^2M^{-1}.
\end{aligned}
$$

Continuing in this manner, we find that $A^3 = MD^3M^{-1}$, and, in general,

$$
A^m = MD^mM^{-1}. \tag{4.55}
$$

Equation 4.44 shows that it is easy to compute the powers of the diagonal matrix D.

Applying the result of Equation 4.55 to the exponential of a diagonalizable matrix yields the useful representation

$$
\begin{aligned}
e^A &= \sum_{k=0}^{\infty}\frac{A^k}{k!} = \sum_{k=0}^{\infty}\frac{1}{k!}(MD^kM^{-1}) = M\left(\sum_{k=0}^{\infty}\frac{D^k}{k!}\right)M^{-1}\\
&= Me^DM^{-1}. \tag{4.56}
\end{aligned}
$$

---

[4]When the matrix is not diagonalizable, it is still possible to show the matrix is similar to a simpler matrix that is "almost" diagonal. For a discussion, see the treatment of Jordan matrices in several of the references in the Annotated Bibliography for this chapter.

Chapter 4 ■ EIGENVALUES AND EIGENVECTORS

☐ **EXAMPLE 4.14**   *Matrix Functions*

To apply the techniques for diagonalization of a matrix, we compute e^A for the matrix

$$A = \begin{bmatrix} 4 & 1 \\ 3 & 2 \end{bmatrix}.$$

The eigenvalues are $\lambda_1 = 1$ and $\lambda_2 = 5$, with corresponding eigenvectors

$$\mathbf{x}_1 = \begin{bmatrix} 1 \\ -3 \end{bmatrix} \quad \text{and} \quad \mathbf{x}_2 = \begin{bmatrix} 1 \\ 1 \end{bmatrix}.$$

The modal matrix is thus

$$M = \begin{bmatrix} 1 & 1 \\ -3 & 1 \end{bmatrix}$$

or this matrix with the columns reversed. Using M as written,

$$M^{-1} = \frac{1}{4} \begin{bmatrix} 1 & -1 \\ 3 & 1 \end{bmatrix}.$$

Performing a similarity transform verifies that A is diagonalized with the eigenvalues on the diagonal, since

$$\Lambda = M^{-1}AM = \begin{bmatrix} 1 & 0 \\ 0 & 5 \end{bmatrix}.$$

According to Equation 4.56,

$$e^A = Me^\Lambda M^{-1},$$

and

$$e^\Lambda = \begin{bmatrix} e^1 & 0 \\ 0 & e^5 \end{bmatrix}$$

using Equation 4.45. Combining these results,

$$\begin{aligned} e^A &= \begin{bmatrix} 1 & 1 \\ -3 & 1 \end{bmatrix} \begin{bmatrix} e^1 & 0 \\ 0 & e^5 \end{bmatrix} \frac{1}{4} \begin{bmatrix} 1 & -1 \\ 3 & 1 \end{bmatrix} \\ &= \frac{1}{4} \begin{bmatrix} e + 3e^5 & -e + e^5 \\ -3e + 3e^5 & 3e + e^5 \end{bmatrix}. \end{aligned} \tag{4.57}$$

This result will be used in Example 4.17 to solve a system of differential equations.

☐

SPECIAL MATRICES AND THEIR EIGENVALUES (OPTIONAL)

In applications, there are a number of special types of matrices that occur frequently when modeling physical systems. Formulating a problem in terms of these matrices generally simplifies the mathematical analysis. A study of their eigenvalues and eigenvectors may also lead to a better physical understanding of the problem. The particular matrices presented here are symmetric matrices, orthogonal matrices, Hermitian matrices, and unitary matrices.

REAL SYMMETRIC MATRICES

A *real symmetric* matrix $(A = A^T)$ has only real numbers as elements. Since such matrices arise so often from the solution of physical problems, such as the resistive network in Example 4.4, the properties of the eigenvalues and eigenvectors of symmetric matrices are of interest.

Many real matrices have complex eigenvalues, such as the rotation matrix previously discussed. However, in the case of a real symmetric matrix, the eigenvalues and corresponding eigenvectors are *real*, as Problem 4.14 asks you to prove.

The eigenvalues and eigenvectors of an $n \times n$ real symmetric matrix A satisfy the following properties:

1. The eigenvalues and eigenvectors of A are real.

2. Eigenvectors corresponding to distinct eigenvalues are orthogonal.

From the last property, we conclude that if A is a real symmetric matrix with n distinct eigenvalues, then the n corresponding eigenvectors are orthogonal and they form a basis for $\mathbf{R}^n$. These comments apply also to real diagonal matrices, which are obviously symmetric with zero off-diagonal elements.

Repeated Eigenvalues It can be proven that an $n \times n$ symmetric matrix has a complete set of n orthonormal eigenvectors. Since the orthonormal eigenvectors are also linearly independent, any symmetric matrix is diagonalizable. The eigenvectors belonging to repeated eigenvalues are linearly independent for symmetric matrices, so they may be used to create an orthonormal set using the Gram-Schmidt orthonormalization process described in Chapter 2. As an extreme case, think of the $n \times n$ identity matrix with n eigenvalues of 1 but any number of eigenvectors. Simply choose n orthonormal examples for the eigenvectors.

ORTHOGONAL
MATRICES

If the matrix A is real and symmetric, its eigenvalues are real and its eigenvectors are orthogonal. These eigenvectors can be constructed to be orthonormal. Then, the modal matrix would contain orthonormal vectors as its columns. This leads to the conclusion that any real symmetric matrix can be diagonalized and the diagonalizing matrix can be an orthogonal matrix Q. Recalling the definition from Chapter 3, the orthogonal matrix has orthonormal columns and $Q^T = Q^{-1}$.

■ THEOREM 4.6

Orthogonal diagonalization

For every $n \times n$ real symmetric matrix A, there exists an $n \times n$ real orthogonal matrix Q such that

$$Q^{-1}AQ = Q^T AQ = \Lambda, \qquad (4.58)$$

where Λ is a diagonal matrix.

When the relationship of Equation 4.58 holds, the matrix A is said to be *orthogonally diagonalizable*. This transformation sometimes is called the *principal axis theorem* in geometry or mechanics. If the eigenvectors define the axes, the axes are perpendicular. For example, in the study of deformable bodies in mechanics, the eigenvectors give the principal directions along which there is simply pure compression or tension. In other directions, there is shear. An advanced but complete description of the principal axis transformation is given in Goldstein's text listed in the Annotated Bibliography for this chapter.

☐ EXAMPLE 4.15

Orthogonally Diagonalizable Matrix

We wish to find the orthogonal modal matrix that transforms the symmetric matrix

$$A = \begin{bmatrix} 3 & 1 \\ 1 & 3 \end{bmatrix}$$

into diagonal form. The eigenvalues are $\lambda_1 = 2$ and $\lambda_2 = 4$, and the orthonormal eigenvectors are

$$\mathbf{x}_1 = \begin{bmatrix} \dfrac{1}{\sqrt{2}} \\ \dfrac{-1}{\sqrt{2}} \end{bmatrix} \quad \text{and} \quad \mathbf{x}_2 = \begin{bmatrix} \dfrac{1}{\sqrt{2}} \\ \dfrac{1}{\sqrt{2}} \end{bmatrix}$$

To verify that the transformation of Equation 4.58 diagonalizes A, choose Q as the modal matrix

$$M = \begin{bmatrix} \dfrac{1}{\sqrt{2}} & \dfrac{1}{\sqrt{2}} \\ \dfrac{-1}{\sqrt{2}} & \dfrac{1}{\sqrt{2}} \end{bmatrix}.$$

This matrix is clearly orthogonal, and $M^{-1} = M^T$.

Forming the product $M^{-1}AM$, we find

$$M^{-1}AM = \begin{bmatrix} \dfrac{1}{\sqrt{2}} & \dfrac{-1}{\sqrt{2}} \\ \dfrac{1}{\sqrt{2}} & \dfrac{1}{\sqrt{2}} \end{bmatrix} \begin{bmatrix} 3 & 1 \\ 1 & 3 \end{bmatrix} \begin{bmatrix} \dfrac{1}{\sqrt{2}} & \dfrac{1}{\sqrt{2}} \\ \dfrac{-1}{\sqrt{2}} & \dfrac{1}{\sqrt{2}} \end{bmatrix}$$

$$= \begin{bmatrix} 2 & 0 \\ 0 & 4 \end{bmatrix}.$$

□

HERMITIAN AND UNITARY MATRICES

The matrix with complex elements that plays the role of a symmetrical matrix is called *Hermitian*. These matrices are not equal to their transpose but to their *conjugate transpose*. A Hermitian matrix is a square $n \times n$ matrix A with the property

$$A^H = A.$$

Thus, for each element of A, $a_{ij} = \bar{a}_{ji}$. This result also implies that the diagonal entries must be real.

A very interesting property of a Hermitian matrix is that it has real eigenvalues. The eigenvectors corresponding to different eigenvalues are orthogonal to one another. Also, the Hermitian matrix is diagonalizable.

The complex equivalent of an orthogonal matrix is the *unitary* matrix U with the property

$$U^H = U^{-1}.$$

The eigenvectors of U are orthogonal, and they can be scaled to unit length to be orthonormal. The diagonalizing transform for a Hermitian matrix A is

$$\Lambda = U^{-1}AU = U^H AU,$$

where Λ is a *real* diagonal matrix.

□ EXAMPLE 4.16   *Hermitian Matrix*
An example Hermitian matrix is

$$\begin{bmatrix} 1 & 3-i & 5i \\ 3+i & 3 & -3-2i \\ -5i & -3+2i & 2 \end{bmatrix}.$$

□

Hermitian operators in physics. Any operator in physics that corresponds to a physically observable property is Hermitian. These operators, which can often be represented by a matrix A, are the most important operators in quantum mechanics. In fact, it is a basic postulate of quantum mechanics that each type of physically observable property, such as momentum, position, and energy, is associated with a Hermitian operator. A measurement of an observable quantity must result in a number that is an eigenvalue of the associated operator.

Another interesting fact is that two operators $\hat{A}$ and $\hat{B}$ with the same eigenfunctions commute, so that $\hat{A}\hat{B} = \hat{B}\hat{A}$. In terms of matrices, we state that *two matrices commute if they have the same eigenvectors*. Measurements associated with commuting operators can be determined with arbitrary accuracy without the effect of one measurement influencing the other. On the other hand, physical variables with noncommuting operators cannot be measured simultaneously with arbitrary accuracy.

For those readers who are familiar with Heisenberg's uncertainty principle, we say that values associated with noncommuting operators are subject to its limitation. For example, the position and momentum operators in quantum mechanics do not commute. The Heisenberg uncertainty principle leads to the conclusion that if Δx is the uncertainty in a measurement of position of a particle and Δp is the uncertainty in momentum, the product of the uncertainties is

$$\Delta x \Delta p \geq h/2\pi$$

where h is Planck's constant, with the tiny value

$$h = 6.62377 \times 10^{-34} \quad \text{joule-second.}$$

SUMMARY
Table 4.3 summarizes the definitions and properties of the eigenvalues for real and complex matrices with special structure. It is interesting that real symmetric matrices and Hermitian matrices both have real eigenvalues and orthonormal eigenvectors from distinct eigenvalues. Also, any symmetric or Hermitian matrix has a complete set of orthonormal eigenvectors, whether or not the eigenvalues are distinct.

TABLE 4.3 *Properties of special matrices*

Type	Definition	Eigenvalues
Real Matrices:		
Symmetric	$A^T = A$	Real eigenvalues
Skew	$A^T = -A$	Imaginary (or zero) eigenvalues
Orthogonal	$Q^T = Q^{-1}$	All $|\lambda_i| = 1$
Complex Matrices:		
Hermitian	$A^H = A$	Real eigenvalues
Skew Hermitian	$A^H = -A$	Imaginary (or zero) eigenvalues
Unitary	$U^H = U^{-1}$	All $|\lambda_i| = 1$

Symmetric and Hermitian matrices can be diagonalized as follows:

$$Q^{-1}A_S Q = \Lambda,$$
$$U^{-1}A_H U = \Lambda,$$

where an orthogonal matrix Q is used to diagonalize the symmetric matrix A_S and a unitary matrix U is used to diagonalize the Hermitian matrix A_H. The columns of Q or U contain a complete set of orthonormal eigenvectors.

APPLICATIONS TO DIFFERENTIAL EQUATIONS

The idea of writing a system of linear algebraic equations in matrix form can be extended to certain systems of linear differential equations. We give examples in this section and continue the discussion of techniques to solve systems of differential equations in Chapter 5.

Consider the system of differential equations

$$\frac{d\mathbf{x}(t)}{dt} = A\mathbf{x}, \tag{4.59}$$

where A is a real constant matrix. The system has the solution

$$\mathbf{x}(t) = e^{At}\mathbf{x}_0, \tag{4.60}$$

with the value at $t = 0$ of $\mathbf{x}(0) = \mathbf{x}_0$. To prove this result is correct, we expand the exponential according to the definition of Equation 4.43,

$$\mathbf{x} = e^{At}\mathbf{x}_0 = \left[I + At + A^2\frac{t^2}{2!} + \cdots \right]\mathbf{x}_0, \tag{4.61}$$

and show that the derivative of the series can be written as $Ae^{At}\mathbf{x}_0$. Since A is a constant matrix,

$$\begin{aligned}
\frac{d}{dt}A^k\frac{t^k}{k!} &= \frac{d}{dt}\frac{t^k}{k!}A^k = \frac{kt^{k-1}}{k!}A^k \\
&= \frac{A^k t^{k-1}}{(k-1)!} = A\left[A^{k-1}\frac{t^{k-1}}{(k-1)!}\right].
\end{aligned} \tag{4.62}$$

Combining these results, we obtain

$$\begin{aligned}
\frac{d\mathbf{x}(t)}{dt} &= \frac{d}{dt}e^{At}\mathbf{x}_0 \\
&= A\left[I + At + A^2\frac{t^2}{2!} + \cdots\right]\mathbf{x}_0 \\
&= Ae^{At}\mathbf{x}_0 = A\mathbf{x}(t).
\end{aligned}$$

Substituting $t = 0$ in the series yields the result

$$\mathbf{x}(0) = Ae^{A0}\mathbf{x}_0 = I\mathbf{x}_0 = \mathbf{x}_0.$$

☐ **EXAMPLE 4.17** **_Differential Equation Solution by Matrix_**

A differential equation system using the A matrix of Example 4.14

$$\frac{dx_1(t)}{dt} = 4x_1 + x_2,$$

$$\frac{dx_2(t)}{dt} = 3x_1 + 2x_2 \qquad\qquad \textbf{(4.63)}$$

can be written in the form

$$\frac{d\mathbf{x}(t)}{dt} = A\mathbf{x}$$

with

$$\mathbf{x} = \begin{bmatrix} x_1(t) \\ x_2(t) \end{bmatrix} \quad \text{and} \quad A = \begin{bmatrix} 4 & 1 \\ 3 & 2 \end{bmatrix}.$$

The solution is $\mathbf{x}(t) = e^{At}\mathbf{x}_0$, according to Equation 4.60. Applying the results of Example 4.14 and Equation 4.57 to evaluate e^B with the substitution $B = At$ yields

$$e^{At} = \frac{1}{4} \begin{bmatrix} e^t + 3e^{5t} & -e^t + e^{5t} \\ -3e^t + 3e^{5t} & 3e^t + e^{5t} \end{bmatrix}.$$

Letting the initial condition at $t = 0$ be $\mathbf{x}_0 = [x_{01}, x_{02}]^T$, the solution is

$$\mathbf{x} = \frac{1}{4} \begin{bmatrix} e^t + 3e^{5t} & -e^t + e^{5t} \\ -3e^t + 3e^{5t} & 3e^t + e^{5t} \end{bmatrix} \begin{bmatrix} x_{01} \\ x_{02} \end{bmatrix}.$$

Performing the matrix multiplication and combining terms, the solution becomes

$$x_1(t) = k_1 e^t + k_2 e^{5t},$$

$$x_2(t) = -3k_1 e^t + k_2 e^{5t},$$

where $k_1 = (x_{01} - x_{02})/4$ and $k_2 = (3x_{01} + x_{02})/4$.

Substituting the results in Equation 4.63 shows that these functions solve the equations and satisfy the conditions at $t = 0$. Thus, the eigenvalues and eigenvectors of A, as computed in Example 4.14, determine the form of the solution. It is a general result that the solutions of the $n \times n$ system of equations

$$\frac{d\mathbf{x}(t)}{dt} = A\mathbf{x}$$

are linear combinations of $e^{\lambda_1}, e^{\lambda_2}, \ldots, e^{\lambda_n}$ if A has distinct eigenvalues

$$\lambda_1, \lambda_2, \ldots, \lambda_n.$$

☐

☐ EXAMPLE 4.18   *System of Differential Equations*

Consider the two carts in Figure 4.4. We assume they move on wheels that are massless and frictionless. The carts are coupled by a shock absorber that acts as a damper to their motion. Since the motion is along one axis only, the velocities can be taken as scalar quantities $v_1(t)$ and $v_2(t)$ in the direction shown in the figure. Motion in the other direction is indicated by a negative value for the velocities.

For each cart, $F = ma$ by Newton's law, where m is a constant representing the cart's mass. In terms of velocity, $F = dv(t)/dt = m\dot{v}$.

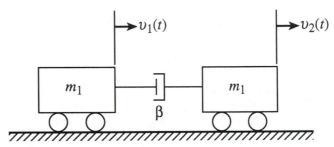

FIGURE 4.4 *Two carts coupled by a damper*

The damper force increases with velocity as $F = \beta v$ in a direction to oppose the motion. In this case, $v = v_2 - v_1$ according to the definition of the directions of the velocities in the figure. Thus,

$$
\begin{aligned}
m_1\dot{v}_1 &= \beta(v_2 - v_1), \\
m_2\dot{v}_2 &= -\beta(v_2 - v_1),
\end{aligned}
$$

which can be rewritten as the matrix equation

$$
\begin{bmatrix} \dot{v}_1 \\ \dot{v}_2 \end{bmatrix} =
\begin{bmatrix} -\dfrac{\beta}{m_1} & \dfrac{\beta}{m_1} \\ \dfrac{\beta}{m_2} & -\dfrac{\beta}{m_2} \end{bmatrix}
\begin{bmatrix} v_1 \\ v_2 \end{bmatrix}.
$$

This system has the form $\dot{\mathbf{v}} = A\mathbf{v}$, in which $\dot{\mathbf{v}}$ represents the vector whose components are derivatives. You should solve for the eigenvalues and eigenvectors of the system and try to interpret the results before looking at the M-file EX4_18.M on our disk. Chapter 5 presents more examples of such systems.

☐

REINFORCEMENT EXERCISES AND EXPLORATION PROBLEMS
---■---

In these problems, do the computations by hand unless otherwise indicated and then check those that yield numerical or symbolic results with MATLAB.

REINFORCEMENT EXERCISES

P4.1. **Eigenvalues** Consider the homogeneous differential equation

$$D^2[y(x)] + 5D[y(x)] + 6y(x) = 0,$$

where $D = d/dx$ is the derivative operator. The solution is of the form

$$y(x) = c_1 e^{\alpha x} + c_2 e^{\beta x}.$$

Show that the exponentials are *eigenvectors* for the differential equation and find the eigenvalues α and β .

P4.2. **Eigenvalues** If $Ax = \lambda x$ represents an eigenvalue equation, find the eigenvalue for the following matrix equation

$$\begin{bmatrix} 7 & -8 & -8 \\ 9 & -16 & -18 \\ -5 & 11 & 13 \end{bmatrix} \begin{bmatrix} 1 \\ 3 \\ -2 \end{bmatrix} = \lambda \begin{bmatrix} 1 \\ 3 \\ -2 \end{bmatrix}.$$

P4.3. **Eigenvalues** Consider the following matrix that depends on the parameter t

$$\begin{bmatrix} t & 2t \\ 2t & -t \end{bmatrix}.$$

Find the eigenvalues.

P4.4. **Eigenvectors** Solve the equation $Ax = \lambda x$ for the eigenvectors of the matrix

$$A = \begin{bmatrix} 3 & 1 \\ 2 & 2 \end{bmatrix}.$$

P4.5. **Eigenvalues** Find the eigenvalues and eigenvectors for the matrix

$$A = \begin{bmatrix} 2 & 1 & 1 \\ 2 & 3 & 2 \\ 1 & 1 & 2 \end{bmatrix}.$$

P4.6. **Eigenvalues** Find the eigenvalues and eigenvectors for the matrix

$$A = \begin{bmatrix} 2 & -12 \\ 8 & 0 \end{bmatrix}.$$

P4.7. **Symmetric matrix** The matrix

$$A = \begin{bmatrix} 3 & 0 & -2 \\ 0 & 2 & 0 \\ -2 & 0 & 0 \end{bmatrix}$$

is a *real*, *symmetric* matrix, since the elements a_{ij} are real and $a_{ij} = a_{ji}$.

    a. Find the eigenvalues and eigenvectors for the matrix.

    b. Form the matrix Q whose columns consist of the eigenvectors (normalized to be unit vectors) and show that $Q^{-1} = Q^T$.

    c. Show that Q is an *orthogonal* matrix in which the rows are orthonormal vectors $\in R^n$.

d. Form the product $D = Q^T A Q$ and show that D is a diagonal matrix.

P4.8. **Zero eigenvalue** Using the characteristic polynomial for a matrix A, prove that A is singular if and only if it has a zero eigenvalue.

P4.9. **Eigenvalues** Assume that A is an $n \times n$ matrix and let $\mathbf{x}$ be a vector in $\mathbf{R}^n$.

a. What vectors $\mathbf{x}$ are parallel to the image vector $A\mathbf{x}$?

b. Given the MATLAB statement `[V,D]=eig(A)`, what is the interpretation of `(A-D(k,k)*eye(n))*V(:,k)=0` for `k=1,2, ..., n`?

c. Using the trace and determinant of

$$A = \begin{pmatrix} 3 & -2 \\ 1 & 2 \end{pmatrix},$$

find the eigenvalues of A.

P4.10. Cayley-Hamilton theorem Using the matrix

$$A = \begin{bmatrix} 3 & -2 \\ 1 & 2 \end{bmatrix},$$

verify the Cayley-Hamilton theorem. Check the result using MATLAB commands.

EXPLORATION PROBLEMS

P4.11. Rotation matrix Given the rotation matrix

$$R_\theta = \begin{bmatrix} \cos \theta & -\sin \theta \\ \sin \theta & \cos \theta \end{bmatrix},$$

show the following:

a. The matrix has eigenvectors and corresponding eigenvalues

$$\lambda = e^{i\theta} : \begin{bmatrix} 1 \\ -i \end{bmatrix},$$

$$\lambda = e^{-i\theta} : \begin{bmatrix} 1 \\ i \end{bmatrix}.$$

b. The eigenvectors in Part a are orthogonal with respect to the complex inner product in $\mathbf{C}^2$.

P4.12. Diagonalizable matrices Support or criticize the assertion that *most matrices are diagonalizable*. Give reasonable arguments to support your position.

Hint: A little experimentation with matrices created with the MATLAB command **rand** may help you decide.

Chapter 4 ■ EIGENVALUES AND EIGENVECTORS

P4.13. Inverse of a matrix Considering the Caley-Hamilton theorem, show that the inverse of a nonsingular $n \times n$ matrix A can be written as a matrix polynomial of degree $n - 1$. Using this approach, determine the inverse of the matrix

$$\begin{bmatrix} 1 & -1 & 4 \\ 3 & 2 & 1 \\ 2 & 1 & -1 \end{bmatrix}.$$

Is this method efficient for computation?

P4.14. Eigenvalues of a symmetric matrix It is not obvious that the eigenvalues of an $n \times n$ real symmetric matrix are real, since the roots of the nth-order characteristic equation with real coefficients are, in general, complex numbers. Prove that the eigenvalues of a real symmetric matrix are real.

Hint: Assume that there are complex eigenvalues λ with complex conjugates λ^* and show that $\lambda = \lambda^*$ so that the eigenvalues are real.

P4.15. Matrix functions with repeated eigenvalues Consider that a function of a matrix can be written as

$$f(A) = p(A)q(A) + r(A)$$

from the equation $f(\lambda) = p(\lambda)q(\lambda) + r(\lambda)$, where $p(\lambda) = |A - \lambda I|$ is the characteristic equation. Since $p(\lambda) = 0$, it follows that

$$f(A) = r(A).$$

If $\lambda = \lambda_0$ is a double root of the characteristic equation $p(\lambda_0) = 0$, then it is also true that $p'(\lambda_0) = 0$. Thus, we can show that $f'(\lambda_0) = r'(\lambda_0)$. In general, for an eigenvalue λ_i of multiplicity k, the equations to determine the coefficients in the polynomial representation of a matrix function are

$$
\begin{aligned}
f(\lambda_i) &= r(\lambda_i), \\
\frac{df(\lambda_i)}{d\lambda} &= \frac{dr(\lambda_i)}{d\lambda}, \\
&\vdots \\
\frac{d^{(k-1)}f(\lambda_i)}{d\lambda^{(k-1)}} &= \frac{d^{(k-1)}r(\lambda_i)}{d\lambda^{(k-1)}}.
\end{aligned}
$$

Use this method to compute A^5 for the matrix

$$\begin{bmatrix} 4 & -1 \\ 1 & 2 \end{bmatrix}.$$

P4.16. Matrix function Compute A^{543} for the matrix

$$A = \begin{bmatrix} -3 & -4 \\ 2 & 3 \end{bmatrix}.$$

P4.17. Orthonormal set of vectors Find an orthonormal set of eigenvectors for the matrix

$$A = \begin{bmatrix} 9 & -2 & -2 & -4 & 0 \\ -2 & 11 & 0 & 2 & 0 \\ -2 & 0 & 7 & -2 & 0 \\ -4 & 2 & -2 & 9 & 0 \\ 0 & 0 & 0 & 0 & 3 \end{bmatrix}.$$

P4.18. Eigenvalues of similar matrices Show that similar matrices A and B have the same eigenvalues. Thus, you must show that

$$\det(A - \lambda I) = \det(B - \lambda I).$$

P4.19. Partitioned matrix Find the determinant and eigenvalues of the matrix in Example 4.10 by partitioning it into 2×2 matrices.

P4.20. Summing-up questions Answering these questions should require little or no calculation.

    a. Find the eigenvalues and eigenvectors for the 3×3 identity matrix. Generalize the result for an $n \times n$ identity matrix.

    b. Using a 2×2 example, show that two matrices with the same eigenvalues are not necessarily similar.

    c. Using a 2×2 example, show that a matrix similar to a diagonal matrix does not necessarily have distinct eigenvalues.

    d. Find the eigenvalues and eigenvectors for the $n \times n$ matrix such that $(a_{ij} = 1)$ for $i, j = 1, 2, \ldots n$.

ANNOTATED BIBLIOGRAPHY

1. Bronson, Richard, *Matrix Operations*, Schaum's Outline Series, McGraw Hill, Inc., New York, 1989. *As with other Outline Series volumes, the text contains many examples. In particular, algorithms and examples are given for the important numerical techniques of matrix algebra, such as the QR decomposition.*

2. Dahlquist, Germund and A. Björck, *Numerical Methods*, Prentice Hall, Inc., Englewood Cliffs, NJ, 1974. *A useful reference that covers important techniques in numerical computing.*

3. Datta, Biswa Nath, *Numerical Linear Algebra and Applications*, Brooks Cole Publishing Company, Pacific Grove, CA, 1995. *The text discusses many numerical algorithms including the calculation of eigenvalues and eigenvectors and potential errors that may arise.*

4. Goldstein, Herbert, *Classical Mechanics*, Addison Wesley, Reading, MA, 1950. *An excellent, but advanced, treatment of classical mechanics with detailed treatments of the use of matrices, eigenvalues, and eigenvectors.*

5. Hill, David R., *Experiments in Computational Matrix Algebra*, Random House, New York, 1988. *The text discusses the calculation of eigenvalues and eigenvectors using MATLAB commands. Various computational problems and errors are also analyzed.*

6. Uspensky, J. V., *Theory of Equations*, McGraw-Hill, Inc., New York, 1948. *This paperback text presents a complete treatment of polynomial equations, including approximate evaluation of the roots.*

ANSWERS

P4.1. Eigenvalues If D is the differential operator d/dt, then $D(e^{\lambda t}) = \lambda e^{\lambda t}$, so it follows that the function $e^{\lambda t}$ is an eigenvector of D associated with the eigenvalue λ. This eigenvector is a solution of the linear homogeneous differential equation $(D - \lambda)y(t) = 0$. For the homogeneous differential equation $D^2[y(x)] + 5D[y(x)] + 6y(x) = 0$, assume the solution is of the form $y(x) = e^{mx}$. Substituting in the differential equation leads to the characteristic equation $m^2 + 5m + 6 = 0$. This equation is factored as $(m + 3)(m + 2)$, so the roots are $m_1 = -3$ and $m_2 = -2$. The corresponding solutions are then

$$y_1(x) = e^{-3x} \quad \text{and} \quad y_2(x) = e^{-2x}.$$

Superimposing the solutions leads to the general solution $y(x) = c_1 e^{-3x} + c_2 e^{-2x}$.

P4.2. Eigenvalues Compute

$$\begin{bmatrix} 7 & -8 & -8 \\ 9 & -16 & -18 \\ -5 & 11 & 13 \end{bmatrix} \begin{bmatrix} 1 \\ 3 \\ -2 \end{bmatrix} = \begin{bmatrix} -1 \\ -3 \\ 2 \end{bmatrix} = (-1) \begin{bmatrix} 1 \\ 3 \\ -2 \end{bmatrix}.$$

Thus, $\mathbf{x} = [1, 3, -2]^T$ is the eigenvector associated with the eigenvalue $\lambda = -1$.

P4.3. Eigenvalues For the matrix

$$A = \begin{bmatrix} t & 2t \\ 2t & -t \end{bmatrix},$$

the characteristic equation $\det(A - \lambda I) = \lambda^2 - 5t^2 = 0$ yields the eigenvalues $\lambda_1 = \sqrt{5}t$ and $\lambda_2 = -\sqrt{5}t$.

P4.4. Eigenvectors Solving the characteristic equation $\lambda^2 - 4\lambda + 4 = 0$ shows that the eigenvalues are $\lambda_1 = 1$ and $\lambda_1 = 4$. The eigenvectors are $[1/\sqrt{2}, 1/\sqrt{2}]$ and $[-1, 2]$. The normalized eigenvectors (to 4 decimal places) are

$$\mathbf{x}_1 = \begin{bmatrix} 0.7071 \\ 0.7071 \end{bmatrix} \quad \text{and} \quad \mathbf{x}_1 = \begin{bmatrix} -0.4472 \\ 0.8944 \end{bmatrix}.$$

P4.5. Eigenvalues For the given matrix A, the characteristic equation to solve is $\lambda^3 - 7\lambda^2 + 11\lambda - 5 = 0$. MATLAB could solve this easily. However, a few observations will lead to a hand solution. The trace of A is $2 + 3 + 2 = 7$, which is the sum of the eigenvalues. Reducing the matrix and computing the determinant yields $\det(A) = 5$. This value is the product of the eigenvalues. Thus, if there is a rational root, it is a factor of 5. Try ± 1 and ± 5. We find that the eigenvalues are $1, 1, 5$.

P4.6. **Eigenvalues** The matrix

$$A = \begin{bmatrix} 2 & -12 \\ 8 & 0 \end{bmatrix}$$

has the characteristic equation $\lambda^2 - 2\lambda + 96$. The roots are complex, so we resort to MATLAB to find that the eigenvalues are $\lambda_1 = 0.5000 + 9.7852i$ and $\lambda_2 = 0.5000 - 9.7852i$. The eigenvectors are

$$\mathbf{x}_1 = \begin{bmatrix} -0.7736 + 0.0395i \\ 0.0 + 0.6325i \end{bmatrix} \quad \text{and} \quad \mathbf{x}_2 = \begin{bmatrix} -0.7736 - 0.0395i \\ 0.0 - 0.6325i \end{bmatrix}.$$

P4.7. **Symmetric matrix** For the matrix A, the eigenvalues are $2, -1, 4$. The matrix of eigenvectors is

$$Q = \begin{bmatrix} 0 & \dfrac{2}{\sqrt{5}} & \dfrac{1}{\sqrt{5}} \\ 1 & 0 & 0 \\ 0 & \dfrac{-1}{\sqrt{5}} & \dfrac{2}{\sqrt{5}} \end{bmatrix}.$$

The inverse is

$$Q^{-1} = Q^T = \begin{bmatrix} 0 & 1 & 0 \\ \dfrac{2}{\sqrt{5}} & 0 & \dfrac{-1}{\sqrt{5}} \\ \dfrac{1}{\sqrt{5}} & 0 & \dfrac{2}{\sqrt{5}} \end{bmatrix}.$$

Diagonalizing A yields

$$Q^{-1}AQ = \begin{bmatrix} 2 & 0 & 0 \\ 0 & 4 & 0 \\ 0 & 0 & -1 \end{bmatrix}.$$

P4.8. **Zero eigenvalue** As shown in the text, the determinant of the matrix A is the product of its eigenvalues. Since $|A| = 0$ if and only if A is singular, it follows that A is singular if and only if it has a zero eigenvalue.

P4.9. **Eigenvalues**

a. Two vectors $\mathbf{x}$ and $\mathbf{y}$ are parallel if one is a scalar multiple of the other. Thus, $\mathbf{x} = \lambda \mathbf{y}$. The solutions to $A\mathbf{x} = \lambda\mathbf{x}$ are thus parallel to the vector $A\mathbf{x}$.

b. D(k,k) is the eigenvalue and V(:,k) is the corresponding eigenvector and the equation is $\det(A - \lambda I)\mathbf{x} = 0$.

c. The trace and determinant yield the equations

$$\begin{aligned} \lambda_1 + \lambda_2 &= 5, \\ \lambda_1 \lambda_2 &= 6, \end{aligned}$$

with solutions $\lambda_1 = 2$ and $\lambda_2 = 3$.

P4.10. **Cayley-Hamilton theorem** The given matrix has the characteristic equation $\lambda^2 - 5\lambda + 8$. Using the Caley-Hamilton theorem leads to the result $A^2 - 5A + 8I = 0$, or

$$\begin{bmatrix} 3 & -2 \\ 1 & 2 \end{bmatrix}^2 - 5\begin{bmatrix} 3 & -2 \\ 1 & 2 \end{bmatrix} + 8\begin{bmatrix} 1 & 0 \\ 0 & 1 \end{bmatrix} = \begin{bmatrix} 0 & 0 \\ 0 & 0 \end{bmatrix}.$$

5

LINEAR DIFFERENTIAL EQUATIONS

PREVIEW_____

Since the eighteenth century, the general laws of the physical universe have often been described by *differential equations*. Such equations contain the derivatives of unknown functions that generally represent quantities of interest. The functions may represent force, pressure, temperature, electromagnetic potential, or similar quantities.

   For engineering and scientific purposes, differential equations are created as mathematical idealizations of reality to simplify complex phenomena encountered in the physical world. For example, experiments with a real object might be replaced with a computer simulation of the motion of the object subject to various forces. The simulation actually solves the equations of motion for the idealized object, perhaps represented by a mass point rather than a solid body.

   In practical cases, the differential equation problems are usually solved by approximation methods. Our approach in this chapter is first to present analytical solutions of simple equations to introduce the basic properties of differential equations. Then, computer solutions of various problems involving differential equations are introduced to illustrate the techniques involved.

After discussing the classification of differential equations, we study linear equations. We then consider linear, homogeneous differential equations. If a physical system is being modeled, the solutions represent the effects only of stored energy without the application of external stimuli, such as a force or voltage. We study nonhomogeneous systems next. These equations may model physical systems that are subject to external applied forces for mechanical systems or voltages in electrical circuits.

The final section treats the transformation of an nth-order differential equation to a system of first-order differential equations. This approach leads to a convenient formulation of the equations for solution using matrix techniques. Examples show the physical significance of the eigenvectors associated with the equations as the normal modes of vibration of a mechanical or electrical system.

Computer Computation This chapter presents the analytical solutions of a restricted class of differential equations whose solutions are easily found for most problems. However, we also wish to use MATLAB's routines, as listed in Table 5.1, to help find the solutions.

TABLE 5.1 *MATLAB commands for differential equations*

Command	Purpose
Symbolic:	
dsolve	Solve symbolic differential equations
ezplot	Plot results
simple	Simplify expressions
solve	Find roots of characteristic equation
subs	Substitute parameters
Numerical:	
ode23,	Solve differential equations
ode45	
roots	Find roots of characteristic equation

Although this chapter deals almost exclusively with linear ordinary differential equations with constant coefficients, other types of equations are also important. Chapter 6 treats more advanced types of ordinary differential equations. Partial differential equations are introduced in Chapter 13.

CLASSIFICATION OF DIFFERENTIAL EQUATIONS
---■---

An equation containing the derivatives of one or more dependent variables, with respect to one or more independent variables, is called a *differential equation*. Differential equations are broadly defined according to the following characteristics:

1. Type (ordinary or partial);

2. Order (defined by the highest derivative present);

3. Linearity (linear or nonlinear).

There may also be a number of additional specific properties of a differential equation that further categorizes it. These properties will be introduced when particular equations are discussed.

The classification of differential equations by *type* considers an equation to be an *ordinary* differential equations or a *partial* differential equations. The distinction is important in that the properties of the solutions, as well as the solution methods, may differ considerably for the two classes of equations.

ORDINARY
DIFFERENTIAL
EQUATION

An ordinary differential equation is any equation containing only a finite number of ordinary derivatives of one or more dependent variables with respect to one independent variable. For example, the equation

$$\frac{d^2 y(t)}{dt^2} + t \frac{dy(t)}{dt} + 5y(t) = e^t \tag{5.1}$$

involves the dependent variable $y(t)$, the independent variable t, and derivatives of $y(t)$ with respect to t. The highest derivative in the differential equation defines the *order* of the equation. Equation 5.1 is thus a second-order, ordinary differential equation. In addition, according to the definition of a linear function given in Chapter 3, this equation is a *linear differential equation*, since the dependent variable y and its derivatives are of the first degree. The coefficients can depend only on the independent variable or be constant.

In contrast to the linear equation, the equation

$$\frac{d^2y}{dt^2} + \sin y = 0 \tag{5.2}$$

is said to be nonlinear, since $\sin y$ is a not a linear function of y. Nonlinear equations are generally more difficult to deal with than linear equations, and nonlinear equations are not discussed in this chapter.

There are numerous applications of ordinary differential equations to physical problems. The differential equations of motion of a system relate position, velocity, and acceleration as functions of time. Many electric and electronic circuits obey laws described by differential equations relating currents or voltages and their time derivatives. Servomechanisms and other control systems can be modeled mathematically by ordinary differential equations that relate mechanical and electrical components of the system.

PARTIAL DIFFERENTIAL EQUATIONS

Partial differential equations involve the derivatives of at least one dependent variable but two or more independent variables. An example would be

$$\frac{\partial^2 z(x,y)}{\partial x^2} + \frac{\partial^2 z(x,y)}{\partial y^2} = 0, \tag{5.3}$$

which is a second-order partial differential equation. Problems involving continuous media, such as elastic bodies, gases, and fluids, lead to partial differential equations. The differential equations describing heat flow and electromagnetic radiation also involve partial derivatives. The fascinating properties of such partial differential equations are studied in Chapter 13.

SOLUTIONS AND INITIAL CONDITIONS

The main issue to be faced when we deal with a differential equation is the problem of finding its solutions. A function is a *solution* of a differential equation if when the function and its derivatives are substituted into the equation, the resulting expression is an identity. Thus, $y(t) = e^t$ is a solution of the equation

$$\frac{dy(t)}{dt} = y(t), \tag{5.4}$$

since

$$\frac{dy(t)}{dt} = e^t = y(t).$$

This function represents a specific solution to Equation 5.4. Notice that the function $y(t) = ce^t$ is also a solution where c is an arbitrary constant.

In algebra, an important theorem called the *fundamental theorem of algebra* asserts that a polynomial of the nth degree always has exactly n roots. The theory of differential equations answers questions about the existence and uniqueness of solutions rather than the number of solutions.

In general, a differential equation has a continuum of solutions but frequently we wish to find the solution to a differential equation that satisfies certain conditions.

The most important types of conditions are *initial conditions* and *boundary conditions*. Boundary conditions are defined in Chapter 6. In this chapter, we solve the *initial value problem*. Consider the differential equation

$$\frac{dy(t)}{dt} = f(t, y) \qquad (5.5)$$

subject to the condition

$$y(t_0) = y_0. \qquad (5.6)$$

This differential equation is a first-order ordinary differential equation in a more general form than that of Equation 5.4. Equation 5.5 with the initial condition of Equation 5.6 is called an initial value problem. For an arbitrary function $f(t, y)$, there is no simple way of solving the equation but various special cases must be considered.

If the function $f(t, y)$ and its partial derivative $\partial f(t, y)/\partial y$ is continuous in a rectangular region R of the ty-plane, then there exists a unique solution to the initial value problem in the region.[1] These are sufficient conditions for the existence of a solution and are useful since they are easy to apply. If the continuity conditions are not met, there may still be a unique solution to the problem.

The solution is defined as the function that satisfies the differential equation in the region R and passes through the point $y(t_0) = y_0$. Another way of stating the result is to say that the coordinates of any point (t_0, y_0) in R are the *initial values* for some solution of Equation 5.5.

Consider Equation 5.5 with $f(t, y) = y(t)$,

$$\frac{dy(t)}{dt} = y(t), \qquad (5.7)$$

which has a solution $y(t) = ce^t$. Let $y_0 = y(t_0)$ and write the initial condition value as $c = y_0 e^{-t_0}$. Then, the function

$$y(t) = ce^t = y_0 e^{(t-t_0)}$$

solves the differential equation and satisfies the initial condition. Since $f(t, y) = y(t)$ and its partial derivative $\partial f(t, y)/\partial y = 1$ are continuous functions of t and y in the entire ty-plane, the region in which $f(t, y)$ is defined is the entire plane. This region where the function is defined is called the *domain of definition*, or simply the *domain*, of $f(t, y)$.

---

[1] Partial derivatives are discussed in Chapter 10.

NOTATION

Primes or dots are often used to denote differentiation. Usually, the prime is used as $f'(x) \equiv df/dx$ with independent variables x, y, or z. The dot is used as $\dot{f} \equiv df/dt$ with the independent variable t. The notation $y^{(n)}$ means the nth derivative of y with respect to the independent variable. However, these conventions are not always applied, and the meaning of the functions and variables is normally assigned when a specific problem is solved.

The unknown functions and independent variables do not necessarily represent physical quantities in mathematical discussions, but they might be assigned meaning in a problem, such as letting t represent time. In some problems, it might be convenient to designate $x(t)$ as the unknown position of a body as a function of time.

The argument x of the function $f(x)$ or t of $f(t)$ is frequently omitted when no confusion would result. When a quantity is a function of two or more variables, the variables are usually explicitly stated. Thus, we write $f(x, y)$ for a function of both x and y.

LINEAR DIFFERENTIAL EQUATIONS

An ordinary linear differential equation of order n is an equation in the form

$$\frac{d^n y(t)}{dt^n} + a_{n-1}(t)\frac{d^{n-1}y(t)}{dt^{n-1}} + \cdots + a_1(t)\frac{dy(t)}{dt} + a_0(t)y(t) = f(t). \quad \textbf{(5.8)}$$

In our discussions, the coefficients $a_{n-1}(t), a_{n-2}(t), \ldots, a_0(t)$, $y(t)$, and the function $f(t)$ are assumed to be defined and continuous in some interval $a \leq t \leq b$ of the t-axis. Unless explicitly stated to the contrary, the term *differential equation* used in this chapter refers to this differential equation and its variations.

The basic questions about Equation 5.8 concern the number and type of solutions. The questions can be answered by theorems of existence and uniqueness, which will not be proved here, but only stated. Several references listed in the Annotated Bibliography at the end of this chapter present the rigorous mathematical details.

UNIQUENESS OF SOLUTION

Suppose there is a solution to the differential equation of Equation 5.8 that satisfies the initial conditions for the solution $y(t)$ and the first $n-1$ derivatives of the solution at some point in the domain of $f(t)$. Then, that solution is the unique solution.

Unique solution of linear differential equation
Let the coefficients in Equation 5.8,

$$a_i(t), i = 0, 1, \ldots, n-1 \quad and \quad f(t),$$

be continuous in some common interval containing the point t_0. If $y(t)$ is found that satisfies the equation and the initial conditions for

$$y(t_0), \dot{y}(t_0), \ldots, y^{(n-1)}(t_0),$$

the resulting solution is unique. ─────────────── ■

Finding solutions to ordinary differential equations is the main topic of this chapter and parts of Chapter 6.

If $f(t) = 0$, Equation 5.8 is said to be *homogeneous* with respect to y and its derivatives. Otherwise, when $f(t) \neq 0$, the equation is *nonhomogeneous*. Some solution methods require that $f(t)$ be set to zero to solve the homogeneous equation as the first step in solving the complete, nonhomogeneous equation.

The coefficients in Equation 5.8 are functions of t as written and the equation is said to have *nonconstant* or *variable coefficients*. When the coefficients are constant, the equation is a linear differential equation with *constant coefficients*. The distinction between the two cases is quite important. For the special case of linear differential equations with *constant coefficients*, the problem of finding the solutions for the homogeneous case has been completely solved.

For the most part, the present chapter is concerned with finding solutions of such equations. Generally speaking, finding the analytic solutions of differential equations with variable coefficients requires special techniques, as discussed in Chapter 6.

OPERATORS   For convenience in notation, we can define an operator

$$\mathcal{L}_n = \frac{d^n}{dt^n} + a_{n-1}(t)\frac{d^{n-1}}{dt^{n-1}} + \cdots + a_0(t), \qquad (5.9)$$

and write Equation 5.8 in "operator" notation in the form

$$\mathcal{L}_n[y(t)] = f(t). \qquad (5.10)$$

Thus, $\mathcal{L}_n$ is the nth-order linear differential operator that operates on the dependent variable in the differential equation to yield the function on the right hand side. In most instances, the subscript indicating the order of the differential operator will be omitted. For example, the equation $\ddot{y} + ty = e^t$ could be abbreviated

$$\mathcal{L}[y(t)] = e^t,$$

with $\mathcal{L} = d^2/dt^2 + t$ in this case. Obviously, care must be taken to define the exact form of the linear differential operator in a given problem.

Returning to Equation 5.8, there is an important theorem that states that there exist n linearly independent solutions of the homogeneous equation $(\mathcal{L}_n[y(t)] = 0)$. Furthermore, a linear combination of these solutions is called the *general solution* of the homogeneous equation. The general solution to the nonhomogeneous equation $\mathcal{L}_n[y(t)] = f(t)$ is a sum of the general solution of the homogeneous equation and one solution of the nonhomogeneous equation.

■ THEOREM 5.2   *Solution of $\mathcal{L}_n[y(t)] = 0$*

Let $\mathcal{L}_n[y(t)] = 0$ represent the nth-order linear differential equation shown in Equation 5.8 with continuous coefficients in the interval $[ab]$. Then,

    a. *There exist n linearly independent solutions of the homogeneous differential equation $\mathcal{L}_n[y(t)] = 0$ in the given interval $a \leq t \leq b$;*

    b. *If $y_1(t), \ldots, y_n(t)$ are linearly independent solutions of $\mathcal{L}_n[y(t)] = 0$ in the given interval $a \leq t \leq b$, then*

$$y_c(t) = c_1 y_1(t) + c_2 y_2(t) + \cdots + c_n y_n(t),$$

where the c_i are constants, not all zero, is the general solution of the homogeneous equation.

━━━━━━━━━━━━━━━━━━━━■

The solution $y_c(t)$ in the theorem is often called the *complementary solution*. In contrast, a solution of Equation 5.8 that corresponds to a specific nonzero $f(t)$ is referred to as a *particular* solution.

Solutions of the Linear Differential Equation   The solution set of the nth-order homogeneous differential equation

$$\mathcal{L}_n[y(t)] = 0$$

forms an n-dimensional vector space. Thus, considering the discussion of vector spaces in Chapter 2, we expect that if

$$y_1(t), y_2(t), \ldots, y_n(t)$$

are solutions of the homogeneous equation, then linear combinations of solutions and scalar multiples of solutions are also solutions, since

    1. $y_1(t) + y_2(t) + \cdots + y_n(t)$ is a solution;

    2. $\alpha y_i(t)$ for $i = 1, \ldots, n$ is a solution for any nonzero number α.

We immediately generalize these results to state that any function that is a linear combination of the solutions written as

$$y_c(t) = c_1 y_1(t) + c_2 y_2(t) + \cdots + c_n y_n(t) \qquad (5.11)$$

---

is also a solution. As was indicated in Theorem 5.2, this is the general solution of the homogeneous equation when the individual solutions $y_i(t), i = 1, 2 \ldots, n$, form a linearly independent set.

The complete solution to $\mathcal{L}_n[y(t)] = f(t)$ can be constructed from the general solution of the homogeneous equation and one particular solution of the nonhomogeneous equation. Let $y_p(t)$ be a particular solution of $\mathcal{L}_n[y(t)] = f(t)$. Then, the complete solution of $\mathcal{L}_n[y(t)] = f(t)$ is $y_p + y_c$, or

$$y(t) = y_p(t) + c_1 y_1(t) + c_2 y_2(t) + \cdots + c_n y_n(t) \tag{5.12}$$

if $y_1(t), y_2(t), \ldots, y_n(t)$ are linearly independent solutions of the homogeneous equation.

Principle of Superposition It can be difficult to find the particular solution to the nonhomogeneous equation $\mathcal{L}_n[y(t)] = f(t)$. To simplify the problem, it is sometimes possible to write $f(t)$ as

$$f(t) = f_1(t) + f_2(t) + \cdots + f_m(t) \tag{5.13}$$

and to find a solution to each equation $\mathcal{L}_n[y(t)] = f_j(t)$, $j = 1, 2, \cdots, m$. Then, the sum of these particular solutions is a particular solution since the equation is linear. The idea here of the *principle of superposition* is to decompose the problem into simpler problems that might be easier to solve than the original problem and then add the individual solutions. The particular solution being sought is thus

$$y_p(t) = y_{p1}(t) + y_{p2}(t) + \cdots + y_{pm}(t), \tag{5.14}$$

where each $y_{pi}(t), i = 1, 2, \ldots, m$, is a solution to $\mathcal{L}_n[y_{pi}(t)] = f_i(t)$.

Although the previous discussions give no clue as how to find the functions $y_1(t), \ldots, y_n(t)$ that solve the homogeneous differential equation, the coefficients c_i in Equation 5.11, or a particular solution $y_p(t)$, quite a bit has been accomplished. We know that if n linearly independent solutions of the homogeneous equation and a particular solution to the nonhomogeneous equation can be found, the complete solution for the linear ordinary differential equation is known.

We shall see that finding the solutions to first- and second-order equations rarely presents any great difficulty. This is fortunate since these equations are of vital importance in many areas of science and engineering. The solution of higher-order differential equations can often be found by converting the differential equation into a system of first-order differential equations, which are then solved using matrix methods. This is the approach taken in later sections of this chapter.

The next section treats the case of first-order linear differential equations. Studying the methods of solution to the first order equations will yield insight about the problems encountered with more complicated equations.

Consider the first-order, linear differential equation

$$\frac{dy(t)}{dt} + p(t)y(t) = f(t), \tag{5.15}$$

which we write as $\dot{y} + p(t)y = f(t)$. Assuming that $p(t)$ and $f(t)$ are continuous in some common interval, the equation can be solved by multiplying each term by an *integrating factor* in the form $e^{\int p(t)dt}$ to yield

$$\dot{y}e^{\int p(t)\,dt} + p(t)ye^{\int p(t)\,dt} = f(t)e^{\int p(t)\,dt}.$$

Notice that the left side of this equation is the derivative of the product $ye^{\int p(t)\,dt}$. Thus,

$$\frac{d}{dt}\left[ye^{\int p(t)\,dt}\right] = f(t)e^{\int p(t)\,dt}$$

can be integrated and solved for $y(t)$, with the result

$$y(t) = e^{-\int p(t)\,dt}\int f(t)e^{\int p(t)\,dt}dt + ce^{\int -p(t)\,dt}, \tag{5.16}$$

where c is the constant of integration. This expression is the general solution to Equation 5.15.

The general solution of Equation 5.16 contains two terms. The first term describes the effect of the function $f(t)$, which is often called the *forcing function* when used in problems that model physical systems. In the study of linear systems in engineering, the function $f(t)$ is also called the *input* or *input function*, and the solution $y(t)$ is termed the *output*. The differential equation describes how the system reacts to the effects of the input function.

The second term in Equation 5.16 contains an arbitrary constant c, which is determined by demanding that the solution meet an *initial condition*. This specifies the value of y at some specific value of t, say, t_0. Mathematically, we write the initial condition as $y(t_0) = y_0$. The differential equation and the initial condition taken together is called an *initial value problem*.

Constant Coefficients In case the function $p(t) = a$ where a is a scalar, Equation 5.15 becomes

$$\frac{dy(t)}{dt} + ay(t) = f(t). \tag{5.17}$$

Assuming that the equation is defined on the interval $t \geq 0$, the integrating factor for the equation is

$$e^{\int_0^t a\,d\tau} = e^{at},$$

where the variable of integration has been changed to τ to emphasize that the integral is a function of t, the upper limit of integration, not the "dummy" variable τ.

The complete solution then takes the form

$$y(t) = e^{-at} \int_0^t f(\tau)e^{a\tau}\,d\tau + ce^{-at}. \tag{5.18}$$

Applications The first-order differential equation is used to *model* a number of phenomena in mathematical terms. Some of the problems treated with first-order equations include

1. Growth and decay,

2. Radioactive half life,

3. Carbon dating,

4. Cooling of a body,

5. Series electrical circuits.

In these applications, the mathematical model is formulated as a first-order differential equation since it has been observed that the rate of change of some variable (e.g., population, mass, temperature, etc.) is proportional to the value of the variable. A few of these problems are given as exercises in this chapter. Many such problems are treated in several of the references in the Annotated Bibliography at the end of the chapter.

□ **EXAMPLE 5.1** ***First-Order Example***
Consider the equation

$$\dot{y}(t) = f(t) \tag{5.19}$$

defined on the interval $t_1 < t < t_2$ with the initial condition $y(0) = y_0$. From Equation 5.18 with $a = 0$, the solution is then

$$y(t) = \int_{t_0}^t f(\tau)\,d\tau + c = \int f(\tau)\,d\tau.$$

Solution. As an example, let the function $f(t)$ be defined as

$$f(t) = \frac{1}{(t-1)}$$

for $t > 1$ as the region of interest. The domain is the half-plane for $t > 1$, so the result of the integration yields

$$y(t) = \int \frac{1}{t-1}\,dt = \ln|t-1| + c, \quad t > 1. \tag{5.20}$$

This is a solution, as can be verified by substituting $y(t)$ in the differential equation. The function also satisfies the conditions of the existence and uniqueness theorem, which indicates that $y(t)$ is the only solution. As c is varied from $-\infty$ to ∞, the solution passes through all the points of the plane.

$\square$

The differential equation in Example 5.1 could be integrated directly. In this case, resort to a computer solution is not necessary. However, MATLAB can be helpful in plotting the results to allow us to compare the solutions for various initial conditions, as in Example 5.2.

$\square$ EXAMPLE 5.2   *MATLAB First-Order Example*

The accompanying MATLAB script plots the solution of Equation 5.20 for three values of the constant $c = y_0 - \ln|t_0 - 1|$. Three example values for c of $-5, 0$, and 5 are shown in Figure 5.1. These correspond to the points (t_0, y_0) as follows

$$
\begin{aligned}
(t_0, y_0) &= (2, -10), & c &= -10.0 \\
(t_0, y_0) &= (2, 0), & c &= 0.0 \\
(t_0, y_0) &= (4, 8), & c &= 6.9014
\end{aligned}
$$

In this case, we know the solution so the analytical form can be used for plotting. The time values are selected to start at $t = 1.1$ so that the infinite value at $t = 1$ is avoided.

MATLAB Script _____

```
Example 5.2
% EX5_2.M Plot solution to ydot=1/(t-1)
% for 3 initial values. y=ln|t-1|+c.
c=[-10 0 6.9014];          % Define constant values
t=[1.005:.1:10];           % Define t to avoid t=1
m=length(t);
n=length(c);
y=zeros(m,n);
for I=1:3
 y(:,I) =log(abs(t-1)') + c(I);
end
clf                        % Clear any figures
plot(t,y(:,1),t,y(:,2),t,y(:,3))
xlabel('t')
ylabel('y(t)')
title('Solution of dy/dt=1/(t-1)')
grid
axis([0,10,-20,20])        % Plot limits
gtext('(t0,y0)=(2,-10)')   % Annotate with mouse clicks
gtext('(t0,y0)=(2,0)')
gtext('(t0,y0)=(4,8)')
```

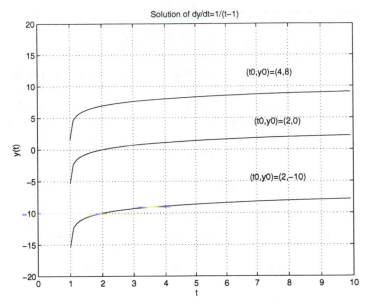

FIGURE 5.1 *Response of the first-order system in Example 5.2*

MATLAB can also be used to solve the differential equation and to serve as a check on the hand solution for the examples in this chapter. The command **dsolve** from the *Symbolic Math Toolbox* can be applied in the forms

```
y=dsolve('Dy=1/(t-1)')
y1=dsolve('Dy=1/(t-1)','y(2)=-10') %Initial values for t=2.
y2=dsolve('Dy=1/(t-1)','y(2)=0')
y3=dsolve('Dy=1/(t-1)','y(4)=8')   %Initial value for t=4.
```

to yield the general symbolic solution and the solutions that match the selected initial conditions.

□

WHAT IF?   Solve the problem in Example 5.2 for several other initial conditions using the symbolic command **dsolve**, and plot the result with **ezplot**. If **ezplot** is called repeatedly, the graphics window is rescaled with each new plot. You may need to use the command **axis** to size the graph properly.

If your version of MATLAB has the command **gtext** as used in Example 5.2, experiment with the command on the plot of Figure 5.1 by moving the annotation for the curves. Type **help graphics** to see other graphics commands.

HIGHER ORDER DIFFERENTIAL EQUATIONS

The nth-order differential equation with constant coefficients is treated in this section. The general solution to the homogeneous equation is easily found in principle. This problem of finding n functions to satisfy the homogeneous equation is reduced to finding the n roots of an algebraic equation. Then, special cases of the equation will be studied.

Consider the nth-order, linear differential equation

$$\frac{d^n y(t)}{dt^n} + a_{n-1}\frac{d^{n-1} y(t)}{dt^{n-1}} + \cdots + a_1 \frac{dy(t)}{dt} + a_0 y(t) = f(t), \qquad \textbf{(5.21)}$$

where the a_i are constants. This equation will be written in operator form as $L_n[y(t)] = f(t)$ to distinguish the operator from $\mathcal{L}$, which applies to equations with variable coefficients, as in Equation 5.9.

HOMOGENEOUS SOLUTION

First, we find the general solution to the homogeneous equation. As discussed in Chapter 4, the function $y(t) = e^{\lambda t}$ is an eigenfunction of each derivative term in Equation 5.21. Then, substitution of $e^{\lambda t}$ in the equation $L_n[e^{\lambda t}] = 0$, where L_n is the nth-order linear differential operator previously defined, yields

$$e^{\lambda t}(\lambda^n + \cdots + a_0) = g(\lambda)e^{\lambda t} = 0.$$

Accordingly, $e^{\lambda t}$ will be a solution provided that λ satisfies the algebraic equation of order n, $g(\lambda) = 0$. The equation $g(\lambda) = 0$ is called the *characteristic equation* associated with Equation 5.21. The roots of the equation

$$g(\lambda) = (\lambda^n + a_{n-1}\lambda^{n-1} + \cdots + a_1\lambda + a_0) = 0 \qquad \textbf{(5.22)}$$

define the valid values of λ for which $e^{\lambda t}$ satisfies the homogeneous equation. There are n solutions to this polynomial equation.

DISTINCT ROOTS

Consider first the case where the roots of the characteristic Equation 5.22 are distinct; that is, no root is repeated. In this case, the general solution of the homogeneous equation with constant coefficients is simply written down as the superposition of exponential terms with the roots of the characteristic equation as the constant in the exponents. The result can be stated as the following theorem.

Solution of $L_n[y(t)] = 0$

If the n roots of the characteristic equation are all distinct and the roots are $\lambda_1, \lambda_2, \ldots, \lambda_n$, then the solution of the nth-order homogeneous differential equation defined by Equation 5.21 with $f(t) = 0$ is

$$y_c(t) = c_1 e^{\lambda_1 t} + c_2 e^{\lambda_2 t} + \cdots + c_n e^{\lambda_n t}, \tag{5.23}$$

where the designation $y_c(t)$ is used for this complementary solution.

─────────────────── ■

REPEATED ROOTS

Suppose it is not possible to find n distinct roots of the characteristic Equation 5.22. In this case, it can be shown that the solutions of the differential equations for a root λ_k of multiplicity m are of the form

$$e^{\lambda_k t}, \quad t e^{\lambda_k t}, \quad t^2 e^{\lambda_k t}, \ldots t^{m-1} e^{\lambda_k t}.$$

To show the result for a root of multiplicity two, we first assume that two roots of the characteristic equation are distinct and then let the roots approach each other to become one root of multiplicity two. Let λ_1 and λ_2 be two distinct real roots of the characteristic equation. Then, the function

$$\frac{e^{\lambda_1 t} - e^{\lambda_2 t}}{\lambda_1 - \lambda_2} \tag{5.24}$$

is a solution of the homogeneous differential equations with constant coefficients. Now assume that the coefficients of the characteristic equation change so that λ_2 tends to λ_1. Equation 5.24 can be written

$$\lim_{\lambda_2 \to \lambda_1} \frac{e^{\lambda_1 t} - e^{\lambda_2 t}}{\lambda_1 - \lambda_2}. \tag{5.25}$$

Remembering the definition of the derivative from calculus, this limit is the derivative of $e^{\lambda_1 t}$ with respect to λ_1.[2] Therefore, one of the solutions of the differential equation is

$$\frac{d}{d\lambda_1} e^{\lambda_1 t} = t e^{\lambda_1 t}$$

when the characteristic equation yields repeated roots.

SUMMARY OF HOMOGENEOUS SOLUTIONS

Table 5.2 lists the various solutions according to the type of roots of the characteristic equation as real or complex numbers. In any case, there will be a total of n functions $y_1(t), y_2(t), \ldots, y_n(t)$ thus obtained. The general solution of $L[y_c(t)] = 0$ will be

$$y_c(t) = c_1 y_1(t) + c_2 y_2(t) + \cdots + c_n y_n(t). \tag{5.26}$$

─────────────────────

[2] The definition of the derivative is reviewed in Chapter 6.

─────────────────────

TABLE 5.2 *Homogeneous solutions*

Roots	Solutions
Simple real root λ	$e^{\lambda t}$
Simple complex root $a \pm ib$	$e^{at}\cos bt, e^{at}\sin bt$
Real root λ_k of multiplicity k	$e^{\lambda_k t}, te^{\lambda_k t}, \ldots, t^{k-1}e^{\lambda_k t}$
Complex root of multiplicity k	$e^{at}\cos bt, e^{at}\sin bt, \ldots$
	$t^{k-1}e^{at}\cos bt, t^{k-1}e^{at}\sin bt$

WHAT IF? Using the results for multiple roots of the characteristic equation of a homogeneous differential equation, show that the solution of the equation

$$\frac{d^2 y(t)}{dt^2} = 0$$

is $y(t) = c_1 + c_2 t$.

Show that the solution of the equation

$$\frac{d^2 y(t)}{dt^2} + 6\frac{dy(t)}{dt} + 9y(t) = 0 \qquad\qquad \textbf{(5.27)}$$

with initial conditions $y(0) = 2$ and $dy(0)/dt = 0$ is $y(t) = 2(1 + 3t)e^{-3t}$.

After solving the equations by hand, experiment with the **dsolve** command from the *Symbolic Math Toolbox*. For example, the command

```
y = dsolve('D2y=0')
```

solves the equation $d^2 y(x)/dx^2 = 0$. Modify the command to solve the equation with t as the independent variable. After using **dsolve** to solve Equation 5.27, use the command **simple** to yield the form given as the answer.

INDEPENDENCE OF THE HOMOGENEOUS SOLUTIONS

Analogous to the fact that any vector in $\mathbf{R}^n$ can be expressed as a linear combination of n linearly independent vectors, any solution to the nth-order homogeneous linear differential equation $\mathcal{L}_n[y(t)] = 0$ can be expressed as a linear combination of n linearly independent solutions. The n constants in the expression for the general solution are found by applying n conditions, as explained later. For the case of the equation with constant coefficients, the linearly independent set of functions has already been mentioned; it is the set of exponential functions previously discussed. A theorem verifies this in the general case.

■ THEOREM 5.4   *Linear Independence of* $t^m e^{\lambda t}$

    Let m be a nonnegative integer and let λ be a real or complex number. Then, any set of functions of the form

$$t^m e^{\lambda t},$$

defined on any common interval, is linearly independent.

─────────────────── ■

The Wronskian determinant treated in Problem 5.11 can be used to test these functions for linear independence.

FORM OF THE HOMOGENEOUS SOLUTIONS

When the coefficients of the homogeneous differential equation are real, complex roots of the characteristic equation occur in conjugate pairs. Thus, if $\lambda_1 = i$ $(i = \sqrt{-1})$ is a root, then another root is $\lambda_2 = -i$. Two solutions are then e^{it} and e^{-it}. Since linear combinations of solutions are also solutions of the homogeneous equation, we can form combinations that are real. The exponential terms can be combined to yield

$$\cos t = \frac{e^{it} + e^{-it}}{2} \quad \text{and} \quad \sin t = \frac{e^{it} - e^{-it}}{2i}, \qquad (5.28)$$

which follow from the identity (Euler's formula)

$$e^{it} = \cos t + i \sin t \qquad (5.29)$$

discussed in Chapter 2. The two solutions belonging to the pair of complex roots can then be written

$$y = c_1 \cos t + c_2 \sin t. \qquad (5.30)$$

When the roots are of the form $a \pm bi$, the functions

$$e^{(a \pm bi)t} = e^{at}(\cos bt \pm i \sin bt) \qquad (5.31)$$

are solutions of the homogeneous differential equation. Forming the linear combinations, we can obtain the real functions

$$e^{at} \cos bt \quad \text{and} \quad e^{at} \sin bt, \qquad (5.32)$$

corresponding to the solutions $y = e^{at}(c_1 \cos bt + c_2 \sin bt)$.

    Using trigonometric identities, the equation with two arbitrary constants $y = c_1 \cos bt + c_2 \sin bt$ can be written in one of the following forms

$$y = C \cos(bt + \alpha) \quad \text{or} \quad D \sin(bt + \beta), \qquad (5.33)$$

where C and α or D and β serve as the two arbitrary constants. Thus, we expect that real solutions to a homogeneous differential equation can be found when the coefficients are real, as summarized in the following theorem.

───

Higher Order Differential Equations

211

Real-Valued Solutions

> *Suppose the constants in the nth-order differential equation*

$$L_n[y(t)] = y^{(n)}(t) + a_{n-1}y^{(n-1)}(t) + \cdots + a_0 y(t) = 0$$

> *are all real. There exists a set of n linearly independent, real-valued, solutions. If a solution satisfies real initial conditions, it is real-valued.*

────────────────────────────────■

Summary for Homogeneous Differential Equations The homogeneous solutions to the linear equation with constant coefficients

$$L_n[y(t)] = 0$$

is easily constructed according to Equation 5.26 if the roots of the characteristic polynomial can be found. The existence of the roots is guaranteed by the fundamental theorem of algebra presented in Chapter 2. The theorem asserts that the polynomial of degree n always has exactly n roots. For equations up to the fourth degree, explicit formulas for the solutions (roots) are known.

Numerical methods may be used to find the roots of polynomial equations. Fortunately, many computer algorithms are available to approximate the roots of the characteristic polynomial for a homogeneous differential equation. Those available with MATLAB were mentioned in Chapter 4.

SECOND ORDER DIFFERENTIAL EQUATIONS
─────────────■─────────────

It is not difficult to answer the question Why is the second order differential equation the most important equation in most areas of engineering and physics? The reason lies in the observations by Newton in mechanics and many contributors to our understanding of electricity that allow us to relate the energy in a system to the associated differential equation.

For example, Newton's second law of motion states that the motion of an object of variable mass $m(t)$ subject to a force F is

$$\frac{d}{dt}(mv) = F,$$

where v is the velocity of the body. In the cases that m is constant, the equation can be written

$$m\ddot{x}(t) = F,$$

where $x(t)$ is position, $\dot{x}(t) = v$, and $\ddot{x}(t)$ is the acceleration. As a simple example, Newton's law describes the free fall of an object. Neglecting air resistance, the equation becomes $m\ddot{y}(t) = mg$, where $y(t)$ is the displacement downward and g is the acceleration of gravity. For motion near the surface of the earth, the force of gravity can be assumed to be constant.

Other forces that are commonly encountered can be proportional to velocity.[3] These are often called *damping forces* because they tend to retard the motion of an object. Air resistance for a falling body or friction due to an object sliding across a surface are forms of damping forces. Assuming that the damping force is proportional to velocity, Newton's law becomes

$$m\ddot{x}(t) = F_d = -\beta\dot{x}(t),$$

where β is a positive damping constant and the negative sign is necessary because the force acts in a direction opposite to the motion.

The third important force is described by Hooke's law that describes the force that acts as a resistance to stretching of a body such as a spring. If an "idealized" spring is displaced by an amount x from its equilibrium position, the force to restore the spring to its original length is $F_s = kx$.

An idealized mechanical system is diagramed in Figure 5.2. The motion of a mass, damping, and a spring force would be governed by the equation

$$m\ddot{x}(t) = F_d + F_s + f(t) = -\beta\dot{x}(t) - kx(t) + f(t) \qquad \text{(5.34)}$$

if a force $f(t)$ acts on the mass.

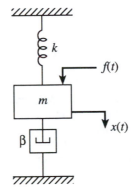

FIGURE 5.2 *Idealized mechanical system*

Rewriting Equation 5.34 as

$$m\ddot{x}(t) + \beta\dot{x}(t) + kx(t) = f(t),$$

we can associate $b = \beta/m$ and $c = k/m$ and write the homogeneous, second-order differential equation of motion as

$$\ddot{x}(t) + b\dot{x}(t) + cx(t) = 0. \qquad \text{(5.35)}$$

---

[3]In some applications, these forces may be proportional to higher powers of the velocity.

Equations in this form are so important in engineering and physics that we will analyze the equation further in this chapter and later chapters. In Chapter 11, we relate this equation to the energy in the system. The kinetic energy due to the motion of the object is

$$\frac{1}{2}mv^2 = \frac{1}{2}m[\dot{x}(t)]^2.$$

The potential energy comes from the spring restoring force and depends on the position of the object but not on its velocity. If the damping force is absent, the sum of these kinetic and potential energies is a constant. The basic electrical or electronic system consisting of resistors, capacitors, and inductors can also be described by second-order differential equations. Problem 5.20 asks you to compare the equations for mechanical and electrical systems.

□ EXAMPLE 5.3   *Second-Order Example*

Consider the second-order homogeneous differential equation

$$\frac{d^2y}{dt^2} + b\frac{dy}{dt} + cy = 0, \tag{5.36}$$

where b and c are real constants. Forming the characteristic equation

$$\lambda^2 + b\lambda + c = 0 \tag{5.37}$$

leads to the solutions

$$
\begin{aligned}
\lambda_1 &= -\frac{1}{2}b + \frac{1}{2}\sqrt{b^2 - 4c}, \\
\lambda_2 &= -\frac{1}{2}b - \frac{1}{2}\sqrt{b^2 - 4c}.
\end{aligned}
\tag{5.38}
$$

The type of solution can be described in terms of the value of the *discriminant*, $b^2 - 4c$. If the discriminant is zero, the roots are real and equal with value $-b/2$. If the discriminant is positive, the roots are real and distinct. Otherwise, the roots are complex conjugate pairs.

The complete solution to the homogeneous differential equation with constant coefficients can be written

$$
\begin{aligned}
y(t) &= c_1 e^{-\frac{b}{2}t} \times \exp(+\frac{1}{2}\sqrt{b^2 - 4c}\,t) \\
&\quad + c_2 e^{-\frac{b}{2}t} \times \exp(-\frac{1}{2}\sqrt{b^2 - 4c}\,t).
\end{aligned}
\tag{5.39}
$$

Equation 5.39 shows that the form of the homogeneous solution to the second-order differential equation with constant coefficients is determined by the relationship between the coefficients b and c when the initial conditions are nonzero.

We will assume that $b > 0$ and $c > 0$ corresponding to a physical system with passive elements.[4] There are three cases to consider based on the ratio of

---

[4]It is possible to design physical systems with negative values of the coefficients. An example is an oscillator using positive feedback.

b^2/c, as follows:

$$b^2 > 4c \qquad \text{overdamped;}$$
$$b^2 = 4c \qquad \text{critically damped;}$$
$$b^2 < 4c \qquad \text{underdamped.} \qquad (5.40)$$

The terms overdamped, critically damped and underdamped refer to the motion of an object modeled by the differential equation of motion previously discussed as Equation 5.35.

When $b^2 > 4c$, the solution decays from any initial value to zero with the form

$$y(t) = c_1 e^{\lambda_1 t} + c_2 e^{\lambda_2 t}$$

since $\lambda_1 < 0$ and $\lambda_2 < 0$ because the coefficients of the equation are positive. This solution is called the overdamped solution, compared to the solution that occurs when $b^2 = 4c$. Physically, all oscillations in the solution are damped out, so only exponential decay is exhibited.

When the coefficients of the equation are such that $b^2 = 4c$, the solution represents a *critically damped* solution, and the graph of $y(t)$ represents eventual exponential decay. The characteristic equation then has two real equal roots leading to a solution of the form

$$y(t) = (c_1 + c_2 t)e^{-bt/2}.$$

If the coefficient b is reduced at all, the solution will contain oscillations.

The term *underdamped* is applied to the solution that allows an oscillatory solution, but the solution is eventually "damped" and approaches zero. The solution to the second-order homogeneous equation in the underdamped case can be written

$$y(t) = C e^{-bt/2} \sin(\omega t + \phi),$$

where $\omega = \sqrt{4c - b^2}/2$ and the constants C and ϕ are determined by the initial conditions. In the extreme case that $b = 0$, the system is said to be undamped, and the solution is pure sinusoidal oscillation called harmonic oscillation. This will be studied in a later example.

In summary, the second-order homogeneous differential equation with positive coefficients and nonzero initial conditions has two distinct types of solution depending on the relationship between the coefficients b and c. If $b^2 < 4c$, an underdamped solution results and the solution displays damped sinusoidal oscillation. If $b^2 > 4c$, the solution exhibits exponential decay. The critically damped solution that divides the two types of solutions occurs when $b^2 = 4c$.

□

MATLAB SOLUTIONS TO DIFFERENTIAL EQUATIONS

In solving a differential equation and indeed any type of equation, MAT-LAB gives us the advantage of being able to vary the parameters in the equation and solve the equation repeatedly with different parameters. The changes in the solutions as the parameters are varied are often best shown by plotting the results.

MATLAB Symbolic Differential Equation Solution

Consider the second-order differential equation

$$\frac{d^2 y(t)}{dy^2} + b\frac{dy(t)}{dt} + y(t) = 0$$

subject to the conditions

$$y(0) = 1 \quad \text{and} \quad \dot{y}(0) = 0.$$

For this equation, $b = 2$ is the critical damping value since $c = 1$ in Equation 5.36.

The accompanying MATLAB script solves the general differential equation and plots the solutions for different values of b with $c = 1$. When $b = 3$, the solution in Figure 5.3 represents overdamped behavior. The underdamped case with $b = 1$ is also shown.

MATLAB Script _____

```
Example 5.4
% EX5_4.M  Solve symbolically the second order equation
%   D2y+b*Dy+c*y=0  and plot for b=1, b=3 with c=1.
%
y = dsolve('D2y+b*Dy+1*y=0','y(0)=1','Dy(0)=0','t');
y=simple(y)                 % Simplify the solution
% Substitute values b=1 and b=3
clf                         % Clear any figures and
hold on                     %  plot multiple graphs
ezplot(subs(y,3.0,'b'),[0,10])
gtext('b=3')                % Annotate text with mouse
ezplot(subs(y,1.0,'b'),[0,10])
gtext('b=1')
title('Solution to D2y+b*Dy+y=0, y(0)=1,Dy(0)=0')
ylabel('y(t)')
grid
hold off                    % Default setting
```

_____

The symbolic MATLAB command **dsolve** is used to solve the differential equation subject to the given initial condition. Notice the use of the **subs** command to substitute various values for b. The command **gtext** is used to annotate the graph at points designated by the mouse cursor position.

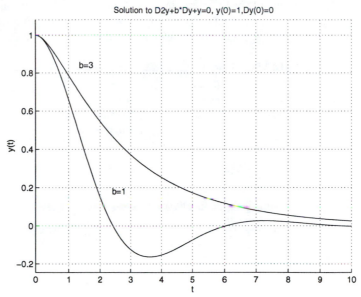

FIGURE 5.3 *Response of the homogeneous second order system with damping b*

☐

MATLAB SOLUTIONS OF CHARACTERISTIC POLYNOMIALS

Once the characteristic polynomial for a differential equation is formed, the MATLAB command **roots** can be used to find the roots of a polynomial defined numerically. The symbolic command **solve** can be used to determine the roots for a symbolic polynomial if symbolic solutions can be found.

☐ EXAMPLE 5.5 **MATLAB Solution**

The characteristic equation for the fifth-order differential equation

$$z^{(5)} + 3z^{(4)} + 3z^{(3)} + z^{(2)} = 0$$

is $\lambda^5 + 3\lambda^4 + 3\lambda^3 + \lambda^2$. Writing this as $\lambda^2(\lambda+1)^3$, the roots are evidently $\lambda_1 = 0$ with multiplicity two and $\lambda_2 = -1$ with multiplicity three. The general solution is thus

$$z = (c_1 + c_2 t) + (c_3 + c_4 t + c_5 t^2)e^{-t}.$$

For the equation

$$\ddot{y}(t) + 2\dot{y}(t) + 5y(t) = 0,$$

the characteristic equation is $\lambda^2 + 2\lambda + 5 = 0$ with the roots $\lambda_{1,2} = -1 \pm 2i$. The general solution written in terms of real functions is

$$y(t) = e^{-t}(c_1 \cos 2t + c_2 \sin 2t),$$

as previously discussed.

The accompanying MATLAB session is from an edited diary file. The command **roots** determines the roots of the characteristic equations just presented. The argument of **roots** in these problems is a row vector with the coefficients of the characteristic polynomial as elements.

MATLAB Script _____

```
Example 5.5
>>format compact
>>p=[1 3 3 1 0 0];    % Define coefficients of 5th order polynomial
r=roots(p)            % Find 5 roots
r =
        0
        0
  -1.0000 + 0.0000i
  -1.0000 - 0.0000i
  -1.0000
>>p1=[1 2 5];         % Second order equation
>>r1=roots(p1)
r1 =
  -1.0000 + 2.0000i
  -1.0000 - 2.0000i
>>quit
```
_____

□

W H A T   I F ?   Try the symbolic command **solve** to solve the quadratic characteristic Equation 5.37. Notice that the MATLAB command **solve** also has no difficulty solving for the roots of the characteristic equations in Example 5.5

PARTICULAR SOLUTIONS OF DIFFERENTIAL EQUATIONS

Various techniques are used to find particular solutions to ordinary differential equations. In this section, the techniques of solution described are the following:

1. Undetermined coefficients,

2. Variation of parameters,

3. MATLAB solution.

METHOD OF UNDETER-MINED COEFFICIENTS

A particular solution to the nth-order linear differential equation

$$\mathcal{L}_n[y_p(t)] = f(t)$$

can be found in principle using a number of techniques. When the equation has constant coefficients, as in Equation 5.21, the simplest approach is the method of *undetermined coefficients*. This amounts to making an educated "guess" as to the form of y_p from the form of the forcing function $f(t)$. This method is appropriate when $f(t)$ is a constant, a polynomial in t, exponential or trigonometric functions of t, or finite sums or products of these functions.

Table 5.3 lists the form of $f(t)$ in the left column and the assumed undetermined coefficient solutions in the right column. The solution method consists of substituting the assumed solution in the equation and finding the unknown constants.

TABLE 5.3 *Undetermined coefficient solutions*

$f(t)$	Choice for y_p
$Ke^{\alpha t}$	$ce^{\alpha t}$
$Kt^n,\ (n = 0, 1, \ldots)$	$c_n t^n + c_{n-1} t^{n-1} + \cdots + c_1 t + c_0$
$K \sin \omega t$ or $K \cos \omega t$	$c_1 \cos \omega t + c_2 \sin \omega t$
$Ke^{\alpha t} \cos \omega t$ or $Ke^{\alpha t} \sin \omega t$	$e^{\alpha t}(c_1 \cos \omega t + c_2 \sin \omega t)$

In the table, it is assumed that $e^{\alpha t}$, $\sin \omega t$, or $\cos \omega t$ are not solutions of the homogeneous equation, as described in Equation 5.26 and Table 5.2. If the chosen particular solution happens to be a solution of the homogeneous equation $L_n[y(t)] = 0$, multiply the trial particular solution by t or powers of t if the homogeneous solution results from multiple roots of the characteristic equation.

Summary The solution approach using undetermined coefficients can be made more rigorous by stating explicitly the conditions under which a particular solution can be found by this method. The following theorem defines the solutions to Equation 5.21,

$$L_n[y_p(t)] = f(t),$$

when $f(t)$ has the form of a polynomial times an exponential function.

Particular solutions

Consider the *n*th-order, linear, nonhomogeneous equation with constant coefficients

$$L_n[y_p(t)] = p(t)e^{\gamma t}, \tag{5.41}$$

in which $p(t)$ is a polynomial of degree r in t and γ is a complex number. In the
following, let $k = 0$ if γ is not a root of the characteristic equation $(g(\gamma) \neq 0)$,
and let k be the multiplicity of the root γ if $g(\gamma) = 0$. Then, there exists a
particular solution of Equation 5.41 of the form

$$y_p(t) = t^k h(t)e^{\gamma t}, \tag{5.42}$$

where $h(t)$ is an *r*th-degree polynomial. The coefficients of $h(t)$ can be found by
the method of undetermined coefficients.

── ■

The theorem covers all the cases just shown in Table 5.3 and also the
case that $f(t)$ would solve the homogeneous equation. The notation $g(\lambda)$
means the characteristic equation

$$g(\lambda) = \lambda^n + a_{n-1}\lambda^{n-1} + \cdots + a_1\lambda + a_0,$$

as in Equation 5.22.

To interpret Theorem 5.6, we consider the cases involving the solutions to the homogeneous equation by the exponential term in the forcing
function as follows:

1. $e^{\gamma t}$ does not solve $L_n[e^{\gamma t}] = 0$, so $k = 0$.

2. $e^{\gamma t}$ solves $L_n[e^{\gamma t}] = 0$ and k is the multiplicity of the root γ in the
   characteristic equation.

For example, consider the equation

$$L_2[y(t)] = \ddot{y}(t) + 2\dot{y}(t) + y(t) = e^{-t}. \tag{5.43}$$

Since the characteristic equation is $(\lambda + 1)^2 = 0$, a root is $\lambda = -1$, but
with multiplicity two. The complementary solution is thus

$$y_c(t) = (c_1 + c_2 t)e^{-t}.$$

The exponential forcing term e^{-t} is a solution of $L_2[y](t) = 0$, $k = 2$ in
the theorem, and a solution is sought in the form

$$y_p(t) = t^2 h(t)e^{-t}.$$

Suppose that the polynomial in the forcing function of Equation 5.41
is of the form

$$p(t) = a_n t^n + a_{n-1}t^{n-1} + \cdots + a_0.$$

Then, we search for a polynomial $h(t)$ in the form

$$h(t) = b_n t^n + b_{n-1}t^{n-1} + \cdots + b_0.$$

It is now possible to substitute the solution of Equation 5.42 into the differential Equation 5.41 and solve for the coefficients of $h(t)$, treating them as unknowns. The result will be a system of linear equations by equating coefficients of like powers on each side of Equation 5.41. In Equation 5.43, $p(t) = 1$ so the degree is $r = 0$. In this case, $h(t) = C$ in the particular solution, where C is a constant to be determined. Thus, the particular solution to the equation

$$\ddot{y}(t) + 2\dot{y}(t) + y(t) = e^{-t}$$

is $y_p(t) = Ct^2 e^{-t}$. You should substitute $y_p(t)$ in the differential equation and show that the undetermined coefficient is $C = 1/2$.

☐ EXAMPLE 5.6   *Undetermined Coefficients*
Consider the equation

$$\ddot{z}(t) + \omega_1^2 z(t) = re^{i(\omega t + \alpha)}, \qquad (5.44)$$

which is the equation of a harmonic oscillator subject to an external force. The function $z(t)$ represents the time evolution of the oscillations. The solutions to the homogeneous equation are the functions $e^{i\omega_1 t}$ and $e^{-i\omega_1 t}$. A linear combination of these functions can be written

$$z_c(t) = r_1 \cos(\omega_1 t + \alpha_1),$$

as previously discussed. The constants can be determined by the initial conditions $z(0)$ and $\dot{z}(0)$. The constant r_1 is called the *amplitude* of the initial oscillation and the constant α_1 is the *phase*.

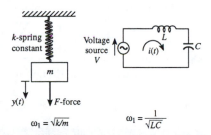

FIGURE 5.4 *Harmonic oscillators: (a) mechanical system; (b) electrical circuit*

In Figure 5.4(a), the variable $y(t)$ represents the vertical displacement of the mass-spring system and force is the input. Figure 5.4(b) shows an electrical oscillator. Equation 5.44 describes both systems when the variable z is the displacement for the mechanical system or current in the electrical circuit. If z represents current in the electrical circuit, the derivative of the voltage is the input function.

The number ω_1, which is assumed to be known, has units of radian *frequency* measured in radians per second. The frequency in cycles per second, or hertz, is $\omega_1/2\pi$. Since this oscillator has no damping, any initial oscillation will continue forever at this radian frequency. The complete solution will be

$$
\begin{aligned}
z(t) &= z_c(t) + z_p(t) \\
&= r_1\cos(\omega_1 t + \alpha_1) + z_p(t),
\end{aligned}
\tag{5.45}
$$

where the particular solution $z_p(t)$ is determined by the forcing function in Equation 5.44.

Two cases will be discussed, since the solution will depend on the frequency of the input function ω. In the first case, $\omega \neq \omega_1$. When $\omega = \omega_1$, the response of the system will not be a constant amplitude sinusoidal but will increase without limit with time.

a. If $\omega \neq \omega_1$, the particular solution has the form $z_p = \rho e^{i\omega t}$, where ρ is a constant not necessarily real. Substituting the trial solution into the original differential equation defined in Equation 5.44 yields the result

$$
\rho = \frac{re^{i\alpha}}{\omega_1^2 - \omega^2},
$$

so that

$$
z_p(t) = \frac{r}{\omega_1^2 - \omega^2} e^{i(\omega t + \alpha)}.
\tag{5.46}
$$

From this result, it is evident that the response has the same characteristics as the input function but is changed in amplitude. Also, since the differential equation is linear and

$$
re^{i(\omega t + \alpha)} = r\cos(\omega t + \alpha) + ir\sin(\omega t + \alpha),
$$

the real part of z_p is the response to the cosine term and the imaginary part of z_p' corresponds to the sine term. This is easily seen by writing $z(t) = x(t) + iy(t)$ and separating the Equation 5.44 into real and imaginary parts.

b. When $\omega = \omega_1$, ω is a root of the characteristic equation. In this case, according to Equation 5.42, a solution is sought in the form

$$
z_p = \sigma t e^{i\omega_1 t},
\tag{5.47}
$$

where σ can be a complex number.

Substituting z_p from Equation 5.47 in Equation 5.44 and solving for σ yields

$$
\sigma = \frac{re^{i\alpha}}{2i\omega_1},
$$

so the particular solution becomes

$$
z_p(t) = \frac{rte^{i(\omega_1 t + \alpha)}}{2i\omega_1} = \frac{rte^{i[\omega_1 t + \alpha - (\pi/2)]}}{2\omega_1},
\tag{5.48}
$$

since $e^{-i\pi/2} = 1/i$.

Thus, for $\omega = \omega_1$, the phenomenon of *resonance* leads to the result that the amplitude $rt/2\omega_1$ becomes variable and increases linearly with time. The system is absorbing energy at its natural frequency ω_1, but there is no dissipative element in the system. For the spring-mass system in Figure 5.4(a), the natural frequency is $\omega_1 = \sqrt{k/m}$, and the vertical displacement is the variable. For the

LC circuit in Figure 5.4(b), $\omega_1 = \sqrt{1/LC}$, and the current is the variable. Note that the input to the circuit is the derivative of the voltage dV/dt, as shown by the solution to Problem 5.20.

Returning to Example 5.3, it can be seen that Equation 5.44 is a special case of the more general second-order differential equation with damping coefficient $b = 0$ and $c = \omega_1^2$. That example gave the solution to the homogeneous equation. Substituting the appropriate values in Equation 5.39, the complementary solution to Equation 5.44 is

$$z_c(t) = c_1 e^{i\omega_1 t} + c_2 e^{-i\omega_1 t}.$$

The complete solution to the harmonic oscillator problem with no damping is, finally,

$$
\begin{aligned}
z(t) &= z_c(t) + z_p(t) \\
&= c_1 e^{i\omega_1 t} + c_2 e^{-i\omega_1 t} + z_p(t).
\end{aligned}
\tag{5.49}
$$

It will be informative for the reader to plot the results for the undamped oscillation model. When the frequency of the forcing function is not the resonance frequency, the solution is given by Equation 5.46. If resonance occurs, Equation 5.48 describes the response of the system described by Equation 5.44. Problem 5.25 asks you to plot both the time response and the amplitude of the response as the input function is changed in frequency. Problem 5.26 adds damping to the system and asks for similar plots of the response to a sinusoidal forcing function.

$\square$

Resonance   The effect of forcing a differential equation in synchronization with the frequency of the response is called *resonance*. In this case, energy is always added to the system by the forcing term and the response tends to grow in amplitude with time unless there is enough damping in the system to dissipate the added energy.

The spectacular destruction of the Tacoma Narrows bridge in 1940 has been attributed to the phenomenon of resonance. According to this explanation, a resonance condition was induced by wind blowing across the roadway, which subjected the structure to periodic vertical forces that eventually caused the bridge to collapse after undergoing "wild oscillations." A film was made of the bridge's destruction that is often shown to warn against resonance effects.[5]

More recent studies of the Tacoma Narrows bridge collapse dispute the resonance explanation. The new analysis blames nonlinear effects as the main factors leading to the large oscillations of the bridge. Proponents of the new theory do not credit resonance as the cause since resonance is a linear phenomenon modeled by linear differential equations. Problem 5.32 provides more information about this interesting debate.

---

[5]Resonance is usually to be avoided in mechanical structures. In electronic oscillators, resonance is a helpful and desired phenomenon.

The complete solution to the second-order differential equation with constant coefficients is of particular importance. If the equation is

$$\frac{d^2y(t)}{dt^2} + b\frac{dy(t)}{dt} + cy(t) = f(t),$$ (5.50)

the roots of the characteristic equation λ_1 and λ_2 were given in Example 5.3. Assuming that the roots are distinct, the solution to Equation 5.50 can be written

$$\begin{aligned} y(t) \;=\; & c_1 e^{\lambda_1 t} + c_2 e^{\lambda_2 t} \\ & + \frac{e^{\lambda_1 t}}{\lambda_1 - \lambda_2} \int e^{-\lambda_1 t} f(t)\,dt \\ & + \frac{e^{\lambda_2 t}}{\lambda_2 - \lambda_1} \int e^{-\lambda_2 t} f(t)\,dt. \end{aligned}$$ (5.51)

This form emphasizes the fact that the particular solution can be derived from the exponential functions of the homogeneous solution by the integration operations shown.

MATLAB
SOLUTION

MATLAB can solve the second-order differential equation defined by Equation 5.50.

☐ EXAMPLE 5.7 **_MATLAB Complete Differential Equation Solution_**
This MATLAB session shows the use of the symbolic command **dsolve** to solve an example of a second-order differential equation.

MATLAB Script
```
>> yeq='D2y+3*Dy+2*y=f(t)';    % Define equation for dsolve
>> y=dsolve(yeq,'t');          % Solve the equation
>> pretty(y)                   %   and simplify the result
y =
    /                              /
    |                              |
    |   exp(t) f(t) dt exp(- t) +  |  - exp(2 t) f(t) dt exp(- 2 t)
    |                              |
    /                              /
    + C1 exp(- t) + C2 exp(- 2 t)
```

We recognize the MATLAB result as

$$\begin{aligned} y(t) \;=\; & e^{-t}\int e^t f(t)\,dt - e^{-2t}\int e^{2t} f(t)\,dt \\ & + c_1 e^{-t} + c_2 e^{-2t}. \end{aligned}$$

Comparing the solution Equation 5.51 for the equation

$$\frac{d^2y(t)}{dt^2} + 3\frac{dy(t)}{dt} + 2y(t) = f(t)$$

shows that MATLAB has found the complete solution in symbolic form.

In fact, the symbolic MATLAB command **dsolve** can solve certain differential equations with variable coefficients and even nonlinear differential equations. However, the MATLAB **dsolve** routine must find the solutions to the homogeneous equation to solve differential equations with variable coefficients. This is demonstrated in Problem 5.29.

□

VARIATION OF PARAMETERS A method designated *variation of parameters* will yield (in principle) a particular solution of $\mathcal{L}[y_p(t)] = f(t)$, even for differential equations with variable coefficients, provided the related homogeneous equation can be solved. The technique is discussed in a later section of this chapter.

SYSTEMS OF DIFFERENTIAL EQUATIONS

The topics in the remaining sections of this chapter will unite our knowledge of matrices, eigenvalues, and eigenvectors with differential equations. Instead of the equations involving single dependent variables previously studied, systems of differential equations will be presented. These equations often arise naturally in physical problems when a model is devised that assumes that the rate of growth or decrease of a variable, $\dot{x}_i(t)$, is a linear combination of all the variables. The rate of change of the ith variable can be written in terms of the other variables as

$$\dot{x}_i(t) = a_{i1}(t)x_1(t) + \cdots + a_{in}(t)x_n(t), \qquad i = 1, 2, \ldots, n.$$

The coefficients $a_{ij}(t)$, for $i, j = 1, \ldots, n$, are determined by physical considerations, and n is the number of variables involved. Thus, n first-order differential equations in the n unknown functions $x_i(t), i = 1 \ldots, n$, have been formed.

Although the equation set just written only involves the first derivative in $x(t)$, it will be shown that nth-order differential equations and higher-order systems of equations can be converted to a system of n first-order equations. Therefore, the solution method for the first-order system of equations is a very general method of solving linear differential equations. The approach is also well suited to matrix methods and modern numerical techniques.

We consider the linear system

$$
\begin{aligned}
\dot{x}_1(t) &= a_{11}(t)x_1(t) + \cdots + a_{1n}(t)x_n(t) + f_1(t), \\
\dot{x}_2(t) &= a_{21}(t)x_1(t) + \cdots + a_{2n}(t)x_n(t) + f_2(t), \\
&\vdots \\
\dot{x}_n(t) &= a_{n1}(t)x_1(t) + \cdots + a_{nn}(t)x_n(t) + f_n(t).
\end{aligned}
\tag{5.52}
$$

This equation set will be written as

$$\dot{\mathbf{x}}(t) = A(t)\mathbf{x}(t) + \mathbf{f}(t), \tag{5.53}$$

where $\dot{\mathbf{x}}(t)$ is the vector of derivatives, $A(t)$ is the $n \times n$ matrix of coefficients, $\mathbf{x}(t)$ is the vector of the x functions, and $\mathbf{f}(t)$ is the vector of forcing functions in Equation 5.53. It will be assumed that the coefficients, $a_{ij}(t)$ for $i, j = 1, \ldots, n$, are continuous in the interval of interest on the t-axis, say, I.

As is usual with linear differential equations, we first consider the solutions of the homogeneous equation $\dot{\mathbf{x}}(t) = A(t)\mathbf{x}(t)$. The set of solutions forms a vector space so that linear combinations of solutions and scalar multiples of solutions are also solutions. The dimension of the solution space is n, and a *fundamental set of solutions* is any set of n linearly independent solutions to the homogeneous equation. Thus, if $\mathbf{x}_1(t), \mathbf{x}_2(t), \ldots, \mathbf{x}_n(t)$ are linearly independent solution vectors of

$$\dot{\mathbf{x}}(t) = A(t)\mathbf{x}(t),$$

the general solution is

$$\mathbf{x}(t) = c_1\mathbf{x}_1(t) + c_2\mathbf{x}_2(t) + \cdots + c_n\mathbf{x}_n(t),$$

where $c_1, \ldots, c_n$ are arbitrary constants. These constants can be determined if the initial conditions are specified. The test for independence of the solution vectors involves forming a determinant whose columns are the solution vectors evaluated at some point in the interval I on which the solution is defined. Thus, $\mathbf{x}_1(t), \mathbf{x}_2(t), \ldots, \mathbf{x}_n(t)$ are linearly independent if and only if, for any t in I,

$$\begin{vmatrix} x_{11}(t) & x_{12}(t) & \cdots & x_{1n}(t) \\ x_{21}(t) & x_{22}(t) & \cdots & x_{2n}(t) \\ \vdots & \vdots & \ddots & \vdots \\ x_{n1}(t) & x_{n2}(t) & \cdots & x_{nn}(t) \end{vmatrix} \neq 0.$$

In the following sections, a number of systems of linear differential equations with constant coefficients in the matrix A are presented. As in the case of the linear differential equation with constant coefficients studied earlier, the solutions to the homogeneous system of equations are easily found in principle.

HOMOGENEOUS SYSTEM WITH CONSTANT COEFFICIENTS

Letting $f(t) = 0$ in Equation 5.52, a homogeneous system of linear first-order differential equations $\dot{\mathbf{x}}(t) = A\mathbf{x}(t)$ with constant coefficients can be written as

$$\begin{bmatrix} \dot{x}_1(t) \\ \dot{x}_2(t) \\ \vdots \\ \dot{x}_n(t) \end{bmatrix} = \begin{bmatrix} a_{11} & a_{12} & \cdots & a_{1n} \\ a_{21} & a_{22} & \cdots & a_{2n} \\ \vdots & \vdots & & \vdots \\ a_{n1} & a_{n2} & \cdots & a_{nn} \end{bmatrix} \begin{bmatrix} x_1(t) \\ x_2(t) \\ \vdots \\ x_n(t) \end{bmatrix}. \tag{5.54}$$

Since A is an $n \times n$ matrix, the general solution to Equation 5.54 consists of n linearly independent vectors $\mathbf{x}_i(t), i = 1, 2, \ldots, n$. In the discussion to follow, we assume that the elements of matrix A are real constants.

Although many of the matrix techniques in this chapter are very general, it is important to realize that the equations being solved here represent a set of *first-order* equations that are linear in the unknowns. Also, the coefficients are *constant*, so the matrix A in Equation 5.54 is independent of time. Equations with nonconstant coefficients are studied in more detail in Chapter 6.

The system of Equation 5.54 will be studied in matrix notation

$$\dot{\mathbf{x}}(t) = A\mathbf{x}(t),\quad (5.55)$$

which has a striking resemblance to the first-order differential equation $\dot{x}(t) = ax(t)$ in which a is a scalar. The theory leading to the solution of the system of equations parallels the approach used to solve the first-order single equation, as previously presented. Immediately, we are lead to try solutions of the form $x(t) = \alpha e^{\lambda t}$, but in vector form $\mathbf{x}(t) = \mathbf{u}e^{\lambda t}$, where $\mathbf{u}$ is a constant vector.

Thus, we assume a solution of the form $\mathbf{x}(t) = \mathbf{u}e^{\lambda t}$ and substitute $\dot{\mathbf{x}}(t) = \lambda \mathbf{u}e^{\lambda t}$ in Equation 5.55. For $\mathbf{x}(t)$ to be a solution, the equation

$$\lambda \mathbf{u}e^{\lambda t} = A\mathbf{u}e^{\lambda t}$$

must be satisfied. Dividing by $e^{\lambda t}$, which is never zero for a finite value of t, yields

$$A\mathbf{u} - \lambda \mathbf{u} = (A - \lambda I)\mathbf{u} = \mathbf{0}.$$

This shows that the solution of the linear system of equations can be found if the eigenvalue problem is solved for the eigenvalues and eigenvectors of the matrix A. This problem was studied extensively in Chapter 4.

The following theorem defines the solutions of the set of differential equations in Equation 5.54 when the eigenvalues and eigenvectors are known.

■ THEOREM 5.7   *Solution of the Linear Homogeneous System*
Let A be an $n \times n$ matrix with eigenvalues

$$\lambda_1, \lambda_2, \ldots, \lambda_n,$$

all of which may not be distinct. Let $\mathbf{u}_i$ be associated with the eigenvalue λ_i. Then, if $\mathbf{u}_1, \mathbf{u}_2, \ldots, \mathbf{u}_n$ are linearly independent as vectors in $\mathbf{R}^n$, the linearly independent solutions of $\dot{\mathbf{x}}(t) = A\mathbf{x}(t)$ are

$$\mathbf{u}_1 e^{\lambda_1 t}, \mathbf{u}_2 e^{\lambda_2 t}, \ldots, \mathbf{u}_n e^{\lambda_n t}. \quad (5.56)$$

When we are able to produce n linearly independent eigenvectors, the general solution can be written as

$$\mathbf{x}(t) = c_1\mathbf{u}_1 e^{\lambda_1 t} + c_2\mathbf{u}_2 e^{\lambda_2 t} + \ldots + c_n\mathbf{u}_n e^{\lambda_n t} \qquad (5.57)$$

even if the eigenvalues are not distinct.

If the eigenvalues are distinct, then the eigenvectors associated with the distinct eigenvalues of the matrix are linearly independent. However, if A does not have distinct eigenvalues, A may or may not have n linearly independent eigenvectors, as previously discussed in Chapter 4. Even in this case, n linearly independent solutions to the system of equations can be found.

□ EXAMPLE 5.8   *Linear System of Equations Example*
Consider the system

$$
\begin{aligned}
\dot{x}_1(t) &= x_1(t) + x_2(t), \\
\dot{x}_2(t) &= 4x_1(t) + x_2(t),
\end{aligned}
\qquad (5.58)
$$

with initial conditions $\mathbf{x}(t_0) = \mathbf{x}_0$, a constant vector. This represents a system of two first-order differential equations with initial values of the functions defined at $t = t_0$. The system in matrix form is $\dot{\mathbf{x}}(t) = A\mathbf{x}(t)$, where

$$A = \begin{bmatrix} 1 & 1 \\ 4 & 1 \end{bmatrix}.$$

The eigenvalues of this system are distinct, so the solution can be written using the form defined in Theorem 5.7. However, we will show a number of the mathematical details in this example.

The assumed solution to the equation set of Equation 5.58 takes the form $\mathbf{x}(t) = \mathbf{u}e^{\lambda t}$ when λ satisfies the characteristic equation $\lambda\mathbf{u}e^{\lambda t} = A\mathbf{u}e^{\lambda t}$. Thus, the assumed solutions are

$$
\begin{aligned}
x_1(t) &= u_1 e^{\lambda t}, \\
x_2(t) &= u_2 e^{\lambda t},
\end{aligned}
$$

with the values of λ and the vector $\mathbf{u} = [u_1, u_2]^t$ to be determined.

Substituting in the system of Equation 5.58 and dividing both sides by $e^{\lambda t}$ (which is nonzero) yields the system of equations

$$
\begin{aligned}
\lambda u_1 &= u_1 + u_2, \\
\lambda u_2 &= 4u_1 + u_2,
\end{aligned}
$$

which is the eigenvalue equation $A\mathbf{u} = \lambda\mathbf{u}$ in the form

$$\begin{bmatrix} 1 & 1 \\ 4 & 1 \end{bmatrix} \begin{bmatrix} u_1 \\ u_2 \end{bmatrix} = \lambda \begin{bmatrix} u_1 \\ u_2 \end{bmatrix}.$$

By methods described in Chapter 4, the eigenvalues are found from the equation $(A - \lambda I) = 0$ to be $\lambda_1 = -1$ and $\lambda_2 = +3$ with eigenvectors

$$\mathbf{u}_1 = \begin{bmatrix} -1/2 \\ 1 \end{bmatrix} \quad \text{and} \quad \mathbf{u}_2 = \begin{bmatrix} 1/2 \\ 1 \end{bmatrix}.$$

The general solution to the homogeneous equation set is then

$$\mathbf{x}(t) = c_1 \begin{bmatrix} -1/2 \\ 1 \end{bmatrix} e^{-t} + c_2 \begin{bmatrix} 1/2 \\ 1 \end{bmatrix} e^{3t}. \tag{5.59}$$

The constants c_1 and c_2 are determined by the initial values of $x_1(t_0)$ and $x_2(t_0)$. In equation form, the solutions of the system in Equation 5.58 are

$$x_1(t) = -\frac{1}{2}c_1 e^{-t} + \frac{1}{2}c_2 e^{3t},$$

$$x_2(t) = c_1 e^{-t} + c_2 e^{3t}.$$

To summarize, once the two distinct eigenvalues and corresponding eigenvectors of A are found, the solution can be written as

$$\mathbf{x}(t) = c_1 \mathbf{u}_1 e^{\lambda_1 t} + c_2 \mathbf{u}_2 e^{\lambda_2 t}$$

in terms of the eigenvalues and eigenvectors of A. The specific result was shown in Equation 5.59.

The constants c_1 and c_2 are determined by the initial conditions $\mathbf{x}(t_0) = \mathbf{x}_0$ by solving the algebraic system

$$\mathbf{x}(t_0) = \mathbf{x}_0 = \sum_{k=1}^{2} c_k \mathbf{u}_k e^{\lambda_k t_0},$$

which can be written as $U\mathbf{c} = \mathbf{x}_0$. The solution is

$$\mathbf{c} = U^{-1}\mathbf{x}_0, \tag{5.60}$$

in which

$$U = [\mathbf{u}_1 e^{\lambda_1 t_0}, \mathbf{u}_2 e^{\lambda_2 t_0}] \quad \text{and} \quad \mathbf{c} = \begin{bmatrix} c_1 \\ c_2 \end{bmatrix}.$$

The inverse U^{-1} is guaranteed to exist since the eigenvectors of A are linearly independent.

Specific initial conditions. Suppose that the initial conditions at $t_0 = 0$ are

$$\mathbf{x}(0) = \begin{bmatrix} 1 \\ 1 \end{bmatrix}.$$

Using Equation 5.60 with $t_0 = 0$, the matrix U becomes

$$U = \begin{bmatrix} -\dfrac{1}{2} & \dfrac{1}{2} \\ 1 & 1 \end{bmatrix}, \tag{5.61}$$

which yields the vector $\mathbf{c}$ as $[-1/2, 3/2]^T$. Substituting c_1 and c_2 in Equation 5.59 leads to the solution

$$\mathbf{x}(t) = -\frac{1}{2} \begin{bmatrix} -\dfrac{1}{2} \\ 1 \end{bmatrix} e^{-t} + \frac{3}{2} \begin{bmatrix} \dfrac{1}{2} \\ 1 \end{bmatrix} e^{3t}. \tag{5.62}$$

Matrix solution. As discussed in Chapter 4, the system of differential equations

$$\frac{d\mathbf{x}(t)}{dt} = A\mathbf{x}, \tag{5.63}$$

where A is a real constant matrix, has the solution

$$\mathbf{x}(t) = e^{At}\mathbf{x}_0 \tag{5.64}$$

with the initial values at $t = 0$ of $\mathbf{x}(0) = \mathbf{x}_0$. Although you are asked to supply the details in Problem 5.30, the solution to the system of Equation 5.58 can be computed as

$$\mathbf{x}(t) = e^{At}\mathbf{x}_0 = e^{M\Lambda M^{-1}t}\mathbf{x}_0, \tag{5.65}$$

where M is the modal matrix and Λ is the diagonal matrix

$$\Lambda = M^{-1}AM.$$

For the present example,

$$M = \begin{bmatrix} -\dfrac{1}{2} & \dfrac{1}{2} \\ 1 & 1 \end{bmatrix}$$

using the notation M for the modal matrix, as in Chapter 4. A comparison with the matrix U of Equation 5.61 shows that U and the modal matrix are the same when the initial conditions are given at $t = 0$.

$\square$

MATLAB SOLUTIONS OF SYSTEMS OF DIFFERENTIAL EQUATIONS

The MATLAB commands **dsolve**, **ode23**, and **ode45** are used to solve ordinary differential equations. The commands can save us a great deal of labor, but as with any computer program, they must be used with care. This section briefly discusses these commands.

Symbolic solutions to differential equations are obviously useful for problems in which parameters can be varied to gain greater understanding of the physical system being modeled. Thus, in a model with damping, the damping constant can be altered, and the various responses to a forcing function can be plotted. Unfortunately, the number of differential equations that allow symbolic solutions is relatively small. In fact, almost all systems of differential equations that arise from practical problems require numerical solution. Also, very few differential equations with variable coefficients can be solved symbolically.

In this section, two of MATLAB's numerical routines to solve differential equations are also introduced. However, more details concerning the algorithms and possible numerical errors are not presented until Chapter 6.

SYMBOLIC SOLUTIONS

For the purposes of this chapter, **dsolve** is the most important command in MATLAB's *Symbolic Math Toolbox*. As described in the *MATLAB User's Guide*, the symbolic problem-solving software called Maple is used to find the solution to differential equations defined symbolically. The command **dsolve** can be used to solve a single differential equation or a system of equations. The command was employed to solve a first-order differential equation in Example 5.2. Example 5.9 demonstrates the use of **dsolve** for a system of differential equations. As we shall see in Chapter 6, **dsolve** can be used to solve equations that are very difficult to solve by hand.

WHAT IF? Suppose you want more information about the **dsolve** command. Of course, the *MATLAB User's Guide* contains a detailed description of each MATLAB command. However, the online help is also very useful and more convenient to use when solving problems with MATLAB. For example, type

```
help dsolve
```

to view the description of **dsolve**.

The help for **dsolve** will provide several screens of information which may scroll by too rapidly to read. To prevent the scrolling, use the command **more on** and then push the space bar to advance page by page on the screen. Use the command **more off** when you are finished.

You will probably notice that the results from **dsolve** are often not in the most convenient form for analysis. The command **pretty** may display the output in a more readable format. Also, try the command **simple** to simplify the algebraic form of the answer.

NUMERICAL SOLUTIONS

MATLAB has the functions **ode23** and **ode45** for computing numerical solutions to ordinary differential equations. Both functions integrate a first-order system of differential equations in the form

$$\frac{d\mathbf{x}(t)}{dt} = \mathbf{f}(t, \mathbf{x}).$$

Details about the algorithms are not presented in this chapter, but **ode23** is used in examples to demonstrate MATLAB's numerical capability to solve differential equations. The two-digit numerical suffix indicates the order of the Runge-Kutta equations used to solve the equations. Thus, **ode23** uses second-order and third-order equations. Chapter 6 presents a description of the modified Runge-Kutta algorithms that MATLAB uses.

The function **ode23p** is a variation of **ode23** that can be used to solve a system of equations and plot the results. For a complete list of MATLAB differential equation solving commands and related functions, use the command **help funfun**.

MATLAB dsolve *Example*

The symbolic command **dsolve** can solve many systems of differential equations. The accompanying MATLAB script shows the M-file to solve the system of Example 5.8. Unfortunately, variables such as x1 are not accepted by **dsolve**, so the variable names have been changed to x and y; thus, the system of equations becomes

$$\dot{x}(t) = x(t) + y(t),$$
$$\dot{y}(t) = 4x(t) + y(t)$$

with the initial conditions $[1,1]^T$.

MATLAB Script _____

```
Example 5.9
% EX5_9.M Solve the system of equations
% Dx1(t)= x1 + x2
% Dx2(t)=4x1 + x2
% Use dsolve with x=x1 and y=x2; Initial [1 1]
xvec='Dx=x+y,Dy=4*x+y'      % Print the results
initc='x(0)=1,y(0)=1'
[x1,x2]=dsolve(xvec,initc)
```

The results from the diary file are given in the following session listing.

MATLAB Script _____

```
Example 5.9
>>ex5_9
xvec =
Dx=x+y,Dy=4*x+y
initc =
x(0)=1,y(0)=1
x1 =
3/4*exp(3*t)+1/4*exp(-t)
x2 =
3/2*exp(3*t)-1/2*exp(-t)
>>quit
```

A comparison of these MATLAB results with Equation 5.62 shows that the **dsolve** solution is the same as the analytical solution. Thus, the system defined in Equation 5.58 can be solved in various ways. By hand, the solution depends on our ability to find the eigenvectors of the system matrix. For small systems and larger systems in special form, a solution by hand is often possible. The use of **dsolve** is very convenient when a symbolic solution can be found.

□

Solving the system of Example 5.9 presents no problem to the **dsolve** command. We now wish to formulate the problem so that a numerical solution can be found. This numerical solution to the equations

$$\frac{d\mathbf{x}(t)}{dt} = \mathbf{f}(t,\mathbf{x}) \tag{5.66}$$

is a sequence of vectors $[\hat{\mathbf{x}}_1, \hat{\mathbf{x}}_2, \ldots \hat{\mathbf{x}}(n)]$ corresponding to an increasing sequence of time points $[t_1, t_2, \ldots t_n]$ such that $\hat{\mathbf{x}}(t_i)$ approximates $\mathbf{x}(t_i)$. When the MATLAB commands are used to solve the equations numerically, the time points will be chosen automatically by the routine, and the user has no control over their selection. The automatic selection of the time points is made by the program to achieve the maximum accuracy, as described in Chapter 6.

A system of differential equations in the form

$$\dot{\mathbf{x}}(t) = A\mathbf{x}(t),$$

where A is a matrix of constants, is a special case of Equation 5.66. This yields a linear relationship between $\dot{\mathbf{x}}$ and $\mathbf{x}$ in the form shown as Equation 5.54. Although the MATLAB differential-solving routines **ode23** and **ode45** do not require time invariance or even linearity of the system of equations, the emphasis in this chapter is on linear equations with constant coefficients.

When using **ode23** or **ode45**, the system of equations must be defined in a separate function M-file. One advantage to defining the system of equations independently of the routine to solve the system is that various differential-solving routines could be used to solve the equations and the results compared. A typical call to **ode23** is

```
[t,x]=ode23('cl2ordf',t0,tf,x0)
```

The parameters for the command are the function defining the equations (`cl2ordf`), the initial time (`t0`), the final time (`tf`), and the vector (`x0`) defining the initial conditions at the initial time. There are other parameters, including one called *tolerance* which defines the desired accuracy of the solution. In this chapter, we accept the default value of the tolerance as 1×10^{-3}.

☐ **EXAMPLE 5.10**   ***MATLAB Linear System Example***
Consider the system of equations solved analytically in Example 5.8:

$$\begin{aligned} \dot{x}_1(t) &= x_1(t) + x_2(t), \\ \dot{x}_2(t) &= 4x_1(t) + x_2(t). \end{aligned} \qquad (5.67)$$

The accompanying M-file CL2ORD.M solves the equation set and plots the result. The command **ode23** actually solves the system of differential equations with given initial conditions. Inputs to the command must include an M-file function that defines the equation set, the starting time and final time for the solution, and a column vector containing the initial values. The outputs are a column vector or time values and the solution vectors with one column vector per time point.

When M-file CL2ORD.M is executed, the final time must be input. In the example, the equations are solved between $t = 0$ and $t = 1$ second. The function CL2ORDF.M defines the system of first-order differential equations. The inputs to the function are the values of t and $\mathbf{x}$ from **ode23** and the system matrix

---

from the calling M-file as a global variable. The output is a vector representing $\dot{\mathbf{x}}$ which **ode23** uses to approximate the solution to the equation system at each time point.

The initial conditions were $\mathbf{x}_0 = [1, 1]^T$, and the results were plotted from $t = 0$ to $t = 1$. Different line types distinguish between the solutions $x_1(t)$ and $x_2(t)$.

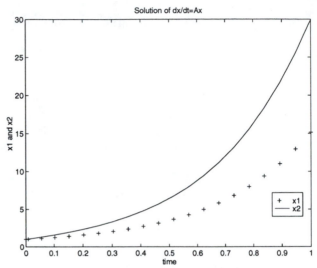

FIGURE 5.5 *Numerical solution of the differential equation system of Example 5.8*

MATLAB Script _____

```
Example 5.10
% CL2ORD.M - Plot solution to differential eq. system xdot=Ax:
% This M-file uses function CL2ORDF.M to define equations
%     and ode23 to solve them.
% INPUT:  Final time; Plot is from t=0 to t=tfinal
%    (System matrix is defined as A=[1 1;4 1]; x(0)=[1 1])
% OUTPUT: A plot of solutions x1 and x2 versus t
%
global A                % This passes A to function CL2ORDF
A=[1 1;4 1];            % Define system matrix
t0=0;
tf = input('Input tfinal - function is exp(-t) + exp(3t) ')
% Initial values x1(0) = 1, x2(0) = 1.  (Define as column vector)
x0=[1 1]';
[t,x]=ode23('CL2ORDF',t0,tf,x0);  % MATLABs Runge-Kutta routine
plot(t,x(:,1),'+',t,x(:,2),'-')   % Plot x1 and x2
title('Solution of dx/dt=Ax')
xlabel('time'); ylabel('x1 and x2')
legend('x1','x2')       % Annotate the graph
```

In terms of the variables defined in M-function CL2ORDF, the relationship between the variables defined in the function and the variables in Equation 5.67 is

$$\begin{aligned} \texttt{x\_prime(1)} &= \dot{x}_1(t), \\ \texttt{x\_prime(2)} &= \dot{x}_2(t). \end{aligned}$$

MATLAB Script _____

```
Example 5.10
function x_prime = cl2ordf(t,x)
% CALL: x_prime = cl2ordf(t,x);  Function to define the system
%    x_prime = A*x at each point x(t); A is a global variable
global A        % A is system matrix passed to function
x_prime=A*x;
```
_____

□

WHAT IF? Suppose you wish to study the operation of **ode23** in more detail. First, see the help screen for the command. Then, solve a problem and see how many points MATLAB uses for the solution. In Example 5.10, the equations were solved at 21 time points between 0 and 1. The array **x** has size 21×2. Also, plot the results for various values of **tf** and see how the numerical accuracy decreases for large values of time. This is not surprising, since the solution increases exponentially with time. A system with such a response is called *unstable*. Compare the solution with the symbolic solution in Example 5.9.

Remember that the MATLAB command **eig** will compute the eigenvalues and eigenvectors for the matrix A if the analytic solution to the problem is desired. In other words, MATLAB can be used to do the algebra in Example 5.8 or to check the results.

HOMOGENEOUS SYSTEMS WITH REPEATED EIGENVALUES
––––––––––––––––––––––■––––––––––

In a previous section, we found solutions of the homogeneous system of equations $\dot{\mathbf{x}}(t) = A\mathbf{x}(t)$ that were combinations of function of the form $\mathbf{u}e^{\lambda t}$. If the $n \times n$ matrix A has n distinct eigenvalues, the associated eigenvectors are linearly independent, and these eigenvectors are a basis for the solutions of the linear system of equations.

When the set of eigenvalues of A contains repeated eigenvalues, there may or may not be n linearly independent eigenvectors. The test based on the rank of $A - \lambda I$ to determine whether n independent eigenvectors exist for an $n \times n$ matrix with repeated eigenvalues was given in Chapter 4. For a system of differential equations $\dot{\mathbf{x}}(t) = A\mathbf{x}(t)$, an eigenvalue of multiplicity m will always provide m linearly independent solutions of the system.

☐ EXAMPLE 5.11   *Linear System Example* $\dot{\mathbf{x}} = I\mathbf{x}$

Consider the system $\dot{\mathbf{x}}(t) = A\mathbf{x}(t)$,

$$\begin{bmatrix} \dot{x}_1(t) \\ \dot{x}_2(t) \\ \dot{x}_3(t) \end{bmatrix} = \begin{bmatrix} 1 & 0 & 0 \\ 0 & 1 & 0 \\ 0 & 0 & 1 \end{bmatrix} \begin{bmatrix} x_1(t) \\ x_2(t) \\ x_3(t) \end{bmatrix}, \qquad (5.68)$$

in which A is the identity matrix I. The eigenvalue equation $\det(A - \lambda I) = 0$ yields $(1 - \lambda)^3 = 0$. Hence, $\lambda = 1$ is the only eigenvalue, but with multiplicity three.[6] The eigenvector equation is

$$(A - 1I)\mathbf{u} = \begin{bmatrix} 0 & 0 & 0 \\ 0 & 0 & 0 \\ 0 & 0 & 0 \end{bmatrix} \begin{bmatrix} u_1 \\ u_2 \\ u_3 \end{bmatrix} = \begin{bmatrix} 0 \\ 0 \\ 0 \end{bmatrix}$$

for any choice of u_1, u_2, u_3. Thus, even though the eigenvalue $\lambda = 1$ is repeated three times, we can find three linearly independent eigenvectors. Choosing them to be the natural basis for $\mathbf{R}^3$ yields

$$\mathbf{u} = u_1 \begin{bmatrix} 1 \\ 0 \\ 0 \end{bmatrix} + u_2 \begin{bmatrix} 0 \\ 1 \\ 0 \end{bmatrix} + u_3 \begin{bmatrix} 0 \\ 0 \\ 1 \end{bmatrix}.$$

Since any nonzero vector $\mathbf{x}_0$ is an eigenvector when $A = I$, the solution to $\dot{\mathbf{x}}(t) = I\mathbf{x}(t)$ of Equation 5.68 with $\mathbf{x}(0) = \mathbf{x}_0$ is $\mathbf{x}(t) = \mathbf{x}_0 e^t$.

☐

Suppose that the $n \times n$ matrix in the equation $\dot{\mathbf{x}}(t) = A\mathbf{x}(t)$ does not yield n linearly independent eigenvectors. Then, if λ_r is a repeated eigenvalue of multiplicity m, solutions are sought in the form

$$(\mathbf{k}_1 + \mathbf{k}_2 t + \cdots + \mathbf{k}_m t^{m-1})e^{\lambda_r t}, \qquad (5.69)$$

where the vectors $\mathbf{k}_i$ are column vectors to be determined.[7]

Thus, the procedure to find n linearly independent solutions to the homogeneous differential equation system consists of the following steps:

1. Find the linearly independent eigenvectors $\mathbf{u}_i$ corresponding to distinct eigenvalues, λ_i, that each yield a solution of the form $c_i \mathbf{u}_i e^{\lambda_i t}$.

2. Find the linearly independent solutions corresponding to each of the multiple eigenvalues. For example, for an eigenvalue of multiplicity m, the solution is of the form

$$(\mathbf{k}_1 + \mathbf{k}_2 t + \cdots + \mathbf{k}_m t^{m-1})e^{\lambda t}.$$

3. Form a linear combination of all of the solutions found in Step 1 and Step 2 to yield the general solution to $\dot{\mathbf{x}}(t) = A\mathbf{x}(t)$.

---

[6]It is not necessary to solve for the eigenvalues of a diagonal matrix. The diagonal elements are the eigenvalues.

[7]For some matrices A, it may be possible to find m linearly independent eigenvectors for an eigenvalue of multiplicity m. In this case, it is not necessary to multiply the column vectors in Equation 5.69 by the terms in t.

When there are not n linearly independent eigenvectors of the $n \times n$ matrix A of the system of equations $\dot{\mathbf{x}}(t) = A\mathbf{x}(t)$, various methods can be applied to find the m vectors in Step 2. Several of these methods are described in textbooks listed in the Annotated Bibliography at the end of this chapter. Unfortunately, every method requires a considerable amount of algebra in most cases. Once the techniques are mastered, the aid afforded by computer solutions will be greatly appreciated. One straightforward approach is presented in Example 5.12.

☐ **EXAMPLE 5.12** *Repeated Eigenvalues*

Consider the system of equations

$$\dot{\mathbf{x}}(t) = \begin{bmatrix} 1 & 1 \\ 0 & 1 \end{bmatrix} \mathbf{x}(t) \quad \text{with} \quad \mathbf{x}(0) = [x_{01}, x_{02}]^T. \qquad (5.70)$$

For the system matrix, $\lambda = 1$ is an eigenvalue of multiplicity 2 and $\mathbf{u} = [1, 0]^T$ is an eigenvector. Since the system is

$$\dot{x}_1 = x_1 + x_2 \quad \text{and} \quad \dot{x}_2 = x_2,$$

the second equation is easily solved to yield $x_2 = e^t$. Thus, $\dot{x}_1 = x_1 + e^t$, which is a first-order nonhomogeneous equation that can be solved by various techniques.

Returning to Equation 5.70, the eigenvector $\mathbf{u}$ belonging to eigenvector $\lambda = 1$ can form one solution. We immediately write

$$\mathbf{x}_1 = \begin{bmatrix} 1 \\ 0 \end{bmatrix} e^t$$

as a solution. Notice that this solution could not serve to satisfy arbitrary initial conditions since $x_{12} = 0$. We search for a second solution that not only satisfies the equation set but also will allow any arbitrary initial condition to be satisfied.

According to the prescription in Equation 5.69, the second solution is of the form

$$\mathbf{x}_2 = (\mathbf{k}_1 + t\mathbf{k}_2)e^{\lambda t},$$

where $\lambda = 1$ is the eigenvalue and $\mathbf{k}_2 = \mathbf{u}$, the known eigenvector. Substitution of this assumed solution in Equation 5.70 results in the equation

$$e^{\lambda t}(t\lambda\mathbf{u} + \lambda\mathbf{k}_1 + \mathbf{u}) = e^{\lambda t}(tA\mathbf{u} + A\mathbf{k}_1).$$

Dividing both sides by $e^{\lambda t}$ and equating coefficients of the polynomial in t yields

$$t(A\mathbf{u}) = t(\lambda\mathbf{u}) \quad \text{and} \quad (A\mathbf{k}_1) = (\lambda\mathbf{k}_1 + \mathbf{u}).$$

The first equation yields only the eigenvalue equation $A\mathbf{u} = \lambda\mathbf{u}$ that was known, and the second equation becomes

$$(A - \lambda I)\mathbf{k}_1 = \mathbf{u},$$

which allows solution for the unknown vector $\mathbf{k}_1$. One approach is to multiply both sides of this equation by $A - \lambda I$ and notice that the right-hand side is again $(A - \lambda I)\mathbf{u} = \mathbf{0}$, so

$$(A - \lambda I)^2 \mathbf{k}_1 = 0.$$

After evaluation, the equation for k_1 becomes

$$\begin{bmatrix} 0 & 0 \\ 0 & 0 \end{bmatrix} k_1 = \begin{bmatrix} 0 \\ 0 \end{bmatrix}.$$

Thus, any vector $k_1 \neq 0$ will satisfy the equation. Choosing k_1 to be $[0, 1]^T$ results in the solution

$$\mathbf{x}(t) = c_1 \begin{bmatrix} 1 \\ 0 \end{bmatrix} e^t + c_2 \left(\begin{bmatrix} 0 \\ 1 \end{bmatrix} + t \begin{bmatrix} 1 \\ 0 \end{bmatrix} \right) e^t,$$

which solves Equation 5.70 and will satisfy any arbitrary initial conditions. The known eigenvector $\mathbf{u}$ is orthogonal to k_1 for the choice of k_1 that was made.

The solution is easily verified with the MATLAB command **dsolve**.

$\square$

NONHOMOGENEOUS SYSTEMS OF DIFFERENTIAL EQUATIONS

This section is concerned with the solution of the equation

$$\dot{\mathbf{x}}(t) = A\mathbf{x}(t) + \mathbf{f}(t), \tag{5.71}$$

where $\mathbf{f}(t) \neq 0$ and A is a matrix of constants. There are a number of ways to find the complete solution to the equation that follow from the methods that have been applied to the scalar equation $\dot{x}(t) = ax(t) + f(t)$. The following discussions cover the techniques of *undetermined coefficients* and *variation of parameters*.

METHOD OF UNDETERMINED COEFFICIENTS

Occasionally it is possible to determine the particular solution to Equation 5.71 by inspection, just as it was for the nth-order differential equation. Table 5.4 lists a few cases.

TABLE 5.4 *Particular solutions*

$\mathbf{f}(t)$	$\dot{\mathbf{x}}_p(t) = A\mathbf{x}_p(t) + \mathbf{f}(t)$
$\mathbf{b}$	$\mathbf{x}_p = A^{-1}\mathbf{b}$
$\mathbf{b}e^{\omega t}$	$\mathbf{x}_p = \mathbf{k}e^{\omega t}$
$\mathbf{b}e^{i\omega t}$	$\mathbf{x}_p = \mathbf{k}e^{i\omega t}$

In the table, $\mathbf{b}$ is a known constant vector and A must be an invertible matrix. If the forcing function is an exponential, vector $\mathbf{k}$ is a constant vector to be determined. The solution technique is to substitute the assumed particular solution $\mathbf{x}_p(t)$ in the equation and determine the constant vector $\mathbf{k}$. The functions of ω are assumed not to be solutions of the homogeneous equation. The complete solution to Equation 5.71 is then the sum of $\mathbf{x}_p(t)$ and the general solution of the homogeneous equation $\dot{\mathbf{x}}_c(t) = A\mathbf{x}_c(t)$.

□ EXAMPLE 5.13   *Undetermined Coefficients*

Consider finding a particular solution for the system

$$\dot{\mathbf{x}}(t) = \begin{bmatrix} 0 & 1 \\ -1 & 0 \end{bmatrix} \mathbf{x} + re^{i\omega t} \begin{bmatrix} 0 \\ 1 \end{bmatrix}.$$

Try a solution of the form $\mathbf{x}_p = \mathbf{k}e^{i\omega t}$, with the result

$$i\omega e^{i\omega t}\mathbf{k} = e^{i\omega t}\begin{bmatrix} 0 & 1 \\ -1 & 0 \end{bmatrix}\mathbf{k} + re^{i\omega t}\begin{bmatrix} 0 \\ 1 \end{bmatrix}.$$

Solving for $\mathbf{k}$ leads to the equation

$$\begin{bmatrix} k_1 \\ k_2 \end{bmatrix} = \frac{1}{1 - \omega^2}\begin{bmatrix} -i\omega & -1 \\ 1 & -i\omega \end{bmatrix}\begin{bmatrix} 0 \\ -r \end{bmatrix} = \frac{r}{1 - \omega^2}\begin{bmatrix} 1 \\ i\omega \end{bmatrix}.$$

If $\omega \neq 1$, a particular solution is

$$\mathbf{x}_p(t) = \frac{r}{1 - \omega^2}\begin{bmatrix} 1 \\ i\omega \end{bmatrix}e^{i\omega t}.$$

□

VARIATION OF PARAMETERS

Remarkably, the general solution of the nonhomogeneous system of equations

$$\dot{\mathbf{x}}(t) = A\mathbf{x}(t) + \mathbf{f}(t), \tag{5.72}$$

where $\mathbf{f}(t)$ is any continuous function of t, can always be determined from the solution to the homogeneous system obtained by setting $\mathbf{f}(t) = \mathbf{0}$. The method of solving the system is known as *variation of parameters*.

Considering the homogeneous equation $\dot{\mathbf{x}}(t) = A\mathbf{x}(t)$, assume that a set of linearly independent eigenvectors $\mathbf{u}_i$ of A have been found such that

$$\dot{\mathbf{x}}(t) = \sum_{k=1}^{n} c_k \mathbf{u}_k e^{\lambda_k t} \tag{5.73}$$

is the general solution to this homogeneous equation. Let the function $\Phi(t)$ be defined as a matrix whose columns are solutions to $\dot{\mathbf{x}}(t) = A\mathbf{x}(t)$. Thus,

$$\Phi(t) = [\mathbf{u}_1 e^{\lambda_1 t}, \mathbf{u}_2 e^{\lambda_2 t} \ldots, \mathbf{u}_n e^{\lambda_n t}].$$

$\Phi(t)\mathbf{c}$ is the general solution to the homogeneous system of equations when $\mathbf{c}$ is the column vector of coefficients of Equation 5.73. The matrix

$\Phi(t)$ is said to be a *fundamental matrix* of the system $\dot{\mathbf{x}}(t) = A\mathbf{x}(t)$. A fundamental matrix is not unique for the system, and any fundamental matrix can be used to find the solutions.

The matrix Φ will be used to find a solution to the nonhomogeneous system of Equation 5.72 by the method of *variation of parameters*. To begin, assume there exists a solution in the form

$$\mathbf{x}_p = \Phi(t)\mathbf{u}(t). \tag{5.74}$$

Forming $\dot{\mathbf{x}}_p$ yields

$$\dot{\mathbf{x}}_p = \dot{\Phi}(t)\mathbf{u}(t) + \Phi(t)\dot{\mathbf{u}}(t). \tag{5.75}$$

However, $\lambda_i \mathbf{u}_i e^{\lambda_i t} = A\mathbf{u}_i e^{\lambda_i t}$ for $i = 1,\ldots,n$ since the vectors $\mathbf{u}_i$ are eigenvectors of A. Thus, $\dot{\Phi}(t) = A\Phi(t)$, and Equation 5.75 becomes

$$\dot{\mathbf{x}}_p = A\Phi(t)\mathbf{u}(t) + \Phi(t)\dot{\mathbf{u}}(t). \tag{5.76}$$

Comparing this equation with the assumed solution $\dot{\mathbf{x}}_p = A\mathbf{x}_p + \mathbf{f}(t)$ of Equation 5.72 leads to the conclusion that if

$$\Phi(t)\dot{\mathbf{u}}(t) = \mathbf{f}(t), \tag{5.77}$$

then $\Phi(t)\mathbf{u}(t)$ is a solution to Equation 5.72.

The vector $\mathbf{u}(t)$ is obtained by writing

$$\dot{\mathbf{u}} = \Phi^{-1}(t)\mathbf{f}(t)$$

and integrating, with the result

$$\mathbf{u}(t) = \int_{t_0}^{t} \Phi^{-1}(\tau)\mathbf{f}(\tau)\,d\tau.$$

Then, the particular solution for the nonhomogeneous equation is found by forming the product $\Phi(t)\mathbf{u}(t)$ according to Equation 5.74.

The complete solution to Equation 5.72 is, finally,

$$\mathbf{x}(t) = \Phi(t)\mathbf{c} + \mathbf{x}_p(t) \tag{5.78}$$

for any choice of $\mathbf{c}$. This is the matrix equivalent of Equation 5.18.

☐ EXAMPLE 5.14   *Variation of Parameters*

The system of equations $\dot{\mathbf{x}}(t) = A\mathbf{x}(t) + \mathbf{f}(t)$ for the specific case

$$\dot{\mathbf{x}}(t) = \begin{bmatrix} 1 & 0 \\ -1 & 3 \end{bmatrix}\mathbf{x}(t) + \begin{bmatrix} e^t \\ 1 \end{bmatrix}$$

will be solved with $\mathbf{x}(0) = [2,1]^T$ using the variation-of-parameters method. The eigenvalues of A are 1 and 3, and two corresponding eigenvectors are $[2,1]^T$ and $[0,1]^T$. Thus, a fundamental matrix for the system is

$$\Phi(t) = \begin{bmatrix} 2e^t & 0 \\ e^t & e^{3t} \end{bmatrix}.$$

Assuming the forcing function can be written in the form $\Phi(t)\dot{\mathbf{u}}(t) = \mathbf{f}(t)$ according to Equation 5.77 leads to the equation

$$\begin{bmatrix} 2e^t & 0 \\ e^t & e^{3t} \end{bmatrix} \dot{\mathbf{u}}(t) = \begin{bmatrix} e^t \\ 1 \end{bmatrix}.$$

Solving for $\dot{\mathbf{u}}(t)$ yields

$$\dot{\mathbf{u}}(t) = \begin{bmatrix} \dfrac{1}{2} \\ e^{-3t} - e^{-2t}/2 \end{bmatrix}.$$

Integrating each term determines $\mathbf{u}(t)$. The solution to the inhomogeneous equation is $\mathbf{x}_p = \Phi(t)\mathbf{u}(t)$ according to Equation 5.74, so multiplying $\mathbf{u}(t)$ by $\Phi(t)$ yields

$$\mathbf{x}_p = \begin{bmatrix} 2e^t & 0 \\ e^t & e^{3t} \end{bmatrix} \begin{bmatrix} \dfrac{t}{2} \\ -\dfrac{e^{-3t}}{3} + \dfrac{e^{-2t}}{4} \end{bmatrix}$$

$$= \begin{bmatrix} te^t \\ \dfrac{te^t}{2} + \dfrac{e^t}{4} - \dfrac{1}{3} \end{bmatrix}.$$

The complete solution is $\mathbf{x}(t) = \Phi(t)\mathbf{c} + \mathbf{x}_p(t)$, with only $\mathbf{c}$ to be determined. The initial conditions applied at $t = 0$ yield the equation for $\mathbf{c}$ as

$$\mathbf{x}(0) = \Phi(0)\mathbf{c} + \mathbf{x}_p(0),$$

which becomes

$$\mathbf{x}(0) = \begin{bmatrix} 2 & 0 \\ 1 & 1 \end{bmatrix} \mathbf{c} + \begin{bmatrix} 0 \\ -\dfrac{1}{12} \end{bmatrix}.$$

Hence,

$$\mathbf{c} = \begin{bmatrix} 1 \\ \dfrac{1}{12} \end{bmatrix}.$$

Using Equation 5.78, the complete solution becomes

$$\Phi(t)\mathbf{c} + \mathbf{x}_p(t) = \begin{bmatrix} 2e^t + te^t \\ \dfrac{te^t}{2} + \dfrac{5e^t}{4} + \dfrac{e^{3t}}{12} - \dfrac{1}{3} \end{bmatrix}.$$

Why are the terms like te^t necessary in the solution?

□

TRANSFORMING DIFFERENTIAL EQUATIONS

In this section, the nth-order linear differential equation

$$\frac{d^n y(t)}{dt^n} + a_{n-1}(t)\frac{d^{n-1}y(t)}{dt^{n-1}} + \cdots + a_1(t)\frac{dy(t)}{dt} + a_0(t)y(t) = f(t) \quad \textbf{(5.79)}$$

with initial conditions

$$y(t_0), \dot{y}(t_0), \ldots, y^{(n-1)}(t_0)$$

will be converted into a set of n first-order differential equations. The main reason for the conversion to a first-order system is that a number of methods are available to solve the problem completely using numerical techniques. Under fairly general conditions, we are certain that a solution to Equation 5.79 exists and is unique.

■ THEOREM 5.8   *Uniqueness*

Let the coefficients $a_i(t), i = 0, 1, \ldots, n-1$ and $f(t)$ be continuous in some common interval containing the point t_0. If $y(t)$ is found that satisfies the equation and the initial conditions for

$$y(t_0), \dot{y}(t_0), \ldots, y^{(n-1)}(t_0),$$

the resulting solution is unique.

As has been the case for the other equations studied in this chapter, the problem will be first formulated assuming the coefficients in Equation 5.79 are variable. Then, the equation with constant coefficients will be studied.

To convert Equation 5.79 into a first order system, we replace the variable $y(t)$ and its derivatives by n new variables defined as follows:

$$
\begin{aligned}
x_1(t) &= y(t), \\
x_2(t) &= \dot{x}_1(t) = \dot{y}(t), \\
x_3(t) &= \dot{x}_2(t) = \ddot{y}(t), \\
&\vdots \\
x_n(t) &= \dot{x}_{n-1}(t) = y^{(n-1)}(t).
\end{aligned}
$$

The key to writing the n first-order equations is to notice that

$$\dot{x}_i(t) = x_{i+1}(t), \qquad i = 1, 2, \ldots, n-1,$$

and that $\dot{x}_n(t) = y^{(n)}(t)$. Rearranging the nth-order equation yields the last equation in terms of the coefficients and the forcing function $f(t)$ in the form

$$
\begin{aligned}
\dot{x}_n &= y^{(n)} \\
&= -a_{n-1}(t)y^{(n-1)} - a_{n-2}(t)y^{(n-2)} - \cdots - a_1(t)\dot{y} - a_0(t)y + f(t) \\
&= -a_{n-1}(t)x_n - a_{n-2}(t)x_{n-1} - \cdots - a_1(t)x_2 - a_0(t)x_1 + f(t).
\end{aligned}
$$

In matrix form, the equations become

$$
\begin{bmatrix} \dot{x}_1(t) \\ \dot{x}_2(t) \\ \vdots \\ \dot{x}_n(t) \end{bmatrix} =
\begin{bmatrix}
0 & 1 & \cdots & 0 \\
0 & 0 & \cdots & 0 \\
\vdots & \vdots & \ddots & \vdots \\
0 & 0 & \cdots & 1 \\
-a_0(t) & -a_1(t) & \cdots & -a_{n-1}(t)
\end{bmatrix}
\begin{bmatrix} x_1(t) \\ x_2(t) \\ x_3(t) \\ \vdots \\ x_n(t) \end{bmatrix}
+
\begin{bmatrix} 0 \\ 0 \\ 0 \\ \vdots \\ 1 \end{bmatrix} f(t).
$$

These equations can be written as

$$
\dot{\mathbf{x}} = A(t)\mathbf{x} + \mathbf{f}(t), \tag{5.80}
$$

where $\mathbf{f}(t)$ is understood to be the scalar function $f(t)$ multiplied by the column vector with zero entries except for the last entry which is 1.

Thus, the system of Equation 5.80 is equivalent to Equation 5.79 in the sense that if $\mathbf{x}$ is a solution of Equation 5.80, then the first component $x_1(t) = y(t)$ is a solution of Equation 5.79. The initial conditions result in the equation $\mathbf{x}(t_0) = \mathbf{c}$, where $\mathbf{c}$ is the constant vector

$$
\mathbf{c} = \begin{bmatrix} c_1 \\ c_2 \\ \vdots \\ c_n \end{bmatrix}
= \begin{bmatrix} y(t_0) \\ \dot{y}(t_0) \\ \vdots \\ y^{(n-1)}(t_0) \end{bmatrix}
= \begin{bmatrix} x_1(t_0) \\ x_2(t_0) \\ \vdots \\ x_n(t_0) \end{bmatrix}.
$$

□ EXAMPLE 5.15   ***Reduction of a Second-Order Equation***
Consider the second-order equation

$$
m\ddot{x}(t) + b\dot{x}(t) + kx(t) = f(t). \tag{5.81}
$$

Using the principles just defined, set $x_1(t) = x(t)$ and $x_2(t) = \dot{x}(t)$, so the first order system becomes

$$
\begin{aligned}
\dot{x}_1(t) &= x_2(t), \\
\dot{x}_2(t) &= -\frac{k}{m}x_1(t) - \frac{b}{m}x_2(t) + \frac{f(t)}{m}.
\end{aligned}
$$

The matrix equation is

$$
\dot{\mathbf{x}}(t) = \begin{bmatrix} 0 & 1 \\ -\dfrac{k}{m} & -\dfrac{b}{m} \end{bmatrix} \mathbf{x} + \begin{bmatrix} 0 \\ 1 \end{bmatrix} \frac{f(t)}{m}.
$$

This method could be applied to convert the second-order equations that were discussed in Example 5.3 and Example 5.6.

If the system represented by Equation 5.81 is a mechanical system, as previously discussed, we associate $x(t) = x_1(t)$ as the position of a mass. Then, the variables $\dot{x}_1(t)$ and $x_2(t)$ represent the velocity. Thus, the first matrix equation defines the velocity of the mass. The second equation is the acceleration.

Assuming that m, b, and k are positive values, the eigenvalues of the matrix determine the frequency of oscillation (or rate of decay) of the complementary solution. If $b = 0$, there is no damping, and the frequency of the harmonic motion caused by any initial conditions would be $\omega = \sqrt{k/m}$.

$\square$

WHAT IF?   Compare Example 5.6 with Example 5.13. These two examples solve the second-order differential equation without damping.

SOLUTION METHODS

If the equation has constant coefficients, the homogeneous solution of $\dot{\mathbf{x}}(t) = A\mathbf{x}(t)$ could be found by using the eigenvalues and eigenvectors for the matrix of coefficients, as previously discussed. A particular solution to the equation $\dot{\mathbf{x}}(t) = A\mathbf{x}(t) + \mathbf{f}(t)$ could also be found by various methods presented earlier in the text.

Another approach is to use an algorithm to perform numerical integration to solve $\dot{\mathbf{x}}(t) = A\mathbf{x}(t) + \mathbf{f}(t)$. As stated before, MATLAB employees a technique called *Runge-Kutta* integration to solve such problems. The details of the algorithm will be discussed in Chapter 6.

Linear differential equations were introduced in this chapter by writing the equation as an nth-order differential equation. When the coefficients are constant, the homogeneous solution was found by first solving for the roots of the nth degree characteristic equation. These roots led to the homogeneous solution as a linear combination of exponential functions with n arbitrary constants. The particular solution, called the *forced solution*, could be found by various methods, such as undetermined coefficients or variation of parameters. The complete solution is formed by adding the homogeneous solution and a particular solution.

It is important to remember that the values of the arbitrary constants that occur in the homogeneous solution must be determined by applying the initial conditions to the *complete* solution of the nonhomogeneous equation and not just to the homogeneous solution.

STATE-SPACE REPRESENTA-TION

The description of a physical system presented in this section is often called the *state-space* description, in which a system is modeled with first-order differential equations. This state-space formulation is well suited to computer solution. Although we treated only linear systems with constant coefficients in examples in this chapter, the state-space approach can be applied to both linear and nonlinear systems, as well as to systems with variable coefficients. It is frequently the basis for simulation of the response of dynamic systems. Several references listed in the Annotated Bibliography at the end of the chapter treat state-space methods in detail.

Models   Consider a physical device or system that is modeled by the nth-order linear differential equation of Equation 5.79:

$$\frac{d^n y(t)}{dt^n} + a_{n-1}(t)\frac{d^{n-1}y(t)}{dt^{n-1}} + \cdots + a_1(t)\frac{dy(t)}{dt} + a_0(t)y(t) = f(t).$$

The function $f(t)$ can be considered the *input* function. This function $f(t)$ is also called the *forcing function*, or *excitation*, to the system. The set of functions

$$y(t), \dot{y}(t), \ldots, y^{(n-1)}(t)$$

is called a set of *state variables* for the system.   The values of these functions at a specific time, say, t_0, is called the *state* of the system at time t_0. The *output*, or *response*, of the system $y(t)$ at any future time is uniquely determined by the state of the system at t_0 and knowledge of the input function $f(t)$ in the interval from t_0 to t.

In terms of the matrix formulation

$$\dot{\mathbf{x}} = A(t)\mathbf{x} + \mathbf{f}(t), \tag{5.82}$$

the vector $\mathbf{x}(t)$ is called the *state vector* and the matrix A is the *system matrix*. Once the state of the system is known at any time t_0, the solution to the state equation is determined for all future time $t \geq t_0$ if the input $f(\tau)$ is known for $t_0 \leq \tau \leq t$.

Notice that in Equation 5.82, $\mathbf{f}(t)$ is a vector. This means that the state-space approach can handle multiple inputs to the system, not just the single input of Equation 5.79. Multiple outputs can also be defined.

State Variables   When modeling a physical system using the state-space approach, it is first necessary to choose the state variables. These are usually chosen as those variables that are associated with energy storage in the system. For example, the potential energy and kinetic energy of a mass are defined by the position and velocity, respectively. So position and velocity of the masses in a system can be chosen as the state variables.

In electrical circuits, inductors store energy in their magnetic fields and capacitors, in their electric fields. The energy is thus defined by the current through an inductor or the voltage across the capacitor. Therefore, all the capacitor voltages and inductor currents are usually chosen as the state variables.

Although the state variables are not unique, an nth-order system requires n state variables. The solution to the equation system describes how the outputs of the physical system evolve in time.

State-Space Representation

Figure 5.6 shows a series circuit consisting of an inductor (L), a resistor (R), and a capacitor (C). The time variation of the current $i(t)$ can be described by a second-order differential equation, as you are asked to show in Problem 5.20. We therefore expect two state variables to completely describe the system. If there is initial energy in the circuit, it is due to any initial voltage across the capacitor and any initial current through the inductor. Thus, the state variables can be chosen as

$$i_L(t) = i(t) = x_1(t) \quad \text{and} \quad v_C(t) = x_2(t).$$

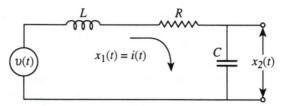

FIGURE 5.6 *Electrical circuit defined by state variables*

The voltages in the circuit written in terms of the circuit elements are

$$
\begin{aligned}
v_R &= Ri \\
v_L &= L\frac{di}{dt} = L\dot{x}_1 \\
v_C &= \frac{1}{C}\int_{-\infty}^{t} i(\tau)\, d\tau.
\end{aligned}
$$

Differentiating the capacitor voltage shows that

$$\frac{dv_C}{dt} = \dot{x}_2(t) = \frac{x_1(t)}{C}. \tag{5.83}$$

Applying Kirchhoff's voltage law to the circuit starting with the inductor voltage leads to the relationship

$$\dot{x}_1(t) = -\frac{R}{L}x_1(t) - \frac{1}{L}x_2(t) + \frac{1}{L}v(t). \tag{5.84}$$

Then, writing Equation 5.84 and Equation 5.83 in matrix form leads to the state equation

$$
\begin{bmatrix} \dot{x}_1(t) \\ \dot{x}_2(t) \end{bmatrix} = \begin{bmatrix} -\dfrac{R}{L} & -\dfrac{1}{L} \\ \dfrac{1}{C} & 0 \end{bmatrix} \begin{bmatrix} x_1(t) \\ x_2(t) \end{bmatrix} + \begin{bmatrix} \dfrac{1}{L} \\ 0 \end{bmatrix} v(t). \tag{5.85}
$$

If $R = 0$, the frequency of the oscillating current and voltage caused by any initial conditions would be $\omega = \sqrt{1/LC}$, as determined by solving the system matrix for its eigenvalues. What is the result if $R \neq 0$?

☐

SECOND-ORDER EXAMPLES

There are many ways to characterize a physical system that is modeled by a differential equation or a system of differential equations. Previous examples have shown that the characteristic equation or the system matrix contains the information about the response of the system to initial stored energy. The eigenvalues and eigenvectors completely characterize the system described by homogeneous differential equations. The resulting complementary solution is often called the *natural*, or *unforced*, response of the system. Even when the unforced response is not of interest in itself, the study of the behavior of the system due to initial conditions often leads to a better understanding of the forced response.

For a nonhomogeneous differential equation or system of equations, the sinusoidal function is widely used for both mathematical and experimental analyses of systems. As a test signal for the physical system or as a forcing function for the model, the sinusoidal input has the advantage of mathematical simplicity. The use of the sinusoid as a forcing function is explored in detail in Chapter 8, where Fourier analysis is presented.

For a system that may be subjected to sudden disturbances, the *step function* is frequently used as a forcing function.[8] Mathematically, the unit step function is defined as

$$U(t - t_0) = \left\{ \begin{array}{ll} 0, & t < t_0, \\ 1, & t \geq t_0. \end{array} \right.$$

This function is zero until $t = t_0$, and then it rises instantaneously to its unit value. For the second-order differential equation with constant coefficients, we shall see that the response to the step function will be classified as an underdamped, critically damped, or an overdamped response. This follows the terminology used to describe the response to initial conditions only in Example 5.3.

☐ **EXAMPLE 5.17**

Solutions of Second-Order Equations

In this example, a second-order differential equation will be solved in several ways. It is good practice to study an equation to be solved before applying numerical techniques. If an analytical solution can be found, this serves as a check on the results of the computer solution. The computer solution then might be used to solve the problem with the variables changed to simulate changes in a physical system or to determine the optimum design for a specified purpose. For example, it may be necessary to assure that oscillations do not occur in response to a step disturbance. By selecting the proper amount of damping, a nonoscillatory response can be achieved.

Figure 5.7 shows a simple physical system consisting of a massless beam supported by a spring and a damper. This system can be modeled by the equation

$$m\ddot{y}(t) + \beta\dot{y}(t) + ky(t) = f(t). \tag{5.86}$$

---

[8]Other test signals, such as the ramp function and impulse function, are used to analyze linear systems. The textbook by Ogata and the one by Kaplan listed in the Annotated Bibliography for this chapter both discuss these functions applied to the analysis of linear systems.

The system is assumed to be at rest before $t = 0$, but a mass is placed on the beam at $t = 0$. The problem is to find the motion of the system measured as the motion of the beam and mass $y(t)$ and the final position of the beam after any oscillations have died out.

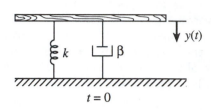
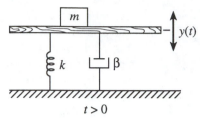

FIGURE 5.7 *Physical system of Example 5.17*

The response of a system to a forcing function can be divided into a *transient solution* and a *steady-state solution*. Mathematically, the steady-state position is defined by the solution $y(t)$ as t approaches infinity. The transient response of a mechanical system will generally exhibit damped vibrations before the system reaches a steady state. In this example, we search for values of damping β that will eliminate the vibrations as the system responds to a step function input.

As a preliminary design, let the physical components have the following properties

$$
\begin{aligned}
m &= 1 \text{ kilogram,} \\
\beta &= 4 \text{ newton-seconds/meter,} \\
k &= 40 \text{ newtons/meter,}
\end{aligned}
$$

where the International System of Units (SI) is used to define the mass m in kilograms, the viscous friction β in newton-seconds/meter, and the spring constant k in newtons/meter. The forcing function is $f(t) = mg$, where g is the force of gravity (9.81 meters per second^2). The solution will be given as displacement measured in meters.

First, we will solve the equation by the traditional methods described in previous sections of this chapter. Dividing the equation by the value of m yields

$$\ddot{y}(t) + 4\dot{y}(t) + 40y(t) = 9.81, \qquad t \geq 0. \tag{5.87}$$

The characteristic equation is $\lambda^2 + 4\lambda + 40 = 0$, with solutions $\lambda = -2 \pm 6i$. Since the real part of the eigenvalues are negative, the complementary solution exhibits damped sinusoidal oscillation (vibration) decaying to zero with time. Such a system is said to be *stable*. Since there is an imaginary part to the eigenvalues, vibration of the beam will occur with this damping constant.

Since the forcing function is constant, assume a particular solution

$$y_p(t) = K.$$

Because the derivatives of y_p are zero, substituting $y_p(t) = K$ in the equation gives the simple result $40K = 9.81$, or $K = 0.2453$ meters. Thus, the complete solution to Equation 5.87 has the form

$$y(t) = 0.2453 + e^{-2t} \left[c_1 \sin(6t) + c_2 \cos(6t) \right], \qquad t \geq 0.$$

Chapter 5 ■ LINEAR DIFFERENTIAL EQUATIONS

Forming $y(0) = \dot{y}(0) = 0$ and solving for the constants gives the final result

$$y(t) = 0.2453 \left[1 - \frac{1}{3}e^{-2t}\sin(6t) - e^{-2t}\cos(6t) \right]. \qquad (5.88)$$

Notice that the particular solution provides a check on the overall solution. It is required that

$$\lim_{t \to \infty} y(t) = 0.2453$$

as the *steady-state* solution. The nonconstant part of the solution in Equation 5.88 is the *transient* solution that decays to zero with time as it must in any real system with damping.

Arbitrary constants. It is important to realize that both the homogeneous and the particular solutions play a role in satisfying the initial conditions in the problem. The arbitrary constants in the solution are not zero here, even though the system is at rest at $t = 0$. In this case, the input function is not zero initially. Placing the mass on the beam gives the system an "impulse" of force at $t = 0$, which excites the natural response of the system, as well as creating a forced solution.

Chapter 6 presents a further discussion of forcing functions, such as the step function, that are discontinuous. For certain forms of $f(t)$ and its derivatives that are zero at the initial time with the system at rest, the homogeneous solution would be zero, and the complete solution would only involve the particular solution. The textbook by Kaplan cited in the Annotated Bibliography for this chapter discusses these special cases.

General form of the equation. Writing Equation 5.86 in the general form

$$\ddot{y}(t) + 2\zeta\omega_n\,\dot{y}(t) + \omega_n^2\,y(t) = \frac{f(t)}{m}. \qquad (5.89)$$

allows the equation to be a model for any second-order differential equation with constant coefficients if the coefficients are interpreted properly. The characteristic equation

$$\lambda^2 + 2\zeta\omega_n\lambda + \omega_n^2 = 0$$

has the roots

$$\lambda_{1,2} = -\zeta\omega_n \pm \omega_n\sqrt{\zeta^2 - 1}. \qquad (5.90)$$

The Greek letter ζ (zeta) is used to designate the damping ratio,

$$\zeta = \text{damping ratio} = \frac{\beta}{2\sqrt{km}},$$

which has the interpretation of the actual damping β coefficient divided by the critical damping value.

Problem 5.20 asks you to define the differential equation for an electrical circuit and relate the coefficients to a mechanical circuit described by an analogous differential equation. The force-voltage analogy is typically used for series connections of the electrical components. In a series RLC circuit shown in Figure 5.6, resistance plays the role of damping and inductance is analogous to mass. For this circuit, the electrical analog of Equation 5.86 takes the form

$$L\ddot{q}(t) + R\dot{q}(t) + \frac{1}{C}q(t) = v(t). \qquad (5.91)$$

In this equation, $q(t)$ is the electrical charge and the current in the circuit is $i(t) = dq(t)/dt$.

Second-order parameters. The parameters listed in Table 5.5 relate mechanical and electrical systems in terms of damping for a second-order differential equation model in which mass and inductance are analogs. In Table 5.5, the parameters apply for the mechanical system we have been studying in this example and the series (RLC) electrical circuit that is analogous.

TABLE 5.5 *Second-order parameters*

Parameter	Mechanical	Electrical
ω_n	$\sqrt{\dfrac{k}{m}}$	$\sqrt{\dfrac{1}{LC}}$
ζ	$\dfrac{\beta}{2\sqrt{km}}$	$\dfrac{R}{2}\sqrt{\dfrac{C}{L}}$
Critical damping	$\beta^2 = 4km$	$R^2 = 4\dfrac{L}{C}$
Analog	m, mass	L, inductance

Between the low-damping condition for a system and the overdamped condition, critical damping occurs for the values given in the table. In the mechanical circuit, $\beta = 2\sqrt{km}$. For a series electrical circuit, the critical value of the resistance must be $R = 2\sqrt{L/C}$.

The critical value ($\zeta = 1$) for our present example of a mechanical system is

$$\beta = 2\sqrt{40} = 4\sqrt{10}.$$

The solution to the differential equation at the critical damping value is

$$y_{cr}(t) = (c_{1cr} + c_{2cr}t)e^{-\omega_n t} + y_p \tag{5.92}$$

using the real and equal roots of the characteristic Equation 5.90. The particular solution is $y_p = 9.81/40$ as before. From the initial conditions, the constants are determined to be $c_{1cr} = -y_p$ and $c_{2cr} = -\omega_n y_p$ so the solution at the critical damping value is

$$y_{cr}(t) = .2453\left[1 - (1 + \omega_n t)e^{-\omega_n t}\right] \tag{5.93}$$

with $\omega_n = \sqrt{40}$ radians/second. For values of $\beta \geq 4\sqrt{10}$, the solution will rise to the steady state value without oscillation or overshooting the final value.

MATLAB solution. Attempting to solve the second-order differential Equation 5.86 directly using **dsolve** leads to a rather complicated expression that is difficult to interpret. For specific values of the coefficients and a step function input as in Equation 5.87, the analytical solution can be found and plotted.

Figure 5.8 shows the step response for two values of ζ as computed and plotted by the accompanying MATLAB script. The MATLAB variable z represents the damping ratio. The critical damping ratio is z=1. A damping value of $\beta = 4$ in the differential equation leads to a damping ratio of z=.3162 and the underdamped response in the figure. In Problem 5.31 you are asked to define the second-order system for this example and plot various responses.

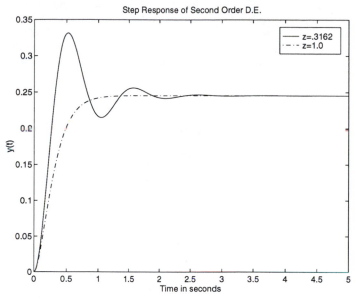

FIGURE 5.8 *Solution of the second-order differential equation of Example 5.17*

MATLAB Script _____

```
Example 5.17
% EX5_17.M Step response of second order differential equation
%   D2y + 2*z*Wn*Dy + Wn^2*y=U(t) for underdamped and critically
%   damped system ; z=2/Wn and z=1, wn=sqrt(40)
clear              % Clear workspace and
clf                %  figures
Wn=sqrt(40)        % Natural frequency - radians/second
z1=2/Wn            % Damping
t=[0:.005:5];      % Time for plotting
y1=.2453*( 1 -(1/3)*exp(-2*t).*sin(6*t)-exp(-2*t).*cos(6*t));
ycrit=.2453*( 1-(1 + Wn*t).*exp(-Wn*t));
plot(t,y1,'-',t,ycrit,'-.')
axis([0,5,0,.5])    % Set plotting limits
xlabel('Time in seconds')
ylabel('y(t)')
title('Step Response of Second Order D.E.')
Legend('z=.3162','z=1.0')
```

□

The matrix approach is often used to solve a set of coupled second order equations by defining a 4×4 system matrix. The eigenvalues and eigenvectors play an important role in determining the form of the homogeneous solution.

☐ EXAMPLE 5.18

MATLAB Solution of Coupled Second Order Equations

Consider the mass and spring system shown in Figure 5.9. The system requires two coordinates to completely describe its motion. Such a system is often called a system with *two degrees of freedom*. In general, the analysis of a system described by ordinary differential equations requires as many simultaneous differential equations as there are degrees of freedom if the system has more than one degree of freedom. The equations of motion for the coupled mass system are

$$
\begin{aligned}
m_1 \ddot{y}_1(t) &= -k_1 y_1(t) &&+ k_2[y_2(t) - y_1(t)], \\
m_2 \ddot{y}_2(t) &= -k_2[y_2(t) - y_1(t)] &&- k_3 y_2(t),
\end{aligned}
$$

where $y_i(t)$ describes the motion of the ith mass, $i = 1, 2$.

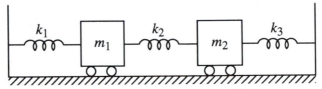

FIGURE 5.9 *Coupled-mass system with two degrees of freedom*

The coupled equations can be written as a first-order system by defining

$$
\begin{aligned}
x_1(t) &= y_1(t), \\
x_2(t) &= y_2(t), \\
x_3(t) &= \dot{y}_1(t), \\
x_4(t) &= \dot{y}_2(t)
\end{aligned}
$$

and substituting the new variables in the original equations. The result is

$$
\begin{aligned}
\dot{x}_1(t) &= \dot{y}_1(t) &=& \quad x_3(t), \\
\dot{x}_2(t) &= \dot{y}_2(t) &=& \quad x_4(t), \\
\dot{x}_3(t) &= \ddot{y}_1(t) &=& \quad -\left(\frac{k_1 + k_2}{m_1}\right) x_1(t) + \frac{k_2}{m_1} x_2(t), \\
\dot{x}_4(t) &= \ddot{y}_2(t) &=& \quad \frac{k_2}{m_2} x_1(t) - \left(\frac{k_2 + k_3}{m_2}\right) x_2(t).
\end{aligned}
$$

The result for the special case of $k_1 = k_2 = k_3 = k$ and $m_1 = m_2 = m$ becomes

$$
\begin{aligned}
\dot{x}_1(t) &= x_3(t), \\
\dot{x}_2(t) &= x_4(t), \\
\dot{x}_3(t) &= -\frac{2k}{m} x_1(t) + \frac{k}{m} x_2(t), \\
\dot{x}_4(t) &= \frac{k}{m} x_1(t) - \frac{2k}{m} x_2(t).
\end{aligned}
\tag{5.94}
$$

Thus, the form of the last equation is $\dot{\mathbf{x}} = A\mathbf{x}$, where

$$A = \begin{bmatrix} 0 & 0 & 1 & 0 \\ 0 & 0 & 0 & 1 \\ -\dfrac{2k}{m} & \dfrac{k}{m} & 0 & 0 \\ \dfrac{k}{m} & -\dfrac{2k}{m} & 0 & 0 \end{bmatrix}$$

Analysis. Letting $k = 1$ and $m = 1$ for simplicity, the eigenvalues of the system matrix were found using the MATLAB command **eig** to be $\pm 1.0i$ and $\pm\sqrt{3}i$. The fundamental, or *natural*, frequencies of oscillation are $\omega_1 = 1$ and $\omega_2 = \sqrt{3}$ radians per second. Examining the eigenvectors (not shown) indicates that there are two modes of oscillation, as you are asked to investigate in Problem 5.19. The eigenvectors associated with the natural frequencies are called the *normal modes* of the system. They describe the relative amplitudes of the displacement of the masses if the system vibrated only at the corresponding natural frequency. However, the absolute amplitudes depend on the initial conditions.

Test cases. The script EX5_18.M solves the equations by calling M-function cldesf to define these equations for **ode23**.

MATLAB Script _____

```
Example 5.18
% EX5_18.M  Use MATLAB ode23 to solve the system
%   y1''=-2y1+ y2
%   y2''=  y1-2y2
%    transformed into the system xdot=Ax where A is 4x4
% INPUTS: Initial time, final time, initial conditions and title
% OUTPUT: A (global variable); Plot of motion y1(t), y2(t)
global A            % Pass A to function
A=[0 0 1 0;0 0 0 1;-2 1 0 0;1 -2 0 0]      % System matrix
t0=input('Initial time=   ')
tf=input('Final time=   ')
x0=input('[y1(t0) y2(t0) doty1(t0) doty2(t0)] =  ')
x0t=x0';               % Transpose of initial conditions for ode23
% Calls function cldesf to define state equations.
[t,x]=ode23('cldesf',t0,tf,x0t);  % Numerical solution of system
% y values
y1=x(:,1);             % Change to physical variables in example
y2=x(:,2);
% Plot y1 and y2, the motion of the masses
titlef=input('Title= ','s')   % Input the title
subplot(2,1,1),plot(t,y1)     % Plot two graphs on one axis
ylabel('Displacement y1')
subplot(2,1,2),plot(t,y2)
ylabel('Displacement y2');xlabel('Time')
title(eval('titlef'))
```
_____

The function cldesf defines the equation set of Equation 5.94.

```
Example 5.18
function xdot=cldesf(t,x)
%  CALL: xdot=cldesf(t,x) This function defines the equations
%     xdot(t)=A*x(t) used by MATLAB commands ode23 and ode45.
%     A is passed to function by a global statement
global A
xdot=A*x;
```

In fact, the function creates the equation set $\dot{\mathbf{x}} = A\mathbf{x}$ for the MATLAB differential solver routines such as **ode23**. The matrix A is passed by the calling program using a **global** command. In this way, any other similar system can be solved by changing the definition of A in the calling script.

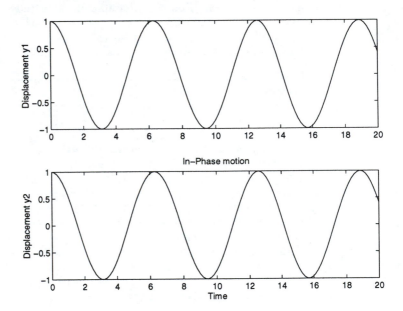

FIGURE 5.10 *In-phase motion in Example 5.18*

The plots in Figure 5.10 show the motion that results from pulling the masses 1 unit from their equilibrium position and releasing them at $t = 0$. The initial conditions are

$$[y_1(0), y_2(0), \dot{y}_1(0), \dot{y}_2(0)] = [1, 1, 0, 0].$$

For this case, the period of oscillation is $T_1 = 2 * \pi/\omega_1$ seconds, or about 6.28 seconds. Notice that the center spring is not compressed or stretched for this in-phase motion.

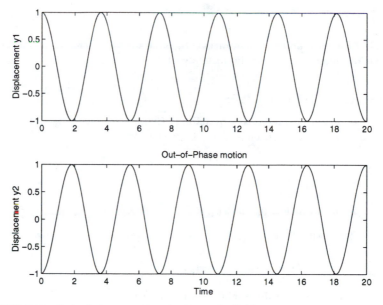

FIGURE 5.11 *Out-of-phase motion in Example 5.18*

In the second case, shown in Figure 5.11, the initial conditions were selected to be

$$[y_1(0), y_2(0), \dot{y}_1(0), \dot{y}_2(0)] = [1, -1, 0, 0].$$

The central spring is compressed as the two masses are brought together. When released, the masses oscillate in opposite directions. The period of oscillation is $T_2 = 2\pi/\sqrt{3} = 3.627$ seconds.

In the first mode, the two masses oscillate together at the lower frequency $\omega_1 = 1$, executing what is called *in phase*-motion. This is the same frequency as for a single mass and spring, where $\omega_1 = \sqrt{k/m} = 1$. In the faster mode, the masses move in the opposite directions, executing *out-of-phase* motion with radian frequency $\omega_2 = \sqrt{3k/m} = \sqrt{3}$.

The general solution for arbitrary initial conditions is a combination of these two modes. The analytical solution consists of combinations of the terms $\sin t$, $\cos t$, $\sin(\sqrt{3}t)$, and $\cos(\sqrt{3}t)$. O'Neil's textbook listed in the Annotated Bibliography describes how to construct the general solution using the eigenvectors and arbitrary initial conditions.

In both cases shown in this example, each oscillation has the same amplitude since the masses and spring constants are the same. If the mass or spring constant values were different, this would not be the case.

□

REINFORCEMENT EXERCISES AND EXPLORATION PROBLEMS

———————————————— ■ ————————————————

In these problems, do the computations by hand unless otherwise indicated, and then check those that yield numerical or symbolic results with MATLAB.

REINFORCEMENT EXERCISES

P5.1. **Classify differential equations** Classify the following equations as to linearity and order, and define the independent and dependent variables:

    a. $y' + 5y = x^2, \quad y = y(x);$

    b. $(x + y)^2 \dfrac{dy}{dx} = 1, \quad y = y(x);$

    c. $\ddot{\theta} + \alpha \sin \theta = 0, \quad \theta = \theta(t);$

    d. $\dfrac{dx}{dt} + 3\dfrac{dy}{dt} + y = e^t, \quad x = x(t), y = y(t);$

    e. $\dfrac{dx}{dy} + y^4 x = \sin y, \quad x = x(y).$

P5.2. **First-order differential equation** Apply the first-order integration formula to solve the equation

$$\dot{u}(t) + 2tu(t) = f(t),$$

with $u(0) = 0$, for the cases:

    a. $f(t) = 2t;$

    b. $f(t) = 1.$

P5.3. **First-order application** Often it can be assumed that the decay rate of a radioactive element is proportional to the amount of the element present. Then the rate of decay is

$$\frac{dA(t)}{dt} = -\alpha A(t),$$

where A is the quantity of material present after t years and α is a positive constant. Assume that 0.5% of an element disappears in 12 years. Then, answer the following questions:

    a. What percentage will disappear in 1000 years?

    b. What is the half-life of the element?

The half-life of an element is defined as the time it takes for 50% of the material to decay.

P5.4. **First-order equation** Solve the equation

$$\dot{y}(t) + 5y(t) = 50 \quad \text{with} \quad y(0) = y_0$$

by the following two methods:

    a. Integrating factor method;

    b. Method of undetermined coefficients.

P5.5. **Assumed solutions of differential equations** What is the assumed solution for the following?

---

a. Particular solution of $\dfrac{d^3 y(t)}{dt^3} = 0$.

b. Homogeneous solution of $y''' - y'' - 12y' = 0$.

c. Homogeneous solution of $\ddot{y}(t) + 4y(t) = 0$.

d. Complete solution of

$$\ddot{y}(t) + 2\dot{y}(t) + y(t) = 3t + 2 + 3e^t.$$

P5.6. Second-order differential equation Solve the second-order differential equation

$$y''(x) - 4y'(x) + 3y(x) = f(x)$$

with initial conditions $y(0) = 1$ and $y'(0) = -3$ in the cases:

a. $f(x) \equiv 0$;

b. $f(x) = 10e^{-2x}$.

Note: The solution of this equation does not approach zero as x approaches infinity.

P5.7. System of differential equations Solve for the constants in the general solution of $\mathbf{x}' = A\mathbf{x}$ as described in Example 5.8, where

$$A = \begin{bmatrix} 1 & 1 \\ 4 & 1 \end{bmatrix},$$

using the initial conditions are $x_1(0.1) = 1$ and $x_2(0.1) = 1$.

P5.8. System of differential equations Solve the initial value problem

$$\mathbf{x}' = \begin{bmatrix} 0 & 1 & 0 \\ 0 & 0 & 1 \\ 2 & 1 & -2 \end{bmatrix} \mathbf{x} \quad \text{with} \quad \mathbf{x}(0) = \begin{bmatrix} 1 \\ 0 \\ 1 \end{bmatrix}.$$

P5.9. System of differential equations Convert the third-order differential equation

$$y'''(x) - 5y''(x) - y'(x) + 5y(x) = 0$$

to a first-order system and find the following:

a. The eigenvalues of the system matrix;

b. The eigenvectors of the system matrix.

P5.10. Step response of second-order equation Solve the equation

$$\ddot{y}(t) + 3\dot{y}(t) + 2y(t) = U(t),$$

in which $U(t)$ is the unit step function

$$U(t) = \begin{cases} 0, & t < 0, \\ 1, & t \geq 0. \end{cases}$$

Convert the equation to a first-order matrix equation. Then, find the fundamental matrix and use the method of variation of parameters. Assume the system is initially at rest.

Reinforcement Exercises and Exploration Problems

P5.11. Wronskian The *Wronskian* $W[\phi_1(t), \phi_2(t), \ldots, \phi_n(t)]$ of n functions having $n-1$ derivatives on an interval I is the determinant function

$$W[\phi_1(t), \phi_2(t), \ldots, \phi_n(t)] = \begin{vmatrix} \phi_1(t) & \cdots & \phi_n(t) \\ \dot{\phi}_1(t) & \cdots & \dot{\phi}_n \\ \vdots & \ddots & \vdots \\ \phi_1^{(n-1)}(t) & \cdots & \phi_n^{(n-1)}(t) \end{vmatrix}.$$

Consider the nth-order differential equation

$$L[y(t)] = \frac{d^n y(t)}{dt^n} + a_{n-1}\frac{d^{n-1}y(t)}{dt^{n-1}} + \cdots + a_1\frac{dy(t)}{dt} + a_0 y(t) = 0,$$

where the a_i are constants. Assume that $\phi_1(t), \phi_2(t), \ldots, \phi_n(t)$ are the solutions of this homogeneous equation on an interval I containing the point t_0. Then, the following theorem defines a test for linear independence of the solutions.

Let $\phi_1(t), \phi_2(t), \ldots, \phi_n(t)$ be n solutions of the homogeneous differential equation $L[y(t)] = 0$ on an interval I containing the point t_0. The solutions are linearly independent if

$$W[\phi_1(t_0), \phi_2(t_0), \ldots, \phi_n(t_0)] \neq 0.$$

For the equation $L[y(t)] = 0$, a useful result is that

$$W[\phi_1(t), \phi_2(t), \ldots, \phi_n(t)] = e^{-a_{n-1}(t-t_0)}W[\phi_1(t_0), \phi_2(t_0), \ldots, \phi_n(t_0)],$$

where a_{n-1} is the coefficient of $y^{(n-1)}$. Solve the differential equation

$$y^{(3)}(t) + \ddot{y}(t) + \dot{y}(t) + y(t) = 0$$

with initial values $y(0) = 0$, $\dot{y}(0) = 1$, $\ddot{y}(0) = 0$ and show that the three solutions are linearly independent. Do the problem by direct evaluation of the Wronskian first and then evaluate at $t = 0$.

P5.12. Differential equation eigenvalues Consider the second-order differential equation

$$\ddot{y}(t) + 2\alpha\dot{y}(t) + {\omega_0}^2 y(t) = 0,$$

which might represent a series RLC circuit or a mechanical system with a mass, spring, and dashpot.

- a. What is the significance of the case when the characteristic equation yields repeated eigenvalues?

- b. Write the homogeneous solution in the case of a double eigenvalue.

P5.13. Differential equation normal modes Consider the mass and spring system shown in Figure 5.12 and do the following:

- a. Find the eigenvalues and eigenvectors.

- b. Describe the fundamental modes of oscillation for small vibrations of the masses in terms of the movements of the three masses.

Chapter 5 ■ LINEAR DIFFERENTIAL EQUATIONS

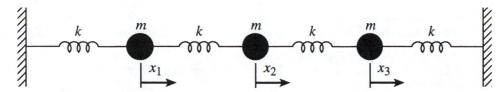

FIGURE 5.12 *Mass-spring system*

P5.14. MATLAB solution of differential equations Write an M-file to solve Problem 5.6 and do the following:

a. Plot the solutions to the equation in Problem 5.6.

b. Plot the analytical and numerical results in Problem 5.6.

c. Plot a curve of the error in Part b on a reasonable scale.

P5.15. MATLAB Solution of differential equations Solve the homogeneous differential equation

$$9.082\ddot{y}(t) + b\dot{y}(t) + 890y(t) = 0$$

with initial values $y(0) = 0.15$ and $\dot{y}(0) = 0$ for the following values of b:

a. $b = 200$;

b. $b = 179.8$;

c. $b = 100$.

Plot all three solutions on the same graph to show overdamped, critically damped, and underdamped responses.

Hint: The command **dsolve** does not handle floating-point numbers. Try rational fractions for the coefficients if you attempt a symbolic solution.

P5.16. MATLAB solution of differential equations Write an M-file to solve Problem 5.10 and plot the results if the initial conditions are $y(0) = 0$ and $\dot{y}(0) = 0$.

P5.17. MATLAB solution of differential equations Given the third-order differential equation:

$$y^{(3)}(t) + 8\ddot{y}(t) + 37\dot{y}(t) + 50y(t) = 4e^{-3t}$$

with initial values $y(0) = 1$, $\dot{y}(0) = 2$ and $\ddot{y}(0) = 1$:

a. Find the analytic solution by hand but try to use MATLAB to help with the calculations.

b. Plot the results for $0 \le t \le 2$.

P5.18. MATLAB solution of differential equations Consider the third-order differential equation

$$y^{(3)}(t) + a_1\ddot{y}(t) + a_2\dot{y}(t) + a_3y(t) = f(t)$$

with initial values $y(0)$, $\dot{y}(0)$, $\ddot{y}(0)$.

a. Create an M-file that allows input of the various variables, solves the equation, and plots the solution.

b. Test the M-file using the equation in Problem 5.17 and compare the plotted results for the analytical and MATLAB solution.

P5.19. MATLAB solution of normal modes In Example 5.18, determine the eigenvectors corresponding to the eigenvalues given. Verify the analysis given in the example for the normal modes of oscillation.

EXPLORATION PROBLEMS

P5.20. Electrical and mechanical analogs For the circuit in Figure 5.13, determine the differential equation for the voltage in the circuit in terms of the charge $q(t)$, where $i(t) = dq(t)/dt$ is the current in the circuit.

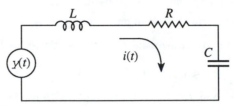

FIGURE 5.13 *Series electrical circuit*

Relate the coefficients and variables of the electrical circuits to the coefficients of the analogous equation for a mechanical system described in terms of force, mass, damping, spring constant, position and velocity. For example, force and voltage are analogous.

P5.21. Solutions that are not unique Consider the differential equation

$$\dot{x} = g(x) = 3x^{2/3}.$$

This equation can be solved by writing $dx/dt = g(x)$ and finding the solution as

$$t - c = \int \frac{dx}{g(x)}.$$

Show the equation does not meet the requirements of the existence and uniqueness theorem and find the solutions for all t. Where do two solutions exist?

Hint: A solution is $x = 0$.

P5.22. First-order equation Solve the equation

$$\dot{x} = \frac{2}{t^2 - 1}.$$

Hint: Expand the function on the right-hand side using partial fractions.

P5.23. Real solutions Write the real solutions of the homogeneous equation

$$y^4(x) + y(x) = 0.$$

This equation arises in the study of deflection of beams.

P5.24. Complete solution Write the complete solution to the differential equation

$$y''(x) - y'(x) - 2y(x) = e^{-x}.$$

P5.25. Forced undamped oscillation Consider the second-order differential equation without damping but with a sinusoidal forcing function,

$$m\ddot{y}(t) + ky(t) = A\cos(\omega t).$$

Let $A = 1$, $m = 1$, and $k = 100$.

    a. Plot the time response for $\omega = 5$.

    b. Plot the time response for $\omega = 12$.

    c. Plot the amplitude of the response as the input frequency is changed from $\omega = 0$ to $\omega = 20$ radians/second.

P5.26. Forced damped oscillations Consider the second-order differential equation with damping,

$$m\ddot{y}(t) + b(\dot{y}) + ky(t) = A\cos(\omega t).$$

Plot the response to a sinusoidal forcing function as the input frequency is changed from $\omega = 0$ to $\omega = 20$ radians/second for the damping ratio

$$\zeta = \frac{b}{2\sqrt{km}}$$

with values $\zeta = 0, 0.2, 0.5, 1, 2$. Let $A = 1$, $m = 1$, and $k = 100$.

P5.27. Particular solution Solve for the particular solution of the differential equation

$$\ddot{z}(t) + z(t) = t\cos t.$$

P5.28. Motion of a rigid mass Consider the motion of a rigid mass in a frictionless environment with the equation of motion

$$f(t) = m\frac{d^2x(t)}{dt^2}.$$

If the system matrix is designated A, write the state equations, compute e^{At}, and solve for $\mathbf{x}(t)$ in terms of $x_1(0)$ and $x_2(0)$.

P5.29. MATLAB solutions to advanced equations Use the MATLAB symbolic command **dsolve** to try to solve the differential equation with variable coefficients

$$\ddot{y}(t) - \frac{2t}{1+t^2}\dot{y}(t) + \frac{2}{1+t^2}y(t) = f(t).$$

Reduce the answer to the simplest form.

P5.30. Matrix solution   Consider the system of equations

$$\dot{x}_1(t) = x_1(t) + x_2(t),$$
$$\dot{x}_2(t) = 4x_1(t) + x_2(t),$$

defined in Example 5.8, and subject to the initial conditions

$$\mathbf{x}(0) = [1, 1]^T.$$

Find the solution using the matrix exponential solution

$$\mathbf{x}(t) = e^{At}\mathbf{x}_0$$

as shown in Equation 5.64.

P5.31. MATLAB second order solution   Write a MATLAB program to solve the general second order differential equation

$$a\frac{d^2y(t)}{dt^2} + b\frac{dy(t)}{dt} + cy(t) = f(t)$$

using the matrix formulation. The MATLAB command **ode23** solves the state space form of these equations. Apply the program to various examples treated in this chapter by varying the values of the coefficients and the forcing function $f(t)$. Specifically, solve the problem of Example 5.17 with a step function input for different values of damping and spring constant.

P5.32. Resonance in bridges   An interesting research project would be to compare the linear theory explanation for the collapse of the Tacoma Narrows suspension bridge with the recent discussions involving nonlinear phenomenon. The articles by Berreby and Lazer and McKenna in the Annotated Bibliography for this chapter are good starting points.

ANNOTATED BIBLIOGRAPHY

1. Abramowitz, Milton, and I. A. Stegun, *Handbook of Mathematical Functions*, Dover Publications, Inc., New York, 1970 (ninth printing). *An invaluable reference for formulas, graphs, and tables of functions important in advanced applied mathematics.*

2. Kaplan, Wilfred, *Operational Methods For Linear Systems*, Addison Wesley Publishing Company, Reading, MA, 1962. *This textbook presents a fairly complete and rigorous treatment of linear systems. Linear differential equations are studied in detail.*

3. O'Neil, Peter V., *Advanced Engineering Mathematics*, Wadsworth Publishing Company, Belmont, CA, 1991. *The text gives very complete treatment of important topics in engineering mathematics. In particular, matrix solutions of systems of differential equations are treated in detail.*

4. Ogata, Katsuhiko, *System Dynamics*, Prentice Hall, Englewood Cliffs, NJ, 1992. *The book presents a solid engineering approach to system analysis and modeling.*

5. Polking, J. C., *Ordinary Differential Equations using MATLAB*, Prentice Hall, Englewood Cliffs, NJ, 1995. *Various aspects of solving differential equations with MATLAB are presented.*

6. Zill, Dennis G., *Differential Equations with Computer Lab Experiments*, PWS Publishing Company, Boston, MA, 1995. *The text treats linear differential equations and presents many examples.*

ANSWERS

P5.1. **Classify differential equations**

   a. Linear; first order; x is independent; y is dependent;

   b. Nonlinear; first order; x is independent; y is dependent;

   c. Nonlinear; second order; t is independent; θ is dependent;

   d. Linear; first order; t is independent; x and y are dependent;

   e. Linear; first order; y is independent; x is dependent.

P5.2. **First-order differential equation** For the equation $\dot{u} + 2tu = f(t)$, the general solution is

$$u(t) = e^{-\int p(t)\,dt} \int f(t) e^{\int p(t)\,dt}\,dt + c e^{\int -p(t)\,dt},$$

with $p(t) = 2t$. Also, $c = 0$ since $u(0) = 0$ was given as the initial condition.

   a. Letting $f(t) = 2t$ leads to the result

$$u(t) = e^{-2\int t\,dt} \int_0^t 2\tau e^{2\int \tau\,d\tau}\,d\tau.$$

     Letting $w = \tau^2$ and $dw = 2\tau d\tau$, the integral with respect to τ is $\int \exp(w)\,dw = \exp(w)$. Thus,

$$u(t) = e^{-t^2}\left[e^{\tau^2} \right]_0^t = 1 - e^{-t^2}.$$

   b. If $f(t) = 1$, the solution becomes

$$u(t) = e^{-t^2} \int_0^t e^{\tau^2}\,d\tau.$$

The integral in Part b is related to the *error function*. This integral is tabulated in mathematical handbooks, such as the one by Abramowitz and Stegun listed in the Annotated Bibliography. Also, MATLAB can be used to compute numerical values of the integral.

P5.3. **First-order application** Let $A = A_0$ at $t = 0$ and

$$A(t) = A_0 \exp(-\alpha t)$$

for $t \geq 0$. Then at $t = 12$ years, $A = 0.995 A_0$ and

$$A_0 e^{-12\alpha} = 0.995 A_0,$$

so that $\alpha = (0.995)^{1/12} = .000418$.

a. At $t = 1000$, $A = A_0 \exp(-0.418) = 0.658A_0$, or 34.2% disappears in 1000 years;

b. The half-life is determined when $A(t) = A_0/2 = A_0 \exp(-\alpha t)$, with the result $t = -\ln 2/.000418 \approx 1660$ years.

P5.4. **First-order equation** The equation $dy/dt + 5y = 50$ with initial condition $y(0) = y_0$ can be solved in a number of ways.

a. Applying the equation for the general solution yields

$$y(t) \quad = \quad e^{-5\int_0^t dt} \int_0^t 50 e^{\int_0^\tau dt} \, dt + y_0 e^{-5\int_0^t dt}$$

$$= \quad y_0 e^{-5t} + 50 e^{-5t} \left(\frac{1}{5} \left[e^{5t} \right]_0^t \right)$$

$$= \quad (y_0 - 10) e^{-5t} + 10.$$

b. The complementary solution is $y_c = c_1 \exp(-5t)$. Using the method of undetermined coefficients, a "guess" is made that $y_p = k$, so $5k = 50$, as determined by substituting in the equation. The solution is $y(t) = 10 + c_1 e^{-5t}$, with $c_1 = y_0 - 10$ as in Part (a).

P5.5. **Assumed solutions of differential equations** Let $D \equiv d/dt$.

a. $y(t) = At^2 + Bt + C$.

b. $D(D^2 - D - 12) = D(D - 4)(D + 3)$, so the roots of the characteristic equation $\lambda(\lambda - 4)(\lambda + 3)$ are $0, 4, -3$ and

$$y(t) = c_1 + c_2 e^{4x} + c_3 e^{-3t}.$$

c. The roots of the characteristic equation $\lambda^2 + 4$ are $\lambda = \pm 2i$ so

$$y(t) = c_1 e^{i2t} + c_2 e^{-i2t}.$$

d. The characteristic equation $(\lambda + 1)^2 = 0$ yields the complementary solution $y_c = (c_1 + c_2 t) \exp(-t)$. By the method of undetermined coefficients, the assumed particular solution is

$$y_p = At + B + Ce^t.$$

The form of the complete solution is thus

$$y(t) = (c_1 + c_2 t)e^{-t} + At + B + Ce^t.$$

P5.6. **Second-order differential equation** For the differential equation

$$y'' - 4y' + 3y = f(t),$$

the characteristic equation is $\lambda^2 - 4\lambda + 3$ with the solutions $\lambda_1 = 1$ and $\lambda_2 = 3$. Since the characteristic roots or eigenvalues are positive, the solution to the differential equation is said to be *unstable* because the solution $y(t)$ does not go to zero as $t \to \infty$. The solution is $y_t = y_p + c_1 e^x + c_2 e^{3x}$, with c_1 and c_2 determined by the initial conditions $y(0) = 1$ and $y'(0) = -3$.

a. For the input function $f(t) = 0$, the initial conditions yield

$$1 \quad = \quad c_1 + c_2,$$
$$-3 \quad = \quad c_1 + 3c_2.$$

The solution is $y(t) = 3e^x - 2e^{3x}$.

b. If $f(t) = 10e^{-2x}$, the assumed solution becomes

$$y(t) = Ce^{-2x} + c_1 e^x + c_2 e^{3x}.$$

Substituting the assumed particular solution in the equation yields $y_p = \frac{2}{3}\exp(-2x)$. Applying the initial conditions

$$
\begin{aligned}
y(0) &= c_1 + c_2 + \frac{2}{3} = 1, \\
y'(0) &= c_1 + 3c_2 - \frac{4}{3} = -3,
\end{aligned}
$$

leads to the result $c_1 = 4/3$ and $c_2 = -1$. The complete solution is

$$y(t) = \frac{4}{3}e^x - e^{3x} + \frac{2}{3}e^{-2x}.$$

P5.7. **System of differential equations** The system of equations

$$
\begin{aligned}
\dot{x}_1(t) &= x_1(t) + x_2(t), \\
\dot{x}_2(t) &= 4x_1(t) + x_2(t),
\end{aligned}
$$

has the solution

$$
\begin{aligned}
x_1(t) &= -\frac{1}{2}c_1 e^{-t} + \frac{1}{2}c_2 e^{3t}, \\
x_2(t) &= c_1 e^{-t} + c_2 e^{3t},
\end{aligned}
$$

as shown in Example 5.8. The equation for the coefficients is

$$
\begin{bmatrix} -\frac{1}{2}e^{-0.1} & \frac{1}{2}e^{0.3} \\ e^{-0.1} & e^{0.3} \end{bmatrix}
\begin{bmatrix} c_1 \\ c_2 \end{bmatrix} =
\begin{bmatrix} x_1(0.1) \\ x_2(0.1) \end{bmatrix} =
\begin{bmatrix} 1 \\ 1 \end{bmatrix}.
$$

Using MATLAB to solve the system yields (to four places)

$$
\begin{bmatrix} c_1 \\ c_2 \end{bmatrix} =
\begin{bmatrix} -0.5526 \\ 1.1112 \end{bmatrix}.
$$

A check on the solution shows that the initial conditions are satisfied.

P5.8. **System of differential equations** Given the system of equations

$$
\mathbf{x}' = \begin{bmatrix} 0 & 1 & 0 \\ 0 & 0 & 1 \\ 2 & 1 & -2 \end{bmatrix} \mathbf{x},
$$

the eigenvalues are $1, -1, -2$ with the corresponding eigenvectors

$$
\mathbf{u}(1) = \begin{bmatrix} 1 \\ -1 \\ 1 \end{bmatrix}, \quad
\mathbf{u}(2) = \begin{bmatrix} 1 \\ 1 \\ 1 \end{bmatrix}, \quad
\mathbf{u}(3) = \begin{bmatrix} 1 \\ -2 \\ 4 \end{bmatrix}.
$$

Thus, the solution is

$$
\mathbf{x}(t) = c_1 \begin{bmatrix} 1 \\ -1 \\ 1 \end{bmatrix} e^{-t} + c_2 \begin{bmatrix} 1 \\ 1 \\ 1 \end{bmatrix} e^t + c_3 \begin{bmatrix} 1 \\ -2 \\ 4 \end{bmatrix} e^{-2t}.
$$

Then,

$$\mathbf{x}(0) = \begin{bmatrix} 1 \\ 0 \\ 1 \end{bmatrix} = \begin{bmatrix} 1 & 1 & 1 \\ -1 & 1 & -2 \\ 1 & 1 & 4 \end{bmatrix} \mathbf{x},$$

with the solution for $\mathbf{c}$

$$\mathbf{c} = \begin{bmatrix} \frac{1}{2} \\ 1 \\ \frac{1}{2} \\ 0 \end{bmatrix}.$$

Substituting the values for the coefficients yields

$$\mathbf{x}(t) = \begin{bmatrix} (e^t + e^{-t})/2 \\ (e^t - e^{-t})/2 \\ (e^t + e^{-t})/2 \end{bmatrix} = \begin{bmatrix} \cosh t \\ \sinh t \\ \cosh t \end{bmatrix}.$$

P5.9. **System of differential equations** The differential equation

$$y''' - 5y'' - y' + 5y = 0$$

can be converted to a system of first-order equations with the substitutions $x_1 = y$, $x_2 = x_1' = y'$, and $x_3 = x_2' = y''$. The result is

$$\begin{bmatrix} x_1' \\ x_2' \\ x_3' \end{bmatrix} = \begin{bmatrix} 0 & 1 & 0 \\ 0 & 0 & 1 \\ -5 & 1 & 5 \end{bmatrix} \begin{bmatrix} x_1 \\ x_2 \\ x_3 \end{bmatrix}.$$

The eigenvalues are $1, -1, 5$. The corresponding eigenvectors are

$$\mathbf{u}_1 = \begin{bmatrix} 1 \\ 1 \\ 1 \end{bmatrix}, \quad \mathbf{u}_2 = \begin{bmatrix} 1 \\ -1 \\ 1 \end{bmatrix}, \quad \mathbf{u}_3 = \begin{bmatrix} 1/25 \\ 1/5 \\ 1 \end{bmatrix}.$$

P5.10. **Step response of second-order equation** The second-order equation $\ddot{y} + 3\dot{y} + 2y = U(t)$ becomes

$$\begin{bmatrix} x_1' \\ x_2' \end{bmatrix} = \begin{bmatrix} 0 & 1 \\ -2 & -3 \end{bmatrix} \begin{bmatrix} x_1 \\ x_2 \end{bmatrix} + \begin{bmatrix} 0 \\ 1 \end{bmatrix} U(t),$$

which has the form $\mathbf{x}' = A\mathbf{x} + \mathbf{f}$. The eigenvalues of A are $-1, -2$, and two eigenvectors are $\mathbf{u}_1 = [1, -1]^T$ and $\mathbf{u}_2 = [-1, 2]^T$. The assumed solution is $\mathbf{x}(t) = \Phi(t)\mathbf{c} + \mathbf{x}_p(t)$. A fundamental matrix is

$$\Phi(t) = \begin{bmatrix} e^{-t} & -e^{-2t} \\ -e^{-t} & 2e^{-2t} \end{bmatrix}.$$

Assuming a particular solution of the original equation is $\mathbf{x}_p(t) = \Phi(t)\mathbf{u}(t)$ leads to the relationship $\Phi(t)\dot{\mathbf{u}}(t) = \mathbf{f}(t)$. In this case,

$$\dot{\mathbf{u}}(t) = \Phi^{-1}(t)\mathbf{f}(t) = \begin{bmatrix} 2e^t & e^t \\ e^{2t} & e^{2t} \end{bmatrix} \begin{bmatrix} 0 \\ 1 \end{bmatrix} = \begin{bmatrix} e^t \\ e^{2t} \end{bmatrix}.$$

Integrating $\dot{\mathbf{u}}(t)$ and multiplying by $\Phi(t)$ yields

$$\mathbf{x}_p(t) = \begin{bmatrix} e^{-t} & -e^{-2t} \\ -e^{-t} & 2e^{-2t} \end{bmatrix} \begin{bmatrix} e^t \\ e^{2t}/2 \end{bmatrix} = \begin{bmatrix} 1/2 \\ 0 \end{bmatrix}.$$

The equation for the initial conditions becomes

$$\mathbf{x}(t) = \begin{bmatrix} 1 & -1 \\ -1 & 2 \end{bmatrix} \mathbf{c} + \begin{bmatrix} 1/2 \\ 0 \end{bmatrix} = \begin{bmatrix} 0 \\ 0 \end{bmatrix},$$

with the solution $\mathbf{c} = [-1, -1/2]^T$. Finally, the complete solution is

$$\mathbf{x}(t) = \Phi(t)\mathbf{c} + \mathbf{x}_p(t) = \begin{bmatrix} -e^{-t} + \frac{1}{2}e^{-2t} + \frac{1}{2} \\ e^{-t} - e^{-2t} \end{bmatrix}.$$

P5.11. Wronskian The solution to the third-order homogeneous equation

$$y^{(3)}(t) + \ddot{y}(t) + \dot{y}(t) + y(t) = 0$$

leads to the characteristic equation $\lambda^3 + \lambda^2 + \lambda + 1 = 0$, with solutions $-1, i, -i$. Thus, the solutions to the homogeneous equation is a linear combination of the functions

$$\phi_1(t) = \cos t, \quad \phi_2(t) = \sin t, \quad \phi_3(t) = e^{-t}.$$

Specifically, the complementary solution is $y_c(t) = c_1 \cos t + c_2 \sin t + c_3 e^{-t}$. The Wronskian is

$$W(x) = \begin{vmatrix} \cos t & \sin t & e^{-t} \\ -\sin t & \cos t & -e^{-t} \\ -\cos t & -\sin t & e^{-t} \end{vmatrix} = 2e^{-t}.$$

Thus, the solutions are linearly independent. Computing $W(0) = 2$ and applying the theorem in the problem shows that

$$W(x) = e^{-t}W(0) = 2e^{-t},$$

which is easier to compute than evaluating $W(x)$. Using the initial values $y(0) = 0$, $\dot{y}(0) = 1$, $\ddot{y}(0) = 0$, the complementary solution becomes $y_c(t) = \sin t$.

P5.12. Differential equation eigenvalues

    a. The eigenvalue equation $\lambda^2 + 2\alpha\lambda + w_0^2 = 0$ yields the eigenvalues $\lambda_{1,2} = \alpha \pm \sqrt{\alpha^2 - w_0^2}$. When $\alpha = w$, the eigenvalues are real and equal and the response is said to be *critically damped*.

    b. The critically damped response for the homogeneous equation is $y(t) = e^{-\alpha t}(c_1 + c_2 t)$.

P5.13. Differential equation normal modes The differential equations of motion for the coupled mass system are

$$\begin{array}{rcl} m\ddot{x}_1(t) & = & -kx_1(t) \quad + \quad k[x_2(t) - x_1(t)], \\ m\ddot{x}_2(t) & = & -k[x_2(t) - x_1(t)] \quad - \quad k[x_3(t) - x_2(t)], \\ m\ddot{x}_3(t) & = & -k[x_2(t) - x_3(t)] \quad - \quad kx_3(t). \end{array}$$

Writing this in matrix form as $\ddot{X} = AX$, leads to the system matrix

$$\begin{vmatrix} -2\alpha & \alpha & 0 \\ \alpha & -2\alpha & \alpha \\ 0 & \alpha & -2\alpha \end{vmatrix},$$

where $\alpha = k/m$. Computing the roots of $|A - \lambda I| = 0$ leads to the characteristic equation $-(\lambda + 2\alpha)^3 + 2\alpha^2(\lambda + 2\alpha) = 0$ with the roots $\lambda_1 = -2\alpha$, $\lambda_2 = -2\alpha + \sqrt{2}\alpha$, and $\lambda_3 = -2\alpha - \sqrt{2}\alpha$. The natural frequencies are given by the square roots of the eigenvalues. The frequencies are

$$w_1 = -2\sqrt{\alpha}, \quad w_2 = \sqrt{(2 + \sqrt{2})\alpha}, \quad w_3 = \sqrt{(2 - \sqrt{2})\alpha}.$$

a. In one mode, the center mass is stationary and the other masses move in opposite directions

$$\mathbf{X}_1 = \begin{bmatrix} \dfrac{1}{\sqrt{2}} \\ 0 \\ -\dfrac{1}{\sqrt{2}} \end{bmatrix} \sin(\omega_1 t + \phi_1).$$

b. The mode in which the masses move in the same direction is

$$\mathbf{X}_2 = \begin{bmatrix} \dfrac{1}{2} \\ \dfrac{1}{\sqrt{2}} \\ \dfrac{1}{2} \end{bmatrix} \sin(\omega_2 t + \phi_2).$$

c. The third mode in one in which the center mass moves opposite the other masses

$$\mathbf{X}_3 = \begin{bmatrix} \dfrac{1}{2} \\ -\dfrac{1}{\sqrt{2}} \\ \dfrac{1}{2} \end{bmatrix} \sin(\omega_3 t + \phi_3).$$

Comment: MATLAB problems are on the disk accompanying this textbook.

6

ADVANCED DIFFERENTIAL EQUATIONS

PREVIEW_____

The previous chapter dealt primarily with linear differential equations with constant coefficients subject to given initial conditions. Furthermore, the solutions were formulated in terms of continuous functions that we often call *elementary functions*, such as e^t or trigonometric functions. This chapter covers ordinary differential equations, which we will call *advanced differential equations*. Such equations may be special forms of the equations studied in Chapter 5, but they also include differential equations with variable coefficients.

This chapter first summarizes many important properties of functions and infinite series of functions that arise in the solution of differential equations. These discussions introduce concepts that will be used in the remainder of the textbook. The Taylor series is an example of an infinite series that plays a particularly important role in advanced mathematics.

Numerical methods to solve differential equations are introduced next, and the methods are first applied to various forms of ordinary differential equations. Then, boundary value problems are studied. Finally, the chapter introduces Bessel and Legendre functions as solutions of differential equations with nonconstant coefficients. These functions also play an important role in the solution of certain partial differential equations, as presented in Chapter 13.

FUNCTIONS AND DIFFERENTIAL EQUATIONS

Functions with discontinuities arise in some applications. Of special interest are functions that are mostly continuous but with a finite number of *jump* discontinuities. These functions are called *piecewise continuous* functions. In solving practical problems, discontinuous functions are often used as idealizations of functions that change rapidly. For example, the *step* function can represent a constant signal that is suddenly applied to a system at rest. In time, the step function is zero until a time t_0 and then it is a nonzero constant value for $t > t_0$. The step function is called a piecewise continuous function since it is continuous at all points except at $t = t_0$.

In this section, we discuss continuous and piecewise continuous functions. Then, we present several examples of the response of differential equations to piecewise continuous forcing functions.

CONTINUOUS FUNCTIONS

Many theorems we use in this textbook state that a function $f(x)$ of a real variable must be continuous in an interval of interest in $\mathbf{R}$ for the theorem to be true. The concepts of continuity, uniform continuity, and piecewise continuity will be discussed for scalar and vector functions.

The function $f(x)$ is said to be *continuous* at the point $x = x_0$ in the domain of $f(x)$ if

$$\lim_{x \to x_0} f(x) = f(x_0).$$

This definition implies that if $f(x_0)$ exists, the limit $\lim_{x \to x_0} f(x)$ exists, and the limit is $f(x_0)$. In other words, if the function $f(x)$ is continuous at x_0, then there are points on the graph of $f(x)$ that are arbitrarily close to the point $(x_0, f(x_0))$. The definition of a continuous function is often also stated in terms of the definition of the limit so that a function can be tested for continuity.

Consider the real function $f(x)$ defined in a closed interval I such that $a \le x \le b$. Then, $f(x)$ is *continuous* at the point $x = x_0$ in I if for every $\epsilon > 0$, there exists a $\delta > 0$ so that

$$|f(x) - f(x_0)| < \epsilon, \tag{6.1}$$

provided that $|x - x_0| < \delta$ and $x \in I$. In words, $f(x)$ is defined to be continuous at $x = x_0$ if $f(x)$ is arbitrarily close to $f(x_0)$ when x is sufficiently close to x_0. If $f(x)$ is not continuous at a point, it is said to be *discontinuous* at the point.

☐ **EXAMPLE 6.1**   ***Continuous Functions***

It is straightforward to determine the continuity of $f(x) = x$ at any point $x_0 \in \mathbf{R}$ using Equation 6.1. We must show that for any $\epsilon > 0$, a δ can be found such that

$$|f(x) - f(x_0)| = |x - x_0| < \epsilon$$

when $|x - x_0| < \delta$. Letting $\delta = \epsilon$, the result shows that $f(x)$ is continuous at any point on the real line.

More generally, every polynomial

$$P(x) = a_0 + a_1 x + a_2 x^2 + \cdots + a_k x^k$$

is continuous at each point in $\mathbf{R}$. Thus, $P(x) = x$ is a special case.

☐

In general, when testing for continuity using the definition previously given, the number δ may depend on both ϵ and the point x_0. However, if δ does not depend on the particular point x_0 in I, the function $f(x)$ is said to be *uniformly continuous* in I. The function $f(x) = x$ is thus uniformly continuous at every point in $\mathbf{R}$ according to the result in Example 6.1.

PIECEWISE CONTINUOUS FUNCTIONS

For the nth-order differential equation

$$\frac{d^n y(t)}{dt^n} + a_{n-1}(t)\frac{d^{n-1}y(t)}{dt^{n-1}} + \cdots + a_1(t)\frac{dy(t)}{dt} + a_0(t)y(t) = f(t) \quad \textbf{(6.2)}$$

introduced in Chapter 5, we have assumed $f(t)$ to be continuous over the interval of interest. In some applications, the forcing function $f(t)$ may have discontinuities at points in the interval. A unique solution for the differential equation can be found if $f(t)$ has only a finite number of *jump discontinuities* in an interval. At the point of a jump discontinuity, say, t_0, $f(t)$ has limits from the right (f_0^+) and the left (f_0^-). The magnitude of the jump is $\left|f_0^+ - f_0^-\right|$, and it must be finite.

A *piecewise continuous* function in an interval I is a function that is continuous on a finite number of subintervals of I such that at each discontinuity the function has finite left-hand and right-hand limits. A piecewise continuous function may have a finite number of jump discontinuities in the interval I.

When $f(t)$ is piecewise continuous in an interval, a solution to Equation 6.2 is defined to be a function $y(t)$ that satisfies the differential equation except at the points of discontinuity of $f(t)$. In solving the initial value problem over the entire interval, the values of $y(t)$ and its derivatives to order $n - 1$ at the end of one subinterval are used as initial values for the solution in the next subinterval. There will be a jump discontinuity in the nth derivative of $y(t)$ at each point of discontinuity.

Piecewise Continuous Forcing Function

Consider the differential equation

$$\ddot{y}(t) + 3\dot{y}(t) + 2y(t) = P_1(t) \tag{6.3}$$

with initial conditions $y(0) = \dot{y}(0) = 0$ and input forcing function

$$P_1(t) = \begin{cases} 0, & t < 0, \\ 1, & 0 \le t \le 1, \\ 0, & t > 1. \end{cases}$$

The function $P_1(t)$ is shown in Figure 6.1(a). This is a piecewise continuous function that is continuous within three regions. Clearly, $y(t) = 0$ for $t < 0$. Thus, the problem could be solved first in the region $0 \le t \le 1$ to yield $y_1(t)$. In this region, the roots of the characteristic equation $\lambda^2 + 3\lambda + 2 = 0$ are -1 and -2. The solution is

$$y_1(t) = \frac{1}{2} - e^{-t} + \frac{1}{2}e^{-2t}, \quad \text{for} \quad 0 \le t \le 1, \tag{6.4}$$

as determined by the techniques in Chapter 5. The solution for $t > 1$ is

$$y_2(t) = c_1 e^{-t} + c_2 e^{-2t} \quad (t > 1) \tag{6.5}$$

with two arbitrary constants since $f(t) = 0$ in this region. Equating values at the point of discontinuity,

$$\begin{aligned} y_1(1) &= y_2(1), \\ \dot{y}_1(1) &= \dot{y}_2(1), \end{aligned}$$

and solving for c_1 and c_2 will yield the solution for $y_2(t)$.

We think a simpler way to solve the problem is to recognize the pulse as the superposition of two unit step functions, where the unit step function is defined as

$$U(t) = \begin{cases} 0, & t < 0, \\ 1, & t \ge 0. \end{cases}$$

The pulse can be constructed by adding a step function $U(t)$ at $t = 0$ and a negative step $-U(t - 1)$ shifted in time to start at $t = 1$. The superposition is shown in Figure 6.1(b). Then, the solution of the linear differential equation is the sum of the solutions due to the individual unit step functions.

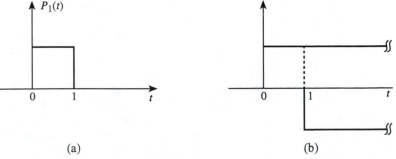

(a)   (b)

FIGURE 6.1 *(a) Pulse input (b) unit step functions*

Since the differential equation is linear and the forcing function $U(t)$ yields the solution of Equation 6.4, then the shifted unit step $-U(t-1)$ yields

$$y_2(t) = \begin{cases} 0, & 0 \le t < 1, \\ -\left[\dfrac{1}{2} - e^{-(t-1)} + \dfrac{1}{2}e^{-2(t-1)}\right], & t > 1. \end{cases}$$

The result is $y(t) = y_1 + y_2$ or

$$y(t) = \begin{cases} \dfrac{1}{2} - e^{-t} + \dfrac{1}{2}e^{-2t}, & 0 \le t \le 1, \\ (e-1)e^{-t} - \dfrac{(e^2-1)}{2}e^{-2t}, & t \ge 1. \end{cases}$$

Thus, the constants in Equation 6.5 are $c_1 = (e-1)$ and $c_2 = -(e^2-1)/2$.

$\square$

□ EXAMPLE 6.3   *MATLAB Pulse Solution*

The accompanying MATLAB scripts solve a second-order differential equation numerically using the command **ode23** introduced in Chapter 5. The equation is

$$a\ddot{y}(t) + b\dot{y}(t) + cy(t) = P(t),$$

with initial conditions specified as inputs. The forcing function $P(t)$ is a pulse of unit height with duration determined by input values to the MATLAB M-file. When the M-file executes, the user must input the following data:

a. The coefficients a, b, c;

b. The starting and ending time of $P(t)$;

c. The initial and final time for the plot;

d. The initial conditions $[y(0), \dot{y}(0)]$;

e. A title and labels for the plot.

The algorithm used by **ode23** attempts to approximate the solution to the equation set $\dot{\mathbf{x}} = A\mathbf{x}$ starting at t_0 and at a sequence of time points

$$t_1 = t_0 + h_1, t_2 = t_1 + h_2, \ldots,$$

where h_i is called the *step size* between points t_i and t_{i+1}. Since **ode23** uses a variable step size based on the characteristics of the solution, it is not possible to predict the values of t at which the equation is evaluated while **ode23** executes.

In particular, the number of time points between $t = 0$ and $t = 1$ will vary with the specific equation, so it is not possible to use a fixed vector of points to represent the pulse. Thus, the function **cl2ordpf** must generate the value of the pulse based on the value of t used by **ode23**. The **if** statements in the M-function supply the value of the pulse as **ode23** executes.

For the specific example, the input values for the coefficients were $a = 1$, $b = 3$, and $c = 2$ and the initial conditions were zero, as in the previous Example 6.2. Using **ode23**, the system of equations corresponding to the equation

$$\ddot{y}(t) + 3\dot{y}(t) + 2y(t) = P_1(t), \tag{6.6}$$

is solved. Figure 6.2 shows the input pulse and the response.

MATLAB Script _____

Example 6.3

```
% CL2ORDP.M To compute solution of ay''+by'+cy=pulse and plot y(t)
%   INPUT:   a (a not 0), b, c, tstart-tend for pulse,
%            t0-tf for plot.
%  Calls function cl2ordfp and ode23
global a_d2y b_dy c_y  tstart_t tend_t     % Pass variables
fprintf('Solve aD2y+bDy+cY=Pulse[Tstart to Tend]\n')
a_d2y=input('Input a=  ')                   % Input coefficients
b_dy=input('Input b=  ')
c_y=input('Input c=  ')
% Input Pulse length to pass to function
tstart_t=input('Input Tstart of pulse ')   % Input time of pulse
tend_t=input('Input Tend of pulse ')
%
t0=input('Initial time for equation =  ')   % Input time of plot
tf=input('Final time=  ')
x0=input('[y(0) Dy(0)] =  ')
x0t=x0'
% ode23 may call function cl2ordpf several times for each t.
[t,x]=ode23('cl2ordpf',t0,tf,x0t);
% Output of ode23 is vector t and matrix x
% y values
y=x(:,1);        % Rename variables: solution y
dy=x(:,2);       % Derivative of y
%
title_x=input('Title =  ', 's')            % Input title and labels
xlabel_1=input('xlabel =  ','s')
ylabel_1=input('ylabel =  ','s')
% Pulse_to_plot - a column vector the length of t
length_t=length(t);                        % Plot the input pulse
clear pulse_to_plot
for i=1:length_t
  if t(i) < tstart_t
      pulse_to_plot(i) = 0;
  elseif (tstart_t <= t(i)) & (t(i) <= tend_t)
      pulse_to_plot(i) = 1;
  elseif t(i) > tend_t
      pulse_to_plot(i) = 0;
  end
end
pulse_to_plot= (pulse_to_plot)';
plot(t,y,'-',t,pulse_to_plot,'*')          % Plot the solution
title(eval('title_x'))
xlabel(eval('xlabel_1'))
ylabel(eval('ylabel_1'))
grid
```

The function cl2ordpf defines the differential equation in state-space form for **ode23**. The coefficients of the equation and the times for the pulse are passed as global variables.

MATLAB Script _____

Example 6.3

```
function xdot=cl2ordpf(t,x)
%   CALL: xdot=cl2ordpf(t,x), defines the second order differential
%     equation xdot(t)=Ax(t)+[0 1]'*P(t).  t,x are scalar inputs
%     from functions such as ode23.m. P(t) is a unit pulse.
%   Elements of A and tstart-tend of the pulse are global variables.
global a_d2y b_dy c_y  tstart_t tend_t  % Coefficients
% Compute pulse_t; a scalar value of pulse at time t
%        If t < tstart_t          pulse_t = 0
%  If tstart_t <= t <= tend_t   pulse_t = 1
%        If t > tend_t           pulse_t = 0
if t < tstart_t
    pulse_t = 0.0;
 elseif  (tstart_t <= t) & (t <= tend_t)
    pulse_t = 1.0 ;
 elseif t > tend_t
    pulse_t =0.0 ;
end
%
% The equation ay''+by'+cy = f(t) as xdot(t)=Ax(t) with x1=y, x2=y'
xdot(1) = 0*x(1) + x(2);
xdot(2) = - (c_y)*x(1)/(a_d2y) -(b_dy)*x(2)/(a_d2y)   ...
    +(pulse_t)/(a_d2y);
```

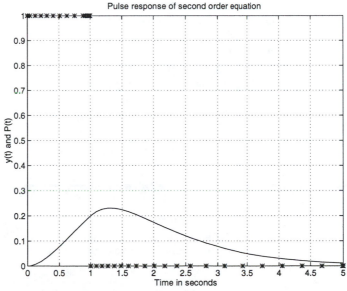

FIGURE 6.2 *MATLAB solution of Equation 6.6*

☐

WHAT IF? Suppose you wish to study the response of a second-order differential equation to a pulse input. The M-file CL2ORDP.M presented in Example 6.3 allows you to change the coefficients of the equation, the initial conditions, and the length of the pulse, as well as plotting parameters. Try a few other examples varying the equation, the initial conditions, or the pulse length.

The function (`cl2ordpf`) defining the differential equation could be modified to study the effect of varying the type of input. Change the function to define a triangular pulse as the input function and compare the results with the response to a rectangular pulse.

SEQUENCES AND SERIES

This section explores the conditions under which expressions of the form

$$a_1 + a_2 + \cdots$$

have meaning, where the terms $a_1, a_2, \ldots$ are real numbers. Such an expression is called an *infinite series*. Although the study of series can be a fascinating topic in itself, we present here only a few results that will be useful for solving advanced differential equations. First, series that consist of constant terms are studied. Then, infinite series of functions are described. Except for a few special cases, the solutions to differential equations with variable coefficients usually involve infinite series. Later sections in this chapter will present more details about series used to solve differential equations.

The concept of an infinite series occurs even in operations involving simple numbers. For example, we can write

$$\frac{1}{3} = 0.33333\ldots,$$

which is another way of writing the series

$$\frac{1}{3} = \frac{3}{10} + \frac{3}{100} + \frac{3}{1000} + \cdots.$$

For computer summation, a finite number of terms can be summed for any series. The example for $1/3$ shows that after the third term is summed the result is the *rational number*

$$\frac{1}{3} = \frac{333}{1000} = 0.333,$$

and we say that the series has been *truncated* to three terms. If this is a sufficiently good approximation, the other terms in the series can be neglected. Thus, 0.333 is said to be a *rational approximation* to the number 1/3. Our intuitive justification for truncating the series is that the remaining terms will not affect the result significantly if three decimal places of accuracy is sufficient. Contrast this result with the series

$$1 + 1 + 1 + \cdots,$$

for which truncation would not be justified if the sum of the series is sought since the sum does not converge to a finite value.

We expect a series to tend toward a finite sum if it represents the solution to a physical problem and hope to be able to calculate the error that results from truncation after n terms. Study of the convergence of a series of interest allows us to decide how many terms are necessary to achieve a certain accuracy. However, before the properties of infinite series are studied, it is useful to consider infinite sequences of numbers.

INFINITE SEQUENCES

An infinite sequence of real numbers is a succession of numbers defined by some rule of formation. The *terms* in a sequence s can be written in subscript notation as

$$s_1, s_2, \ldots$$

so that to each positive number n there corresponds a real number s_n in the sequence. If a formula exists for the terms, the sequence is usually defined by giving the formula for the nth term. Thus, the sequence defined as $s_n = 1/n$ has the terms

$$1, \frac{1}{2}, \frac{1}{3}, \cdots.$$

As another example, consider the sequence designated F_n such that the first two terms are $F_0 = 1$ and $F_1 = 1$ and the remaining terms are calculated as

$$F_{n+1} = F_n + F_{n-1}, \qquad n \geq 1.$$

This is called the Fibonacci sequence, in which each new term is the sum of the previous two terms. The terms are $1, 1, 2, 3, 5, \ldots$. Since each term after the first is defined in terms of its predecessors, this is said to be a *recursively defined sequence*.

The *limit* of a sequence is defined in terms of the behavior of its nth term according to the following definition:

Let L be a real number. A sequence s is said to have limit L if for every number $\epsilon > 0$ there exists a number N such that

$$|s_n - L| < \epsilon$$

for every integer $n > N$.

If the limit of the sequence is the number L, we write

$$\lim_{n \to \infty} s_n = L,$$

and the sequence is said to *converge* to L. Otherwise, the sequence is said to *diverge*.

Using the definition of the limit, it is easily shown that

$$\lim_{n \to \infty} \frac{1}{n} = 0$$

by choosing $N > 1/\epsilon$ so that $n \geq N$ implies that

$$\frac{1}{n} \leq \frac{1}{N} < \epsilon.$$

Sequences and their limits are important in the study of infinite series. In particular, the sum of an infinite series is defined in terms of a sequence.

INFINITE SERIES   An *infinite series* is an expression of the form

$$a_1 + a_2 + \cdots = \sum_{n=1}^{\infty} a_n. \tag{6.7}$$

This notation used in the text indicates a series of real numbers with each term a_i being a constant. Writing the series in the form of Equation 6.7 does not imply that the series has a finite sum. In fact, the term *sum of an infinite series* must be defined since the usual operation of addition applies only to a finite set of numbers. However, the sum of a finite number of terms in the infinite series is a well-defined number. As the limit of the number of terms in this sum approaches infinity, the sum may approach a number. In this case, the infinite series is said to *converge*.

Convergence of Series   Consider the infinite series

$$a_1 + a_2 + \cdots = \sum_{n=1}^{\infty} a_n,$$

and associate with it the sequence of *partial sums* S_n, so that the nth sum has the form

$$S_n = a_1 + a_2 + \cdots + a_n = \sum_{j=1}^{n} a_j.$$

Writing out several of the sums shows that $S_1 = a_1, S_2 = a_1 + a_2, \ldots,$ are the individual sums.

The infinite series $\sum_{n=1}^{\infty} a_n$ is said to *converge* if the sequence of partial sums is convergent. If the partial sums converge, there exists a number S such that

$$\lim_{n \to \infty} S_n = S$$

and number S is the *sum* of the infinite series. If the series does not converge, it *diverges*.

Tests for Convergence   In the theory of series, a number of tests have been developed to determine if a series is convergent or not. Several of these tests are listed in the Table 6.1.

TABLE 6.1 *Tests for convergence of series*

Test	Definition
Limit	If $\lim\limits_{n\to\infty} a_n \neq 0$, the series $\sum\limits_{n=1}^{\infty} a_n$ diverges.
Ratio	If $a_n \neq 0$ for $n = 1, 2, \ldots$ and the limit $$\lim_{n\to\infty} \left\lvert \frac{a_{n+1}}{a_n} \right\rvert = L, \text{ then}$$ If $L < 1$, $\sum\limits_{n=1}^{\infty} a_n$ converges; If $L = 1$, the test yields no information; If $L > 1$, $\sum\limits_{n=1}^{\infty} a_n$ diverges.

The *limit test* is a consequence of the following theorem.

■ THEOREM 6.1   **Limit test**
  If the infinite series $\sum_{n=1}^{\infty} a_n$ converges, then

$$\lim_{n\to\infty} a_n = 0.$$

Thus, the nth term of a convergent series must approach zero as n approaches infinity. However, even if the limit test shows that the nth term approaches zero as n goes to infinity, it does not mean that the series converges. We can conclude only that if

$$\lim_{n\to\infty} a_n \neq 0,$$

the infinite series $\sum_{n=1}^{\infty} a_n$ diverges.

The *ratio test* compares two terms in the series in the form

$$\lim_{n\to\infty} \left\lvert \frac{a_{n+1}}{a_n} \right\rvert.$$

If this limit is a number L less than 1, the series is convergent. If the limit is ∞ or a real number $L > 1$, the series is divergent. When the limit $L = 1$, the ratio test is inconclusive.

☐ **EXAMPLE 6.4**   *Infinite Series*

Table 6.2 shows the results of the ratio and limit test for various series.

TABLE 6.2 *Examples of tests for convergence*

Series	Test	Result
$\displaystyle\sum_{n=1}^{\infty} n$	$\displaystyle\lim_{n\to\infty} n = \infty$	The series diverges.
$\displaystyle\sum_{n=1}^{\infty} \frac{1}{n}$	$\displaystyle\lim_{n\to\infty} \frac{1}{n} = 0$	The limit test is inconclusive.
$\displaystyle\sum_{n=1}^{\infty} \frac{1}{2^n}$	$\displaystyle\lim_{n\to\infty} \frac{1}{2^n} = 0$	The limit test is inconclusive.
$\displaystyle\sum_{n=1}^{\infty} \frac{2^n}{n!}$	$\displaystyle\lim_{n\to\infty} \frac{2}{n+1} = 0$	The series converges.
$\displaystyle\sum_{n=1}^{\infty} \frac{1}{n^p}$	$\displaystyle\lim_{n\to\infty} \left(\frac{n}{n+1}\right)^p = 1$	The ratio test is inconclusive.

The limit test shows that the series

$$\sum_{n=1}^{\infty} n$$

diverges since the nth term does not approach zero as n approaches infinity. In the case of the series

$$\sum_{n=1}^{\infty} \frac{1}{n} \quad \text{and} \quad \sum_{n=1}^{\infty} \frac{1}{2^n},$$

the limit test indicates that the nth terms of both series approach zero in the limit. Therefore, the test is inconclusive. In fact, the series

$$\sum_{n=1}^{\infty} \frac{1}{n} = 1 + \frac{1}{2} + \cdots$$

is divergent. Thus, the sum of fractions of the form $1/n$ does not have a limit. However, the sum of terms such as $1/2^n$ does converge to a limit. It is straightforward to show that the series

$$\sum_{n=1}^{\infty} \frac{1}{2^n} = \frac{1}{2} + \frac{1}{4} + \frac{1}{8} + \cdots$$

converges to 1 since the partial sums are

$$S_1 = \frac{1}{2}, \quad S_2 = \frac{3}{4}, \quad S_n = \frac{2^n - 1}{2^n}, \cdots$$

and

$$S = \lim_{n \to \infty} S_n = \lim_{n \to \infty} \frac{2^n - 1}{2^n} = \lim_{n \to \infty} \left(1 - \frac{1}{2^n}\right) = 1.$$

The other examples in Table 6.2 use the ratio test

$$\lim_{n \to \infty} \left| \frac{a^{n+1}}{a_n} \right|.$$

The series

$$\sum_{n=1}^{\infty} \frac{2^n}{n!}$$

converges by the ratio test since

$$\lim_{n \to \infty} \left| \frac{a^{n+1}}{a_n} \right| = \lim_{n \to \infty} \left| \frac{2}{n+1} \right| = 0.$$

One common series cannot be tested for convergence by the limit or ratio test. The series is called a *p-series* and has the form

$$\sum_{n=1}^{\infty} \frac{1}{n^p} = 1 + \frac{1}{2^p} + \cdots. \tag{6.8}$$

The ratio test is inconclusive since the limit of the ratio of the terms $|a_{n+1}/a_n|$ is 1. A test called the *integral test* can be used to determine the convergence of the *p*-series. This series converges for $p > 1$ and diverges for $p \leq 1$. Any textbook that treats calculus will discuss various tests for convergence of series in detail.

□

Geometric Series A *geometric series* is a series with each term after the first being a fixed multiple of the preceding term. The multiplier is a real number r, called the *ratio*, so that $a_{n+1} = ra_n$. If the sum is taken from $n = 0$, the geometric series is represented as

$$\sum_{n=0}^{\infty} ar^n = a + ar + \cdots \quad (a \neq 0). \tag{6.9}$$

We will show that the series converges to the sum

$$\sum_{n=0}^{\infty} ar^n = \frac{a}{1 - r} \tag{6.10}$$

if $-1 < r < 1$ but diverges if $|r| > 1$. Furthermore, if the infinite series with nth partial sum S_n converges and has sum S, then for every number $\epsilon > 0$, there exists a number N such that

$$|S - S_n| < \epsilon$$

for every $n > N$. Thus, we can approximate the sum as closely as desired by taking more terms in the series if necessary.

The nth partial sum for the geometric series is found by subtracting the terms

$$
\begin{aligned}
S_n - rS_n &= a + ar + \cdots + ar^n - \left(ar + ar^2 + \cdots + ar^{n+1}\right) \\
&= a - ar^{n+1},
\end{aligned}
$$

so that $S_n - rS_n = a(1 - r^{n+1})$. Thus, solving for S_n leads to the result

$$
S_n = \frac{a(1 - r^{n+1})}{1 - r}
$$

for the sum of the first $n + 1$ terms. Taking the limit as n goes to infinity with $|r| < 1$ shows that the sum of the series is $a/(1 - r)$, as shown in Equation 6.10.

Consider the fraction $1/3$ represented as the series

$$
\frac{1}{3} = \frac{3}{10} + \frac{3}{100} + \frac{3}{1000} + \cdots .
$$

Substituting $a = 3/10$ and $r = 1/10$ in Equation 6.9 leads to the result

$$
\sum_{n=0}^{\infty} \frac{3}{10}\left(\frac{1}{10}\right)^n = \frac{3}{10}\frac{1}{1 - 1/10} = \frac{1}{3}.
$$

Considering the partial sums of this series, we are confident that taking more terms in a truncated series leads to a better approximation for $1/3$.

INFINITE SERIES OF FUNCTIONS
An infinite series of functions can be written as

$$
f_1(x) + f_2(x) + \cdots + f_n(x) + \cdots = \sum_{n=1}^{\infty} f_n(x), \tag{6.11}
$$

indicating that each term $f_i(x)$ is a function of the real variable x. For an infinite series of functions, as in Equation 6.11, the series may converge for a range of values x and diverge otherwise. Of course, for a particular choice of x, the series $\sum f_n(x)$ is just a series of constants so that many of the results derived previously, such as the ratio test, can be applied to series of functions.

Convergence   Consider a sequence of real-valued functions defined for all x in some interval I. If the series

$$
\sum_{n=1}^{\infty} f_n(x)
$$

converges for each $x \in I$, the series is said to *converge pointwise* in I. Letting the limit be $f(x)$ implies

$$f(x) = \sum_{n=1}^{\infty} f_n(x) = \lim_{N \to \infty} \sum_{n=1}^{N} f_n(x).$$

Stated another way, if the series converges pointwise on the interval I for each x, there is a number $f(x)$ such that, given $\epsilon > 0$, there is an integer K such that

$$\left| \sum_{n=1}^{N} f_n(x) - f(x) \right| < \epsilon \qquad \text{(6.12)}$$

whenever $N \geq K$. Thus, the series can approximate $f(x)$ at the point x as closely as possible, provided that enough terms in the series are summed. At other points in the interval, the series converges, but the number of terms to keep the error in approximating $f(x)$ from exceeding ϵ may vary from one x to another in the interval. In other words, K can depend on the value of x in the interval.

Uniform Convergence A stronger form of convergence called *uniform convergence* is sometimes obtained. The series

$$\sum_{n=1}^{\infty} f_n(x)$$

is said to *converge uniformly* to a function in the interval I if the relationship in Equation 6.12 is true for *all* $x \in I$ and for *all* $N \geq K$. Notice that uniform convergence implies pointwise convergence, but not conversely.

When a series is uniformly convergent, it is sometimes said that the series converges at the same rate for each point in I. One way to test a series for uniform convergence is to apply a test known as the *Weierstrass test*. To apply the test, terms in the series are compared with the terms in a known constant series.

■ THEOREM 6.2   ***Weierstrass test***

Let $\sum_{n=1}^{\infty} f_n(x)$ be a series of real-valued functions defined for a set of values $x \in I$. If there is a convergent constant series $\sum_{n=1}^{\infty} p_n$ such that

$$|f_n(x)| \leq p_n$$

for all $x \in I$ and $n = 1, 2, \ldots$, then the series $\sum_{n=1}^{\infty} f_n(x)$ is uniformly convergent in I.

Uniform Convergence

According to the Weierstrass test, the trigonometric series

$$\sum_{k=1}^{\infty} \frac{\sin kx}{k^2}$$

converges uniformly for all real x because

$$\left| \frac{\sin kx}{k^2} \right| \le \frac{1}{k^2},$$

and the constant series converges according to the previous discussion of this p-series of order 2 in Equation 6.8. We will study this trigonometric series again in Chapter 8 in connection with Fourier series.

☐

POWER SERIES   For applications, important infinite series are those whose terms are constants times successive powers of the independent variable with the form

$$\sum_{n=0}^{\infty} c_n x^n. \tag{6.13}$$

Such a series is called a *power series* in x. It is often convenient to expand the series in powers of $(x - a)$, which leads to the series

$$c_0 + c_1(x - a) + c_2(x - a)^2 + \cdots = \sum_{n=0}^{\infty} c_n(x - a)^n, \tag{6.14}$$

where a and the coefficients $c_0, c_1, \ldots,$ are constants. Equation 6.13 is a special case of Equation 6.14 with $a = 0$.

The power series has a *radius of convergence* $R \ge 0$ such that the series converges when $|x - a| < R$ and diverges when $|x - a| > R$. The possibilities are as follows:

1. If the number $R = 0$, the series converges only at $x = a$.

2. If $R = \infty$, the series converges for all x.

3. If $R \ne 0$ and r is such that $0 < r < R$, then the series converges uniformly for $|x - a| \le r$.

In the last case, if R is a finite number, the series may converge or diverge at the endpoints where $r = R$.

The radius of convergence R is

$$R = \lim_{n \to \infty} \left| \frac{c_n}{c_{n+1}} \right|,$$

as you are asked to show in Problem 6.2. The *interval of convergence* for the power series is thus a symmetric interval extending a distance R on each side of $x = a$.

In the case of the series

$$\sum_{n=1}^{\infty} \frac{x^n}{n!},$$

where x is any real number, the ratio test shows that the series converges for any choice of x, since

$$\lim_{n \to \infty} \frac{|x|}{n+1} = 0.$$

Representation of Functions   Power series are used extensively for computing or approximating values of functions. Suppose that a power series converges to the value of $f(x)$, so that

$$f(x) = \sum_{n=0}^{\infty} a_n x^n = a_0 + a_1 x + \cdots + a_n x^n + \cdots. \qquad \textbf{(6.15)}$$

Such a series is called a *power series representation*, or a *power series expansion*, of $f(x)$. Alternatively, the function is said to be *represented*, or *expressed*, by the series. Although many functions of interest can be represented by a power series, there are many functions that cannot. If $f(x)$ can be represented by a power series in an interval, the function is termed *analytic* in the interval. The properties of functions that can be represented by power series will be explored further after Taylor series are introduced.

TAYLOR SERIES

We now turn to the *Taylor series* representation of a function. Taylor series are important series representations for both analytical and numerical studies. These series are the basis for many numerical algorithms to approximate functions and solve differential equations. Briefly described, a Taylor series is a power series with the coefficients determined from the derivatives of the function being represented according to Taylor's formula. The Taylor series yields the power series expansion for a function $f(x)$ together with an expression for the remainder after n terms.

Let $f(x)$ be represented by a power series within its interval of convergence as

$$
\begin{aligned}
f(x) &= c_0 + c_1(x - a) + c_2(x - a)^2 + \cdots \\
&= \sum_{n=0}^{\infty} c_n(x - a)^n, \qquad a - R < x < a + R, \qquad \textbf{(6.16)}
\end{aligned}
$$

where R is the convergence radius. This series is the *Taylor series* of $f(x)$ at $x = a$ if the coefficients are derived as

$$c_0 = f(a), \ c_1 = \frac{f'(a)}{1!}, \ c_2 = \frac{f''(a)}{2!}, \ldots, \ c_n = \frac{f^{(n)}(a)}{n!}, \ldots,$$

where the notation for the coefficient c_n means that the nth derivative of $f(x)$ is to be formed and then evaluated at $x = a$ before division by $n!$. Substituting these coefficients in Equation 6.16 yields the result

$$f(x) = f(a) + \frac{f'(a)}{1!}(x - a) + \cdots + \frac{f^{(n)}(a)}{n!}(x - a)^n + \cdots. \tag{6.17}$$

This formula is known as Taylor's formula, and the series is the Taylor series. When $a = 0$, the series is often called the *Maclaurin series* for $f(x)$.

Notice that the Taylor series for $n + 1$ terms is the nth-degree polynomial

$$P_n(x) = \sum_{k=0}^{n} \frac{f^{(k)}(a)}{k!}(x - a)^k, \tag{6.18}$$

which approximates the function $f(x)$ at the point $x = a$. Thus, $P_n(x)$ is a polynomial in powers of $(x - a)$. To compute the Taylor series and approximate $f(x)$ near $x = a$, it is necessary to calculate the value $f(a)$ and the derivatives $f'(a), f''(a), \cdots, f^{(n)}(a)$.

An important question is which functions can be represented by the Taylor series just presented in Equation 6.17. Before exploring this issue, we consider the application of the Taylor series to functions for which the Taylor series yields useful approximations.

TAYLOR SERIES REMAINDER

Consider the application of the Taylor series to represent a function by directly applying Equation 6.17. Unless the function $f(x)$ is very simple, computation of the higher derivatives of $f(x)$ may become impractical. Generally, the infinite series must be truncated after only a relatively few terms. Thus, the question arises as to how well the truncated series approximates the function.

Let the function $f(x)$ be represented by a Taylor series truncated after the term involving the nth derivative. The function $f(x)$ is then represented as

$$f(x) = f(a) + \frac{f'(a)}{1!}(x - a) + \cdots + \frac{f^{(n)}(a)}{n!}(x - a)^n + R_n(x),$$

where $R_n(x)$ is the *remainder* and measures the error made when $f(x)$ is replaced by a truncated Taylor series. The following theorem defines the remainder in terms of the $(n + 1)$st derivative of $f(x)$.

Taylor series remainder

Let $f(x)$ be defined and continuous with continuous derivatives up to order $(n+1)$ on an interval I. Then, for each $x \in I$,

$$f(x) \;=\; f(a) + \frac{f'(a)}{1!}(x-a) + \cdots + \frac{f^{(n)}(a)}{n!}(x-a)^n$$
$$+ \frac{f^{(n+1)}(x_1)}{(n+1)!}(x-a)^{n+1}$$

for some $x_1 \in I$.

The Taylor series *remainder* term is thus

$$R_n = \frac{f^{(n+1)}(x_1)}{(n+1)!}(x-a)^{n+1}. \tag{6.19}$$

Because x_1 is not explicitly given, the remainder is not known exactly. However, the remainder can often be used to obtain an upper estimate of the error in the approximation.

☐ EXAMPLE 6.6   ***Taylor Series***

If $f(x) = e^x$, then $f^{(k)}(x) = e^x$ for all $k \geq 0$ in the Taylor series expansion of Equation 6.17. Thus, expanding at $a = 0$, $f^{(k)}(0) = 1$, and the Taylor series becomes the Maclaurin series

$$e^x = 1 + x + \frac{x^2}{2!} + \cdots + \frac{x^n}{n!} + \frac{e^{x_1} x^{n+1}}{(n+1)!},$$

for some x_1 between 0 and x. In terms of the Taylor polynomials defined in Equation 6.18, the first few polynomials for e^x are

$$\begin{aligned}
P_0 &= 1, \\
P_1 &= 1 + x, \\
P_2 &= 1 + x + \frac{1}{2}x^2, \\
P_3 &= 1 + x + \frac{1}{2}x^2 + \frac{1}{6}x^3.
\end{aligned} \tag{6.20}$$

The series converges since

$$\lim_{n \to \infty} \left| \frac{a_{n+1}}{a_n} \right| = \lim_{n \to \infty} \frac{|x|}{n+1} \to 0,$$

and also the remainder $R_n \to 0$ as n goes to infinity. The magnitude of the remainder is

$$0 \;<\; |R_n(x)| < \frac{|x|^{n+1}}{(n+1)!}, \qquad x < 0,$$

$$0 \;<\; |R_n(x)| < \frac{e^{x_1} x^{n+1}}{(n+1)!}, \qquad x > 0,$$

where $0 < x_1 < x$. Note that if $x > 0$, $e^{x_1} < e^x$ since $0 < x_1 < x$. If $x < 0$, $e^{x_1} < 1$.

To compute e, let $x = 1$ and write the Taylor series truncated after $N + 1$ terms. Then,

$$e = \sum_{n=0}^{N} \frac{1}{n!} = 1 + \frac{1}{1!} + \cdots + \frac{1}{N!}$$

with an error of less than $\dfrac{e^1}{(N+1)!}$ since $0 < x_1 < 1$.

□

WHAT IF?   To see the errors involved, plot the function $y = e^x$ and the approximations P_n from Equation 6.20 for $n = 1, 2, 3$. Using an interval $[-2, 2]$ for x shows how the accuracy of the approximation varies away from the origin.

TAYLOR SERIES EXAMPLES

Table 6.3 presents the Taylor series and gives a few examples of functions that have Taylor series representation. For the function

$$f(x) = \frac{1}{1-x},$$

the series can always be obtained by applying Taylor's formula to $f(x)$ using the point $x = 0$ for expansion of the series. In this case, notice that the geometric series

$$\sum_{n=0}^{\infty} x^n$$

converges for $-1 < x < 1$ with the sum $1/(1 - x)$, as shown by Equation 6.10. Recognizing the power series as a geometric series with sum $1/(1 - x)$ leads to a simple way of creating the power series expansion of $f(x)$. It can be shown that a valid power series expansion obtained by any method must coincide with the Taylor series.

For the exponential, sine, and cosine functions, the expansions shown in the table are valid for all x. Notice that the Taylor series for $\cos x$ can be derived by differentiating the series for $\sin x$. It is no coincidence that

$$\frac{d}{dx} \sin x = \cos x$$

and the Taylor series expansions have the same relationships. The following theorem relates the Taylor series expansion of a function to the derivative and integral of the function.

■ **THEOREM 6.4**   *Differentiation and integration*

A power series may be differentiated (or integrated) term by term in any interval interior to its interval of convergence. The resulting series has the same interval of convergence as the original series and represents the derivative (or integral) of the function to which the original series converges.

——————————————————————————■

TABLE 6.3 *Examples of Taylor series*

Definition and examples

Taylor series:

$$f(x) = f(a) + \frac{f'(a)}{1!}(x - a) + \cdots + \frac{f^{(n)}(a)}{n!}(x - a)^n + \cdots$$

Examples:

$$\frac{1}{1 - x} = 1 + x + x^2 + \cdots, \quad |x| < 1$$

$$e^x = 1 + x + \frac{x^2}{2!} + \cdots$$

$$\sin x = x - \frac{x^3}{3!} + \frac{x^5}{5!} - \frac{x^7}{7!} + \cdots$$

$$\cos x = 1 - \frac{x^2}{2!} + \frac{x^4}{4!} - \frac{x^6}{6!} + \cdots$$

MATLAB and Taylor Expansions   A MATLAB program can be used to sum the terms in a power series expansion that have been calculated and plot the result over a numerical range of x. The command **taylor** returns the Taylor series of a function defined symbolically. The number of terms to be calculated can be specified.

W H A T   I F ?   Verify the Taylor series examples in Table 6.3 using the symbolic command **taylor**. For example, the commands

```
>>f2='exp(x)'
>>T2=taylor(f2)
```

compute the first six terms of the Taylor expansion of e^x and also define the order of the error.

Suppose you want more terms in the symbolic series and a numerical result. Evaluate the Taylor series computed by **taylor** at various values of the argument and compare it to the numerical result computed by MATLAB.

Uniqueness of Taylor series   Whenever a Taylor series converges to a function, the series is unique. This can be stated as a theorem.

Power and Taylor series

> *Every power series with nonzero convergence radius is the Taylor series of its sum.*

———————————————————————————■

Functions that can be represented by a Taylor series in an interval are termed *analytic* in the given interval.[1] For a function of a real variable, it is a necessary condition that the function be continuous and have derivatives of all orders in an interval to be analytic in the interval. Fortunately, almost all the functions arising in physical problems are analytic. Most of the common functions, such as polynomials, exponentials, and trigonometric functions, are analytic in every interval where the functions are continuous. Yet even a function such as $|x|$ is not analytic in any interval that includes $x = 0$. The function is continuous for all x but has a discontinuous derivative at $x = 0$, so it cannot be represented by a Taylor series there.

Considering the form of the Taylor series in Equation 6.17, it is tempting to say that a continuous function that has continuous derivatives of all orders in an interval I could be represented accurately by a Taylor series in the interval. Unfortunately, there are real functions that have finite derivatives of all orders but cannot be represented by a power series. Thus, the continuity of the function and its derivatives is *not sufficient* to guarantee the existence of a Taylor series that represents the function.

In summary, even when the terms in a Taylor series for a function $f(x)$ can be computed, the series may fail to converge to $f(x)$ as its sum. Also, when the Taylor series represents a function within the interval of convergence, the series and the function may not agree outside this interval. However, for most cases of interest in physical problems, the Taylor series will serve well.

NUMERICAL METHODS FOR DIFFERENTIAL EQUATIONS

———————————————■———————————————

The study of numerical (computer) solutions of differential equations has produced a large and rich body of literature. In contrast, this introduction to numerical methods is brief and covers only a few of the methods available to solve differential equations. Our purpose is to understand the terminology and to create simple programs that solve certain types of differential equations.

This section deals with equations of the form

$$\frac{dy(t)}{dt} = f[t, y(t)] \tag{6.21}$$

---

[1] The theory of analytic functions is an important part of the study of functions of a complex variable.

with initial condition $y(a) = c$. This is the initial value problem studied extensively in Chapter 5, although there we concentrated on the special case when f was a linear function of $y(t)$. Many of the methods to solve Equation 6.21 can be extended to represent a system of first-order differential equations. Also, with the methods presented here, nonlinear differential equations or differential equations with variable coefficients can be solved.

Since any numerical method involves some approximation, the accuracy of the result is always a matter of concern. Error analysis is difficult in all but the simplest cases, and the reader should consult several of the references in the Annotated Bibliography at the end of this chapter for a detailed discussion.

This section begins with a few simple examples to indicate the general techniques and problems that are involved in numerical approximations. A useful algorithm is then presented for solving differential equations. This Runge-Kutta method is widely used, easy to program, and accurate for many problems of interest.

APPROXIMATIONS FOR THE DERIVATIVE

The expression for the derivative of a function $f(t)$ is a fundamental formula in calculus. The definition is

$$f'(t) = \lim_{h \to 0} \frac{f(t+h) - f(t)}{h}, \qquad (6.22)$$

where

$$f'(t) \equiv \frac{df(t)}{dt}$$

is the exact derivative at the point t. For computer calculation, h will be a finite value that cannot be taken as zero. Also, roundoff error limits the accuracy of the calculation, even if h is chosen to be very small. This error will be ignored in the following discussion. However, the error involved can be shown to depend on h by expanding $f(t)$ in a Taylor series.

Consider a function $f(t)$ that is analytic in the neighborhood of a point t, as shown in Figure 6.3. This function can be expanded in a Taylor series at t to find the value $f(t+h)$.

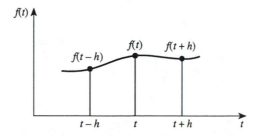

FIGURE 6.3 *A function near the point t*

The Taylor formula yields

$$f(t+h) = f(t) + hf'(t) + \frac{h^2}{2!}f''(t)\cdots, \qquad (6.23)$$

which is easily solved for $f'(t)$ as

$$f'(t) = \frac{f(t+h) - f(t)}{h} - \frac{h}{2!}f''(t_1), \qquad (6.24)$$

where $t \leq t_1 \leq t + h$ in the remainder term. Comparing this expression to that of Equation 6.22 shows that the error in the approximation is determined by the remainder term

$$-\frac{h}{2!}f''(t_1). \qquad (6.25)$$

When the Taylor series is truncated, the sum of the omitted terms is called the *truncation*, or *discretization*, error. If each term in the series is smaller that the preceding term, the magnitude of the first truncated term defines the *order of magnitude* of the error. Since the error in Equation 6.25 is assumed proportional to h, the form of error term leads us to the conclusion that decreasing the value of h decreases the error linearly.[2]

Comment: In the Taylor series for the derivative, the justification for neglecting higher-order terms containing $h^2, h^3, \ldots$ is that these terms in h are assumed to become smaller with increasing powers of h. However, suppose the time scale time for a physical problem is such that the time duration, say, T seconds, of the phenomenon being studied is very large. In such cases, it may be necessary to choose an increment $h \gg 1$. To apply the error analysis just discussed, it is necessary to *normalize* the independent variable so the phenomenon occurs over a duration of order unity. Thus, the quantity h/T could be used in the approximation and to determine the order of the error.

☐ EXAMPLE 6.7   *Approximate Solution of a Differential Equation*
In this example, we attempt to solve the differential equation

$$\frac{dy(t)}{dt} = f[t, y(t)], \qquad a \leq t \leq b,$$

with initial condition $y(a) = c$ using the Taylor series approximation to the derivative derived in Equation 6.24. The simplest approach is to first divide the interval $[a, b]$ into subintervals of length h, so that the function $y(t)$ is evaluated at the points

$$t_0 = a, t_1 = a + h, t_2 = a + 2h, \ldots, t_{M-1} = b.$$

---

[2]More complete discussions of the errors involved in Taylor series approximation to the derivative are presented in several of the references in the Annotated Bibliography for this chapter.

Notice that there are M points in the interval, but there are $M-1$ subintervals. The length of the interval is $h(M-1) = |b-a|$. Since many numerical techniques solve a differential equation "step by step" starting at t_0 and advancing the solution one point at a time, the interval between steps on the t axis is called the *step size*. The magnitude of the step size h in this example determines the error involved in the approximation, as discussed previously in connection with the Taylor remainder term of Equation 6.25.

Letting the approximate values be designated $y_i(t_i) = y_i$, and using the approximation for the derivative, the differential equation becomes a *difference equation*

$$\frac{y_{n+1} - y_n}{h} = f(t_n, y_n). \tag{6.26}$$

Solving for y_{n+1} yields the recursion formula

$$y_{n+1} = y_n + hf(t_n, y_n), \tag{6.27}$$

subject to $y_0 = c$, a constant. This formulation is sometimes called *Euler's method*.

Although no restrictions have been put on $f(t_n, y_n)$ in Equation 6.26, we will solve a very simple example, so that the errors will become evident. The equation

$$\frac{dy(t)}{dt} = y(t), \qquad y(0) = 1, \tag{6.28}$$

has the exact solution $y(t) = e^t$. The approximation of Equation 6.27 leads to the relationship

$$y_{n+1} = y_n + hy_n = (1+h)y_n$$

with $y_0 = 1$. Solving for $y_i, i = 0, 1, \ldots, 10$, with $h = 0.1$ using the MATLAB M-file EX6_7.M on our disk leads to the following results:

```
h =   0.1000
```

t	yn	exp	error
0	1.0000	1.0000	0
0.1000	1.1000	1.1052	-0.0052
0.2000	1.2100	1.2214	-0.0114
0.3000	1.3310	1.3499	-0.0189
0.4000	1.4641	1.4918	-0.0277
0.5000	1.6105	1.6487	-0.0382
0.6000	1.7716	1.8221	-0.0506
0.7000	1.9487	2.0138	-0.0650
0.8000	2.1436	2.2255	-0.0820
0.9000	2.3579	2.4596	-0.1017
1.0000	2.5937	2.7183	-0.1245

The exact solution and the Euler solution are plotted in Figure 6.4. The error column in the table of MATLAB results and Figure 6.4 both show that the error increases as t increases. This is because the errors are accumulating across the interval. Solving the equation with $h = 0.2$ will show that the error is approximately doubled when the step size is doubled as you are asked to show in Problem 6.19. Reducing the value of h leads to a more accurate solution.

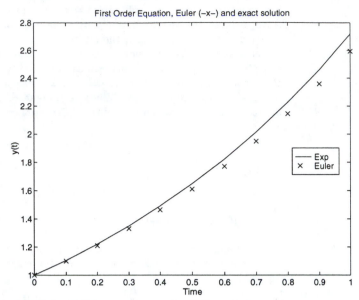

First Order Equation, Euler (–x–) and exact solution

FIGURE 6.4 *Euler solution to first-order differential equation*

☐

The Euler method is a special case of the Taylor series method that determines the solution at point t_{k+1} as

$$y_{k+1} = y_k + \alpha_1 h + \alpha_2 \frac{h^2}{2!} + \cdots + \alpha_N \frac{h^N}{N!}, \qquad (6.29)$$

where $\alpha_i = y^{(i)}(t_k)$ for $i = 1, 2, \ldots, N$ at each step $k = 0, 1, \ldots, M - 1$. It can be proven that the error involved in such a solution at each point is proportional to h^{N+1} and that the overall *global* error is proportional to h^N. The formula of Equation 6.29 can be abbreviated as

$$y_{k+1} = y_k + h T_N[t_k, y(t_k)]. \qquad (6.30)$$

Although the Taylor series method is simple in principle to implement, it is normally not used for solving the initial value problem. There are more accurate and useful techniques commonly employed. One such method is the Runge-Kutta method described in the following section.

RUNGE-KUTTA
METHODS

The initial value problem of Equation 6.21,

$$\frac{dy(t)}{dt} = f[t, y(t)] \quad \text{with} \quad y(a) = c,$$

can be solved using the Taylor series method as previously described. However, this solution method requires the calculation of higher derivatives of the function f if more terms in the series are added to improve the order of the approximation. The solution techniques called *Runge-Kutta* methods attempt to imitate the Taylor series method, but without requiring calculation of higher derivatives.

Basically, the classical Runge-Kutta approach is to replace the Taylor series terms hT_N in Equation 6.30 by a formula that gives as accurate an approximation, but defined in terms of constants that can be determined. The *Runge-Kutta method of order 4* is one of the most popular formulas. This approximation has the form

$$y_{k+1} = y_k + \left[\frac{1}{6}(k_1 + 2k_2 + 2k_3 + k_4) \right], \tag{6.31}$$

with the definitions

$$
\begin{aligned}
k_1 &= hf(t_k, y_k), \\
k_2 &= hf\left(t_k + \frac{h}{2}, y_k + \frac{k_1}{2} \right), \\
k_3 &= hf\left(t_k + \frac{h}{2}, y_k + \frac{k_2}{2} \right), \\
k_4 &= hf\left(t_k + h, y_k + k_3 \right). \tag{6.32}
\end{aligned}
$$

These equations show that the value of the solution $y(t_{k+1})$ can be calculated without forming derivatives but the function $f[t_k, y(t_k)]$ must be evaluated four times at each point. The overall error is proportional to h^5. The reader will find references that derive the Runge-Kutta equations and discuss the error in the Annotated Bibliography at the end of the chapter.

PROGRAMMING
NUMERICAL
SOLUTIONS

Programs to solve differential equations numerically can contain a few fairly simple instructions or consist of many hundreds of program statements. In any case, most algorithms to solve the initial value problem

$$\frac{dy(t)}{dt} = f[t, y(t)], \qquad y(a) = c,$$

in the interval $a \le t \le b$ at M points can be described in terms of the steps defined in Table 6.4.

TABLE 6.4 *Design of a program for numerical solution*

Definition of Algorithm

Define the initial values of the variables involved.
Calculate the solution at the first point in the interval.
Calculate the solution at the other points.
Meet a criterion for stopping the algorithm.
Output a table of values or a plot describing the solution.

Figure 6.5 shows the general structure of the MATLAB program that will be presented to solve the initial value problem.

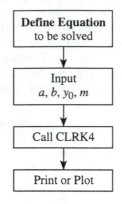

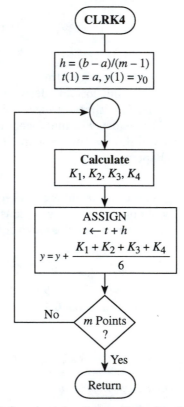

FIGURE 6.5 *MATLAB flowchart for Runge-Kutta fourth-order algorithm.*

The function $f[t, y(t)]$ will be defined by one function. Another function called `clrk4` is used to implement the fourth-order Runge-Kutta algorithm using the formula of Equation 6.31. The function is stored on the disk as file CLRK4.M, and the text is shown in the accompanying MATLAB script.

Initial values for variables include the number of points M, the endpoints a and b, and $y(a)$. The calculations are unique to the specific method used. Our criterion for stopping the calculation in the cases we have discussed thus far is simply that the solution has been calculated at M points. Other criteria are discussed in later sections. Solutions to the equation may be presented as a table of values or as a plot of the solution $y(t)$ versus t. The list of components in a program emphasizes that the first calculation and the last usually require special consideration. Also, the definitions of variables that belong to the problem should be separated from the equations of the algorithm, both for ease of testing and to make the program general in nature.

☐ EXAMPLE 6.8 **MATLAB Runge-Kutta Routine**

The clrk4 routine is written to solve any first-order differential equation of the type discussed in this section. Therefore, the function $f[t, y(t)]$ may change from problem to problem. The MATLAB command **feval** is used to evaluate the function that is passed to the clrk4 routine. The call to clfk4 would be

[T,Y]=clrk4('function',a,b,m,y0)

where 'function' is the name of a MATLAB function file (FUNCTION.M) that defines the input function to the differential equation. Example 6.9 demonstrates the use of the clrk4 function.

MATLAB Script _____

```
Example 6.8
function [T,Y]=clrk4(f,a,b,m,y0)
% CALL: [T,Y]=clrk4(f,a,b,m,y0) solves Ydot=f(T,Y) on t=[a,b].
% INPUTS: f M-file defining input function; interval a,b;
%         m points in interval; y0 initial value
% OUTPUTS: T abscissa, Y solution by 4th order Runge-Kutta method
h=(b-a)/(m-1);  % Step size
T=zeros(1,m);   % Vector of time points
Y=zeros(1,m);   % Vector of solution points
T(1)=a;
Y(1)=y0;
for I=1:m-1;     % There are m-1 steps and m points
 tI=T(I);        % - step through m-1 intervals
 yI=Y(I);
 k1=h*feval(f,tI,yI);            % Runge-Kutta coefficients
 k2=h*feval(f,tI+h/2,yI+k1/2);
 k3=h*feval(f,tI+h/2,yI+k2/2);
 k4=h*feval(f,tI+h,yI+k3);
 Y(I+1)=yI+(k1+2*k2+2*k3+k4)/6;
 T(I+1)=a + h*I;  % Next time step
end
```
_____

☐

The truncation errors arising from the Taylor series and Runge-Kutta methods of solving the first-order differential equation have involved the step size h. However, no method of selecting the appropriate step size was presented. This is because it is not generally possible to select a step size that keeps the error in the solution within a preassigned limit.

Another problem is that the total error associated with numerical solutions is a function of both the truncation error and roundoff errors. Although the truncation error is reduced by reducing the step size, the net effect could be an increase in the total error due to the increase in roundoff error. This is expected since decreasing the step size leads to an increased number of calculations. Thus, covering the interval of interest for the problem leads to a corresponding increase in the cumulative effect of roundoff at each step. Figure 6.6 illustrates the general relationship between step size and total error when roundoff errors are considered. Clearly, the ideal step size depends on a number of factors that may not be known for a given problem.

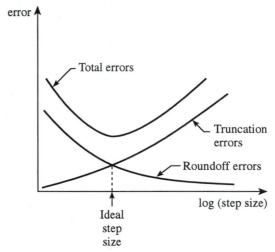

FIGURE 6.6 *Total error associated with numerical solutions.*

Since roundoff error is a function of the computer representation of numbers and is not normally under program control, the step size is the most important parameter in determining the accuracy of a solution for a given numerical method. The simplest attempt to derive a more accurate solution to a problem is to use an approach such as the following:

1. Select a step size h and compute the solution.

2. Change the step size to $h/2$ and recompute the solution.

3. Compare the results to decide on the next step size.

If the results with $h/2$ do not differ significantly from the original solution, then the step size h is probably acceptable. However, if the solution from the smaller step size is much different from the solution using h, the step size should be further reduced and another comparison made. In some cases, the step size could actually be increased, perhaps to $2h$, in order to reduce the time of computation. Then, the tests must be repeated. The characteristic of this method and similar methods to improve the accuracy is that a *fixed step size* is used throughout the entire interval of interest each time the program computes a solution.

Advanced numerical algorithms include *automatic step-size control*. In these algorithms, the step size is adjusted to find an optimal step size at each point in the interval of interest. The practical question is how to select the criterion that causes a change in the step size when necessary as the program executes. One such approach used in the MATLAB routine **ode45** will be discussed in the next section.

MATLAB ROUTINE ODE45 MATLAB provides the functions **ode23** and **ode45** for numerical solution of the initial value problem. For both routines, the step size is automatically calculated until an error parameter is reduced below a given value. This section describes only **ode45**, which has the calling format

```
[t,y]=ode45('function',t0,tf,y0,tol)
```

The arguments of **ode45** define the function f in the equation

$$\frac{dy(t)}{dt} = f[t, y(t)],$$

the interval of interest [t0,tf], and the initial condition $y(t_0)$ =y0. The parameter **tol** is the desired accuracy of the solution. The default value is 1.0×10^{-6} if **tol** is not specified.[3] The algorithm of **ode45** uses a fourth- and fifth-order Runge-Kutta approximation to compare the error at each point and adjust the step size to keep within the error tolerance. Generally, since it uses a higher-order formula, **ode45** will take fewer integration steps than **ode23** and thus produce a solution more rapidly.

The actual **ode45** algorithm is called the *Runge-Kutta-Fehlberg* algorithm. It is described in detail in the textbook *Computer Methods for Mathematical Computations*, listed in the Annotated Bibliography for this chapter.

Comment: The M-files ODE23.M and ODE45.M are stored in ASCII format and thus can be displayed and examined using the MATLAB command **type**. They can also be read with any word processor.

---

[3]The MATLAB commands **ode23** and **ode45** actually solve a system of first-order differential equations. The solution of systems of equations will be treated in another section of this chapter.

Comparison of Methods

The accompanying MATLAB script shows the M-file to call function `clrk4` and **ode45** to compare the error of each method. The differential equation to be solved is

$$\frac{dy(t)}{dt} = y(t)$$

on the interval $[a, b] = [0, 1]$ with initial condition $y(0) = 1$. The errors are compared in Figure 6.7, which shows the logarithm of the magnitude of the difference between the two solutions and the exact solution $y(t) = e^t$. The program also plots the exact solution and the calculated solutions, but the plot is not shown here. The input parameters are $a = 0$ and $b = 1$ to define the interval, $y(0) = 1$ as the initial condition, and $m = 8$ as the number of points in the interval.

The program first prompts the user to input the parameters. Then, the function `clrk4` is called to compute the fixed step-size Runga-Kutta solution of the equation. This function is described in Example 6.8. The differential equation set is defined by function `clrk4exf`. Notice the use of the string variable `'clrk4exf'` enclosed in single apostrophes to define the function name.

The MATLAB function **ode45** uses only eight points in the interval to yield a solution that is considerably more accurate than the fixed-step-size function `clrk4`.

MATLAB Script _____

```
Example 6.9
% CLRK4EX.M Runge-Kutta solution of ydot=f(t,y); t=[a,b] to
%   compare clrk4 and ode45 solutions.
% INPUTS: a,b interval endpoints; m points; y0 initial value
%   Calls function clfrk4exf to define f(t,y)
a=input('Input starting point a= ')
b=input('Input end point b= ')
m=input('Number of points m= ')
y0=input('Initial value y0= ')
% Solve with fixed step size
[Trk4,Yrk4]=clrk4('clrk4exf',a,b,m,y0);    % CLRK4 solution
% ode45 solution
[Tode45,Yode45]=ode45('clrk4exf',a,b,y0); % ODE45 solution
%
% Compare solution with clrk4 and ode45
Yexrk4=exp(Trk4);
Yexode=exp(Tode45);
erk4=abs(Yrk4-Yexrk4);
eode45=abs(Yode45-Yexode);
plot(Trk4,Yexrk4,'+',Trk4,Yrk4,'x')
hold on
plot(Tode45,Yode45)
pause
hold off
% Compare error
semilogy(Trk4,erk4,'--',Tode45,eode45,'-')
xlabel('Time in seconds')
```

```
ylabel('Log10 Error')
title('Errors comparing CLRK4 (- - -) and ODE45 (solid line)')
grid
```

The function `clrk4exf` defines the differential equation

$$\frac{dy(t)}{dt} = \dot{y} = y(t)$$

using the MATLAB variable ydot. To change the differential equation or the forcing function $y(t)$, it is necessary only to modify this function. The file is CLRK4EXF.M on the disk accompanying this text.

MATLAB Script _____

```
Example 6.9
function ydot=clrk4exf(t,y)
% CALL: ydot=clrk4exf(t,y) define ydot=f(t,y) for each t
ydot=y;     % Define first order system
```

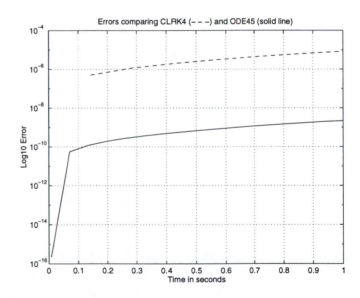

FIGURE 6.7 *Total error for* **clrk4** *and* **ode45**

VECTOR EQUATIONS

The Runge-Kutta method is easily extended to solve vector equations of the form

$$\dot{\mathbf{x}}(t) = \mathbf{f}[t, \mathbf{x}(t)], \tag{6.33}$$

which is a set of first-order differential equations. In Chapter 5, the initial value problem for this set of equations was discussed in detail. This system can describe virtually any dynamic system, including those systems described by higher-order ordinary differential equations.

The vector form of the fourth-order Runge-Kutta formulas are

$$
\begin{aligned}
\mathbf{k}_1 &= h\mathbf{f}(t_k, \mathbf{x}_k), \\
\mathbf{k}_2 &= h\mathbf{f}\left(t_k + \frac{h}{2}, \mathbf{x}_k + \frac{\mathbf{k}_1}{2}\right), \\
\mathbf{k}_3 &= h\mathbf{f}\left(t_k + \frac{h}{2}, \mathbf{x}_k + \frac{\mathbf{k}_2}{2}\right), \\
\mathbf{k}_4 &= h\mathbf{f}\left(t_k + h, \mathbf{x}_k + \mathbf{k}_3\right).
\end{aligned}
\tag{6.34}
$$

The solution at the point $\mathbf{x}(t_{k+1})$ is then

$$\mathbf{x}_{k+1} = \mathbf{x}_k + \left[\frac{1}{6}(\mathbf{k}_1 + 2\mathbf{k}_2 + 2\mathbf{k}_3 + \mathbf{k}_4)\right]. \tag{6.35}$$

Using the programming model of Example 6.9, the system of equations represented by Equation 6.33 can be defined by a MATLAB function and then the Runge-Kutta routine can be called to solve the system. The MATLAB commands **ode23** or **ode45** can be used to solve the system on the interval $[t_0, t_f]$ when $\mathbf{x}(t_0)$ is known.

For the initial value problems previously treated in Chapter 5 and in this chapter, the solution begins at time t_0 and continues step by step until the final time t_f. When the final time is not known, another criterion must be used to terminate the calculations.

PROJECTILE MOTION

In some problems, a criterion other than reaching the final time must be used to terminate the solution. In fact, in problems involving the flight of a projectile, the time when the projectile strikes the ground is not known until the trajectory is determined. Thus, the time of flight is one of the parameters to be determined. A simple modification of the Runge-Kutta algorithm allows such problems to be solved.

In the following, we assume that the motion is in two dimensions of the xy-plane, so y is the height and x is the horizontal distance. Thus, we consider the flight of a projectile without an engine that starts with an initial velocity from a point $[x, y]$ on the earth's surface. The projectile could be any object, such as a cannonball, a ballistic rocket, or a baseball. The assumption in this example is that the flight of the projectile is affected only by gravity and perhaps air resistance. A solution to the problem would trace the position vector $\mathbf{r}(t)$ and velocity vector $\mathbf{v}(t)$ of the object as it rises in the air and then falls back to the earth.

The equations of motion for the projectile are

$$
\begin{aligned}
\frac{d\mathbf{v}(t)}{dt} &= \frac{1}{m}\mathbf{F}_a(\mathbf{v}(t)) - g\mathbf{n}_y, \\
\frac{d\mathbf{r}(t)}{dt} &= \mathbf{v}(t),
\end{aligned}
\tag{6.36}
$$

where m is the mass of the projectile, $\mathbf{F}_a(\mathbf{v}(t))$ is the force due to air resistance, g is the acceleration due to gravity, and $\mathbf{n}_y$ is a unit vector in the y direction. In terms of the standard unit vectors in two dimensions discussed in Chapter 2, $\mathbf{n}_y = \mathbf{j}$.

For our 2D problem, the components of the vectors in Equation 6.36 are

$$
\begin{aligned}
\mathbf{v}(t) &= [v_x(t), v_y(t)] = [\dot{x}(t), \dot{y}(t)], \\
\mathbf{r}(t) &= [r_x(t), r_y(t)] = [x(t), y(t)].
\end{aligned}
$$

A common approximation assumes that the air resistance is proportional to the square of the velocity. Since this force is in the direction opposite to motion, the vector expression becomes

$$
\mathbf{F}_a(\mathbf{v}(t)) = -\alpha\,\|\mathbf{v}\|\,\mathbf{v}.
\tag{6.37}
$$

The coefficient α varies with the shape and area of the projectile, as well as with the air density. Although the coefficient may vary with velocity for real projectiles, we assume that it is constant in the present example.

Before programming the solution to the system of Equation 6.36, it is necessary to make several choices for the numerical routines. First, it is evident that the solution should stop when the projectile's flight is terminated by striking the ground. This is at the time when $y(t) = 0$ but $t > 0$ if the projectile starts at time $t = 0$. The precise time of impact is one of the unknowns in the problem, so the solution algorithm must be constructed to take into account this physical constraint. The initial position is $\mathbf{r}(0) = [x_0, y_0]$ and the final position is $\mathbf{r}(t_f) = [x_f, 0]$, where x_f is the horizontal distance traveled and t_f is the time of impact.

Also, an estimate of the initial step size must be made for a numerical solution. This is possible considering the physics of the problem. The solution for the projectile problem when air resistance is negligible serves as a check on the numerical algorithm and also gives some indication of a reasonable stepsize to use. This will be called the *ideal* solution. The trajectory defined by $\mathbf{r}(t)$ is a parabola as determined by ignoring air resistance in Equation 6.36 and solving the system of equations

$$\begin{aligned} \dot{v}_x &= \ddot{x}(t) = 0, \\ \dot{v}_y &= \ddot{y}(t) = -g, \end{aligned}$$

(6.38)

with $\mathbf{v}(0) = [v_x(0), v_y(0)]$ and assumed initial position $\mathbf{r}(0) = [0, 0]$.

Using the techniques of Chapter 5, the solutions for $x(t)$ and $y(t)$ are found to have the form

$$\begin{aligned} x(t) &= c_1 t + c_2, \\ y(t) &= c_3 t^2 + c_4 t + c_5. \end{aligned}$$

(6.39)

To satisfy initial position $x(0) = 0$ and $y(0) = 0$, it is clear that the coefficients $c_2 = c_5 = 0$. Let the initial velocity be defined as

$$\mathbf{v}(0) = v_x(0)\,\mathbf{i} + v_y(0)\,\mathbf{j} = v_0 \cos\theta\,\mathbf{i} + v_0 \sin\theta\,\mathbf{j},$$

which indicates that the projectile is shot, thrown, hit, or fired into the air with an initial velocity $\|\mathbf{v}(0)\| = v_0$ at an angle of θ degrees measured up from the horizontal axis. From elementary physics, $x(t) = v_x t$, so $c_1 = v_0 \cos\theta$ in Equation 6.39, since there is no force in the x direction. Setting $\dot{y}(0) = v_0 \sin\theta$ yields $c_4 = v_0 \sin\theta$. The equation for acceleration yields $\ddot{y} = -2c_3 = -g$ so $c_3 = -g/2$. The result is

$$\begin{aligned} x(t) &= (v_0 \cos\theta)t, \\ y(t) &= -\frac{g}{2}t^2 + (v_0 \sin\theta)t, \end{aligned}$$

with the solution for the time of impact t_f as

$$t_f = \frac{2v_0}{g}\sin\theta.$$

This is the maximum time of flight for a projectile, since adding the effect of air resistance will always shorten the flight. A trial step size could be

$$h = \frac{t_f}{(M-1)},$$

where M points are taken during the ideal flight.

☐ EXAMPLE 6.10   *MATLAB Projectile Motion*

The accompanying MATLAB scripts present a program to solve the projectile equations with air resistance (drag). M-file CLPROJEX.M is executed so the user can input important parameters and then function `clrk4pj` is called to solve the system of equations defined by function `clprojf`.

After the initial values are defined, the initial state vector $X(0)$ is a 1×4 row vector passed to function `clprojf`. In that function, the state equations are defined. The second-order differential equations for $\ddot{x}(t)$ and $\ddot{y}(t)$ are converted into first-order equations, as previously described in Chapter 5. Function `clrk4pj` uses two arrays to implement the Runga-Kutta algorithm. Array T is a column vector with m elements and X is an $m \times 4$ matrix with the columns

$$X(:,1) = x(t) \quad X(:,2) = y(t) \quad X(:,3) = v_x(t) \quad X(:,4) = v_y(t).$$

The Runga-Kutta algorithm terminates when the **break** command indicates that $y(t) < 0$.

Figure 6.8 shows the trajectory for the following parameters

$$
\begin{aligned}
\mathbf{v}(0) &= [40, 40] \text{ (meters/second)}, \\
\text{drag} &= 0.0052 \text{ (1/meter)}, \\
M &= 50 \text{ points}, \\
h &= 0.2 \text{ seconds}.
\end{aligned}
$$

In this example, the initial time is $t_0 = 0$ and the initial position is $\mathbf{r}(0) = [0, 0]$, although the program is easily modified to allow other values.

The parameters just listed correspond to a baseball hit with an initial velocity of 50.6 meters/second (about 125 miles per hour) at an angle of $45°$.[4] Notice in the plot that the final y-point of the solution is slightly negative. This error can be improved somewhat by taking more points. A better approach is to interpolate at least the last few values of y [X(:,2)]. The MATLAB function **spline** or **interp1** described in Chapter 7 could be used. Alternatively, a simple linear interpolation on the last two points would yield the exact time t_f when $y(t) = 0$. Then, before plotting, it would be necessary to adjust the final t value in the T array and let $y = 0$ in the X array.

MATLAB Script _____

```
Example 6.10
% CLPROJEX.M Solve the equations of flight for a projectile
%    with air resistance and plot trajectory
% INPUTS:  drag, initial velocity [Vox Voy] = [X0(1),X0(2)],
%          M number of points in interval, h step size.
% Fixed Parameter Values g, t(0), r(0)
% Calls functions clrk4pj to solve system of equations and
%   clprojf to define the system
global drag g
g=9.81;          % Gravity  9.81 m/sec-sec
t0=0.0;          % Initial time
r=[0 0];         % Initial [x y]
% Input Parameters
drag=input('drag= ');
% drag=0.0 is ideal case
v=input('Enter initial velocity v=[Vx(0) Vy(0)] = ');
M=input('Number of points= ');
h=input('Step in time= ');
%
Vx=v(1);         % Initial x velocity
```

_____

[4] Alejandro Garcia, evidently a baseball fan, discusses the flight of a baseball in more detail in *Numerical Methods for Physics* listed in the Annotated Bibliography of this chapter.

_____

```
Vy=v(2);         % Initial y velocity
X0=[r(1) r(2) Vx Vy];
% Let X(:,1)=x; X(:,2)=y; X(:,3)=x'; X(:,4)=y' state vector
% Create system matrix and solution
%  Last is the index of the last nonzero value of X(:,2), i.e. y
[T,X,Last]=clrk4pj('clprojf',t0,M,h,X0);
%
fprintf('Strike a key to plot results\n')
pause
% Plot range versus height
plot(X(1:Last,1),X(1:Last,2))
title(['Projectile-Tfinal is ',num2str(T(Last)),'sec',';  Drag= ',...
num2str(drag)])
xlabel('Range in meters')
ylabel('Height in meters')
hold off
```

Function `clprojf` defines the equations of motion for the projectile as a system of first-order equations.

MATLAB Script _____

```
Example 6.10
function xdot = clprojf(t,x)
% CALL: xdot = clprojf(t,x) to define equations of motion
%    with drag for Example 6.10
%  x''=-drag*v*x'      x1=x, xdot1=x'=x3  xdot3=x''
%  y''=-drag*v*y'-g    x2=y, xdot2=y'=x4  xdot4=y''
global drag g
V=sqrt(x(3)^2 + x(4)^2);
xdot(1)=x(3);                    % represents x'
xdot(2)=x(4);                    % represents y'
xdot(3)=-drag*V*x(3);           % represents x''
xdot(4)=-drag*V*x(4)-g;         % represents y''
```

Function `clrk4pj` implements a fourth-order Runge-Kutta algorithm to solve the system of differential equations. The calculations continue until the y-value is less than zero.

MATLAB Script _____

```
Example 6.10
function [T,X,Last]=clrk4pj(f,t0,m,h,x0)
% CALL: [T,X,Last]=clrk4pj(f,t0,m,h,x0) Solve Xdot=f(T,X)
% INPUTS: function f; m points; h step size; t0,x0 initial values
% Output: T abscissa, X solution of differential equation system
%  Last is index of last nonzero value of X(:,2)
T=zeros(1,m);               % T array m columns (1xm)
X=zeros(m,length(x0));      % X array m rows, 4 columns (mx4)
T(1)=t0;
X(1,:)=x0;
%There are m-1 steps and m points maximum
```

```
for I=1:m-1;
  tI=T(I);
  xI=X(I,:);
  k1=h*feval(f,tI,xI);
  k2=h*feval(f,tI+h/2,xI+k1/2);
  k3=h*feval(f,tI+h/2,xI+k2/2);
  k4=h*feval(f,tI+h,xI+k3);
  X(I+1,:)=xI+(k1+2*k2+2*k3+k4)/6;
  T(I+1)=t0 + h*I;
 if (X(I+1,2) < 0),
 break;  % Quit upon impact
 end
end
Last=(I+1) % Index of last nonzero value
```

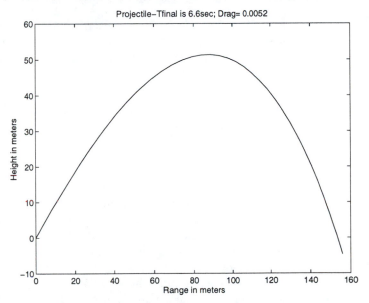

FIGURE 6.8 *Projectile trajectory*

□

W H A T   I F ?   Suppose you wish to improve the solution to the projectile problem as shown in Figure 6.8. Plot a ground line at $y(x) = 0$ and stop the solution more precisely when the projectile hits the ground. Also, try other drag coefficients to determine the changes caused by making the projectile more or less aerodynamic that the coefficient used in the example.

BOUNDARY VALUE PROBLEMS

■

The differential equations studied thus far have led to an *initial value problem* in which the subsidiary conditions were given at one point. Under very general conditions, the initial value problem has a unique solution, as described in Chapter 5. In this section, differential equations with the solution specified at two different points are considered. The locations of the points normally specify the physical boundaries of the problem under consideration. Thus, such problems are called *boundary-value problems*.

Although any order differential equation is theoretically possible, we restrict the discussion of boundary value problems to those with second-order differential equations. This is not a great restriction since many of the important boundary value problems in physics and engineering lead to second-order equations. More general differential equations and conditions for the existence and uniqueness of the solution to the boundary value problem are discussed in several of the textbooks listed in the Annotated Bibliography at the end of the chapter.

Consider the linear differential equation

$$y''(x) + p(x)y'(x) + q(x)y(x) = f(x) \tag{6.40}$$

subject to the conditions

$$y(a) = \alpha \quad \text{and} \quad y(b) = \beta$$

at the ends of the interval of interest where $x = a$ and $x = b$. Let the values α and β be constant and the coefficients and $f(x)$ be continuous in the interval $a \leq x \leq b$. The problem of finding $y(t)$ that satisfies the differential equation and the conditions at the endpoints of the interval where the functions are defined is called a *boundary value problem*.

Suppose that $y_1(x)$ and $y_2(x)$ are linearly independent solutions of the homogeneous differential equation corresponding to Equation 6.40 and $y_p(x)$ is a particular solution. Then, the general solution to the differential equation of Equation 6.40 is

$$y(x) = c_1 y_1(x) + c_2 y_2(x) + y_p(x). \tag{6.41}$$

Thus, applying the boundary conditions leads to the system of algebraic equations

$$
\begin{aligned}
c_1 y_1(a) + c_2 y_2(a) + y_p(a) &= \alpha, \\
c_1 y_1(b) + c_2 y_2(b) + y_p(b) &= \beta.
\end{aligned}
\tag{6.42}
$$

The boundary value problem of Equation 6.40 will have as many solutions as the system of Equation 6.42. The result will be either one solution, no solutions, or an infinite number of solutions, as discussed in Chapter 3, which presented the various cases for the solution of a system of linear equations.

Writing Equation 6.42 as the matrix equation $A\mathbf{c} = \mathbf{b}$ yields the results

$$\begin{bmatrix} y_1(a) & y_2(a) \\ y_1(b) & y_2(b) \end{bmatrix} \begin{bmatrix} c_1 \\ c_2 \end{bmatrix} = \begin{bmatrix} \alpha - y_p(a) \\ \beta - y_p(b) \end{bmatrix}. \tag{6.43}$$

To discuss the solutions, we consider the determinant of the matrix of y values

$$\begin{vmatrix} y_1(a) & y_2(a) \\ y_1(b) & y_2(b) \end{vmatrix} = y_1(a)y_2(b) - y_2(a)y_1(b). \tag{6.44}$$

If the determinant is nonzero, a unique solution exists and the boundary value problem has one and only one solution in the interval $a \leq x \leq b$. If the determinant is zero, the equation may have no solution or an infinite number of solutions.

□ EXAMPLE 6.11   ***Boundary Value Problem***

The homogeneous boundary value problem

$$y''(x) + ky(x) = 0, \quad y(0) = 0, \quad y(\pi) = 0, \tag{6.45}$$

where k is a positive constant, has the solution

$$Y(x) = c_1 \sin(\sqrt{k}x) + c_2 \cos(\sqrt{k}x).$$

This differential equation can be written as the eigenvalue equation

$$L[y(x)] = -ky(x), \tag{6.46}$$

with $L \equiv d^2/dx^2$.

Applying the boundary condition at $x = 0$ leads to the result that $c_2 = 0$. Thus, either

$$c_1 = 0 \quad \text{or} \quad \sin(\sqrt{k}x) = 0.$$

If $c_1 = 0$, the trivial solution $y(x) = 0$ is obtained as a legitimate solution. If $c_1 \neq 0$, then

$$y(x) = c_1 \sin(\sqrt{k}x),$$

with the requirement that $\sqrt{k}x = n\pi$. Only the values

$$k = n^2, \qquad n = 1, 2, \ldots$$

are allowed for k. These are the eigenvalues in Equation 6.46.

The constant c_1 must be determined by other conditions. For example, if it is known that $y(x_1) = a$, $0 < x_1 < \pi$, then

$$c_1 \sin \sqrt{k}x_1 = a$$

and the solution becomes

$$y(x) = a\frac{\sin(\sqrt{k}x)}{\sin(\sqrt{k}x_1)}.$$

□

Eigenvalue problems arise frequently in engineering and physics. In many cases, an ordinary differential equation represents a boundary value problem in the form

$$y''(x) + \lambda y(x) = 0 \tag{6.47}$$

over an interval $0 \leq x \leq L$. A typical condition might be $y(0) = 0$ and $y(L) = 0$. The form of the solution to Equation 6.47 depends on the value of λ. The equation is called an *eigenvalue* equation because it can be written as

$$y''(x) = -\lambda y(x). \tag{6.48}$$

We seek the values of λ that provide nontrivial solutions to the boundary value problem expressed in Equation 6.47. Different solutions are determined by the three cases in which $\lambda = 0$, $\lambda < 0$, or $\lambda > 0$.

Physical Meaning When the eigenvalue equation describes a physical system, the eigenvalues will be determined by the physical properties of the system being considered. In vibration problems, the eigenvalues define the frequencies of vibration. When solving the problem of the buckling of a column, the eigenvalues determine the load at which the column buckles. In the Schrödinger equation of quantum mechanics, the eigenvalues represent the measurable values of energy of a physical system. The solutions of many other problems are also formulated in terms of the eigenvalues (and eigenfunctions) of a differential equation. Chapter 13 presents examples of the use of eigenvalue equations to solve partial differential equations.

Eigenvalue Solution Consider a situation in which a physical system described by Equation 6.47 is constrained at $x = 0$ and $x = L$ so that no motion is possible at these points. As you are asked to prove in Problem 6.8, the only nonzero solution to Equation 6.47 satisfying the conditions $y(0) = 0$ and $y(L) = 0$ occurs when $\lambda > 0$. The solution is

$$y(x) = c_1 \cos(\sqrt{\lambda}\, x) + c_2 \sin(\sqrt{\lambda}\, x). \tag{6.49}$$

Applying the boundary conditions leads to the equations

$$y(0) \;=\; 0 = c_1,$$
$$y(L) \;=\; 0 = c_2 \sin(\sqrt{\lambda}\, L),$$

so that $\sqrt{\lambda} L = n\pi$ for any integer n. Thus, the appropriate eigenvalues and corresponding eigenfunctions are

$$\lambda_n = \frac{n^2 \pi^2}{L^2} \quad \text{and} \quad y_n(x) = c_n \sin\left(\frac{n\pi x}{L}\right). \tag{6.50}$$

Since the equation is linear, the sum of the solutions $y_n(x)$ for $n = 1, 2, \ldots$ is also a solution to the differential equation.

The set of eigenfunctions form an orthogonal set since

$$\langle \sin(\frac{n\pi x}{L}) \sin(\frac{m\pi x}{L}) \rangle = \int_0^L \sin(\frac{n\pi x}{L}) \sin(\frac{m\pi x}{L}) \, dx = 0, \qquad n \neq m,$$

as discussed in Chapter 2. The squared norm of the sinusoid is

$$\int_0^L \left[\sin(\frac{n\pi x}{L}) \right]^2 dx = \frac{L}{2}.$$

Therefore, the set

$$\left\{ \sqrt{\frac{2}{L}} \sin(\frac{n\pi x}{L}) \right\}, \qquad n = 1, 2, \ldots$$

is an *orthonormal set*. We will return to a discussion of such trigonometric sets in Chapter 8, when Fourier series are presented.

Thus, if λ is a parameter in the second-order differential equation subject to boundary conditions, there are an infinite number of possible solutions. If λ is fixed in a problem, the homogeneous solution consists of two elementary functions, as previously discussed in Chapter 5.

□ EXAMPLE 6.12   *Eigenvalue Problem*

The eigenvalue problem of Equation 6.47 has a number of physical applications. A simple one will be presented here to show the meaning of eigenvalues and eigenfunctions when a thin vertical column of length L is subject to an axial, compressive force P, as shown in Figure 6.9(a). The differential equation used to model the bending of the column will be

$$EI\frac{d^2 y(x)}{dx^2} = -Py(x), \qquad 0 < x < L, \tag{6.51}$$

where E is Young's modulus of elasticity and I is the moment of inertia of the column around a vertical line through its center. For steel, $E = 30.0 \times 10^6$ pounds per square inch. The moment of inertia of a square column with cross-sectional area b^2 is $I = b^4/12$.

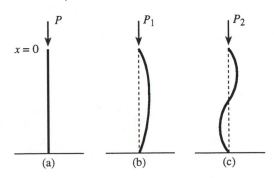

FIGURE 6.9 *Vertical column: (a) unbent beam; (b),(c) bending modes*

Assuming the column is constrained at both ends to not move laterally, the boundary conditions are $y(0) = 0$ and $y(L) = 0$. We expect that the column will not bend (or collapse) until the force P reaches some critical value. It is interesting that the column only bends when P has certain discrete values and that the deflections $y(x)$ can only assume specific shapes. This is a consequence of the simple model in Equation 6.51 and the boundary conditions. In a real situation, the column may collapse before it bends.

Writing $\lambda = P/EI$, Equation 6.51 becomes the eigenvalue equation

$$y''(x) + \lambda y(x) = 0,$$

which has been already solved with the results shown in Equation 6.50. Thus, the possible solutions become

$$\lambda_n = \frac{P_n}{EI} = \frac{n^2\pi^2}{L^2} \quad \text{and} \quad y_n(x) = c_n \sin(\frac{n\pi x}{L}).$$

This leads to the conclusion that the minimum force to bend the column is

$$P_1 = \pi^2 \frac{EI}{L^2}.$$

The shape of the column when it bends is

$$y_1(x) = c_1 \sin(\frac{\pi x}{L})$$

which is the half-wavelength sine wave shown in Figure 6.9(b). The next bending mode represents a full-wave sine wave at a force $P_2 = 4P_1$, as in Figure 6.9(c). This mode might be exhibited if the column is pinned at $x = L/2$ so the first mode is not possible.

This application is typical in the sense that the eigenvalues define the allowed values of a parameter to yield a solution to a given differential equation, and the eigenfunctions describe the allowed physical motion corresponding to the eigenvalues. Solutions to partial differential equations often lead to similar eigenvalue problems.

$\square$

INHOMOGENEOUS BOUNDARY VALUE PROBLEM

The inhomogeneous boundary value problem

$$y''(x) + p(x)y'(x) + q(x)y(x) = f(x),$$
$$y(a) = \alpha, \quad y(b) = \beta \tag{6.52}$$

can be solved analytically by a number of techniques, assuming a valid solution exists. There are also numerical techniques that have been developed to solve boundary value problems. Other boundary value problems are considered in Chapter 13.

EQUATIONS WITH VARIABLE COEFFICIENTS

———————————— ■ ————————————

In Chapter 5, we developed methods for solving linear, ordinary differential equations with constant coefficients. The solution of the homogeneous equation was expressed in terms of elementary functions such as exponentials or trigonometric functions. Linear differential equations with variable (nonconstant) coefficients, such as

$$\frac{d^n y(x)}{dx^n} + a_{n-1}(x)\frac{d^{n-1}y(x)}{dx^{n-1}} + \cdots + a_1(x)\frac{dy(x)}{dx} + a_0(x)y(x) \;\;=\;\; f(x),$$

$$(6.53)$$

were also discussed in Chapter 5. For these equations, the important general properties of linear equations, such as the superposition property, are applicable. In particular, the initial value problem for Equation 6.53 has a unique solution under the conditions described for linear equations in Chapter 5.

Mathematically, the difference between the differential equation with variable coefficients and the equation with constant coefficients lies primarily in the *methods of solution* that must be applied. In this section, the general method of *series solutions* is described. Although other methods of solution have been developed, the series solution technique can be applied to many of the most important equations in engineering and physics.

Linear differential equations with variable coefficients arise in many applications. When the independent variable is time, the equations describe *time-varying* linear systems. These systems are characterized by the fact that the response to a given input depends on the time at which the input is applied. Thus, the time shift or translation property that was useful in studying differential equations with constant coefficients does not apply to time varying systems. If the independent variable is position, the point at which an input excitation is applied determines the solution.

It is perhaps strange that a significant difficulty concerning differential equations with variable coefficients is determination of the solution of the homogeneous equation. In fact, one author says, "That which was a problem of algebra for the equations with constant coefficients becomes a problem demanding the most elaborate tools of mathematical analysis."[5]

However, certain equations with variable coefficients can be easily solved, although these cases represent the exceptions. For example, the first-order differential equation with variable coefficient

$$\frac{dy(t)}{dt} + p(t)y(t) = f(t) \qquad\qquad (6.54)$$

---

[5]Wilfred Kaplan, *Operational Methods for Linear Systems*, Addison Wesley, Reading, MA, 1962.

was discussed in Chapter 5. This equation has the complete solution

$$y(t) = e^{-\int p(t)\,dt}\int f(t)e^{\int p(t)\,dt}\,dt + ce^{-\int p(t)\,dt}, \qquad (6.55)$$

where c is the constant of integration. Also, considering previous examples using MATLAB to solve differential equations, it should be clear that equations with variable coefficients present no special problems for numerical analysis. The initial value problem involving an nth-order differential equation can be reduced to a system of first-order equations and solved numerically with the MATLAB functions **ode23** or **ode45**. The symbolic command **dsolve** is used to solve differential equations defined symbolically.

☐ EXAMPLE 6.13   *First-Order Equation*
The differential equation with constant coefficients

$$y' + (\cos x)y = \sin x \cos x$$

can be solved using Equation 6.55 or the MATLAB symbolic command **dsolve**. Applying the formula for solution,

$$y(x) = e^{-\int \cos x\,dx}\int \sin x \cos x e^{\int \cos x\,dx}\,dx + ce^{-\int \cos x\,dx}.$$

Performing the integration leads to the result

$$y(x) = (\sin x - 1) + ce^{-\sin x}.$$

Problem 6.28 asks you to work out the details.

☐

SERIES
SOLUTIONS

For a very general class of differential equations, a power series solution about the point $x = 0$ is possible in the form

$$y(x) = \sum_{n=0}^{\infty} a_n x^n. \qquad (6.56)$$

The solution technique is to substitute a series in the form of Equation 6.56 into the differential equation and to solve for the coefficients. A theorem presented later defines the conditions for which a series solution to a second-order differential equation is valid.

Power series were discussed in detail in a previous section of this chapter. However, we present here several properties of power series that are useful in the solution of differential equations. The following theorem presents a basic result that defines the relationship between the coefficients of two power series that are equal in the range of their convergence.

■ THEOREM 6.6   *Equality of Power Series*

If two power series

$$\sum_{n=0}^{\infty} a_n(x - x_0)^n = \sum_{n=0}^{\infty} b_n(x - x_0)^n$$

have nonzero convergence radii and have equal sums wherever both series converge, then the series are identical, so that

$$a_n = b_n, \qquad n = 0, 1, 2, \ldots.$$

A corollary of this theorem is that every coefficient is zero if a convergent power series has a sum that is identically zero.

Before proceeding with the solution of differential equations, two other theorems are presented to define certain allowed operations on power series. The first theorem defines the convergence condition for a power series that is differentiated term by term. The other specifies the convergence interval for two series that are added together. Also, the effect of shifting the index in the summation of a series is studied.

■ THEOREM 6.7   *Differentiation of Power Series*

If a power series

$$y(x) = \sum_{n=0}^{\infty} a_n\, x^n$$

is differentiated term by term any number of times, the resulting series is convergent within the interval of convergence of the original series.

For example, differentiating the series in the theorem once yields

$$y'(x) = \sum_{n=1}^{\infty} n a_n\, x^{n-1}.$$

It is also true that a power series may be integrated term by term within its radius of convergence to determine the integral of the function it represents.

■ THEOREM 6.8   *Addition of Power Series*

If two power series are both convergent in some interval, then the sum converges at all points in that interval. Thus, if the series are

$$\sum_{n=0}^{\infty} a_n\, x^n \quad and \quad \sum_{n=0}^{\infty} b_n\, x^n,$$

the series

$$\sum_{n=0}^{\infty} (a_n + b_n)\, x^n$$

---

that results from summing the series term by term converges at all points where both of the individual series converge.

———————————————————■

Shifting Indices   Given two series, it is sometimes useful to change the summation indices so that the coefficients of common powers of x or $(x - x_0)$ can be easily combined. As an example, suppose it is necessary to represent $y''(x) + y(x)$ as a series. Starting with the series

$$y(x) = \sum_{n=0}^{\infty} a_n x^n \qquad (6.57)$$

and differentiating to find $y'(x)$ and $y''(x)$ yields

$$y'(x) = \sum_{n=1}^{\infty} n a_n x^{n-1},$$

$$y''(x) = \sum_{n=2}^{\infty} n(n-1) a_n x^{n-2}. \qquad (6.58)$$

Notice the change in index of the summation as the original series is differentiated. These series arise frequently when differential equations are solved by the series method, as described later in this section.

In order to represent the equation $y''(x) + y(x)$, we add the series for $y''(x)$ and $y(x)$ to form the series

$$y''(x) + y(x) = \sum_{n=2}^{\infty} n(n-1) a_n x^{n-2} + \sum_{n=0}^{\infty} a_n x^n. \qquad (6.59)$$

To collect coefficients of equal powers of x in the sum defined in Equation 6.59, it is necessary to *shift the index* of one of the series so that both series start at the same index. Since the index n is just a dummy summation index, it can be changed as desired, as long as the series contains the same terms as powers of x. In this case, let $n \to n + 2$ in the series for $y''(x)$. Changing the index yields

$$\sum_{n=2}^{\infty} n(n-1) a_n x^{n-2} = \sum_{n=0}^{\infty} (n+2)(n+1) a_{n+2} x^n.$$

Using this result, the sum of series in Equation 6.59 becomes

$$y''(x) + y(x) = \sum_{n=0}^{\infty} (n+2)(n+1) a_{n+2} x^n + \sum_{n=0}^{\infty} a_n x^n$$

$$= \sum_{n=0}^{\infty} [(n+2)(n+1) a_{n+2} + a_n] x^n. \qquad (6.60)$$

Chapter 6 ■ ADVANCED DIFFERENTIAL EQUATIONS

□ **EXAMPLE 6.14**   *Series Solution*

Suppose we seek a series solution to the equation

$$y'(x) = y, \qquad y(0) = 1. \tag{6.61}$$

Assume a power series solution of the form in Equation 6.56 exists and further that term-by-term differentiation of the series is valid in the interval of interest. Writing the assumed solution to Equation 6.61 as

$$y(x) = \sum_{n=0}^{\infty} a_n x^n = a_0 + a_1 x + \cdots + a_n x^n + \cdots$$

and substituting the series in the differential equation leads to the equality

$$\sum_{n=1}^{\infty} n a_n x^{n-1} = \sum_{n=0}^{\infty} a_n x^n.$$

Shifting the index of the series for the derivative to $n + 1$ to sum from zero allows this series to be summed in terms of x^n so that

$$\sum_{n=0}^{\infty} (n+1) a_{n+1} x^n = \sum_{n=0}^{\infty} a_n x^n.$$

Collecting coefficients of like powers of x shows that

$$\sum_{n=0}^{\infty} [(n+1) a_{n+1} - a_n] x^n \equiv 0.$$

According to the corollary to Theorem 6.6, every coefficient in this power series is zero in its interval of convergence. The relationship between the coefficient thus becomes

$$(n+1) a_{n+1} - a_n = 0, \qquad n = 0, 1, 2, \ldots.$$

Writing this as a *recurrence relation*, the $(n+1)$st coefficient can be derived as

$$a_{n+1} = \frac{a_n}{(n+1)}$$

for all n. The coefficient $a_0 = 1$ satisfies the initial condition $y(0) = 1$. Substituting in the recurrence relationship yields the coefficients

$$a_1 = \frac{1}{1!}, \quad a_2 = \frac{1}{2} a_1 = \frac{1}{2!}, \ldots$$

The solution series is then

$$y(x) = \left(1 + x + \frac{x^2}{2!} + \cdots + \frac{x^n}{n!} + \cdots \right). \tag{6.62}$$

In this case, the series can be recognized as the Taylor series for e^x, which converges for all x. We conclude the solution to the differential equation in Equation 6.61 is $y(x) = e^x$.

□

Consider the second-order differential equation with variable coefficients

$$y''(x) + p(x)y'(x) + q(x)y(x) = f(x). \tag{6.63}$$

If the coefficients $p(x)$, $q(x)$, and the function $f(x)$ have power series representations, then this second order equation has a power series solution. When these conditions are met at any point x_0, the coefficients, function, and the solution are said to be *analytic* at the point $x = x_0$. In fact, when a series solution exists, there exists a pair of series solutions to the homogeneous equation.

■ **THEOREM 6.9**

Homogeneous Solution

If the coefficients $p(x)$ and $q(x)$ of the homogeneous differential equation

$$y''(x) + p(x)y'(x) + q(x)y(x) = 0 \tag{6.64}$$

are analytic at $x = x_0$, then there exists a pair of solutions

$$y_1(x) = \sum_{n=0}^{\infty} a_n(x - x_0)^n \quad and \quad y_2(x) = \sum_{n=0}^{\infty} b_n(x - x_0)^n.$$

Note that if the coefficients $p(x)$ and $q(x)$ are polynomials, the series solutions will converge for all x. Second-order equations with coefficients that are polynomials in x form an important class of equations for physical applications.

There are similar theorems for linear first-order differential equations and for higher-order differential equations. Thus, if a differential equation is linear and the coefficient of the highest derivative is 1, we can expect power series solutions at any point where the coefficients are analytic.

Once the solutions to the homogeneous equation are found, the variation of parameters technique will always generate a particular solution $y_p(x)$. Another approach is to assume that the particular solution is also represented as a power series and substitute in the equation to find the coefficients. The complete solution to the equation

$$y''(x) + p(x)y'(x) + q(x)y(x) = f(x)$$

is thus

$$y(x) = c_1 y_1(x) + c_2 y_2(x) + y_p(x).$$

□ **EXAMPLE 6.15**

Homogeneous Solution

To solve the differential equation

$$y''(x) + xy'(x) + y(x) = 0,$$

assume a solution of the form

$$y(x) = \sum_{n=0}^{\infty} C_n x^n.$$

Then, substituting the series in the differential equation yields

$$y''(x) + xy'(x) + y(x) = \sum_{n=2}^{\infty} n(n-1)C_n x^{n-2}$$

$$+ x \sum_{n=1}^{\infty} nC_n x^{n-1} + \sum_{n=0}^{\infty} C_n x^n.$$

It is necessary to shift the summation indices, as discussed in the text following Equation 6.58. Collecting the coefficients of x^n and setting the sum of these coefficients to zero leads to the recursion relationship

$$(1+n)C_n + (n+2)(n+1)C_{n+2} = 0,$$

which shows that

$$C_{n+2} = -\frac{1}{n+2} C_n, \qquad n = 0, 1, 2, \ldots.$$

Since the related indices differ by 2, the coefficients $C_2, C_4, \ldots$ are determined once C_0 is specified. Similarly, the coefficients with odd indices can be derived from C_1. Thus, in the solution of the second-order linear differential equation, the coefficients C_0 and C_1 can be considered arbitrary constants. Writing out the first few coefficients,

$$
\begin{aligned}
C_2 &= -\frac{1}{2}C_0, & C_3 &= -\frac{1}{3}C_1, \\
C_4 &= -\frac{1}{4}C_2 = \frac{1}{2\cdot4}C_0, & C_5 &= -\frac{1}{5}C_3 = \frac{1}{3\cdot5}C_1, \\
&\;\;\vdots & &\;\;\vdots
\end{aligned}
$$

yields the pattern

$$C_{2n} = \frac{(-1)^n}{2\cdot4\cdots2n} C_0, \qquad C_{2n+1} = \frac{(-1)^n}{1\cdot3\cdots(2n+1)} C_1.$$

Thus, the solution can be expressed as the sum of two series with the result

$$y(x) = C_0 + C_0 \sum_{n=1}^{\infty} \frac{(-1)^n}{2\cdot4\cdots2n} x^{2n} + C_1 \sum_{n=0}^{\infty} \frac{(-1)^n}{1\cdot3\cdots(2n+1)} x^{2n+1}.$$

The series corresponding to even-numbered coefficients is even, and the series for the odd-numbered coefficients is odd.

$\square$

BESSEL AND LEGENDRE EQUATIONS

There are a number of functions that arise in the solution of physical problems. The elementary functions of calculus serve as solutions to homogeneous ordinary differential equations with constant coefficients. The derivative of an elementary function is again an elementary function. Also, an elementary function of a real variable $f(x)$ can be extended to a function of a complex variable by replacing x with $z = x + iy$. Then, evaluating the function with $y = 0$ reduces the function to the original real function. However, the integral of an elementary function may not be an elementary function and the solutions of advanced differential equations will lead to the study of nonelementary functions. Several examples are the Bessel and Legendre functions introduced in this section.

Functions other than elementary functions are required to solve differential equations with variable coefficients and certain partial differential equations. These functions are sometimes called *special functions*.[6] Such functions are usually defined by an infinite series whose values are tabulated in various references or can be computed by programs such as MATLAB.

Table 6.5 lists the name of several of these special functions and gives an example of the differential equation solved by the function. The functions are named after the mathematician or physicist who originally proposed or solved the equation. One common feature is that each equation may be solved by the method of series solutions. Also, these equations belong to a class of differential equations that are described by a *Sturm-Liouville* equation. In the table, the parameters ν, λ, and n are constants.

TABLE 6.5 *Examples of special functions*

Function	Differential equation
Bessel	$x^2 y'' + xy' + (x^2 - \nu^2)y = 0$
Hermite	$y'' - 2xy' + 2\lambda y = 0$
Laguerre	$xy'' + (1 - x)y' + \lambda y = 0$
Legendre	$(1 - x^2)y'' - 2xy' + n(n + 1)y = 0$

Although the equations are defined in the xy-plane, many of these functions are useful in cylindrical and spherical coordinates with the ap-

---

[6]Some authors define elementary functions as those that are algebraic such as polynomials. Using this definition, the term special functions would include transcendental functions, such as sinusoids or exponentials.

propriate change of coordinates. Also, each type of equation usually has a number of variations that are suited to fit specific physical problems. In some cases, slightly different notation may be employed. Notice that each of these differential equation contains a parameter. The value or values will be determined by the problem being solved.

Bessel's equation is the most widely applied differential equation of those listed in the table. To mention only a few applications, Bessel's equation arises in problems involving electromagnetic fields and waves, fluid flow, heat conduction, and vibrations of circular membranes. The Bessel functions often arise when partial differential equations are being solved in cylindrical coordinates.

The Legendre functions are used to solve the same types of differential equations as the Bessel functions, but in spherical coordinates. Our justification for studying them here is to present the mathematical features of the equations without emphasizing the physical applications. With a few exceptions, the applications are better explained after partial differential equations are studied in Chapter 13.

The Hermite and Laguerre equations have application in solving various problems in fields of physics such as quantum mechanics. The reader is referred to the list of references in the Annotated Bibliography for more details about these specific equations. Only the Bessel and Legendre equations will be studied in detail in the present section.

ORDINARY AND SINGULAR POINTS

Consider the differential equations in Table 6.5. Except for Hermite's equation, they have the form

$$R(x)y''(x) + p(x)y'(x) + q(x)y(x) = 0 \qquad \textbf{(6.65)}$$

so Theorem 6.9 does not apply directly. The theorem, which defined the conditions for a series solution, may not apply to the equations for which the coefficient of the second derivative term is not 1 but $R(x)$. However, the theorem will apply at points where $R(x) \neq 0$ in Equation 6.65, since the equation can be divided by $R(x)$ to yield a coefficient of 1 for the second derivative term.

To describe the solutions of Equation 6.65 at points where $R(x)$ may be zero, it is useful to form the functions

$$\frac{p(x)}{R(x)} \quad \text{and} \quad \frac{q(x)}{R(x)}$$

and make the following definitions:

1. If the two functions are analytic at the point x_0, the point x_0 is called an *ordinary point* of the differential equation.

2. If at least one of the functions is not analytic at the point x_0, then x_0 is called a *singular point* of the differential equation.

For those differential equations with polynomials as coefficients, the functions $p(x)/R(x)$ and $q(x)/R(x)$ are analytic at every point except where the denominator vanishes. Such points are singular points, and all other real values of x are ordinary points. At ordinary points, the differential equation of Equation 6.65 has a power series solution. Bessel's equation has a singular point at $x = 0$, and Legendre's equation has singular points at $x = \pm 1$.

The theory of solution of differential equations with singular points is quite involved and will not be explored in great detail in this text. Fortunately, the solutions to Bessel's equation and other equations of practical importance have been derived. This chapter presents selected results from the extensive study of these equations that is found in textbooks treating advanced differential equations.

BESSEL'S DIFFERENTIAL EQUATION

Even though Bessel's differential equation is a second-order differential equation, the equation

$$x^2 y''(x) + xy'(x) + (x^2 - \nu^2)y(x) = 0 \tag{6.66}$$

is called the *Bessel equation of order* ν. The parameter $\nu \geq 0$ is a positive number that is usually an integer or a rational number. Bessel differential equations of integral or half-integral order are particularly important in applications. Various solutions to the Bessel equation are created according to the value of ν.

The solutions to Equation 6.66 are presented in many books treating advanced differential equations or partial differential equations. The usual technique to derive series solutions uses the *Frobenius method*, in which a series of the form

$$y(x) = \sum_{k=0}^{\infty} a_k x^{k+\nu} \tag{6.67}$$

is substituted in the equation. The term x^k is required because the Bessel equation has a singular point at $x = 0$ as is seen by dividing Equation 6.66 by x^2. If ν is not a positive integer, the Frobenius series need not be a power series. Also, the Frobenius method may not yield two linearly independent solutions to the differential equation.

We will state several solutions to Bessel's equation as a function of the parameter ν and give an example of a series solution for the case $\nu = 0$. Table 6.6 lists one solution of Bessel's equation based on the value of ν. For convenience, the solutions are listed for $\nu = 0$, $\nu = n$, where n is an integer and ν is an arbitrary real number. When ν is not an integer, the Bessel function is usually written with the gamma function (Γ) in the denominator.

TABLE 6.6 *Solutions of Bessel's equation*

Order	Equation
$\nu = 0$	$J_0(x) = \displaystyle\sum_{k=0}^{\infty} \frac{(-1)^k}{(k!)^2} \left(\frac{x}{2}\right)^{2k}$
$\nu = n$	$J_n(x) = \displaystyle\sum_{k=0}^{\infty} \frac{(-1)^k}{k!(n+k)!} \left(\frac{x}{2}\right)^{2k+n}$
ν	$J_\nu(x) = \displaystyle\sum_{k=0}^{\infty} \frac{(-1)^k}{k!\Gamma(k+\nu+1)} \left(\frac{x}{2}\right)^{2k+\nu}$

The *gamma function* $\Gamma(x)$ is actually defined as an integral as shown in Problem 6.14. It has the important property that

$$\Gamma(x+1) = x\Gamma(x)$$

if $x > 0$. If $x = n$, an integer, then $\Gamma(n+1) = n!$. Thus, the gamma function in the denominator of the formula for $J_\nu(x)$ in Table 6.6 becomes $(n+k)!$ when $\nu = n$, an integer.

For $\nu \geq 0$, but not necessarily an integer,

$$\Gamma(n+\nu+1) = (n+\nu)(n+\nu-1)\cdots(1+\nu)\Gamma(1+\nu).$$

To compute the gamma function for half-integral arguments, we use the fact that

$$\Gamma\left(\frac{1}{2}\right) = \sqrt{\pi}, \quad \Gamma\left(1+\frac{1}{2}\right) = \frac{1}{2}\sqrt{\pi}, \ldots,$$

which are useful equalities to simplify Bessel functions such as $J_{n/2}(x)$, where n is an integer.

☐ **EXAMPLE 6.16** **Bessel Function**

Consider Bessel's equation in Equation 6.66 with $\nu = 0$. Thus, we wish to solve the equation

$$x^2 y''(x) + xy'(x) + x^2 y(x) = 0.$$

Let an assumed solution be $y(x) = \sum_{k=0}^{\infty} c_k x^k$, since $\nu = 0$ in the Frobenius solution of Equation 6.67, and substitute in the differential equation to find

$$\sum_{k=2}^{\infty} k(k-1)c_k x^k + \sum_{k=1}^{\infty} kc_k x^k + \sum_{k=0}^{\infty} c_k x^{k+2} = 0.$$

Converting the indices to sum the coefficients of x^k (from 2 to ∞) gives the result

$$\sum_{k=2}^{\infty} k(k-1)c_k x^k + c_1 x + \sum_{k=2}^{\infty} kc_k x^k + \sum_{k=2}^{\infty} c_{k-2} x^k = 0.$$

Since the coefficient of each power of x must be zero, $c_1 = 0$. The other coefficients can be defined by the recursion relationship

$$[k(k-1)c_k + kc_k + c_{k-2}] x^k = \left[k^2 c_k + c_{k-2}\right] x^k = 0,$$

which shows that $c_k = -c_{k-2}/k^2$. Writing out the coefficients shows that $c_k = 0$ for k an odd number.

Thus, one solution to the Bessel equation of order zero, designated J_0, has the expansion

$$
\begin{aligned}
y(x) &= c_0 \left(1 - \frac{x^2}{2^2} + \frac{x^4}{2^2 4^2} - \cdots\right) \\
&= c_0 \sum_{k=0}^{\infty} \frac{(-1)^k}{(k!)^2} \left(\frac{x}{2}\right)^{2k} = c_0 J_0(x).
\end{aligned}
\tag{6.68}
$$

Comparison of this series with the Taylor series for $\cos x$ given in a previous section shows a similarity. However, the Bessel functions $J_0(x), J_1(x), \ldots$ are not periodic but are "almost" periodic, and they decay to zero with increasing x. In Problem 6.22 you are asked to plot $J_0(x)$ and $J_1(x)$.

$\square$

Second Solution The second-order homogeneous differential equation in Equation 6.66 requires two independent solutions. A general solution to Bessel's equation for all values of ν is

$$y(x) = c_1 J_\nu(x) + c_2 Y_\nu(x),
\tag{6.69}$$

where the Y_ν function is called the Bessel function of the *second kind* of order ν, or *Neumann's function* of order ν. These functions become unbounded as x approaches zero.

If ν is a nonintegral real number, then the Bessel differential equation of order r has the solution

$$y(x) = c_1 J_r(x) + c_2 J_{-r}(x).
\tag{6.70}$$

For more details, the reader is referred to the text by Watson listed in the Annotated Bibliography for this chapter.

Properties of Bessel Functions Many useful relationships involving Bessel functions have been derived, and only a few will be presented here. For example, the derivative of $J_0(x)$ and the function $J_1(x)$ are related as

$$\frac{d}{dx} J_0(x) = -J_1(x).$$

For Bessel functions of the second kind,

$$\frac{d}{dx} Y_0(x) = -Y_1(x).$$

A recurrence relationship for Bessel functions of orders $r, r - 1, r + 1$ is

$$\frac{2r}{x} J_r(x) = J_{r-1}(x) + J_{r+1}(x), \quad x > 0. \tag{6.71}$$

MATLAB Bessel Functions   There are MATLAB commands used for computing various Bessel functions and the gamma function. Table 6.7 lists the most important commands. For the Bessel functions, α is the order and x is a scalar or vector argument.

TABLE 6.7 *MATLAB commands for Bessel functions*

Name	Equation
besselJ	$J_\alpha(x) = \mathbf{besselj}(\alpha, x)$
bessely	$Y_\alpha(x) = \mathbf{bessely}(\alpha, x)$
gamma	$\Gamma(x) = \mathbf{gamma}(x)$

If $x = n$, an integer, $\mathbf{gamma}(n) = n!$

LEGENDRE POLYNOMIALS

The Legendre differential equation

$$(1 - x^2)y'' - 2xy' + n(n + 1)y = 0 \tag{6.72}$$

has as one solution the *Legendre polynomial*

$$P_n(x) = \frac{1}{2^n n!} \frac{d^n}{dx^n} (x^2 - 1)^n, \tag{6.73}$$

where n is an integer. The first six Legendre polynomials are

$$P_0(x) = 1, \qquad\qquad P_3(x) = \frac{1}{2}(5x^3 - 3x),$$

$$P_1(x) = x, \qquad\qquad P_4(x) = \frac{1}{8}(35x^4 - 30x^2 + 3),$$

$$P_2(x) = \frac{1}{2}(3x^2 - 1), \quad P_5(x) = \frac{1}{8}(63x^5 - 70x^3 + 15x).$$

The even-numbered Legendre polynomials are even functions, and the odd-numbered polynomials are odd functions. The polynomials can also be calculated by the recursion relation

$$(n + 1)P_{n+1}(x) = (2n + 1)x P_n(x) - nP_{n-1}(x). \tag{6.74}$$

This relationship is useful for the purpose of numerical calculations, as in Problem 6.25.

As described in Chapter 2, the Legendre polynomials are orthogonal on the interval $[-1, 1]$. They satisfy the relationships

$$\int_{-1}^{1} P_n(x)P_m(x)dx = 0, \qquad m \neq n,$$

$$\int_{-1}^{1} P_n^2(x)dx = \frac{2}{2n + 1}. \tag{6.75}$$

The factor $2/(2n + 1)$ is the normalization factor used to create an orthonormal set of Legendre polynomials.

Legendre's differential equation in Equation 6.72 requires an independent second solution. The second solution is called a *Legendre function of the second kind*, designated $Q_n(x)$. The general solution is thus

$$y(x) = c_1 P_n(x) + c_2 Q_n(x).$$

The Legendre polynomials of the second kind, $Q_n(x)$, are unbounded at $x = \pm 1$.

□ EXAMPLE 6.17   **MATLAB Legendre Functions**

The differential equation

$$(1 - x^2)y'' - 2xy' + \left[n(n + 1) - \frac{m^2}{1 - x^2} \right] y = 0$$

is called *Legendre's associated differential equation*. If $m = 0$, this equation reduces to Legendre's differential equation in Equation 6.72. Assuming that n and m are nonnegative integers, one solution is the associated Legendre polynomial,

$$P_n^m(x) = (1 - x^2)^{m/2} \frac{d^m}{dx^m} P_n(x).$$

If $m > n$, $P_n^m(x) = 0$.

The MATLAB command **legendre**(n, x) computes the *associated* Legendre functions of degree n and order $m = 0, 1, \ldots, n$, evaluated at x. As usual with MATLAB, x can be either a scalar or a vector of values. The first row of the computed matrix contains the Legendre polynomial values $P_n(x)$.

□

W H A T   I F ?   MATLAB's symbolic command **dsolve** can help solve certain differential equations with variable coefficients. In particular, the symbolic algorithms of Maple recognize Bessel functions, and MATLAB commands will compute numerical values of the solutions.

Experiment with the equations in this section by trying to solve them with the commands in MATLAB's *Symbolic Math Toolbox*.

It was shown by J. C. F. Sturm and J. Liouville that Bessel's and Legendre's equations, as will as other differential equations that arise in engineering and physics, can be considered as a class of equations with common general properties. One important result of Sturm-Liouville theory is that each set of functions that solves such an equation forms an orthogonal set.

The Sturm-Liouville equation has the form

$$[r(x)y'(x)]' + [q(x) + \lambda p(x)]y(x) = 0, \tag{6.76}$$

subject to various boundary conditions that depend on the specific equation and the physical constraints for a particular problem being solved.

Bessel's differential equation for integer n,

$$x^2 y''(x) + xy'(x) + (x^2 - n^2)y(x) = 0, \tag{6.77}$$

is easily converted to Sturm-Liouville form by first substituting $\sqrt{\lambda}x$ for x to yield

$$x^2 y''(\sqrt{\lambda}x) + xy'(\sqrt{\lambda}x) + (\lambda x^2 - n^2)y(\sqrt{\lambda}x) = 0. \tag{6.78}$$

Then, divide the equation by x and notice that

$$[xy'(x)]' = xy''(x) + y'(x). \tag{6.79}$$

Thus, Equation 6.77 becomes

$$xy''(\sqrt{\lambda}x) + y'(\sqrt{\lambda}x) + \left(-\frac{n^2}{x} + \lambda x\right)y(\sqrt{\lambda}x) = 0, \tag{6.80}$$

which has the Sturm-Liouville form with

$$r(x) = x, \quad q(x) = -\frac{n^2}{x}, \quad p(x) = x.$$

The function $J_n(\sqrt{\lambda}x)$ is a solution to this transformed equation.

Legendre's differential equation (6.72) has the Sturm-Liouville form

$$[(1 - x^2)y'(x)]' + \lambda y(x) = 0, \quad [\lambda = n(n + 1)], \tag{6.81}$$

with $r(x) = 1 - x^2$, $q(x) = 0$, and $p(x) = 1$.

Orthogonal Eigenfunctions Functions $f(x)$ and $g(x)$ are said to be orthogonal on the interval $[a, b]$ with respect to a weight function $p(x)$ if

$$\int_a^b p(x)f(x)g(x)\, dx = 0. \tag{6.82}$$

It is assumed that $p(x)$ is continuous on $[a, b]$ and that $p(x) \geq 0$ on the interval without being the zero function. This is an extension of the definition of orthogonality for functions given in Chapter 2, where the weighting function $p(x) = 1$.

Assume that the Sturm-Liouville equation

$$[r(x)y'(x)]' + [q(x) + \lambda p(x)]y(x) = 0$$

has solutions $y_m(x)$ and $y_n(x)$ corresponding to the values λ_m and λ_n. Then, if the solutions or their derivatives vanish at a or b,

$$\int_a^b p(x)y_m(x)y_n(x)\,dx = 0. \qquad (6.83)$$

Thus, the solutions or eigenfunctions corresponding to different eigenvalues are orthogonal with respect to the weight function $p(x)$. Although it is possible to solve differential equations with more general boundary conditions, we do not consider such problems until Chapter 13 where partial differential equations are discussed.

Considering Equation 6.81, the Legendre polynomials solve a Sturm-Liouville problem and they are orthogonal on the interval $[-1, 1]$ with a weight function of $p(x) = 1$. The Bessel functions that solve Equation 6.80 are orthogonal with a weight function $p(x) = x$. Table 6.8 lists a number of polynomials that are orthogonal with respect to the weight function shown on the given interval.

TABLE 6.8 *Orthogonal polynomials*

Name	Equation, interval, and weight
Chebyshev	$T_n(x) = \cos(n \arccos x)$
	$[-1, 1]; \quad p(x) = 1/\sqrt{1 - x^2}$
Hermite	$H_n(x) = (-1)^n e^{x^2/2} \dfrac{d^n}{dx^n}(e^{-x^2/2})$
	$(-\infty, \infty); \quad p(x) = e^{-x^2/2}$
Laguerre	$L_n(x) = \dfrac{e^x}{n!} \dfrac{d^n}{dx^n}(x^n e^{-x})$
	$0 \le x < \infty; \quad p(x) = e^{-x}$
Legendre	$P_n(x) = \dfrac{1}{2^n n!} \dfrac{d^n}{dx^n}(x^2 - 1)^n$
	$[-1, 1]; \quad p(x) = 1$

□ EXAMPLE 6.18   *A famous Problem (Optional)*

Consider a particle of mass m that is not subject to external forces. In one dimension, the equation of motion for this *free* particle is

$$m\frac{d^2x(t)}{dt^2} = 0, \qquad (6.84)$$

with solution $x(t) = x_0 + v_x t$. The integration constants are determined by the initial position $x_0 = x(t_0)$ and the initial velocity $v_x = v_x(t_0)$. The momentum of the particle is $p_x = mv_x$ and the total energy is the particle's kinetic energy

$$E = \frac{m}{2}v_x{}^2. \qquad (6.85)$$

The energy of the particle may take any positive value. For purposes of comparison with the quantum mechanical solution to be presented, these results will be called the *classical* solution for the motion of a particle.

In 1926, Erwin Schrödinger introduced the concept of a *wave equation* to describe the motion of a particle based on the findings of the newly emerging field of *quantum mechanics*. In one space dimension, the Schrödinger equation is the differential equation

$$\frac{d^2\psi(x)}{dx^2} + \frac{2m}{\hbar^2}[E - V(x)]\psi(x) = 0. \qquad (6.86)$$

Physically, m is the mass of the particle, $V(x)$ is the potential energy and E is the total energy. The quantity $\hbar = h/2\pi$ is Plank's constant divided by 2π. In SI (mks) units, $\hbar = 1.054 \times 10^{-34}$ joule-second. This is a minute value and becomes significant only when considering particles of atomic dimensions, such as electrons. Thus, the sometimes surprising results of quantum mechanics can be ignored for most physical objects, such as dust particles or baseballs. The wave function $\psi(x)$ is not a measurable quantity, but the energy E of the system can be determined experimentally. The function $\psi(x)$ is used to determine the probability of the particle being at a certain position along the x axis.

First, we consider the case of a free particle for which $V(x) = 0$ and $E > 0$. The Schrödinger equation becomes

$$\frac{d^2\psi(x)}{dx^2} + \frac{2m}{\hbar^2}E\psi(x) = 0. \qquad (6.87)$$

The simplest Sturm-Liouville equation is

$$y''(x) + \lambda y(x) = 0, \qquad (6.88)$$

in which $r(x) = p(x) = 1$ and $q(x) = 0$. Schrödinger's equation fits this form exactly if $V(x) = 0$ and

$$\lambda = \frac{2m}{\hbar^2}E.$$

This problem was discussed previously as Equation 6.47. The solution is

$$\psi(x) = c_1 \cos(\sqrt{\lambda}x) + c_2 \sin(\sqrt{\lambda}x), \qquad (6.89)$$

which is valid for all $\lambda > 0$, since the total energy must be positive. For the free particle, the energy of the particle may be any positive value.

Bessel and Legendre Equations   329

Now consider a particle "trapped" in a region with a potential energy, as shown in Figure 6.10. The confining region is called a *potential well*. The particle cannot escape from the region defined by the well since $V(x)$ is infinite at the boundaries. The classical solution is that the particle will move with velocity $v = \sqrt{2E/m}$ between $x = -a$ and $x = a$, reversing direction when it collides with the potential barrier at $x = \pm a$.

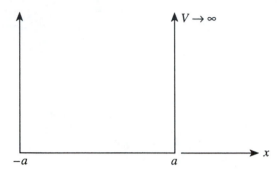

FIGURE 6.10 *Potential energy versus x*

The Schrödinger equation is again given by Equation 6.87,

$$\frac{d^2\psi(x)}{dx^2} + \frac{2m}{\hbar^2}E\psi(x) = 0, \quad |x| < a,$$

within the well since $V(x) = 0$ inside. At the boundaries, $\psi(-a) = \psi(a) = 0$ and $\psi(x) = 0$ for $|x| > a$. Thus, the differential equation is a boundary value problem with distinct eigenvalues and solutions shown in Equation 6.89. Applying the boundary conditions to the solution yields $\psi(x)$ in the form

$$\psi(x) = c_2 \sin(\sqrt{\lambda}x), \quad \lambda = \frac{2m}{\hbar^2}E,$$

with $\sin(\sqrt{\lambda}x) = n\pi$ if a nonzero solution is to be found. This requires that

$$\lambda_n = \frac{n^2\pi^2}{a^2} = \frac{2m}{\hbar^2}E.$$

The allowable energy values are then

$$E_n = \frac{\hbar^2}{2m}\left(\frac{n^2\pi^2}{a^2}\right), \tag{6.90}$$

and the nonzero solutions for the wave function are

$$\psi_n(x) = c_n \sin(\frac{n\pi x}{a}) \qquad n = 1, 2, \dots. \tag{6.91}$$

These functions were previously shown to be orthogonal over the interval, as would be expected from Equation 6.83. The solutions can be normalized if c_n is chosen so that the functions are orthonormal. For example, let $c_n = \sqrt{2/a}$.

Equation 6.90 shows that the allowed energy values for a particle confined to a potential well have discrete values. The energy levels are said to be *quantized*, which is a fundamental result of quantum mechanics. For the reader unfamiliar with quantum mechanics, this may seem to contradict common sense. However, the quantum result does not manifest itself on a macroscopic scale because as the mass of a particle increases beyond that of elementary particles or the potential well becomes larger than atomic dimensions, the differences between energy levels $E_{n+1} - E_n$ become so small that the possible energy levels become indistinguishable from a continuous set.

$\square$

REINFORCEMENT EXERCISES AND EXPLORATION PROBLEMS

In these problems, do the computations by hand unless otherwise indicated, and then check those that yield numerical or symbolic results with MATLAB.

REINFORCEMENT EXERCISES

P6.1. **Defining intervals** The absolute value is useful to define ranges of a variable that extends from negative to positive values. Thus, using the definition, the notation $| t | \leq a$ states that

$$0 \leq t \leq a \text{ when } t > 0, \quad \text{but} \quad 0 \leq -t \leq a \text{ when } t < 0.$$

Stated in another manner, the statement $| t | \leq a$ means that $-a \leq t \leq a$, and the real variable t is defined on the interval $[-a, a]$ including the end points $-a$ and a. What is the interval on which $| t + 2 | \leq 5$ is defined?

P6.2. **Power series** For the power series of Equation 6.14, prove the following:

a. The radius of convergence is $R = \lim_{n \to \infty} \left| \dfrac{c_n}{c_{n+1}} \right|$ when this limit exists.

b. The coefficients are computed by the formula $c_n = \dfrac{f^{(n)}(a)}{n!}$.

P6.3. **Taylor series** Find the radius of convergence for the series

$$\sum_{n=1}^{\infty} \frac{(x - 3)^n}{2^n \, n}.$$

P6.4. **Taylor series** Approximate $\sqrt[3]{e}$ to five decimal places. Use MATLAB to help with the calculations.

P6.5. **Taylor series** Integrate the function

$$\int_0^x \cos t^2 \, dt$$

using a Taylor series expansion. This integral is the *Fresnel cosine integral* used in optics.

P6.6. **Runge-Kutta solution** Solve $\dot{y} = y$ by hand with $y(0) = 1$ using $h = 1$.

P6.7.   **Boundary value problem**   Solve the boundary value problem

$$y''(x) + y(x) = x, \quad 0 \le x \le \frac{\pi}{2},$$

with $y(0) = 2$ and $y(\pi/2) = 1$.

P6.8.   **Eigenvalue problem**   Solve the eigenvalue problem

$$y''(x) = -\lambda y(x), \quad y(0) = 0, \quad y(L) = 0,$$

and show that the only nonzero solutions for $y(x)$ occur for $\lambda > 0$.

P6.9.   **Eigenvalue problem**   Solve the eigenvalue problem

$$y''(x) = -\lambda y(x),$$

subject to the initial conditions

    a. $y'(0) = 0, \quad y'(L) = 0$;

    b. $y(0) = y(2L), \quad y'(0) = y'(2L)$.

P6.10.   **Temperature in a rod**   The temperature in a long rod held at a constant temperature at the ends can be described approximately by the differential equation

$$\frac{d^2 u(x)}{dx^2} - \mu^2 u(x) = 0, \quad 0 \le x \le L,$$

with the boundary conditions $u(0) = u(L) = T_0$. The constant μ depends on the geometric and thermal properties of the rod. Solve for the temperature $u(x)$ in the rod, and write the answer in terms of hyperbolic functions

$$\sinh x = \frac{e^x - e^{-x}}{2} \quad \text{and} \quad \cosh x = \frac{e^x + e^{-x}}{2}.$$

P6.11.   **Series solution**   Solve the initial value problem

$$(1 - x)y''(x) - y'(x) + xy(x) = 0,$$

with $y(0) = 1$ and $y'(0) = 1$.

P6.12.   **Series solution**   Find the series solution to the second-order differential equation

$$y''(x) - 2x^2 y'(x) + 8y(x) = 0,$$

with initial values $y(0) = 0$, $y'(0) = 1$.

P6.13.   **Bessel's equation**   Given the differential equation

$$y''(t) + \beta e^{\alpha t} y(t) = 0,$$

convert this to Bessel's equation with the substitution $u(t) = e^{\alpha t/2}$ and write the solution of the equation. If $\alpha < 0$, this equation with $\beta = k/m$ could model a mass connected by a spring that has a nonlinear, decreasing restoring force.

P6.14.   **Gamma function**   The gamma function with argument $\lambda + 1$ is defined as

$$\Gamma(\lambda + 1) = \int_0^\infty e^{-t} t^\lambda \, dt.$$

a. Show that $\Gamma(\lambda + 1) = \lambda\Gamma(\lambda)$.

b. Show that $\Gamma(1) = 1$.

c. Show that

$$\Gamma(1 + 1/2 + n) = \frac{(2n+1)!}{2^{2n+1}n!}\sqrt{\pi}$$

where n is an integer.

Hint: Integrate by parts in (a)

P6.15. Sturm-Liouville equation   Work out the details and show that Equation 6.80 is indeed the Sturm-Liouville form of Bessel's differential equation with argument $\sqrt{\lambda}x$.

P6.16. MATLAB piecewise continuous function   A column vector $\mathbf{T}$ of arbitrary length from T_0 to T_{final} is to be created. The vector is to contain a pulse of length $T_{\text{end}} - T_{\text{start}}$ such that

$$\text{Pulse}(t) = \begin{cases} 0, & t < T_{\text{start}}, \\ 1, & T_{\text{start}} \leq t \leq T_{\text{end}}, \\ 0, & t > T_{\text{end}}, \end{cases} \qquad (6.92)$$

with $T_{\text{start}} \geq T_0$ and $T_{\text{end}} \leq T_{\text{final}}$. Test the function by plotting the pulse for the values

$$T_0 = 0, \quad T_{\text{final}} = 5, \quad T_{\text{start}} = 0, \quad T_{\text{end}} = 1,$$

where all values are in seconds.

P6.17. MATLAB differential equation solution   Solve and plot the solution to the equation

$$\ddot{y}(t) + y(t) = \begin{cases} 1 - \dfrac{t^2}{\pi^2}, & \text{for } 0 \leq t \leq \pi, \\ 0, & \text{for } t > \pi, \end{cases}$$

with $\dot{y}(0) = y(0) = 0$.

P6.18. MATLAB square wave   Without typing in all the elements, use MATLAB to create the array for the square wave:

$$\text{square}(t) = \begin{cases} 1, & \text{for } i = 1, 10, \\ 0, & \text{for } i = 11, 20. \end{cases}$$

Create two cycles of the square wave.

P6.19. MATLAB Euler solution   Solve the differential equation of Example 6.7 using a step size of $h = 0.2$, and compare the errors with those in the example.

P6.20. MATLAB computer error   For the equation

$$\dot{y} = y, \quad \text{with} \quad y(0) = 1, \quad 0 \leq t \leq 1$$

plot the $\log_{10}$ of the maximum error versus $\log_2$ of the step size as the step size is reduced from $h = 1$ in steps of $h/2$.

P6.21. MATLAB central difference The central difference approximation for the second derivative is written

$$\frac{d^2y}{dx^2} \approx \frac{y_{i+1} - 2y_i + y_{i-1}}{h^2}.$$

Solve the boundary value problem

$$\frac{d^2y(x)}{dx^2} + \nu^2 y(x) = 0,$$

with $y(-\pi) = y(\pi) = 0$. Using MATLAB, set up the matrix eigenvalue equation and find the smallest eigenvalue.

Hint: The matrix is a band matrix, which is easily created using the MATLAB commands **eye**(n) and **diag**(y,k).

P6.22. MATLAB Bessel functions Plot J_0 and J_1 versus x for the range $0 < x < 15$, and estimate the first five zeros of the functions.

P6.23. MATLAB Bessel functions For the equation of Problem 6.13, plot the solution if $y(0) = 1$ and $y'(0) = 0$. Try various values of α and β.

P6.24. MATLAB Bessel functions Bessel functions arise in a wide variety of applications. Plot the function

$$I(x) = \left[2\frac{J_1(\pi x)}{\pi x} \right]^2$$

around $x = 0$ to the first three zeros of the function. This is the Fraunhofer diffraction pattern (light intensity) caused by a light wave passing through a circular aperture.

P6.25. MATLAB Legendre polynomials Compute and plot the Legendre polynomials $P_n(x)$ for $n = 1, \ldots, 5$ on the interval $[-1, 1]$.

EXPLORATION PROBLEMS

P6.26. Limit of a sequence Define $f_n = x^n$ for $0 \le x \le 1$, where $n = 1, 2, \ldots$ is a natural number. Find $\lim_{n\to\infty} f_n$.

P6.27. First-order equation. Solve the equation

$$\frac{dy(t)}{dt} + 2ty = 0.$$

P6.28. Variable coefficients Solve the differential equation

$$y' + (\cos x)y = \sin x \cos x$$

by working out the integrals in the solution by hand. Then, use the MATLAB symbolic command **dsolve** to verify the solution.

P6.29. Falling objects Solve the given equations and plot the resulting velocities versus time for comparison:

a. $m\dfrac{dv(t)}{dt} = mg - bv, \quad \dfrac{b}{m} = 0.1\,\text{second}^{-1};$

b. $m\dfrac{dv(t)}{dt} = mg - bv^2, \quad \dfrac{b}{m} = 0.001\,\text{second}^{-1}.$

Chapter 6 ■ ADVANCED DIFFERENTIAL EQUATIONS

In both problems, $v(0) = 0$ and $g = 9.81 \, \text{meters/second}^2$. If $v(t)$ represents the vertical velocity of a falling object, approximately how long does it take for each object to reach its terminal velocity?

P6.30. Series solution Find the differential equation solved by the series

$$f(x) = \sum_{k=0}^{\infty} \frac{(-1)^k}{(2k)!} x^{2k}.$$

P6.31. Variable coefficients Find the solutions of the equation $y''(x) + e^x y(x) = 1$.

P6.32. Nonlinear equation Find the solution of the equation

$$\frac{dy}{dx} = y^2$$

with $y(0) = 1$.

P6.33. MATLAB step response (Optional) Consider the equation

$$\dot{y}(t) + 0.5y(t) = 0.5U(t) \tag{6.93}$$

defined on the interval $t \geq 0$ with the initial condition $y(0) = y_0$.

Let $U(t) = 1$ for $t \geq 0$ and zero for $t < 0$. This is called a *unit step function* input to the equation. Solve the problem by analytical means.

MATLAB Solution. The command **step** from the *Signals and Systems Toolbox* calculates the step response of a linear system whose equations are of the form

$$\dot{x} = Ax + Bu$$
$$y = Cx + Du \tag{6.94}$$

where A, B, C, D are matrices in the general case. This is called the *state-space* representation and u and y are the input and output functions, respectively. The state space model was discussed in Chapter 5. In this case, the differential equation can be changed to a first-order differential equation in state space representation by the following change of variables:

$$y(t) = x(t),$$
$$\dot{y}(t) = \dot{x}(t).$$

Using these substitutions, Equation 6.93 becomes

$$\dot{x}(t) = \dot{y}(t) = -0.5y(t) + 0.5U(t)$$
$$= Ax(t) + Bu(t).$$

Thus, in Equation 6.94 the coefficients become $A = -0.5$, $B = 0.5$, $C = 1$, and $D = 0$.

Write a MATLAB script to solve the equation and plot the solution $y(t)$ of the equation $\dot{y}(t) + 0.5y(t) = 0.5U(t)$ with initial condition $y(0) = 0$.

If you do not have the **step** function in the version of MATLAB you are using, **ode23** can be used to solve differential equations with a unit step input.

P6.34. Differentiable functions As a research project, answer the following questions:

a. Are there continuous functions that fail to be differentiable?

b. Can you find a function such that $f^{(n)}(0) = 0$ and the derivatives of all orders exist but that is not analytic?

ANNOTATED BIBLIOGRAPHY

1. Fehlberg, Erwin, *Low-Order Classical Runge Kutta Formulas With Step size Control and Their Application to Some Heat Transfer Problems*, NASA Technical Report, NASA TR R-315, National Aeronautics and Space Administration, George C. Marshall Space Flight Center, Marshall, Alabama, July 1969. *This technical report presents first- to fourth-order Runge-Kutta formulas.*

2. Forsythe, George E., M. A. Malcolm, and C. B. Moler, *Computer Methods for Mathematical Computations*, Prentice Hall, Englewood Cliffs, NJ, 1977. *A classic text, which treats the RKF45 (Runge Kutta Fehlberg) algorithm in detail. The MATLAB function* **ode45** *is based on this method.*

3. Goldstein, Herman H., *A History of Numerical Analysis from the 16th through the 19th Century*, Springer Verlag, New York, 1977. *This textbook takes a serious approach to the history of numerical methods. The history as well as the mathematics show the contributions to numerical analysis by famous mathematicians.*

4. Garcia, Alejandro L., *Numerical Methods for Physics*, Prentice Hall, Englewood Cliffs, NJ, 1994. *The textbook applies MATLAB solutions to many numerical problems in physics.*

5. Henrici, P., *Elements of Numerical Analysis*, John Wiley & Sons, New York, 1964. *A classic book that covers many techniques for the solution of differential equations.*

6. Mathews, John M., *Numerical Methods for Mathematics, Science, and Engineering*, Prentice Hall, Englewood Cliffs, NJ, 1992. *This text covers many numerical algorithms. There is a supplement that gives MATLAB implementation of many of these algorithms.*

7. Press, William H., B. P. Flannery, S. A. Teukolsky, and W. T. Vetterling, *Numerical Recipes The Art of Scientific Computing*, Cambridge University Press, Cambridge, England, 1968. *As the title implies, this book presents numerical algorithms with a description and analysis of each one.*

8. Watson, G. N., *Theory of Bessel Functions*, Macmillan, New York, 1944. *There is not much anyone would want to know about the theory of Bessel functions that is not contained in this book.*

ANSWERS

∎

P6.1. **Defining intervals** The statement $|t + 2| \leq 5$ means $-5 \leq t + 2 \leq 5$, which becomes $-7 \leq t \leq 3$. So t is defined on the interval $[-7, 3]$ in this case.

P6.2. **Power series** The power series is $c_0 + c_1(x - a) + c_2(x - a)^2 + \cdots = \sum_{n=0}^{\infty} c_n(x - a)^n$.

a. The radius of convergence is $|x - a| < R$. The ratio test shows that for convergence

$$\lim_{n \to \infty} \left| \frac{c_{n+1}}{c_n} (x - a) \right| = \rho |x - a| < 1,$$

so $|x - a| < 1/\rho = R$. Thus, $R = \lim_{n \to \infty} |c_n / c_{n+1}|$.

b. The second part is easily proven by differentiating each side of the equation and equating $f(a) = c_0$, $f'(a) = c_1$, etc.

P6.3. **Taylor series** The radius of convergence is

$$R = \lim_{n \to \infty} \left| \frac{2^{n+1}(n + 1)}{2^n n} \right| = 2,$$

so that $|x - 3| < 2$ or $1 < x < 5$. In fact, at $x = 1$, the series is an alternating series that converges. At $x = 5$, the series is $\sum_{n=1}^{\infty} (1/n)$, which diverges. The radius of convergence is thus the interval $[1, 5)$, in which $x = 5$ is not included.

P6.4. **Taylor series** Try 5 places in the Taylor series to find $|R_n (1/3)| < 10^{-5}$. Thus, $\sqrt[3]{e} \approx 1.39561$.

P6.5. **Taylor series** Since the Taylor expansion of $\cos t$ is

$$\cos t = 1 - \frac{t^2}{2!} + \cdots + \frac{(-1)^n t^{2n}}{2n!},$$

substitute t^2 and integrate term by term yielding

$$\int_0^x \cos t^2 \, dt = t - \frac{t^5}{5 \cdot 2!} + \cdots + \frac{(-1)^n t^{4n+1}}{(4n + 1)2n!}.$$

P6.6. **Runge-Kutta solution** Compute k_1, k_2, k_3, k_4 using $f(t, y) = y$.

P6.7. **Boundary-value problem** There is a unique solution to this problem. The solution is $y(x) = 2 \cos x + (1 - \pi/2) \sin x + x$.

P6.8. **Eigenvalue problem**

a. For $\lambda = 0$, the equation is $y''(x) = 0$, with solution $y(x) = c_1 x + c_2$. The boundary conditions yield $c_1 = c_2 = 0$.

b. For $\lambda < 0$, the solution cannot match the boundary values. The only possibility is $\lambda > 0$.

P6.9. **Eigenvalue problem** The eigenvalues in each case are $\lambda_n = n^2 \pi^2 / L^2$.

a. The eigenfunctions are $y_n(x) = c_n \cos(n\pi x / L)$, $n = 0, 1, \ldots$.

b. The periodic boundary conditions lead to the solutions

$$y_n(x) = a_n \cos\left(\frac{n\pi x}{L}\right) + b_n \sin\left(\frac{n\pi x}{L}\right).$$

P6.10. **Temperature in a rod** Solutions to the equation $(D^2 - \mu^2)u(x) = 0$ are $e^{+\mu x}$ and $e^{-\mu x}$, so the general solution can be written $u(x) = c_1 \cosh \mu x + c_2 \sinh \mu x$. The boundary conditions yield

$$u(0) = T_0 = c_1,$$
$$u(L) = T_0 = T_0 \cosh \mu L + c_2 \sinh \mu L,$$

so solving for c_2 the solution becomes

$$u(x) = T_0 \left(\cosh \mu x + \frac{1 - \cosh \mu L}{\sinh \mu L} \sinh \mu x \right).$$

P6.11. Series solution The solution is $y(x) = e^x$. Although the equation has a singular point at $x = 1$ (divide by $1 - x$ to see this), the radius of convergence of the solution is infinity.

P6.12. Series solution The series solution $\sum_{n=0}^{\infty} a_n x^n$ is

$$y(x) = x - \frac{4}{3}x^3 + \frac{1}{6}x^4 + \cdots,$$

since $a_2 = a_0 = 0$ and $a_1 = 1$ from the initial conditions. The recurrence relationship is

$$(n+2)(n+1)a_{n+2} - 2(n-1)a_{n-1} + 8a_n = 0$$

for $n = 2, 3, \ldots$. This becomes

$$a_{n+2} = \frac{2(n-1)a_{n-1} - 8a_n}{(n+2)(n+1)}.$$

P6.13. Bessel's equation To change variables, let $u = e^{\alpha t/2}$, $\dot{u} = \alpha u/2$, and $y(t) = y[u(t)]$. Using the chain rule for differentiation,

$$\frac{dy}{dt} = \left(\frac{dy}{du}\frac{du}{dt}\right) = \frac{\alpha}{2}u\frac{dy}{du}.$$

The second derivative becomes

$$
\begin{aligned}
\frac{d^2y}{dt^2} &= \frac{d}{dt}\left(\frac{dy}{dt}\right) = \frac{d}{dt}\left(\dot{u}\frac{dy}{du}\right) = \ddot{u}\frac{dy}{du} + \dot{u}\frac{d}{dt}\left(\frac{dy}{du}\right) = \ddot{u}\frac{dy}{du} + \dot{u}\left[\frac{d}{du}\left(\frac{dy}{du}\right)\frac{du}{dt}\right] \\
&= \ddot{u}\frac{dy}{du} + (\dot{u})^2\frac{d^2y}{du^2}
\end{aligned}
$$

Finally, substitute $v = (2/\alpha)\beta^{1/2}u$ so that

$$u^2\frac{d^2y}{du^2} = v^2\frac{d^2y}{dv^2}, \quad u\frac{dy}{du} = v\frac{dy}{dv}, \quad \frac{4\beta}{\alpha^2}u^2 y = v^2 y.$$

After substituting for the derivatives of y, the original equation becomes

$$v^2\frac{d^2y}{dv^2} + v\frac{dy}{dv} + v^2 y = 0.$$

The solution is

$$y(t) = c_1 J_0\left(\frac{2}{\alpha}\sqrt{\beta}e^{\frac{\alpha t}{2}}\right) + c_2 Y_0\left(\frac{2}{\alpha}\sqrt{\beta}e^{\frac{\alpha t}{2}}\right).$$

P6.14. Gamma function Integrate the gamma function by parts using $u = t^\lambda$ and $dv = e^{-t}dt$ to find that $\Gamma(\lambda + 1) = \lambda\Gamma(\lambda)$. When $\lambda = 0$, $\Gamma(1) = \int_0^\infty e^{-t}\,dt = 1$.

P6.15. Sturm-Liouville equation Substitute $x \to \sqrt{\lambda}x$ in Bessel's equation and use the fact that

$$\frac{dy(\sqrt{\lambda}x)}{dx} = \sqrt{\lambda}\frac{dy}{dx}, \quad \frac{d^2y(\sqrt{\lambda}x)}{dx^2} = \lambda\frac{d^2y}{dx^2}$$

to change the argument and derive Equation 6.78.

Comment: MATLAB problems are on the disk accompanying this textbook.

7

APPROXIMATION OF FUNCTIONS

This chapter presents interpolation and least-squares curve fitting as techniques of fitting particular functions to discrete data and approximation of continuous functions by orthogonal functions.

The first section explains polynomial interpolation over a set of N points. Perhaps the best known example of interpolation is linear interpolation in which a straight line is selected to pass between two consecutive data values in a table. Intermediate values not in the table can be found or *interpolated* using the equation of the line. We discuss this case and other applications of interpolation.

The second section presents another approach to interpolation using *spline functions*. Splines are smooth curves, often used for graphics applications, that pass through a set of data points.

The *least squares* method is presented next. This technique often solves the problem in experimental work of fitting a curve to data. Typically, the curve is known or believed to be of a certain form. Curve fitting is the process of finding the parameters of a function that fits a set of data points according to some criterion even when the function that results does not necessarily pass through any of the given data points. The method of least squares is used to select the parameters in such a way that the sum of the squared error between the function values and the data points is minimized. Since there are typically many more data points than parameters to determine, we shall see that the basic least-squares problem can be formulated and solved as an *overdetermined* system of linear equations.

Continuous functions can be approximated in various ways, such as by using Taylor's series. This chapter presents the expansion of functions in terms of *orthogonal functions*. In addition to the mathematical simplification afforded by orthogonal function representation, orthogonal functions arise in connection with eigenvalue problems for differential equations, as explained in Chapter 6, and in the study Fourier series to be presented in Chapter 8.

MATLAB COMMANDS

Table 7.1 lists a few of the commands in MATLAB that are useful for approximation of functions. The listed MATLAB commands will be explained as they are used in examples in this chapter. The command **help polyfun** will list other MATLAB polynomial and interpolation functions.

TABLE 7.1 *MATLAB approximation functions*

Operation	Form	Result
Polynomial	**polyfit(x,y,n)**	Polynomial that fits the data
	polyval(p,S)	Polynomial evaluation
	polyder(p)	Derivative of a polynomial
Interpolation	**interp1(x,y,xi,'method')**	Interpolate using linear, spline or cubic functions
Least squares	**nnls(A,b)**	Solve $Ax = b$ for $\mathbf{x}$, $x_i \geq 0$
	\ or **polyfit**	Solve $Ax = b$ for an overdetermined system
Spline	**spline(x,y,xi)**	Cubic spline interpolation

This chapter is concerned with functions of a single variable in the form $y = f(x)$. Two-dimensional interpolation will be described in Chapter 10.

POLYNOMIAL INTERPOLATION

■

The results of experimental measurements or numerical computation often yield values of a function only at discrete points of the independent variable. In such cases, values of a function $f(x)$ may be tabulated for discrete values of x. The process of computing the values in between the known data points for $f(x)$ is called *interpolation*.

This section is concerned with methods for finding the value of $f(x)$ between the tabulated points. Polynomial interpolation is introduced first in this section. Then, MATLAB commands that perform interpolation are introduced and used in an example.

Consider the following table of values $(x_i, f(x_i))$:

x_i	x_0	x_1	$\ldots$	x_N
$f(x_i)$	$f(x_0)$	$f(x_1)$	$\ldots$	$f(x_N)$

This table defines $N + 1$ points of an unknown function. Assuming the function can be expressed as a Taylor series, we could approximate the function with a series of the form

$$f(x) \approx a_0 + a_1 x + \cdots + a_N x^N = \sum_{k=0}^{N} a_k x^k, \qquad (7.1)$$

representing an Nth-degree polynomial. In fact, it can be proven that if the points $x_i, i = 0, N$ are distinct, then there is a unique polynomial $p(x)$ of degree $\leq N$ such that $p(x_i) = f(x_i)$ $(0 \leq i \leq N)$. This polynomial is said to *interpolate* the $N + 1$ data points. It is possible, in principle, to solve for the coefficients in Equation 7.1 and to determine the interpolating polynomial. For example, the MATLAB command **polyfit** will determine the polynomial.

Although the Nth-degree polynomial that interpolates the $N + 1$ points can be found, we will not pursue this approach in great detail. Unless the table contains a small number of points, fitting a polynomial to the data points may lead to numerical problems. Also, a high-order polynomial may deviate considerably from the actual function between the tabulated points. Every book treating numerical analysis warns of this possibility, and the effect will be shown in Example 7.1. Considering a physical problem, one would not expect a mathematical model fitting the data resulting from 100 hundred measurements to be a polynomial of degree 99.

A more reasonable approach to interpolation is to fit a low-order polynomial, such as a first-, second-, or third-degree polynomial, between two, three, or four points, respectively. For example, a first-degree polynomial approximation results in a set of straight lines between each two points in the table. Then, points in between the tabulated points can be determined using the linear approximation in the interval of interest. The linear polynomial between x_i and x_{i+1} is

$$P_1(x) = y_i + \left(\frac{y_{i+1} - y_i}{x_{i+1} - x_i} \right) (x - x_i) \qquad (7.2)$$

for $x_i \leq x \leq x_{i+1}$. In this equation, the designation $y_i = f(x_i)$ is used for simplicity.

To interpolate over an interval, a sequence of linear (P_1) or cubic (P_3) polynomials is often used for the subintervals. This avoids the oscillations exhibited by a polynomial of high degree that fits all the points. The oscillations are caused by the large values of the higher derivatives of the polynomial.

□ EXAMPLE 7.1   *MATLAB Interpolation*

This example applies the MATLAB commands **interp1**, **polyfit**, as well as **polyval** to the problem of interpolation. The command **polyfit** will compute the Nth-degree polynomial that fits $N + 1$ points. A polynomial with its coefficients defined in the vector **p** is evaluated at the points specified in the vector **x** by the command **polyval(p,x)**. The command **interp1** is used to fit a sequence of linear or cubic polynomials over the interval where data are given.

To compare several types of interpolation methods, the function

$$y(x) = \frac{1}{1 + x^2}, \qquad -5 \leq x \leq 5, \qquad (7.3)$$

will be used. This is an example of a *Runge function*, named after C. Runge, who pointed out the dangers of polynomial interpolation. Functions of this form are often used as so-called bad examples since the function is smooth but the interpolation polynomial deviates from the function by a relatively large amount toward the endpoints. The point is that even for a very smooth function, the interpolating polynomial may not provide a good approximation. If the data being interpolated come from a function with numerous inflection points, the polynomial approximation could be even more in error between the points.

Interpolation using a linear polynomial between the data points is appropriate in many cases. The accompanying MATLAB script first computes a polynomial that interpolates 11 data points of the function defined in Equation 7.3. Then, **interp1** is used with linear interpolation between each two data points. Refer ahead to Figure 7.2 to see the Runge function interpolated with spline functions that match the function very well.

MATLAB Script _____

```
Example 7.1
% EX7_1.M Interpolate the function y=1/(1+x^2)
%   at N+1 points with an Nth degree polynomial,
%   then compare with straight-line interpolation.
%   Test case is N=10 over the interval x=[-5 5].
N=10;                    % Choose 10th degree polynomial
x=[-5:1:5];             % 11 points
y=1./(1+x.^2);          % Runge function
```

```
% Polynomial Fit Nth degree
p=polyfit(x,y,N);         % p holds coefficients
xplot=[-5:.1:5];          % Define finer grid (101 points)
f=polyval(p,xplot);       % Evaluate at points xplot
% Straight Line interpolation
ystl=interp1(x,y,x,'linear');
% Plot
clf                       % Clear any figures
subplot(2,1,1), plot(x,y,'o',xplot,f,'-')
title('Polynomial Interpolation Figure 7.1')
axis([-6 6 -.5 2.5])      % Set axis limits
subplot(2,1,2), plot(x,y,'o',x,ystl)
title('Straight Line Interpolation')
axis([-6 6 -.5 2.5])
%
```

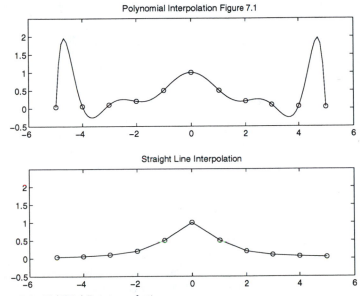

FIGURE 7.1 *MATLAB interpolation*

OTHER INTERPOLATING POLYNOMIALS

It is possible that an Nth-degree polynomial that interpolates functional values at $N + 1$ equally spaced points in an interval suffers from oscillations that increase near the endpoints. The polynomial interpolation in Figure 7.1 shows the effect dramatically. References in the Annotated Bibliography at the end of this chapter treat other methods of interpolation such as Lagrange interpolation, which can be used to interpolate a function defined at unequally spaced points in an interval.

INTERPOLATION BY SPLINE FUNCTIONS

The term *spline* refers to a flexible drafting device that is used to draw a smooth curve through a set of plotted data points. Mathematically, spline interpolation consists of finding a set of polynomials that passes smoothly through data points given as $f(x_i), i = 0, 1, 2, \ldots, N$. Although the theory of splines is quite extensive, we restrict our discussion to cubic splines since these are the most widely used in engineering applications.

A spline function of degree n can be defined in the interval $[a, b]$ with subintervals defined by the $N + 1$ points

$$a = x_0 < x_1 < \ldots < x_N = b.$$

The spline function $s(x)$ has the following properties:

1. In each subinterval $[x_i, x_{i+1}]$, $s(x)$ is a polynomial of degree n.

2. $s(x)$ and its first $(n - 1)$ derivatives are continuous on $[a, b]$.

The *spline function* is constructed of polynomial pieces joined together at the end points of each subinterval. In cubic spline interpolation, cubic polynomials are joined together so that the function and the first and second derivatives agree at each point within the interval. Thus, in each subinterval $[x_i, x_{i+1}]$, the spline function is

$$s_i(x) = a_{i0} + a_{i1}(x - x_i) + a_{i2}(x - x_i)^2 + a_{i3}(x - x_i)^3 \qquad \textbf{(7.4)}$$

for $1 \leq i \leq N$. Since there are $N + 1$ points, a cubic equation with four parameters fits each of the N intervals. Thus, $4N$ parameters are to be determined.

One condition on the polynomials is that they interpolate the points $f(x_i), i = 0, 1, \ldots, N$, leading to $N + 1$ equations. Also, the cubics and the first two derivatives are continuous at each interior point. These conditions yield $3(N - 1)$ equations, for a total of $4N - 2$ equations. The two remaining conditions can be defined by the values of the first or second derivatives at the endpoints. For example, the *natural cubic splines* have a free boundary condition $s''(x_0) = 0$ and $s''(x_N) = 0$.

The $4N$ equations can be solved to yield the coefficients of the spline function in the N intervals. The details are presented in many textbooks treating splines, such as the textbook by John Mathews listed in the Annotated Bibliography for this chapter

☐ EXAMPLE 7.2   *MATLAB Interpolation*
Given the equation from Example 7.1,

$$y(x) = \frac{1}{1 + x^2}, \qquad -5 \leq x \leq 5, \qquad \textbf{(7.5)}$$

a spline function will be used to approximate the function. The MATLAB command **spline** performs the spline fit. In the form **spline(x,y,xi)**, the function interpolates the points (x,y) but uses cubic spline interpolation to create a set of points (xi,yi). Generally, the vector xi specifies a finer spaced abscissa than that given by the x points.

The accompanying MATLAB script shows the M-file that creates and plots the spline function approximation to the function. Eleven points are used for the interpolation, but arbitrarily 101 points are used to create a more finely spaced abscissa for plotting. The spline interpolation in Figure 7.2 produces a very smooth curve compared with the approximations shown in Figure 7.1.

MATLAB Script ──

```
Example 7.2
% EX7_2.M Spline interpolation of the function y=1/(1+x^2)
%    with a cubic polynomial
x=[-5:1:5];              % N+1=11 points
y=1./(1+x.^2);          % Runge function
% Spline function
xspline=[-5:.1:5];      % Finer spacing in x
yspline = spline(x,y,xspline);
% Plot
clf                     % Clear any figures
plot(x,y,'o',xspline,yspline,'-')
title('Spline Function Interpolation Figure 7.2')
axis([-6 6 -.5 2])
```

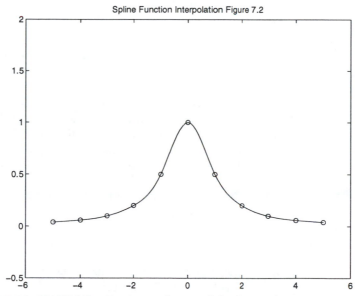

FIGURE 7.2 *MATLAB spline interpolation of the Runge function*

LEAST-SQUARES CURVE FITTING

In previous sections of this chapter, functions were sought that interpolated data points. The data points might have arisen as values at specific points of a given function or as sample values determined experimentally. In the latter case, it is possible that the experimental data contain errors. Borrowing from communications terminology, the data contain "noise," and we seek an approximation that *smooths* the data. Also, in many experiments, the functional relationship that models a process may be known but specific parameters are to be determined. The technique of *least-squares curve fitting* is useful in these situations.

As first discussed in Chapter 2, it is frequently convenient to represent a function $f(x)$ on an interval $[a, b]$ by a linear combination of other functions as

$$f^*(x) = a_0\phi_0(x) + \cdots + a_n\phi_n(x) = \sum_{k=0}^{n} a_k\phi_k(x), \qquad \textbf{(7.6)}$$

where the notation $f^*(x)$ indicates that the sum is an approximation to $f(x)$. The question is how to define an error criterion so that the $\hat{n} = n+1$ coefficients $a_0, a_1, \ldots, a_n$ can be determined.

In this section, the function $f(x)$ is assumed to be defined by N points $(x_1, y_1), \ldots, (x_N, y_N)$ where the abscissas are distinct. The least-squares error criterion is to minimize the square of the Euclidean norm

$$e = ||f^*(x_i) - f(x_i)||^2 = \sum_{i=1}^{N} |f^*(x_i) - f(x_i)|^2. \qquad \textbf{(7.7)}$$

The approximation functions $\phi_i(x)$ in Equation 7.6 could be of any form, but low-degree polynomials are frequently used. This case will be developed using a least-squares line to illustrate the technique. The resulting equations will then be written in matrix notation in a form that is suitable for MATLAB solution.

LEAST-SQUARES STRAIGHT LINE

Consider the straight-line approximation

$$f^*(x) = a_0 + a_1 x, \qquad \textbf{(7.8)}$$

with $\phi_0(x) = 1$ and $\phi_1(x) = x$ in the notation of Equation 7.6. The error term from Equation 7.7 is thus

$$e = \sum_{i=1}^{N} [a_0 + a_1 x_i - f(x_i)]^2. \qquad \textbf{(7.9)}$$

Hence, there are two parameters, a_0 and a_1, that can be varied to minimize the error. The minimum conditions are found by setting the partial derivatives with respect to each of the coefficients to zero. This results in the equations

$$\frac{\partial e}{\partial a_0} = 0,$$

$$\frac{\partial e}{\partial a_1} = 0.$$

Due to the form of the error equation, the second derivative test for maxima and minima from elementary calculus will indicate that a minimum has been found.

Taking the partial derivatives with respect to the coefficients in Equation 7.0 leads to the two equations

$$N a_0 + \left(\sum_{i=1}^{N} x_i \right) a_1 = \sum_{i=1}^{N} f(x_i),$$

$$\left(\sum_{i=1}^{N} x_i \right) a_0 + \left(\sum_{i=1}^{N} x_i^2 \right) a_1 = \sum_{i=1}^{N} x_i f(x_i). \qquad \textbf{(7.10)}$$

In matrix form, Equation 7.10 becomes

$$\begin{bmatrix} N & \sum x_i \\ \sum x_i & \sum x_i^2 \end{bmatrix} \begin{bmatrix} a_0 \\ a_1 \end{bmatrix} = \begin{bmatrix} \sum f(x_i) \\ \sum x_i f(x_i) \end{bmatrix}, \qquad \textbf{(7.11)}$$

where $\sum$ signifies $\sum_{i=1}^{N}$. It is convenient to write the equations by creating a matrix whose columns are derived from the approximating functions evaluated at the data points.

Another way to view the least-squares problem for fitting a straight line is to recognize that the system of equations that must be solved is

$$a_0 + a_1 x_1 = f(x_1),$$
$$a_0 + a_1 x_2 = f(x_2),$$
$$\vdots$$
$$a_0 + a_1 x_N = f(x_N), \qquad \textbf{(7.12)}$$

which is an *overdetermined* system in which there are more equations than unknowns. The matrix form is $A\mathbf{x} = \mathbf{b}$ with

$$A = \begin{bmatrix} 1 & x_1 \\ 1 & x_2 \\ \vdots & \vdots \\ 1 & x_N \end{bmatrix}, \quad \mathbf{x} = \begin{bmatrix} a_0 \\ a_1 \end{bmatrix}, \quad \text{and} \quad \mathbf{b} = \begin{bmatrix} f(x_1) \\ \vdots \\ f(x_N) \end{bmatrix}. \qquad \textbf{(7.13)}$$

The equations will be inconsistent unless all the points $f(x_i)$ lie exactly on a straight line.

Notice that the left-hand side of Equation 7.11 can be written as the product $A^T A \mathbf{x}$ if the product is defined as

$$A^T A \mathbf{x} = \begin{bmatrix} 1 & 1 & \cdots & 1 \\ x_1 & x_2 & \cdots & x_N \end{bmatrix} \begin{bmatrix} 1 & x_1 \\ 1 & x_2 \\ \vdots & \vdots \\ 1 & x_N \end{bmatrix} \begin{bmatrix} a_0 \\ a_1 \end{bmatrix}.$$

Also, the vector on the right-hand side of Equation 7.11 can be written as $A^T \mathbf{b}$

$$\begin{bmatrix} 1 & 1 & \cdots & 1 \\ x_1 & x_2 & \cdots & x_N \end{bmatrix} \begin{bmatrix} f(x_1) \\ \vdots \\ f(x_N) \end{bmatrix}.$$

In general, with N data points and $\hat{n}$ coefficients in the approximating function of Equation 7.6, A is the $N \times \hat{n}$ matrix with elements

$$a_{ij} = \phi_{j-1}(x_i), \quad j = 1, 2, \ldots, \hat{n}, \quad i = 1, 2, \ldots, N.$$

For the straight line approximation, there are $\hat{n} = 2$ coefficients.

Thus, the least-squares equations defined by Equation 7.10 become

$$\left(A^T A \right) \mathbf{X}_{\text{lsq}} = A^T \mathbf{b}, \tag{7.14}$$

where $\mathbf{b}$ is a column vector whose elements are the values of the function $f(x_i), i = 1, 2, \ldots, N$. When this equation is solved, the vector $\mathbf{X}_{\text{lsq}}$ contains the coefficients of the least-squares line. It can be shown that minimizing the norm square $(\|A\mathbf{x} - \mathbf{b}\|)^2$ with the definitions in Equation 7.13 leads to the least-squares formulation of Equation 7.14.

A unique solution of the matrix equation for the least-squares line is possible if not all of x_i's are equal, since the columns of A are linearly independent in that case. Other cases are discussed in several of the references in the Annotated Bibliography at the end of this chapter. In particular, the text by David R. Hill gives a more complete discussion of least-squares curve fitting and presents MATLAB examples.

☐ EXAMPLE 7.3   *MATLAB Least Squares*

It is desired to fit a least-squares straight line to the following data set:

x_i	0.0	1.0	2.0	3.0	5.0
y_i	0.0	1.4	2.2	3.5	4.4

Let the approximating function be

$$f1(x) = a_0 + a_1 x.$$

The least-squares equation is

$$\left(A^T A \right) \mathbf{X}_{lsq} = A^T \mathbf{b},$$

where

$$A = \begin{bmatrix} 1 & 0 \\ 1 & 1 \\ 1 & 2 \\ 1 & 3 \\ 1 & 5 \end{bmatrix} \quad \text{and} \quad \mathbf{b} = \begin{bmatrix} 0.0 \\ 1.4 \\ 2.2 \\ 3.5 \\ 4.4 \end{bmatrix}.$$

The accompanying MATLAB script shows two methods of computing the coefficients of the approximating function. After forming the least-squares equation, the backslash operator \ in the command Xlsq1=Als\bls yields the two coefficients for the line. Also, the command **polyfit** in the form

```
Xlsq2=polyfit(x,y,n)
```

will fit a polynomial of degree n to the data in a least-squares sense and return the coefficients. The command **polyval** is used to evaluate the polynomial at the x-values for plotting. The result by both approaches to the accuracy shown is the function

$$f1(x) = 0.3676 + 0.8784x,$$

since the least-squares line is unique in this case. Figure 7.3 shows the straight-line approximation for the data.

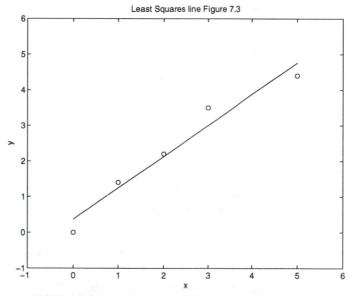

FIGURE 7.3 *MATLAB least-squares curve fit*

The results from executing the MATLAB M-file are displayed in tabular form following the M-file script. The vector Xlsq1 holds the coefficients computed using the MATLAB backslash operator, and Xlsq2 was computed using **polyfit**. The array table holds the data points and the values of $f1(x)$ from the least-squares line evaluated at the abscissas. Also, the error at each point is shown as y-f1. The error is the *vertical distance* between the data value and the least squares line at each x point in Figure 7.3. The least-squares error would be the sum of the squares of these values.

MATLAB Script _____
Example 7.3

```
% EX7_3.M Least-squares curve fit with a line using
%  \ operator and polyfit. The results are displayed
%  and plotted.
x=[0 1 2 3 5];          % Define the data points
y=[0 1.4 2.2 3.5 4.4];
A1=[1 1 1 1 1 ]';       % Least squares matrix
A=[A1 x'];
Als=A'*A;
bls=A'*y';
% Compute least squares fit
Xlsq1=Als\bls;
Xlsq2=polyfit(x,y,1);
f1=polyval(Xlsq2,x);
error=y-f1;
disp('        x         y          f1        y-f1')
table=[x' y' f1' error'];
disp(table)
fprintf('Strike a key for the plot\n')
pause
% Plot
clf
plot(x,y,'o',x,f1,'-')
axis([-1 6 -1 6])
title('Least Squares line Figure 7.3')
xlabel('x')
ylabel('y')
%--------------------------------
Results:
Xlsq1 =
     0.3676
     0.8784
Xlsq2 =
     0.8784    0.3676
```

x	y	f1	y-f1
0	0	0.3676	-0.3676
1.0000	1.4000	1.2459	0.1541
2.0000	2.2000	2.1243	0.0757
3.0000	3.5000	3.0027	0.4973
5.0000	4.4000	4.7595	-0.3595

□

WHAT IF? It may not be obvious that the straight line generated by the least-squares method leads to the minimum squared error. In Problem 7.14, you are asked to compute the error in the approximation of Example 7.3 and experiment with other straight lines to compare the errors.

Suppose the nth-degree polynomial,

$$f^*(x) = a_0 + a_1 x + \cdots + a_n x^n = \sum_{k=0}^{n} a_k x^k, \qquad (7.15)$$

is used to fit N points. Then, the least-squares matrix A becomes

$$A = \begin{bmatrix} 1 & x_1 & \cdots & x_1^n \\ 1 & x_2 & \cdots & x_2^n \\ \vdots & \vdots & & \vdots \\ 1 & x_N & \cdots & x_N^n \end{bmatrix}. \qquad (7.16)$$

There are $\hat{n} = n + 1$ columns and N rows in the matrix A. The resulting least-squares equation is again that of Equation 7.14. When $n = N-1$, the polynomial determined by the least-squares method is the interpolating polynomial.

ORGHOGONAL FUNCTIONS

The least-squares approximation described in the previous section involved the approximation of a function by a linear combination of other functions. In particular, polynomials were selected for the least-squares curve fit. This approximation leads to least-squares equations that could be solved by the methods of linear algebra. Such equations may be difficult to solve numerically if a polynomial of high degree is used in the approximation. One method of simplifying the equations is to use *orthogonal polynomials* as the approximating functions.

Orthogonal systems of functions allow great mathematical simplification in many problems. As we have seen in Chapter 6, orthogonal functions occur frequently when solving eigenvalue problems involving differential equations. In this section, we describe the expansion of a function in terms of orthogonal functions. This technique will be useful in a number of applications presented later.

Consider a set of *orthogonal functions* $\phi_0(x), \phi_1(x), \ldots, \phi_n(x)$ that have been normalized by requiring that

$$\langle \phi_k(x), \phi_k(x) \rangle = \|\phi_k(x)\|^2 = \int_a^b [\phi_k(x)\phi_k(x)]\, dx = 1 \qquad (7.17)$$

for $k = 0, 1, 2, \ldots, n$.

These functions are called *orthonormal* functions on the interval $[a, b]$ since they have the property

$$\langle \phi_i(x), \phi_j(x) \rangle = \int_a^b \phi_k(x)\, \phi_l(x)\, dx = \begin{cases} 0, & i \neq j, \\ 1, & i = j. \end{cases} \qquad (7.18)$$

We seek to approximate $f(x)$ by a linear combination of these orthonormal functions as

$$f^*(x) = a_0\phi_0(x) + \cdots + a_n\phi_n(x) = \sum_{k=0}^{n} a_k\phi_k(x) \tag{7.19}$$

in such a way that the error is a minimum. The resulting series is called the *expansion* of $f(x)$ in terms of the orthogonal functions.

The most common error criterion is to minimize the *square* error. This is the same as the least-squares criterion of Equation 7.7 that was used for the discrete case. For continuous functions, the square error is defined as

$$e = \int_a^b [f(x) - \sum_{k=0}^{n} a_k\phi_k(x)]^2 \, dx. \tag{7.20}$$

The approach to the approximation is to find the appropriate coefficients in Equation 7.19 that minimize this square error integral.[1]

Expanding the terms in Equation 7.20 and using the orthonormal properties of the approximating functions leads to the result

$$e = \int_a^b [f(x)]^2 dx - \sum_{k=0}^{n} c_k{}^2 + \sum_{k=0}^{n} (a_k - c_k)^2, \tag{7.21}$$

in which

$$c_k = \int_a^b f(x)\phi_k(x) \, dx \tag{7.22}$$

are called the *Fourier coefficients* of $f(x)$ relative to $\phi_k(x)$. Since the first two terms in Equation 7.21 are positive, the choice $a_i = c_i$ minimizes the square error integral. Thus, the Fourier coefficients yield a smaller square error than any other choice of coefficients in the approximation using a linear combination of the ϕ_i's.

■ THEOREM 7.1　　*Fourier coefficients*

The Fourier coefficients c_k as defined in Equation 7.22 give the best least-squares fit when a function $f(x)$ is approximated in terms of an orthonormal set of functions $\phi_k(x)$.

————————————————————————————■

One important aspect of the approximation using orthonormal functions is that each Fourier coefficient is determined independently of all the others. Adding terms such as $c_{n+1}\phi_{n+1}(x)$ does not require the recalculation of the previous terms $\phi_k(x), i = 1, \ldots n$, as shown in Problem 7.6.

---

[1]If the integral in Equation 7.20 is divided by $(b - a)$, it is called the *mean-square error*.

Before discussing the properties of orthogonal functions and the conditions for orthogonal expansions, we will give an important example of an orthogonal set. These will be polynomials designated $P_n(x)$, with the following properties:

1. $P_n(x)$ is a polynomial of degree n.

2. $P_n(1) = 1$ for each n.

3. The set of polynomials is orthogonal on the interval $[-1, 1]$.

It can be shown that these properties uniquely determine the $P_n(x)$ and that the resulting polynomials are the *Legendre polynomials* introduced in Chapter 6 as the solutions to Legendre's differential equation. The results presented there showed that $P_n(x)$ can be derived as

$$P_n(x) = \frac{1}{2^n n!} \frac{d^n}{dx^n}(x^2 - 1)^n, \tag{7.23}$$

where n is an integer. The first six Legendre polynomials are

$$P_0(x) = 1, \qquad\qquad P_3(x) = \frac{1}{2}(5x^3 - 3x),$$

$$P_1(x) = x, \qquad\qquad P_4(x) = \frac{1}{8}(35x^4 - 30x^2 + 3),$$

$$P_2(x) = \frac{1}{2}(3x^2 - 1), \qquad P_5(x) = \frac{1}{8}(63x^5 - 70x^3 + 15x).$$

The Legendre polynomials with even-numbered subscripts are even functions and the polynomials with odd subscripts are odd functions. The polynomials can be calculated by the recursion relation

$$(n+1)P_{n+1}(x) = (2n+1)xP_n(x) - nP_{n-1}(x), \tag{7.24}$$

and the square of the norm of $P_n(x)$ is

$$||P_n(x)||^2 = \int_{-1}^{1} P_n^2(x)dx = \frac{2}{2n+1}. \tag{7.25}$$

If the interval is taken to be $[a, b]$ rather than $[-1, 1]$, the generating formula in Equation 7.23 becomes

$$P_n(x) = \frac{1}{(b-a)^n(n!)} \frac{d^n}{dx^n}(x-a)^n(x-b)^n. \tag{7.26}$$

Legendre Series   The *Legendre series*, sometimes called the Fourier-Legendre series, to approximate $f(x)$ on the interval $[-1, 1]$ has the form

$$f(x) = \sum_{n=0}^{\infty} c_n P_n(x). \tag{7.27}$$

The coefficients are computed as the Fourier coefficients defined in Equation 7.22. Using Equation 7.25 and the orthonormal properties of the polynomials results in the Fourier (Legendre) coefficients,

$$c_n = \frac{2n+1}{2} \int_{-1}^{1} f(x) P_n(x)\, dx \qquad (7.28)$$

If $f(x)$ is square integrable on the interval $[-1, 1]$, the series will converge to $f(x)$. The next section presents convergence and other properties of such series.

□ **EXAMPLE 7.4**   *Legendre Series*

Consider expanding the function $f(x)$ defined as

$$f(x) = \begin{cases} -1, & -1 \leq x < 0, \\ 1, & 0 < x \leq 1, \end{cases}$$

in a Legendre series of the form

$$f(x) \approx c_0 + c_1 P_1(x) + c_2 P_2(x) + c_3 P_3(x).$$

We assume that $f(0) = [f(0^-) + f(0^+)]/2 = 0$. This is convergence in the mean at a point of discontinuity.

Since $f(x)$ is odd, we immediately set $c_0 = c_2 = 0$. Then, integrating in Equation 7.28 over the half-integral yields

$$c_1 = 2 \cdot \frac{3}{2} \int_0^1 1 \cdot x\, dx = \frac{3}{2},$$

$$c_3 = 2 \cdot \frac{7}{2} \int_0^1 1 \cdot \left(\frac{5}{2}x^3 - \frac{3}{2}x \right) dx = -\frac{7}{8}.$$

The resulting series is thus

$$\begin{aligned} f(x) &\approx \frac{3}{2} P_1(x) - \frac{7}{8} P_3(x) \\ &= \frac{45}{16}x - \frac{35}{16}x^3. \end{aligned}$$

In Problem 7.12, you are asked to plot the Legendre polynomials and display the approximation to this $f(x)$. With only two terms in the series, don't expect too accurate a fit to the function.

□

PROPERTIES OF ORTHOGONAL FUNCTIONS   There are a number of other orthogonal functions and orthogonal expansions commonly used. Orthogonal expansions are possible using Bessel functions, Hermite polynomials, and Laguerre polynomials mentioned in Chapter 6 as well as many other functions. For example, the important Fourier trigonometric series will be treated in the next chapter. In fact, a set of orthogonal functions to suit a particular problem can be constructed.

Orthogonalization Given a sequence $f_n(x)$ of functions that are continuous and linearly independent on the interval $[a, b]$, it is possible to construct linear combinations of the functions that form an orthogonal set. The procedure is called the *Gram-Schmidt orthogonalization process* and was described in Chapter 2. Problem 7.8 asks you to find an orthonormal basis for the vector space of polynomials over the interval $[-1, 1]$.

Convergence of Orthogonal Series Consider the minimum error involved in the least-squares approximation of $f(x)$ by an expansion using a finite sum of orthogonal functions

$$E_n = \int_a^b [f(x) - \sum_{k=0}^n c_k \phi_k(x)]^2 \, dx, \tag{7.29}$$

where the coefficients c_k are the Fourier coefficients previously defined. An important question is whether E_n converges to zero as the number of terms $n \to \infty$.

If $f(x)$ is piecewise continuous in the interval $[a, b]$ and $f(x)$ can be approximated as closely as possible by the series

$$f(x) \approx \sum_{k=0}^n c_k \phi_k(x),$$

the set of orthogonal functions $\{\phi_k(x)\}$ is termed *complete*. Restating this in terms of Equation 7.29 assuming that the set of approximating functions forms a complete orthonormal set, leads to the relationship

$$E_n = \int_a^b f^2(x) \, dx - \sum_{k=0}^n c_k^2 \geq 0. \tag{7.30}$$

As $n \to \infty$, the expression becomes the equality

$$\int_a^b f^2(x) \, dx = \sum_{k=0}^\infty c_k^2. \tag{7.31}$$

This relationship is called *Parseval's equality*. Thus, the validity of Parseval's equality is equivalent to completeness. Parseval's equality has important physical as well as mathematical consequences, as discussed in Chapter 8.

REINFORCEMENT EXERCISES AND EXPLORATION PROBLEMS

In these problems, do the computations by hand unless otherwise indicated, and then check those that yield numerical or symbolic results with MATLAB.

REINFORCEMENT EXERCISES

P7.1.   Interpolation   Given the data in the following table, determine the polynomial of degree 6 that interpolates the data.

x_i	-3	-2	-1	0	1	2	3
$f(x_i)$	0.6	0.2	0.6	3.0	0.6	0.2	0.6

P7.2.   Least squares   Suppose that f is the force applied to a spring and x is the resulting length of the spring. According to Hooke's law, we expect the relationship to be

$$f = k(x - x_0),$$

where x_0 is the unstretched length of the spring. If the following lengths are measured with the given applied force, find the spring constant k.

x_i in.	6.1	7.6	8.7	10.4
$f(x_i)$ lb.	0.0	2.0	4.0	6.0

Assume the data contain experimental errors.

P7.3.   Least squares   Given the data set

$$(1, 1.7), \ (2, 1.8), \ (3, 2.3), \ (4, 3.2),$$

plot the values and decide the degree of a polynomial that might provide a reasonable least-squares fit to the data. Compute the coefficients of the polynomial.

P7.4.   Least squares   Given the following table for the Bessel function $J_0(x)$, fit the least-squares line to approximate $J_0(6.3)$.

x	6.0	6.2	6.4	6.6	6.8
$J_0(x)$	0.15065	0.20175	0.24331	0.27404	0.29301

P7.5.   Square error integral   Expand the square error integral for approximation by orthogonal functions in Equation 7.20 and show that Equation 7.21 results.

P7.6.   Fourier coefficients   Consider the approximation

$$f_1(t) \approx c_{12} f_2(t)$$

for $t_1 < t < t_2$.

   a. Find the coefficient c_{12} that minimizes the square error.

   b. If $f_1(t) \approx c_{12} f_2(t) + c_{13} f_3(t)$, how is c_{12} now calculated if f_2 and f_3 are orthogonal functions?

P7.7. **Orthogonal functions** Suppose that f and g are orthogonal functions. Prove the Pythagorean theorem for functions

$$\|f + g\|^2 = \|f\|^2 + \|g\|^2.$$

P7.8. **Gram-Schmidt process** Consider the vector space of polynomials

$$f(x) = a_0 + a_1 x + \cdots + a_n x^n$$

defined on the interval $[-1, 1]$. Find an orthonormal basis for these polynomials from the set $1, x, x^2, \ldots, x^n$. The result is the normalized Legendre polynomials. The inner product for the polynomials is

$$\langle f, g \rangle = \int_{-1}^{1} f(x) g(x) \, dx.$$

Hint: Apply the Gram-Schmidt process described in Chapter 2, replacing the dot product by the inner product.

P7.9. **MATLAB interpolation** Fit a spline function to the data in Problem 7.1 and plot the results for the spline fit and the polynomial approximation previously determined. Compare the two results.

P7.10. **MATLAB least squares** Write a MATLAB function to compute and plot the least-square polynomial of degree n given the data points $(x_i, y_i), i = 1, \ldots, N$.

P7.11. **MATLAB least squares** Using the values of the Bessel function in Problem 7.4, compare a straight-line and a quadratic least-squares fit to the data. Find $J_0(6.3)$ by each method.

P7.12. **MATLAB Legendre series** Plot the function $f(x)$ and the Legendre approximation to $f(x)$ in Example 7.4.

EXPLORATION PROBLEMS

P7.13. **Vandermonde matrix** Construct the interpolation polynomial

$$\sum_{k=0}^{N} a_k x^k$$

for the data set $D = \{(x_i, y_i) | i = 0, 1, \ldots, N, \; x_i \neq x_j, \text{ for } i \neq j\}$ by forming the linear system $V\mathbf{a} = \mathbf{y}$. The vector $\mathbf{a}$ holds the coefficients to be determined and $\mathbf{y}$ contains the y values in the data set. The matrix V is called the *Vandermonde matrix*. Write the matrix in terms of the powers of the data points x_i.

For large values of N, the Vandermonde matrix tends to be ill-conditioned. Consider uniformly spaced data so that $x = x_0 + kh$ for $k = 0, 1, \ldots, N$. For the special case $x_0 = 1$ and $h = 1$, investigate the condition number of the Vandermonde matrix for $N = 3$ to $N = 15$.

Hint: The condition number was described in Chapter 3. The MATLAB command **vander** computes the Vandermonde matrix.

P7.14. MATLAB least-squares experiment Using the data from Example 7.3 that led to the least-squares straight line

$$f^*(x) = 0.3676 + 0.8784x,$$

compute the total error from the least squares approximation as defined in Equation 7.9. Then, modify the slope and the intercept of the straight line $f^*(x)$ by small amounts and recompute the error.

P7.15. Vector form of least squares Show that the least-squares straight-line approximation to the data points $(x_1, y_1), (x_2, y_2), \ldots, (x_n, y_n)$ can be written as $\mathbf{y} = A\mathbf{u}$, where

$$\mathbf{u} = \left[\begin{array}{c} b \\ m \end{array} \right]$$

holds the coefficients of the line $y = mx + b$ to be fit to the data. Then, the vector form of the least-squares problem can be stated as finding the vector $\mathbf{u}$ such that the Euclidean norm of $\| \mathbf{y} - A\mathbf{u} \|$ is minimized. Show that the solution can be written as

$$\mathbf{u} = \left(A^T A \right)^{-1} A^T \mathbf{y}.$$

ANNOTATED BIBLIOGRAPHY

1. Forsythe, George E., M. A. Malcolm, and C. B. Moler, *Computer Methods for Mathematical Computations*, Prentice Hall, Englewood Cliffs, NJ, 1977. *The text treats many aspects of interpolation, including spline functions and least-squares curve fitting.*

2. Hill, David R., *Experiments in Computational Matrix Algebra*, Random House, New York, 1988. *The text discusses the least-squares problem in some detail. It contains examples using MATLAB commands.*

3. Mathews, John M., *Numerical Methods for Mathematics, Science, and Engineering*, Prentice Hall, Englewood Cliffs, NJ, 1992. *This text covers many numerical algorithms for interpolation, splines, and least squares. There is a supplement that gives MATLAB implementation of many of these algorithms.*

ANSWERS

P7.1. Interpolation Notice that without additional information, there is no way to determine the accuracy of the various approximations. However, there is no reason to suppose that the true function varies a great deal near the endpoints, as is indicated by the polynomial approximation. This behavior is typical of polynomial interpolation using equally spaced points. The interpolating polynomial is $f(x) = -0.06x^6 + 0.8667x^4 - 3.2067x^2 + 3$. See the books listed in the Annotated Bibliography for more discussion of the errors in polynomial approximation.

P7.2. Least squares The spring constant is $k = 1.4$, found by a least squares fit of the line $f = -8.6 + 1.4x$.

P7.3. Least squares Assuming $f^*(x) = a_0 + a_1x + a_2x^2$ leads to the least-squares quadratic fit, with $a_0 = 2, a_1 = -0.5$, and $a_2 = 0.2$.

P7.4. Least squares $J_0(6.3) = 0.22381$. The linear approximation gives 0.21471. A quadratic approximation would yield 0.22282.

P7.5. Square error integral The mean-square error integral becomes

$E = \int_a^b f^2 \, dx - 2 \int_a^b [a_0\phi_0 + \cdots + a_n\phi_n]f \, dx + \int_a^b [a_0\phi_0 + \cdots + a_n\phi_n]^2 \, dx$. Letting $c_k = \int_a^b f\phi_k \, dx$,

the second integral becomes $\sum_{k=0}^n a_ic_i$. The third integral becomes

$$\int_a^b a_i^2\phi_i^2 \, dx + \text{ product terms in } \phi_i\phi_j = \sum_{k=0}^n a_i^2,$$

since the products $\phi_i\phi_j$ are zero for $i \neq j$.

P7.6. Fourier coefficients

a. As shown in the text, the Fourier coefficient is

$$c_{12} = \frac{\int_{t_1}^{t_2} f_1(t)f_2(t) \, dt}{\int_{t_1}^{t_2} f_2^2(t) \, dt}.$$

b. The least-squares error is

$$E = \int_{t_1}^{t_2} [f_1(t) - c_{12}f_2(t) - c_{13}f_3(t)]^2 \, dt,$$

and it is minimized with respect to c_{12} by setting the derivative $\partial E/\partial c_{12}$ to zero. The result is the same as in (a), so that adding other terms that are orthogonal does not change the previously calculated coefficients.

P7.7. Orthogonal functions The norm squared is
$\|f + g\|^2 = \langle f + g, g + g \rangle = \langle f, f \rangle + \langle g, f \rangle + \langle f, g \rangle + \langle g, g \rangle = \|f\|^2 + 0 + 0 + \|g\|^2$.

P7.8. Gram-Schmidt process To compute the orthonormal polynomials, choose

$$u_1(x) = \frac{1}{\langle 1, 1 \rangle^{1/2}} = \frac{1}{\sqrt{2}}.$$

Then,

$$y_2(x) = x - \langle x, \frac{1}{\sqrt{2}} \rangle \frac{1}{\sqrt{2}} = x,$$

yielding

$$u_2(x) = \frac{1}{\langle x, x \rangle^{1/2}} = \sqrt{\frac{3}{2}} x.$$

The function $u_3(x)$ is formed by creating an orthogonal function from
$y_3(x) = x^2 - \langle x^2, u_1(x) \rangle u_1(x) - \langle x^2, u_2(x) \rangle u_2(x) = x^2 - 1/3$, with the result

$$u_3(x) = \sqrt{\frac{5}{8}}(3x^2 - 1).$$

Comment: MATLAB solutions to many of these problems are on the disk accompanying this textbook.

8   *FOURIER ANALYSIS*

PREVIEW_____

The next two chapters are devoted primarily to the study of Fourier techniques. Collectively, these techniques form a branch of applied mathematics called *Fourier analysis*. The field of study is named for Jean Baptiste Fourier (1768–1830), who showed that any periodic function can be represented as the sum of sinusoids with integrally related frequencies. This observation leads to the study of *Fourier series*, which will be presented first in this chapter. The purpose of the Fourier series is to express a given function as a linear combination of sine and cosine *basis functions*. In many cases, the series is simpler to analyze than the original function. Most importantly for some applications, the components of the series allow physical interpretation of the function in terms of its *frequency spectrum*.

The *Fourier transform* provides an extension of Fourier series to the analysis of nonperiodic functions. As with the series, the point of the transform is to represent a function in a manner that is easier to analyze and understand. Properties and applications of this important transform will be considered in the chapter.

The present chapter concentrates on techniques and transforms that apply to a continuous function $f(t)$. These include Fourier series and Fourier transforms. In Chapter 9, the discrete Fourier transform for functions $f(t_i)$ defined at discrete points and other transforms are considered.

It may be helpful to review the sections on orthogonal functions in Chapter 2 and Chapter 7 before studying the details of Fourier series in this chapter.

FOURIER SERIES

In 1807, Fourier astounded many of his contemporary mathematicians and scientists by asserting that an arbitrary function could be expressed as a linear combination of sines and cosines. These linear combinations of the trigonometric sine and cosine functions, now called a *Fourier trigonometric series*, are applied to the analysis of *periodic* phenomena including vibrations and wave motion.

FOURIER SERIES FORMULA The Fourier series approximates a function $f(t)$ by using a *trigonometric polynomial* of degree N as follows:

$$f(t) \approx \frac{a_0}{2} + \sum_{n=1}^{N} [a_n \cos(nt) + b_n \sin(nt)] = s_N(t), \qquad (8.1)$$

where $s_N(t)$ denotes the nth partial sum. Assuming that $f(t)$ is continuous on the interval $-\pi \leq t \leq \pi$, the coefficients a_n and b_n can be computed by the formulas

$$a_0 = \frac{1}{\pi} \int_{-\pi}^{\pi} f(t) \, dt \qquad (8.2)$$

for the constant term and

$$a_n = \frac{1}{\pi} \int_{-\pi}^{\pi} f(t) \cos(nt) \, dt, \qquad b_n = \frac{1}{\pi} \int_{-\pi}^{\pi} f(t) \sin(nt) \, dt, \qquad (8.3)$$

for $n = 1, 2, \ldots, N$. If $f(t)$ is continuous on the interval and the derivative of $f(t)$ exists, the series converges to $f(t)$ at the point t when $N \rightarrow \infty$. Although the convergence properties of the series in Equation 8.1 will be discussed later, we will use the equality sign in Fourier series expansions, as is commonly done when the sum contains an infinite number of terms.

The series $s_N(t)$ is the *Fourier approximation* to the function $f(t)$ on the interval $[-\pi, \pi]$. From the periodicity of the trigonometric terms, it follows that

$$s_N(t + 2k\pi) = s_N(t) \qquad (8.4)$$

for all t and all integers k.

Notice that the constant term $a_0/2$ in the series of Equation 8.1 is the average value of $f(t)$ on the interval $-\pi \leq t \leq \pi$ since a_0 calculated by Equation 8.2 is twice the average value of $f(t)$ over the interval. The integrals in Equation 8.3 are twice the average value of $f(t)\cos(nt)$ and $f(t)\sin(nt)$, respectively. When the series is written using $a_0/2$ as the constant term, Equation 8.3 can be used for all the coefficients a_n by letting n vary from 0 to N.

☐ EXAMPLE 8.1　　*Fourier Series Example*

Consider the periodic function

$$f(t) = \begin{cases} 0, & -\pi < t < 0, \\ t, & 0 < t < \pi, \end{cases}$$

with period 2π, as shown in Figure 8.1.

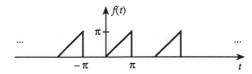

FIGURE 8.1 *Periodic function $f(t)$ for Example 8.1*

Using Equation 8.2, the Fourier coefficient a_0 is

$$a_0 = \frac{1}{\pi} \int_{-\pi}^{\pi} f(t)\, dt = \frac{1}{\pi} \int_{0}^{\pi} t\, dt = \frac{\pi}{2},$$

which yields a constant term of $a_0/2 = \pi/4$. The coefficients of the cosine terms are computed from Equation 8.3 as

$$\begin{aligned} a_n &= \frac{1}{\pi} \int_{-\pi}^{\pi} f(t)\cos(nt)\, dt \\ &= \frac{1}{\pi} \int_{-\pi}^{0} 0\cos(nt)\, dt + \frac{1}{\pi} \int_{0}^{\pi} t\cos(nt)\, dt. \end{aligned}$$

Integrating by parts yields

$$a_n = \frac{1}{\pi} \left[\frac{t}{n}\sin(nt) + \frac{1}{n^2}\cos(nt) \right]_0^{\pi} = \frac{1}{\pi n^2}\left[\cos(n\pi) - 1 \right] \qquad \textbf{(8.5)}$$

for $n = 1, 2, \ldots$.

Similarly, the sine terms are computed as

$$\begin{aligned} b_n &= \frac{1}{\pi} \int_{-\pi}^{\pi} f(t)\sin(nt)\, dt \\ &= \frac{1}{\pi} \int_{-\pi}^{0} 0\sin(nt)\, dt + \frac{1}{\pi} \int_{0}^{\pi} t\sin(nt)\, dt, \end{aligned}$$

which yields

$$b_n = \frac{1}{\pi}\left[-\frac{t}{n}\cos(nt) + \frac{1}{n^2}\sin(nt)\right]_0^\pi = -\frac{1}{n}[\cos(n\pi)] \qquad \textbf{(8.6)}$$

for $n = 1, 2, \ldots$.

The series approximation can be rewritten using the identity relationship $\cos(n\pi) = (-1)^n$ and noticing that $a_n = 0$ when n is an even integer. The result is

$$f(t) = \frac{\pi}{4} - \frac{2}{\pi}\sum_{n=1}^{\infty}\frac{\cos(2n-1)t}{(2n-1)^2} - \sum_{n=1}^{\infty}(-1)^n\frac{\sin(nt)}{n}, \qquad \textbf{(8.7)}$$

where the $(2n-1)$ is introduced in the first sum to assure that only odd terms are included in that summation.

Writing out a few terms yields the series for $f(t)$ as

$$f(t) = \frac{\pi}{4} \quad - \quad \frac{2}{\pi}\cos(t) - \frac{2}{9\pi}\cos(3t) - \cdots$$

$$+ \quad \sin(t) - \frac{1}{2}\sin(2t) + \frac{1}{3}\sin(3t) + \cdots, \qquad \textbf{(8.8)}$$

where the equality holds at points at which $f(t)$ is continuous. At points of discontinuity $t = n\pi$, the series converges to $\pi/2$, as explained later.

□

□ EXAMPLE 8.2   *MATLAB Fourier Series Example*

The accompanying MATLAB script and plots show the Fourier approximation to the function of Example 8.1 for 5 and 20 terms of the series.

MATLAB Script ‗‗

```
Example 8.2
% EX8_2.M Plot the Fourier series of the function f(t)
%    f(t)=0  -pi < t < 0
%    f(t)=t   0  < t < pi
% Plot f(t) for 5 and 20 terms in the series
clear
t =[-pi:.031:pi];                % Time points for plotting
sizet=size(t);
fn = pi/4*(ones(sizet));         % Fourier approximation at each t
yplt=zeros(sizet);               %  for plot of f(t)
% 5 terms
for n=1:5
 fn=fn+ (1/pi)*(-2*cos((2*n-1)*t)/(2*n-1)^2)-((-1)^n*sin(n*t)/n);
end
%
for k=1:length(t)                % Create f(t)
  if t(k) < 0
   yplt(k)=0;
  else
   yplt(k)=t(k);
  end
end
```

```
clf                              % Clear any figures
subplot(2,1,1),plot(t,fn,t,yplt,'--');
xlabel('t')
ylabel('f(t)')
title('Fourier series approximation to f(t) - Figure 8.2')
legend(['N=',num2str(n)],'f(t)') % Annotate the graph
% Add 15 more terms
for n=6:20
 fn=fn+ (1/pi)*(-2*cos((2*n-1)*t)/(2*n-1)^2)-((-1)^n*sin(n*t)/n);
end
subplot(2,1,2),plot(t,fn,t,yplt,'--');
xlabel('t')
ylabel('f(t)')
legend(['N=',num2str(n)],'f(t)')
```

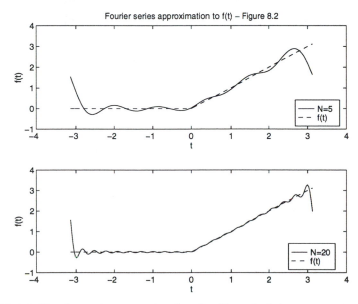

FIGURE 8.2 *Fourier series approximation for Example 8.1*

□

Five terms shown in the first plot of Figure 8.2 do not give a very good approximation, particularly near the endpoints of the function. The second plot showing a 20-term series is much closer to the actual function. Both approximations are converging to $\pi/2$ at the endpoints.

ORTHOGONAL FOURIER TERMS From the discussion of orthogonal functions in Chapter 2, Chapter 6, and Chapter 7, the integrals for the Fourier coefficients in Equation 8.3 can be viewed as the normalized inner product of the function $f(t)$ and the sine and cosine trigonometric functions.

Recall that in Chapter 2, the *inner product* of the trigonometric functions on the interval $[-\pi, \pi]$ was defined as

$$\langle f, g \rangle = \frac{1}{\pi} \int_{-\pi}^{\pi} f(t)g(t)\, dt, \tag{8.9}$$

where the factor $1/\pi$ was introduced to normalize the inner product for the Fourier trigonometric functions.

Then, the trigonometric terms in the Fourier series consist of functions that form an orthonormal set, since for integers k and m

$$\langle \cos(kt), \cos(mt) \rangle = \begin{cases} 1, & k = m \neq 0, \\ 0, & k \neq m, \end{cases}$$

$$\langle \sin(kt), \sin(mt) \rangle = \begin{cases} 1, & k = m \neq 0, \\ 0, & k \neq m, \end{cases}$$

$$\langle \cos(kt), \sin(mt) \rangle = 0, \quad \text{for all } k, m. \tag{8.10}$$

Thus, the Fourier coefficients in the expansion of a function $f(t)$ from Equation 8.3 can be written as

$$\begin{aligned} a_k &= \langle f(t), \cos(kt) \rangle, & k = 0, 1, \ldots, \\ b_k &= \langle f(t), \sin(kt) \rangle, & k = 1, 2, \ldots. \end{aligned} \tag{8.11}$$

Notice that the constant term a_0 is computed as

$$a_0 = \langle f(t), 1 \rangle = \frac{1}{\pi} \int_{-\pi}^{\pi} f(t)\, dt, \tag{8.12}$$

which is the inner product of $f(t)$ and the $\cos(kt)$ term in Equation 8.11 for $k = 0$.

EVEN AND ODD FUNCTIONS

The Fourier approximation of even and odd functions can be computed with significantly less effort than that needed for functions without such symmetry. The properties that define an even or odd function are as follows:

1. An *even* function has a graph that is symmetric with respect to the vertical axis $(t = 0)$ and satisfies the equation

$$f(-t) = f(t). \tag{8.13}$$

2. An *odd* function is symmetric with respect to the origin and satisfies the equation

$$f(-t) = -f(t). \tag{8.14}$$

For the *even* function $f_e(t)$, a range of integration that is symmetrical about the vertical axis, where $t = 0$, yields the result

$$\int_{-\pi}^{\pi} f_e(t)\, dt = 2 \int_{0}^{\pi} f_e(t)\, dt.$$

The integral over a symmetrical range about the vertical axis for an *odd* function $f_o(t)$ is zero; that is,

$$\int_{-\pi}^{\pi} f_o(t)\, dt = 0.$$

Based on these results, if $f(t)$ is an *even function*,

$$f(t) = \frac{a_0}{2} + \sum_{n=1}^{\infty} [a_n \cos(nt)], \qquad (8.15)$$

where

$$a_0 = \frac{2}{\pi} \int_{0}^{\pi} f(t)\, dt, \qquad (8.16)$$

$$a_n = \frac{2}{\pi} \int_{0}^{\pi} f(t) \cos(nt)\, dt. \qquad (8.17)$$

If $f(t)$ is an *odd function*,

$$f(t) = \sum_{n=1}^{\infty} [b_n \sin(nt)], \qquad (8.18)$$

where

$$b_n = \frac{2}{\pi} \int_{0}^{\pi} f(t) \sin(nt)\, dt. \qquad (8.19)$$

☐ **EXAMPLE 8.3**   *Fourier Series Odd Function Example*

Consider the Fourier series for the odd periodic function

$$f(t) = t, \qquad -\pi < t < \pi,$$

shown in Figure 8.3. Each term $a_i = 0$, since there is no constant term or terms in $\cos(nt)$.

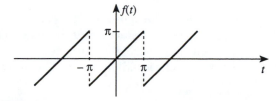

FIGURE 8.3 *Odd periodic function for Example 8.3*

Applying Equation 8.19, the coefficients of the sine terms are

$$b_n = \frac{2}{\pi} \int_0^\pi t \sin(nt)\, dt, \quad n = 1, 2, \ldots,$$

which can be integrated by parts to yield

$$b_n = \frac{2}{\pi} \left[\frac{-t}{n} \cos(nt) + \frac{1}{n^2} \sin(nt) \right]_0^\pi = -\frac{2}{n} \cos(n\pi).$$

Using the fact that $\cos(n\pi) = (-1)^n$, the resulting series can be written as

$$f(t) = -2 \sum_{n=1}^\infty \frac{(-1)^n}{n} \sin(nt).$$

The result is a series of sine terms as expected for the approximation of an odd function. Note that $f(0) = 0$, as required, and $f(n\pi) = 0$ for any integer n. At the discontinuities, the series converges to the midpoint.

$\square$

[-T/2, T/2]
INTERVAL

On the interval $[-T/2, T/2]$, the limits of integration for the Fourier series can be changed from $[-\pi, \pi]$ by assigning to the integration variable t the value $2\pi t/T$. The period of the function is thus T.

Assuming that $f(t)$ is continuous on the interval $-T/2 \le t \le T/2$, the coefficients a_n and b_n can be computed by the formulas

$$a_0 = \frac{2}{T} \int_{-T/2}^{T/2} f(t)\, dt,$$

$$a_n = \frac{2}{T} \int_{-T/2}^{T/2} f(t) \cos\left(\frac{2n\pi t}{T}\right) dt,$$

$$b_n = \frac{2}{T} \int_{-T/2}^{T/2} f(t) \sin\left(\frac{2n\pi t}{T}\right) dt, \qquad \text{(8.20)}$$

where $n = 1, 2, \ldots$ is any positive integer.

The Fourier series on the interval $[-T/2, T/2]$ is thus

$$f(t) = \frac{a_0}{2} + \sum_{n=1}^\infty \left[a_n \cos\left(\frac{2n\pi t}{T}\right) + b_n \sin\left(\frac{2n\pi t}{T}\right) \right]. \qquad \text{(8.21)}$$

Frequency Components   Assuming the variable t represents time, the function $f(t)$ repeats every T seconds. The *frequency* associated with the fundamental sinusoid in the series of Equation 8.21 is $f_0 = 1/T$, measured in cycles per second, or hertz. The parameter

$$\omega_0 = 2\pi f_0 = \frac{2\pi}{T}$$

is the frequency in radians per second.

Since $2n\pi/T = 2n\pi f_0 = n\omega_0$, the series in Equation 8.21 can be written

$$
\begin{aligned}
f(t) &= \frac{a_0}{2} + \sum_{n=1}^{\infty} [a_n \cos(2\pi n f_0 t) + b_n \sin(2\pi n f_0 t)] \\
&= \frac{a_0}{2} + \sum_{n=1}^{\infty} [a_n \cos(n\omega_0 t) + b_n \sin(n\omega_0 t)], \qquad \text{(8.22)}
\end{aligned}
$$

which emphasizes the components in terms of their frequencies.

The first term in cosine or sine is called the *fundamental* component, and the other terms are the *harmonics* with frequencies that are integer multiples of the fundamental component's frequency. Thus, the frequencies of the Fourier series terms are

$$
f_0, 2f_0, 3f_0, \ldots,
$$

although some of the components may be zero for a particular Fourier series. However, $f(t)$ is a continuous function of time, and this aspect of the Fourier series is emphasized when the series is used to approximate $f(t)$. In other applications, the frequencies of the components are of primary interest.

Sometimes a function of a spatial variable x is of interest. If the function has period λ meters, the function repeats as

$$
f(x + \lambda) = f(x).
$$

Then, the variable t in Equation 8.22 is replaced by x, and the frequency components are defined by replacing f_0 with $1/\lambda$. The spatial equivalent of ω_0 is

$$
k = \frac{2\pi}{\lambda},
$$

measured in inverse units of length. Such a formulation of Fourier series is used frequently in problems involving optics. In optical applications, the values nk are called *spatial frequencies*. Thus, λ represents the wavelength of the light wave being analyzed.

□ EXAMPLE 8.4   *Fourier series square wave example*

A square wave of amplitude A and period T shown in Figure 8.4 can be defined as

$$
f(t) = \begin{cases} A, & 0 < t < \dfrac{T}{2}, \\ -A, & -\dfrac{T}{2} < t < 0, \end{cases}
$$

with $f(t) = f(t + T)$, since the function is periodic.

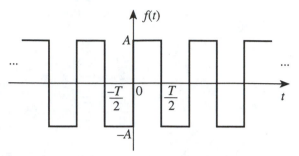

FIGURE 8.4 *Square wave of Example 8.4*

The first observation is that $f(t)$ is odd, which yields the result that $a_0 = 0$ and $a_i = 0$ for every coefficient of the cosine terms. Letting $\omega_0 = 2\pi/T$, the coefficients b_n are

$$b_n = 2 \left(\frac{2}{T}\right) \int_0^{T/2} A \sin(n\omega_0 t) \, dt.$$

The result is

$$f(t) = \frac{4A}{\pi} \sum_{n=1}^{\infty} \frac{\sin[(2n-1)\omega_0 t]}{(2n-1)},$$

where $(2n-1)$ is introduced to assure that only odd terms are included in the summation. The sine waves that make up the Fourier series for the odd square wave are

$$f(t) = \frac{4A}{\pi} \left[\sin(\omega_0 t) + \frac{\sin(3\omega_0 t)}{3} + \cdots\right],$$

so the series consists not only of sine terms, as expected, but also odd harmonics appear. This is due to the rotational symmetry of the function since the wave shapes on alternate half-cycles are identical in shape but reversed in sign. Such waveforms are produced in certain types of rotating electrical machinery.

□

SYMMETRIES

Symmetry in a function $f(t)$ should be exploited to reduce the computational effort of finding the Fourier coefficients. Generally, the symmetry exists either about a vertical line or a horizontal line. Several types of symmetry are presented in Table 8.1. Even and odd symmetry were discussed previously. Rotational symmetry exists about the zero axis, and the waveshape of alternate half-cycles is identical but reversed in sign. This symmetry is also called *half-wave symmetry* since the integrals for the Fourier coefficients need be taken over only half a period.

There is no constant term when rotational symmetry exists. If the function is also odd, only odd-harmonic sine terms will appear in the Fourier series, as was the case for the odd square wave of Example 8.4. An even function with rotational symmetry will have a Fourier series consisting of odd-harmonic cosine terms.

TABLE 8.1 *Symmetry of the function $f(t)$*

Symmetry	Fourier Series
Even function $f(-t) = f(t)$	Cosine terms only
Odd function $f(-t) = -f(t)$	Sine terms only
Rotational $f(t) = -f(-t + T/2)$	Odd harmonics only

COMPLEX SERIES

The Fourier series of Equation 8.1 contains a series of sines and cosines and thus involves real functions. It is often convenient to write the series for a function $f(t)$ with period T as a sum of exponential functions in the form

$$f(t) = \sum_{n=-\infty}^{\infty} \alpha_n e^{in\omega_0 t}, \tag{8.23}$$

where $\omega_0 = 2\pi/T$ as before and the coefficients α_n are the complex Fourier coefficients.

By substituting the identities

$$\cos(n\omega_0 t) = \frac{e^{in\omega_0 t} + e^{-in\omega_0 t}}{2},$$

$$\sin(n\omega_0 t) = \frac{e^{in\omega_0 t} - e^{-in\omega_0 t}}{2i}, \tag{8.24}$$

in the trigonometric form of the series in Equation 8.21, the relationship between the trigonometric and exponential coefficients is found to be

$$\alpha_0 = \frac{a_0}{2},$$

$$\alpha_n = \frac{a_n - ib_n}{2} \quad \text{for } n > 0,$$

$$\alpha_{-n} = \frac{a_n + ib_n}{2}. \tag{8.25}$$

Notice that α_{-n} is the complex conjugate of the term α_n.[1] Thus, the series in Equation 8.23 becomes

$$f(t) = \alpha_0 + \sum_{n=1}^{\infty} [\alpha_n e^{in\omega_0 t} + \alpha_{-n} e^{-in\omega_0 t}]. \tag{8.26}$$

---

[1]These results hold when $f(t)$ is a real-valued function. For complex-valued functions, the real and imaginary parts can be treated separately as real functions.

Orthogonality   To find the coefficients in Equation 8.23, each side is multiplied by $e^{-im\omega_0 t}$ and integrated over the period to yield

$$\int_{-T/2}^{T/2} f(t)e^{-im\omega_0 t}\, dt = \sum_{n=-\infty}^{\infty} \alpha_n \int_{-T/2}^{T/2} e^{i(n-m)\omega_0 t}\, dt. \qquad (8.27)$$

Since the terms with different exponents are orthogonal, all terms but that for which $m = n$ are zero for the integral on the right-hand side. Thus,

$$\int_{-T/2}^{T/2} f(t)e^{-im\omega_0 t}\, dt = \int_{-T/2}^{T/2} e^{-in\omega_0 t}e^{in\omega_0 t}\, dt = \alpha_n T,$$

so that dividing both sides T yields the coefficients

$$\alpha_n = \frac{1}{T}\int_{-T/2}^{T/2} f(t)e^{-in\omega_0 t}\, dt. \qquad (8.28)$$

□ EXAMPLE 8.5   ***Complex Series Square Wave Example***

Consider the odd square wave of Example 8.4 and the complex Fourier coefficients

$$\alpha_n = \frac{1}{T}\int_{-T/2}^{0}(-A)e^{-in\omega_0 t}\, dt + \frac{1}{T}\int_{0}^{T/2}(A)e^{-in\omega_0 t}\, dt, \qquad (8.29)$$

which leads to the series

$$f(t) = \frac{2A}{i\pi}\sum_{n=-\infty}^{\infty}\frac{e^{i(2n-1)\omega_0 t}}{(2n-1)}, \qquad (8.30)$$

as defined in Equation 8.23.

This form contains complex coefficients, but the series can be written in terms of sine waves by combining the corresponding terms for positive and negative arguments. To determine the coefficients, the amount of difficulty is about the same for the trigonometric series and the complex series. However, the complex series perhaps has an advantage when the magnitude of the coefficients are of interest.

Each coefficient has the form

$$\alpha_n = \frac{2A}{in\pi} = \frac{2A}{n\pi}e^{-i\pi/2}, \qquad n = \pm 1, \pm 3, \ldots,$$

and the coefficients for even values, $n = 0, \pm 2, \ldots$, are zero. Notice that the coefficients decrease as the index n increases. The use of these coefficients to compute the *frequency spectrum* of $f(t)$ is considered later.

The trigonometric series is derived from the complex series by expanding the complex series of Equation 8.30 as

$$\begin{aligned}
f(t) &= \sum_{n=-\infty}^{\infty}\alpha_n e^{in\omega_0 t} \\
&= \cdots - \frac{2A}{3\pi i}e^{-i3\omega_0 t} - \frac{2A}{\pi i}e^{-i\omega_0 t} + \frac{2A}{\pi i}e^{i\omega_0 t} + \frac{2A}{3\pi i}e^{i3\omega_0 t} + \cdots
\end{aligned}$$

and recognizing the sum of negative and positive terms for each n as $2\sin(n\omega_0 t)$. The trigonometric series becomes

$$f(t) = \frac{4A}{\pi}\left(\sin(\omega_0 t) + \frac{\sin(3\omega_0 t)}{3} + \cdots\right) = \frac{4A}{\pi}\sum_{n=1}^{\infty}\frac{\sin[(2n-1)\omega_0 t]}{(2n-1)},$$

which is the result of Example 8.4.

□

FOURIER SPECTRUM FOR PERIODIC FUNCTIONS

In general, the Fourier series of a periodic function with period T seconds contains the fundamental sinusoid and numerous harmonics some of which may be zero. The plot of the magnitude of the frequency components is called the *amplitude*, or *frequency spectrum*. The frequency components are spaced $f_0 = 1/T$ hertz apart. On a graph, the spectrum is a series of points (or lines) and is called a *discrete spectrum*. A discrete Fourier spectrum is characteristic of all periodic functions.

Spectrum of Trigonometric Series An alternative method of expressing the trigonometric Fourier series of Equation 8.22 is to write the series with terms of the form

$$c_n\cos(2\pi n f_0 t + \theta_n), \tag{8.31}$$

which we will call a *shifted cosine series* due to the phase shift θ_n in each term. To derive the relationship between the coefficients c_n and the coefficients of the complete cosine and sine series, we set the nth term in the cosine expansion equal to the nth component of the original series as follows:

$$c_n\cos(nt + \theta_n) = a_n\cos(nt) + b_n\sin(nt), \tag{8.32}$$

where a_n, b_n, and θ_n are known.

Using the identity

$$c_n\cos(nt + \theta_n) = c_n\cos(nt)\cos(\theta_n) - c_n\sin(nt)\sin(\theta_n), \tag{8.33}$$

and expanding the left-hand side of Equation 8.32 leads to the relationships,

$$\begin{aligned}c_n\cos(\theta_n) &= a_n,\\ c_n\sin(\theta_n) &= -b_n,\end{aligned}$$

to be solved for c_n and θ_n. Squaring these equations and adding them together yields the solution for c_n, and taking the ratio determines θ_n. The result is

$$c_n = \sqrt{a_n^2 + b_n^2} \quad\text{and}\quad \theta_n = \tan^{-1}\left(-\frac{b_n}{a_n}\right).$$

Notice that the sign of the argument of the tangent is negative.

In terms of the fundamental frequency f_0, the shifted cosine series equivalent to Equation 8.22 can be written

$$f(t) = \frac{a_0}{2} + \sum_{n=1}^{\infty} [c_n \cos(2\pi n f_0 t + \theta_n)], \tag{8.34}$$

in which the numerical value of c_n is the *amplitude* and the angle θ_n is the *phase* of the nth harmonic. In many problems, physical units are associated with the coefficients. For example, if $f(t)$ represents a voltage wave that varies with time, the components yield the values of the voltages at each frequency that must be combined to reproduce $f(t)$. The phase indicates how the individual components must be shifted from the $t = 0$ axis before they are combined.

When plotted versus frequency, the magnitudes of the constant $(a_0/2)$ and the set of harmonic amplitudes, $c_n, n = 1, 2, \ldots$, are referred to as the *amplitude spectrum*, and the plot of phase shifts (θ_n) is the *phase spectrum*. The constant $a_0/2$ corresponding to zero frequency is often called the dc (direct current) component in problems involving electrical circuits.

Spectrum of Complex Series The complex series presented as Equation 8.23,

$$f(t) = \sum_{n=-\infty}^{\infty} \alpha_n e^{in\omega_0 t} = \sum_{n=-\infty}^{\infty} \alpha_n e^{i2\pi n f_0 t},$$

has coefficients α_n that are in general complex. Thus, the actual real sinusoidal terms are composed of terms that have both positive and "negative" frequency components. This use of the negative frequencies should not be disturbing and is given no physical interpretation. After all, the basis functions as complex exponentials have no physical meaning either. The spectrum is usually plotted as α_n and θ_n versus n or frequency $n f_0$ for both negative and positive values of n.

Since the complex series coefficients are complex, they can be written

$$
\begin{aligned}
\alpha_n &= |\alpha_n| e^{i\theta_n}, \\
\alpha_n^* &= |\alpha_n| e^{-i\theta_n}.
\end{aligned}
$$

The sum of two terms for the nth harmonic yields the term f_n as

$$
\begin{aligned}
f_n &= |\alpha_n| e^{i\theta_n} e^{in\omega_0 t} + |\alpha_n| e^{-i\theta_n} e^{-in\omega_0 t} \\
&= |\alpha_n| \left[e^{i(n\omega_0 t + \theta_n)} + e^{-i(n\omega_0 t + \theta_n)} \right] \\
&= 2 |\alpha_n| \cos(2\pi n f_0 t + \theta_n).
\end{aligned}
$$

Comparing this result with the coefficients for the shifted cosine series of Equation 8.34 shows that $c_n = 2 |\alpha_n|$ if the complex series is to be converted to the shifted cosine series. Thus, the one-sided amplitude spectrum ($n = 1, 2, \ldots$) for the complex series consists of the terms $2\,|\alpha_n|$. The phase is given by the terms θ_n just as for the trigonometric series. If both positive and negative frequency components are shown, the terms $|\alpha_n|$ are plotted. The zero frequency term is the same for any representation.

Real Functions If $f(t)$ is a real and even function of t, the coefficients α_n are real and even functions of n. If $f(t)$ is a real and odd function of t, then the coefficients α_n are imaginary and odd functions of n. In either case, plotting $|\alpha_n|$ results in a real and even discrete frequency spectrum for the amplitudes.

Power in the signal It is an important result of alternating current theory that the power associated with a periodic wave of voltage or current $f(t)$ is proportional to the mean-square value of $f(t)$. The mean-square formula is

$$\overline{f^2(t)} = \frac{1}{T} \int_{-T/2}^{T/2} [f(t)]^2 \, dt, \qquad (8.35)$$

which is seen to be the average of the square of $f(t)$. For a pure sinusoid, the average value of its square is one-half the peak value. Thus, for the nth harmonic $c_n \cos(n\omega_0 t)$, the result is $c_n^2/2$.

Applying the mean-square formula to any of the Fourier series representations leads to the power spectrum for the function. The power is computed by squaring the appropriate series and dividing the integral over the period by the period itself to compute the mean-square value of each component. All the cross terms average to zero since the trigonometric functions are orthogonal.

The average power in the time signal must equal the power computed by the Fourier series. A rigorous statement of this fact is called Parseval's theorem. This important result was presented in Chapter 7.

Comment: If the function $f(t)$ represents a voltage signal (volts) or a current (amperes), the power is not strictly given by Equation 8.35 when the signal is applied to a circuit. For example, the average power dissipated in a resistor of R ohms would be $\overline{P(t)} = \overline{f(t)^2}/R$ watts when $f(t)$ represents the voltage across the resistor. Generally, when no confusion would result, the power in a periodic signal is considered to be given by Equation 8.35.

For a periodic signal $f(t)$ with period T, the various forms of the Fourier series and the power associated with the signal are shown in Table 8.2.

TABLE 8.2 *Fourier Series Representation*

Series	*Power*
Sine and cosine series:	
$\dfrac{a_0}{2} + \displaystyle\sum_{n=1}^{\infty} [a_n \cos(2n\pi f_0 t) + b_n \sin(2n\pi f_0 t)]$	$\left(\dfrac{a_0}{2}\right)^2 + \dfrac{1}{2}\displaystyle\sum_{n=1}^{\infty}(a_n^2 + b_n^2)$
where	
$a_n = \dfrac{2}{T}\displaystyle\int_{-T/2}^{T/2} f(t)\cos(2n\pi f_0 t)\,dt$	
$b_n = \dfrac{2}{T}\displaystyle\int_{-T/2}^{T/2} f(t)\sin(2n\pi f_0 t)\,dt$	
Shifted cosine:	
$\dfrac{a_0}{2} + \displaystyle\sum_{n=1}^{\infty} c_n \cos(2\pi n f_0 t + \theta_n)$	$\left(\dfrac{a_0}{2}\right)^2 + \dfrac{1}{2}\displaystyle\sum_{n=1}^{\infty} c_n^2$
where	
$c_n = \sqrt{a_n^2 + b_n^2}, \quad \theta_n = \tan^{-1}\left(-\dfrac{b_n}{a_n}\right)$	
Complex series:	
$\displaystyle\sum_{n=-\infty}^{\infty} \alpha_n e^{i2\pi n f_0 t}$	$\displaystyle\sum_{n=-\infty}^{\infty} |\alpha_n|^2$
where	
$\alpha_n = \dfrac{1}{T}\displaystyle\int_{-T/2}^{T/2} f(t) e^{-i2n\pi f_0 t}\,dt$	

Comparing the power relations in the table, the coefficients of the shifted cosine series are related to those for the sine and cosine series as

$$c_n^2 = a_n^2 + b_n^2,$$

with $c_0 = a_0/2$. The complex series coefficients are related to the coefficients of the trigonometric series as

$$\alpha_n^2 = \frac{1}{2}(a_n^2 + b_n^2)$$

with $\alpha_0 = a_0/2$.

Chapter 8 ■ FOURIER ANALYSIS

MATLAB Fourier Spectrum Example

The spectrum for the square wave of Figure 8.4 can be derived from the complex series of Equation 8.30 in Example 8.5,

$$f(t) = \frac{2A}{i\pi} \sum_{n=-\infty}^{\infty} \frac{e^{i(2n-1)\omega_0 t}}{(2n-1)}.$$

The coefficients of the series consists of the terms

$$\alpha_n = \frac{2A}{n\pi} e^{-i\pi/2}, \quad n = \pm 1, \pm 3,$$

since $i^{-1} = e^{-i\pi/2}$. Thus, the amplitude is $(2A/n\pi)$ with phase $+90°$ for each negative frequency component and $-90°$ for each positive frequency components. Figure 8.5 shows the one-sided amplitude spectrum as computed by the M-function CLSPEC1.M. This MATLAB script creates the spectrum. To specify the power in the signal, the series

$$P = |\alpha_0|^2 + 2 \sum_{n=1}^{\infty} |\alpha_n|^2$$

would be computed. The function `clptdscf` actually plots the line spectrum.

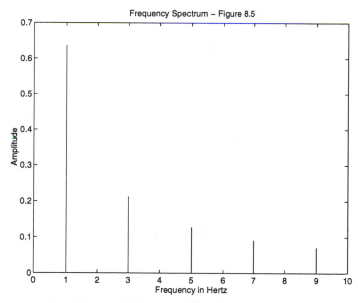

FIGURE 8.5 *Spectrum of square wave of Example 8.4*

The MATLAB command **stem** also plots discrete data sequences. We created `clptdscf` because it gives more control over the output format.

MATLAB Script

Example 8.6

```
% CLSPEC1.M  Plot positive frequency spectrum of square wave
%  The components are 2/(n pi); n odd.
%  Plot 10 components of the discrete spectrum [f F]
%   by calling function clptdscf
%
clear
xunit='Hertz';           % Units of frequency
f=[0:1:10];              % Frequency scale
Fn=zeros(1,11);          % Row vector of 11 elements
% Frequency spectrum
for n=1:5 % Compute 5 positive components
 Fn(2*n-1)=2/((2*n-1)*pi);
end
Fn=[0 Fn];               % Add the zero value
%
clptdscf(f,Fn,xunit)     % Call for plot
```

The function clptdscf plots a discrete function (f,F) in f units specified by the input xunit and the units will be displayed. The title of the graph must be input from the keyboard when the function executes.

MATLAB Script

Example 8.6

```
function clptdscf(f,F,xunit)
% CALL: clptdscf(f,F,xunit) Plot a discrete spectrum [f F]
%  Input to function is
% f  - frequencies
% F  - spectral values
% xunit - units of frequency (Hz or rad/sec)
%  Input title of graph from keyboard
nl=length(f);            % Number of f points
fmin=min(f);             %  and range
fmax=max(f);
Fmax=max(F);
% Plotting range, lengthen axes by 10%
Fmaxp=Fmax+.1*Fmax;
fminp=fmin-.1*fmax;
fmaxp=fmax+.1*fmax;
%
title1=input('Title ', 's' );
clf                      % Clear any figures
axis([fminp fmaxp 0 Fmaxp]) % Manual scaling
for I=1:nl,
 fplots=[f(I) f(I)];
 Fplot=[0 F(I)];
 plot(fplots,Fplot)      % Plot one line at a time
hold on
end
```

```
title(title1)
ylabel('Amplitude')
xlabel(['Frequency in ', xunit])
```

□

□ **EXAMPLE 8.7**   *Fourier spectrum*

Consider the even, periodic pulse train in Figure 8.6. This is an important test signal in electronics and the signal is also of interest because of the characteristics of its Fourier components. The period is T, the amplitude is A, and the pulse has duration τ in each period. The function is an even function with average value over a period of $A\tau/T$.

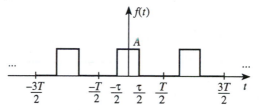

FIGURE 8.6 *Periodic train of rectangular pulses*

Letting $\omega_0 = 2\pi/T$, the coefficients of the complex series are

$$\alpha_n = \frac{1}{T}\int_{-T/2}^{T/2} Ae^{-in\omega_0 t}\, dt = \frac{1}{T}\int_{-\tau/2}^{\tau/2} Ae^{-in\omega_0 t}\, dt.$$

Integrating and substituting $-2\sin(n\omega_0\tau/2)$ for the resulting exponentials and then multiplying and dividing by the term $\omega_0\tau/2$ yields

$$\alpha_n = \frac{A\tau}{T}\frac{\sin(n\omega_0\tau/2)}{n\omega_0\tau/2}.$$

The coefficient α_0 is determined as $A\tau/T$ by l'Hôpital's rule. Notice that the coefficient

$$\alpha_0 = \frac{A\tau}{T} = \frac{\text{area of pulse}}{\text{period}}.$$

Defining the function

$$\operatorname{sinc} x \equiv \frac{\sin x}{x} \tag{8.36}$$

with $x = n\omega_0\tau/2$ leads to the series

$$f(t) = \frac{A\tau}{T} + 2\frac{A\tau}{T}\sum_{n=1}^{\infty}\operatorname{sinc}(n\omega_0\tau/2)\cos(n\omega_0 t)$$

since the coefficients of the cosine series are twice the values of those in the complex series.

Comment: The sinc function is frequently defined as

$$\operatorname{sinc} t = \frac{\sin \pi t}{\pi t}.$$

This is the convention used in the MATLAB *Signal and Systems Toolbox* for the function **sinc**. Letting $t = x/\pi$ converts the MATLAB version to the form in Equation 8.36.

The sinc function determines the magnitude of the frequency components when it is evaluated at the points $n\omega_0\tau/2$. This function will be studied later as it plays an important role in Fourier analysis using Fourier transforms.

As a specific example, let $A = 1$, $\omega_0 = 1$ radian/second and $\tau = \pi/2$ seconds for the pulse train of Figure 8.6. Thus, the period $T = 2\pi$ seconds and $A\tau/T = 1/4$. The accompanying MATLAB script computes and plots the spectrum and the Fourier series representation of pulse train for 20 components in the series.

MATLAB Script _____

```
Example 8.7
% CLSPEC2.M Plot the positive amplitude spectrum and the
%  Fourier series representation for the pulse train with
%  A=1, period T=2*pi, and pulse width tau=pi/2.
n=1:10;                         % Number of components
Wn=zeros(size(n));
Wn=2*(1/pi)*(sin(n*pi/4)./n);   % Frequency spectrum n=1,2,...
Wn=[1/4,Wn];                    % Add dc term
n=[0,n];
%
t=[-3*pi:.02:3*pi];             % Range of t
f=zeros(size(t));
for k=1:1:20;                            % f(t) with 20 terms
f=f+(2/(k*pi))*sin(k*pi/4)*cos(k*t); % in series
end
f=1/4+f;                        % Add dc value A*tau/T
% Put in a zero line and plot frequency
fzero=zeros(size(n));
clf                             % Clear any figures
subplot(2,1,1),plot(n,Wn,'*',n,fzero,'-');
xlabel('w radians per second')
ylabel('(2/pi)*sinc(n*pi/4)')
title('Fourier Series of Pulse Train - Figure 8.7')
% Plot f(t)
subplot(2,1,2),plot(t,f)
xlabel('t time in seconds')
ylabel('f(t)')
```

_____

Figure 8.7 shows the Fourier series spectrum for positive frequencies and the approximation to the pulse train using 20 terms. If a two-sided spectrum were plotted, the terms $\alpha_1, \ldots, \alpha_n, \ldots$, would be divided by 2.

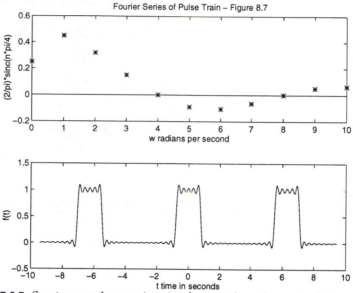

FIGURE 8.7 *Spectrum and approximation for periodic pulse train of Example 8.7*

You should analyze the effect on the spectrum of varying T and the effect on the reconstructed $f(t)$ of varying the number of terms in the series.

□

PROPERTIES OF FOURIER SERIES

This section treats properties of Fourier series that are useful for practical and theoretical purposes. Topics include the time shift property, conditions for convergence, and the Gibbs phenomenon. Theorems define the relationship between the Fourier series and its derivative and integral.

TIME SHIFT

For a periodic function $f(t)$ with period T, the complex Fourier series is

$$f(t) = \sum_{n=-\infty}^{\infty} \alpha_n e^{in\omega_0 t}, \tag{8.37}$$

with $\omega_0 = 2\pi/T$. If the function is shifted to the right along the t-axis by an amount $\tau > 0$, the shifted function can be represented as

$$g(t) = f(t - \tau). \tag{8.38}$$

Thus, a shifted version of $f(t)$ is obtained by substituting the variable $(t - \tau)$ for t. Let the coefficients of the Fourier series expansion of $g(t)$ be $\hat{\alpha}_n$, so that

$$g(t) = \sum_{n=-\infty}^{\infty} \hat{\alpha}_n e^{in\omega_0 t}.$$

With the substitution of $t - \tau$, Equation 8.37 becomes

$$g(t) = f(t - \tau) = \sum_{n=-\infty}^{\infty} \alpha_n e^{in\omega_0(t-\tau)}$$

$$= \sum_{n=-\infty}^{\infty} \alpha_n e^{-in\omega_0\tau} e^{in\omega_0 t}. \qquad (8.39)$$

Thus, the Fourier coefficients for the shifted function are obtained as

$$\hat{\alpha}_n = \alpha_n e^{-in\omega_0\tau},$$

indicating the coefficients of the shifted function have the same magnitude as those for the original function, but the phase of each component is shifted by $-n\omega_0\tau$ radians if τ is in seconds. Also, if $\tau < 0$, the function is considered to be shifted to the left from the origin $t = 0$. In effect, in either case of a shift left or right, τ becomes the new origin for $f(t)$ since $g(\tau) = f(0)$ in Equation 8.38. Figure 8.14 of Problem 8.5 shows an example of a shifted pulse.

CONVERGENCE OF FOURIER SERIES

The general conditions for existence and convergence of the Fourier series,

$$f(t) = \frac{a_0}{2} + \sum_{n=1}^{\infty} [a_n \cos(nt) + b_n \sin(nt)], \qquad (8.40)$$

are fairly complicated, and several references listed in the Annotated Bibliography at the end of the chapter present a rigorous discussion of Fourier series. For example, a strong condition for convergence is that if $f(t)$ is a periodic function with continuous derivatives through the second order for all t, the Fourier series of $f(t)$ converges uniformly to $f(t)$ for all t. However, many functions of interest are not so smooth, and we even wish to consider functions with discontinuities.

For engineering applications, the sufficient conditions for convergence of the Fourier series considering only the function and its first derivative are usually appropriate. Assume that $f(t)$ is piecewise continuous and at t_0, $f'(t_0)$ exists. Then, the series converges to $f(t_0)$ when $t = t_0$. At each point of discontinuity, assume the function takes the average value

$$f(t) = \frac{1}{2}[f(t+) + f(t-)], \qquad (8.41)$$

where $f(t+)$ means the limit of the function from the right and $f(t-)$ is the limit from the left. Then, the Fourier series for $f(t)$ converges to $f(t)$ at every value of t.

In summary, the Fourier series for the functions $f(t)$ considered in this text will converge to the function at points in an interval where $f(t)$ is continuous and $f'(t)$ exists. At points of discontinuity where there is a finite *jump* in the function, the series converges to the average value of the right-hand and left-hand limits as defined in Equation 8.41. This behavior is obvious in the Fourier series approximation of Figure 8.2.

Gibbs Phenomenon   Convergence of the Fourier series as previously discussed assumed that infinitely many terms in the Fourier series were used to approximate a function. In practice, the Fourier series of Equation 8.40 is summed with a finite number of terms as

$$f(t) = \frac{a_0}{2} + \sum_{n=1}^{N} [a_n \cos(nt) + b_n \sin(nt)]. \tag{8.42}$$

When the Fourier series converges, we expect the series to approximate the function in the sense that the square error is minimized by the Fourier coefficients, as discussed in Chapter 7. This error tends to zero as the number of terms $n \to \infty$ if $f(t)$ is continuous in the interval of interest.

An interesting and apparently strange behavior, called the *Gibbs phenomenon* of the Fourier series, occurs near a point of discontinuity. When a finite number of terms of the series is summed, the series shows an oscillatory error, particularly near the discontinuity. This is evident in Figure 8.2 and Figure 8.7. Taking more terms in the series shows that the error does not greatly decrease very close to the points of discontinuity. However, as more terms are taken, the oscillations move closer to the point of discontinuity. In theory, the Gibbs phenomenon would occur at the point of discontinuity, even if an infinite number of terms is taken in the series.[2]

One other feature of a Fourier series that approximates a function with discontinuities is that the nth coefficient is divided by n. For continuous functions, the coefficients decrease as $1/n^2$, as stated in Problem 8.8. In general, the smoother the function, the more rapidly the coefficients decrease with increasing n.

Integration and Differentiation of Fourier Series   Consider the integral of $f(t)$ in terms of its Fourier series

$$\int_{t_1}^{t_2} f(t) \, dt = \frac{a_0}{2} \int_{t_1}^{t_2} dt + \int_{t_1}^{t_2} \sum_{n=1}^{\infty} [a_n \cos(nt) + b_n \sin(nt)] \, dt. \tag{8.43}$$

---

[2]The Gibbs phenomenon is not just a mathematical curiosity. Any real system attenuates the high-frequency components of an input signal higher in frequency than a certain value. For example, a rapidly switched signal approximating a square wave viewed on an oscilloscope that cannot pass all the frequency components of the signal as determined by Fourier analysis will appear to exhibit the Gibbs phenomenon in the displayed signal.

If $f(t)$ is piecewise continuous on the interval $[t_1, t_2]$, the Fourier series of $f(t)$ can be integrated term by term and the resulting series will converge to the integral of $f(t)$ in the interval. Notice that integration will have the effect of dividing the coefficients in the original series by n and hence increases the rate of convergence. In terms of frequency, integration tends to smooth the function by reducing the magnitudes of higher harmonics.

Differentiation of the Fourier series, in effect, multiplies the original series by n and thus increases the magnitude of the coefficients. Using the complex series, the derivative is

$$\frac{df(t)}{dt} = \frac{d}{dt} \sum_{n=-\infty}^{\infty} \alpha_n e^{in\omega_0 t}$$

$$= \sum_{n=-\infty}^{\infty} (in\omega_0) \alpha_n e^{in\omega_0 t}. \tag{8.44}$$

At higher frequencies as n increases, each term in the series is multiplied by a large number and the higher frequency components increase rapidly. However, if $f(t)$ is continuous and $f'(t)$ is piecewise continuous, then the Fourier series for $f'(t)$ can be obtained by differentiating the series for $f(t)$.

APPLICATIONS OF FOURIER SERIES TO DIFFERENTIAL EQUATIONS

Consider the nth-order differential equation with constant coefficients subject to a harmonic series of sinusoids input as the forcing function. The differential equation is thus

$$\frac{d^n y(t)}{dt^n} + a_{n-1} \frac{d^{n-1} y(t)}{dt^{n-1}} + \cdots + a_1 \frac{dy(t)}{dt} + a_0 y(t) = \sum_{n=1}^{N} A_n e^{in\omega_0 t}. \tag{8.45}$$

Following the discussion in Chapter 5, if the frequencies $n\omega_0$ are not those of the characteristic equation, the assumed solution is

$$y(t) = \sum_{n=1}^{N} \alpha_n e^{in\omega_0 t}, \tag{8.46}$$

using the method of undetermined coefficients. Substituting the kth term in the differential equation leads to the relationship

$$\alpha_k [(i\omega_k)^n + a_{n-1}(i\omega_k)^{n-1} + \cdots + a_1(i\omega_k) + a_0] = A_k.$$

Letting $H^{-1}(ik\omega_0)$ designate the term in brackets, the solution for the kth coefficient of the solution is

$$\alpha_k = H(ik\omega_0) A_k,$$

and the complete particular solution is

$$y(t) = \sum_{n=1}^{N} H(in\omega_0) A_n e^{in\omega_0 t}. \tag{8.47}$$

☐ EXAMPLE 8.8   *Fourier Series DE Example*

Consider the simple circuit of Figure 8.8 consisting of a resistor R and capacitor C. The input voltage is designated $f(t)$ and the output voltage across the capacitor is $y(t)$.

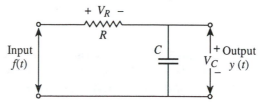

FIGURE 8.8 *RC circuit*

To derive the differential equation for the circuit, apply Kirchhoff's voltage law, with the result

$$f(t) = V_R(t) + V_C(t).$$

Using Ohm's law, $V_R = Ri(t)$, where $i(t)$ is the current through the circuit. Since the current is proportional to the change in voltage across the capacitor,

$$i(t) = C\frac{dV_C(t)}{dt},$$

and Kirchhoff's law can be written

$$f(t) = RC\frac{dV_C(t)}{dt} + V_C(t).$$

Letting $y(t) = V_C(t)$ leads to the resulting equation

$$\frac{dy(t)}{dt} + \frac{1}{RC}y(t) = \frac{1}{RC}f(t).$$

If the input voltage can be written as

$$f(t) = \sum_{n=1}^{N} A_n e^{in\omega_0 t},$$

the solution according to Equation 8.47 is

$$y(t) = \sum_{n=1}^{N} H(in\omega_0) \times A_n e^{in\omega_0 t},$$

and the function $H(in\omega_0)$ is

$$H(in\omega_0) = \frac{1}{1 + in\omega_0 RC}.$$

In this context, the function $H(in\omega_0)$ for a given frequency $\omega = n\omega_0$ is called the *frequency response function* for the circuit being modeled by the differential equation. The magnitude $|H(in\omega_0)|$ defines the amplification (or attenuation) factor at frequency $n\omega_0$. The angle of $H(in\omega_0)$, designated $\arg[H(in\omega_0)]$, gives the phase shift of the input. For this circuit, the results are

$$
\begin{aligned}
|H(in\omega_0)| &= \frac{1}{1 + (n\omega_0 RC)^2}, \\
\arg[H(in\omega_0)] &= -\tan^{-1}(n\omega_0 RC).
\end{aligned}
\tag{8.48}
$$

Problem 8.9 at the end of the chapter further explores this approach to determining the response of a system to a Fourier series input.

The RC circuit can be used to model (approximately) the input circuits of many instruments. Given the values of R and C, a frequency analysis of the circuit shows that at the radian frequency $\omega = 1/RC$, the input frequency component is attenuated by the factor $1/\sqrt{2}$. The value $f = \omega/2\pi$ hertz defines the *bandwidth* of the instrument. Although an input circuit does not completely cut off frequencies above a specific value, frequency components higher in frequency than the bandwidth are considered to be eliminated from the signal for the purposes of a simple analysis.

□

FOURIER TRANSFORMS

Fourier series, as previously presented, were used to represent arbitrary periodic functions as a linear combination of a set of *complete*, *orthogonal* functions, namely, the set of harmonic sines and cosines or exponentials. These functions are ideal for periodic functions. The frequencies present in the series are integer multiples of $2\pi/T$, where T is the period of the function being expanded. However, the application of Fourier series goes much further, since the harmonic functions are *eigenfunctions* of derivative and integral operators. Thus, the Fourier series can be used to solve differential equations and to describe the response of time-invariant linear systems.

We now turn to functions that are not periodic. A Fourier analysis of such functions requires all frequencies, so the sum of functions of discrete frequencies must be replaced by an integral. The result is the *Fourier transform*. As with Fourier series, the transform has a wide range of applications. A few are listed in Table 8.3. Even if the reader is not familiar with every area of application, it is clear from the examples in the table that Fourier transforms play a very important role in engineering and physics.

TABLE 8.3 *Applications of Fourier transforms*

Area	Application
Linear systems	The Fourier transform of the output of a linear system is the product of the system transfer function and the Fourier transform of the input signal.
Optics	A Fourier transform relationship exists between the light distribution on the object and image focal planes of a lens.
Random process	The power density spectrum of a random process is the Fourier transform of the auto correlation function of the process.
Quantum mechanics	The momentum and position of a particle are related by the Fourier transform.
Partial differential equations	Fourier series and Fourier transforms are used to solve various equations.

DEFINITION OF THE FOURIER TRANSFORM

Let the function $f(t)$ be *piecewise continuous* for $-\infty < t < \infty$, and let $\int_{-\infty}^{\infty} |f(t)|\, dt$ exist in the sense that the result is finite. This latter condition is called *absolute convergence* of the integral. Then, the Fourier transform of $f(t)$ exists and is defined as

$$\mathcal{F}[f(t)] = F(i\omega) = \int_{-\infty}^{\infty} f(t)e^{-i\omega t}\, dt. \tag{8.49}$$

The transform $F(i\omega)$ represents the *frequency spectrum* of $f(t)$, and it may be complex even though $f(t)$ is real. The magnitude $|F(i\omega)|$ is called the amplitude spectrum of $F(i\omega)$. Using the notation $F(i\omega)$ emphasizes the fact that the Fourier transform is a function of a complex variable.

□ **EXAMPLE 8.9** **Fourier Transform Example**

Consider the piecewise continuous function f(t) defined as

$$f(t) = \begin{cases} 0, & t < 0, \\ Ae^{-\alpha t}, & t \geq 0, \end{cases}$$

with $\alpha > 0$.

The function has a Fourier transform since

$$\int_{-\infty}^{\infty} |f(t)|\, dt = \int_{0}^{\infty} |A|e^{-\alpha t}\, dt = \frac{|A|}{\alpha}$$

is finite for $\alpha > 0$.

The Fourier transform is thus

$$\mathcal{F}[f(t)] \quad = \quad \int_{-\infty}^{\infty} |A|e^{-\alpha t}e^{-i\omega t}\, dt$$

$$= \int_0^\infty |A|e^{-(\alpha+i\omega)t}\, dt$$

$$= -|A|\frac{e^{-(\alpha+i\omega)t}}{\alpha+i\omega}\Bigg|_0^\infty = \frac{|A|}{\alpha+i\omega} = F(i\omega). \qquad \textbf{(8.50)}$$

The amplitude spectrum is determined by computing $|F(i\omega)|$, and the phase spectrum is $\tan^{-1}(\omega/\alpha)$. For this example,

$$|F(i\omega)| = \frac{|A|}{|\alpha+i\omega|} = \frac{|A|}{\sqrt{\alpha^2+\omega^2}}.$$

$\square$

The Fourier transform represents a function of time in terms of its frequency content. The energy associated with the signal can also be determined from its Fourier transform.

☐ EXAMPLE 8.10   *MATLAB Fourier Transform Example*
The Fourier spectrum of

$$f(t) = \begin{cases} 0, & t < 0, \\ e^{-t}, & t \geq 0. \end{cases}$$

is

$$|F(i\omega)| = \frac{1}{\sqrt{1+\omega^2}},$$

as shown in Example 8.9, with $A = 1$ and $\alpha = 1$. Figure 8.9 shows the plot of $f(t)$ and $|F(i\omega)|$ created by the accompanying MATLAB script. The spectrum is always symmetric around $\omega = 0$ when $f(t)$ is real.

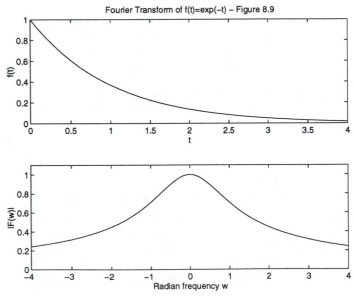

FIGURE 8.9 *Fourier transform for Example 8.10*

MATLAB Script ⎯⎯⎯⎯⎯⎯⎯⎯⎯⎯⎯⎯⎯⎯⎯⎯⎯⎯⎯⎯⎯⎯⎯⎯⎯⎯⎯⎯⎯⎯⎯⎯

Example 8.10

```
% EX8_10.M Plot f(t)=exp(-t), t=[0,4] and
%    the magnitude of the Fourier transform
%
t=[0:.1:4];                % Range in time
foft=exp(-t);              %
% Plot magnitude of F(w)
w=[-4:.1:4];
Fw=1./(sqrt(1+w.^2));      % Fourier transform
%
clf                        % Clear any figures
subplot(2,1,1), plot(t,foft)
xlabel('t')
ylabel('f(t)')
title('Fourier Transform of f(t)=exp(-t) - Figure 8.9')
subplot(2,1,2), plot(w,Fw)
axis([-4 4 0 1.1])
xlabel('Radian frequency w')
ylabel('|F(w)|')
```

Try the MATLAB symbolic command **fourier** or **int** to perform the Fourier integration of $f(t)$.

☐

Energy in a Signal The energy E in a signal is given by the relationship $E = P \times T$ where P is the power and T is the time interval. For a periodic signal $f(t)$ with period T, the power is given by

$$P = \frac{1}{T} \int_{-T/2}^{T/2} [f(t)]^2 \, dt = \sum_{n=-\infty}^{\infty} |\alpha_n|^2,$$

as discussed previously in connection with Fourier series. The Fourier series also provides a direct measure of the power by virtue of Parseval's theorem. In this case, the α_n are the coefficients for the complex series.

For the Fourier transform as previously defined to exist, the signal must have a finite energy over the interval of interest. Thus, if the signal is defined on the interval $[-\infty, \infty]$, the power would be zero since $P = E/T$ and T goes to infinity. Instead of power, the energy in the signal is calculated for signals with Fourier transforms. The relationship between the signal energy in the time domain and the frequency domain is again given by *Parseval's theorem* in the form

$$E = \int_{-\infty}^{\infty} [f(t)]^2 \, dt = \frac{1}{2\pi} \int_{-\infty}^{\infty} |F(i\omega)|^2 \, d\omega.$$

Integrating $|F(i\omega)|^2$ between ω_1 and ω_2 gives the energy contributed by the frequency components between ω_1 and ω_2 to the total energy of the signal.

Properties of the Fourier transform   Various important properties of the Fourier transform will be stated as theorems.

■ THEOREM 8.1   *Linearity*
   Assume that $\mathcal{F}[f(t)] = F(i\omega)$ and $\mathcal{F}[g(t)] = G(i\omega)$. The Fourier transform is linear since

$$\mathcal{F}[\alpha f(t) + \beta g(t)] = \alpha F(i\omega) + \beta G(i\omega).$$

■ THEOREM 8.2   *Time Shifting*
   Shifting $f(t)$ in time simply changes the phase of the Fourier transform, so that

$$\mathcal{F}[f(t - t_0)] = e^{-i\omega t_0} F(i\omega).$$

Notice that the Fourier spectrum of the time-shifted function is not changed from that of the unshifted function.

■ THEOREM 8.3   *Time and Frequency Scaling*
   Scaling the time axis by a nonzero axis yields

$$\mathcal{F}[f(at)] = \frac{1}{|a|} F\left(\frac{\omega}{a}\right)$$

for the Fourier transform, where a is a real, nonzero constant.

If $a < 1$, the time function is spread out on the time axis, and the spectrum of $F(i\omega)$ is compressed by a similar amount.

□ EXAMPLE 8.11   *Fourier Pulse Example*
   The even rectangular pulse of height A and width τ is defined as

$$P(t) = \begin{cases} A, & -\dfrac{\tau}{2} \le t \le \dfrac{\tau}{2}, \\ 0, & |t| > \dfrac{\tau}{2}. \end{cases}$$

The Fourier transform is

$$\begin{aligned} \mathcal{F}[P(t)] &= \int_{-\infty}^{\infty} P(t) e^{-i\omega t}\, dt = A \int_{-\tau/2}^{\tau/2} e^{-i\omega t}\, dt \\ &= \left. -A \frac{e^{-i\omega t}}{i\omega} \right|_{-\tau/2}^{\tau/2} = -A \frac{e^{-i\omega\tau/2} - e^{i\omega\tau/2}}{i\omega}. \end{aligned}$$

This result expressed in terms of the sine function is $(2A/\omega)\sin(\omega\tau/2)$. Multiplying the numerator and denominator by $\tau/2$ yields the Fourier transform as

$$P(i\omega) = A\tau \frac{\sin(\omega\tau/2)}{\omega\tau/2} = A\tau \operatorname{sinc}(\omega\tau/2). \tag{8.51}$$

This is the product of the area under the pulse times $\sin(\omega\tau/2)/(\omega\tau/2)$.

□

MATLAB Pulse Example

Figure 8.10 shows the Fourier transform for two pulses as described in Example 8.11. The positive frequencies of the transform are shown for different pulse widths. The accompanying MATLAB script was used to plot the transforms for the two pulses. Each pulse has amplitude $A = 1$. One pulse has a pulse width of 16 seconds and the other 4 seconds.

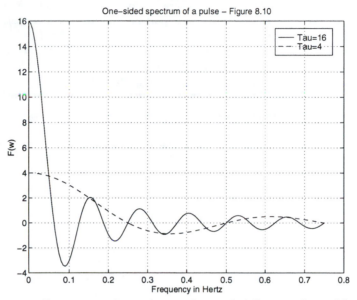

FIGURE 8.10 *Fourier transform of two pulses with different pulse widths*

A number of conclusions about the Fourier transform can be drawn from comparing the two transforms in the figure. Using Equation 8.51 with $A = 1$, the transform of a pulse is

$$P(i\omega) = \tau\frac{\sin(\omega\tau/2)}{\omega\tau/2},$$

and the argument of the sine is $\omega\tau/2 = \pi f\tau$. First, considering the definition of the Fourier transform, it is obvious that at $f = 0$,

$$P(0) = \int_{-\infty}^{\infty} f(t)\,dt,$$

which is the area under the pulse.

Also, at the points

$$\frac{\omega\tau}{2} = \pi f\tau = n\pi, \qquad \text{or} \qquad f = n\left(\frac{1}{\tau}\right),$$

the transform has zeros. For the 16-second pulse, the first zero of the transform occurs at $f = 1/16$ or 0.0625 hertz. The transform of the 4 second pulse crosses zero first at $f = 0.25$ hertz. Another important feature is that as the time pulse becomes shorter, the Fourier transform spreads out in frequency and decreases in amplitude. Although it can be proven rigorously for certain functions, we simply state that if W_t is the width of the time function and W_f is the width of the frequency spectrum in hertz, then

$$W_t W_f \geq K,$$

where K is a constant that would depend on the function being considered. The width of a function that is not a pulse is usually taken to be the distance between points where the function has a value of at least $1/\sqrt{2}$, or 0.707, of its peak value. More important than the exact details, which must be worked out for an individual problem, is the fact that if the function in time or frequency becomes more narrow, the function in the other domain becomes broader. In the present example, as the pulse is shortened by a factor of 4, we expect the range of significant frequencies in the transform to be increased by approximately 4. This is shown in the figure.

MATLAB Script _____
Example 8.12

```
% EX8_12.M Plot the Fourier transform (w>0) of a pulse for various
% widths.  Pulse width is tau = 16 and 4 seconds.
f=[0:.005:.75];          % Frequency
f=f + eps;               % Avoid a divide by zero
F1=zeros(size(f));
F2=zeros(size(f));
F3=zeros(size(f));
tau=16                   % Pulse width in seconds
F1=(1/pi)*(sin(pi*f*tau))./f;
tau=tau/4;               % Pulse width = 4 seconds
F2=(1/pi)*(sin(pi*f*tau))./f;
plot(f,F1,'-',f,F2,'--')
title('One-sided spectrum of a pulse - Figure 8.10')
xlabel('Frequency in Hertz')
Ylabel('F(w)')
grid
legend('Tau=16','Tau=4')
```
_____

□

RELATIONSHIP TO FOURIER SERIES

Comparing the coefficients of the Fourier series of Example 8.7 for a periodic pulse train of rectangular pulses and the Fourier transform of Example 8.11 for a single pulse shows that the series coefficients are

$$\alpha_n = \frac{1}{T} \int_{-\tau/2}^{\tau/2} f(t) e^{-in\omega_0 t}\, dt = \frac{A\tau}{T} \frac{\sin(n\omega_0\tau/2)}{n\omega_0\tau/2}$$

and the transform is

$$\mathcal{F}[f(t)] = F(i\omega) = \int_{-\tau/2}^{\tau/2} f(t) e^{-i\omega t}\, dt = A\tau \frac{\sin(\omega\tau/2)}{\omega\tau/2}.$$

By comparing the two results, it is clear that designating the transform $F(i\omega) = \mathcal{F}[f(t)]$,

$$\frac{F(n\omega_0)}{T} = \frac{A\tau}{T}\frac{\sin(n\omega_0\tau/2)}{n\omega_0\tau/2}.$$

Thus, we conclude that the Fourier series coefficients are obtained by *sampling* the Fourier transform at the points $n\omega_0$ and dividing by the period T. However, the Fourier series itself is a continuous function of time, but the Fourier transform is a function of ω in the frequency domain.

APPLICATIONS TO DIFFERENTIAL EQUATIONS

The Fourier transform is often used to solve differential equations and also to analyze linear time-invariant systems. As the following theorem states, the Fourier transform of the derivative of a function is simply a constant times the Fourier transform of the function.

■ THEOREM 8.4

Differentiation

Assume that $f(t)$ is piecewise continuous and that $f(t)$ and its derivative $f'(t)$ have absolutely convergent integrals for all t. Then,

$$\mathcal{F}[\frac{df}{dt}] = (i\omega)F(i\omega).$$

By extension, the nth derivative $f^{(n)}(t)$ has the Fourier transform

$$\mathcal{F}[f^{(n)}(t)] = (i\omega)^n F(i\omega)$$

if $f(t)$ is piecewise continuous and the derivatives of $f(t)$ have absolutely convergent integrals.

───────────────────────────── ■

The theorem is proven by substituting df/dt in the Fourier transform and integrating by parts using the fact that $f(t) \to 0$ as $t \to \pm\infty$ since we assume that the Fourier transform exists.

Applying the differentiation theorem to the differential equation,

$$\frac{d^n y(t)}{dt^n} + a_{n-1}\frac{d^{n-1}y(t)}{dt^{n-1}} + \cdots + a_1\frac{dy(t)}{dt} + a_0 y(t) = f(t) \qquad \textbf{(8.52)}$$

by forming the Fourier transform of both sides of the equation, we find that

$$[(i\omega)^n + a_{n-1}(i\omega)^{n-1} + \cdots + a_1(i\omega) + a_0]Y(i\omega) = F(i\omega).$$

The solution for $y(t)$ could be found by solving the transformed equation as

$$Y(i\omega) = \frac{F(i\omega)}{[(i\omega)^n + a_{n-1}(i\omega)^{n-1} + \cdots + a_1(i\omega) + a_0]}$$

and taking the inverse Fourier transform. An example will demonstrate this application to the solution of differential equations.

☐ **EXAMPLE 8.13**   *Fourier Solution of Differential Equation*

Consider the second-order equation

$$m\ddot{y}(t) + \beta\dot{y}(t) + ky(t) = f(t), \tag{8.53}$$

for which it is desired to compute the output frequency spectrum. The equation can be rewritten as

$$\ddot{y}(t) + 2\zeta\omega_n\dot{y}(t) + \omega_n{}^2 y(t) = \frac{f(t)}{m} \tag{8.54}$$

using the substitutions $\omega_n{}^2 = k/m$ and $2\zeta\omega_n = \beta/m$. This form of the equation is written in terms of the *natural frequency* of oscillation ω_n that would be obtained if the damping factor β were zero. The factor ζ is a measure of the damping and thus can range from zero in an ideal undamped system to any positive value in a physical system.

Taking the Fourier transform of Equation 8.54 yields the result

$$(-\omega^2 + i2\zeta\omega_n\omega + \omega_n{}^2)Y(i\omega) = \frac{F(i\omega)}{m}.$$

Dividing both sides by ω_n^2 to normalize the frequency and solving for $Y(i\omega)$ gives the *frequency response* of the output as

$$Y(i\omega) = \frac{F(i\omega)/\omega_n^2 m}{[1 - (\omega/\omega_n)^2] + i[2\zeta\omega/\omega_n]}.$$

Using the relationship $\omega_n^2 m = k$, the amplitude of the output spectrum is thus

$$|Y(i\omega)| = \frac{1/k}{\sqrt{[1 - (\omega/\omega_n)^2]^2 + [2\zeta\omega/\omega_n]^2}} |F(i\omega)|. \tag{8.55}$$

If $\phi_F(i\omega)$ is the phase of $F(i\omega)$, the phase of the output is

$$\arg(|Y(i\omega)|) = -\tan^{-1}\left[\frac{2\zeta\omega/\omega_n}{1 - (\omega/\omega_n)^2}\right] + \phi_F(i\omega).$$

☐

Frequency Response (Optional)   The term *frequency response* is often used to describe the steady-state response to a sinusoidal input by a linear, time-invariant system with zero initial conditions. Such systems are described by linear differential equations with constant coefficients, as previously discussed in Chapter 5. The frequency response at one frequency is computed as the ratio of the amplitudes of the Fourier transform of the output signal to the Fourier transform of the input signal at the specified frequency. If the frequency of the input signal is varied from zero to the highest frequency of interest, the equation or plot of the output responses versus frequency is often called the *frequency response* for the system.

The importance of the frequency response stems from the fact that sinusoidal inputs or combinations of sinusoidal inputs to systems arise in many applications. In particular,

1. Any periodic signal of interest can be represented as a linear combination of sinusoidal components (Fourier series).

2. Many natural phenomena are sinusoidal such as the simple harmonic motions or signals generated in mechanical and electrical systems.

3. Sinusoidal signals are important in communications and in the generation of electrical power.

The frequency response of a system governed by a differential equation can be determined by applying the Fourier transform to the equation and taking the ratio of the output and input transforms. If the input function is $f(t)$ with transform $F(i\omega)$ and the response is $y(t)$ with transform $Y(i\omega)$, the input/output relationship $Y(i\omega) = H(i\omega)F(i\omega)$ is shown in Figure 8.11. The function $H(i\omega)$ is called the *transfer function* for the system. As a function of ω, the transfer function defines the frequency response of the system. If $F(i\omega)$ and $Y(i\omega)$ are known, then the transfer function can be determined as

$$H(i\omega) = \frac{Y(i\omega)}{F(i\omega)},$$

where $H(i\omega)$ is in general a complex function of ω that describes the amplitude and phase of the system response at each radian frequency ω to a sinusoidal input at that frequency.

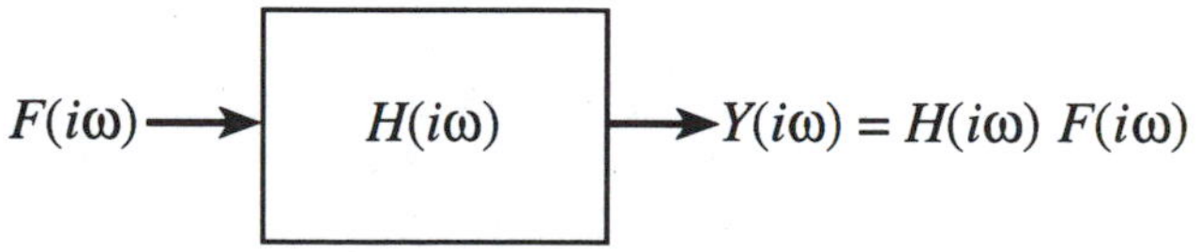

$$F(i\omega) \longrightarrow \boxed{H(i\omega)} \longrightarrow Y(i\omega) = H(i\omega)\,F(i\omega)$$

FIGURE 8.11 *Frequency analysis of a linear system*

□ EXAMPLE 8.14   **MATLAB Frequency Response of Differential Equation**
Writing the differential equation of Example 8.13 as

$$\ddot{y} + 2\zeta\omega_n\dot{y} + \omega_n{}^2 y = \frac{f(t)}{m},$$

the magnitude of the transfer function for the system is

$$|H(i\omega)| = \frac{|Y(i\omega)|}{|F(i\omega)|}$$

$$= \frac{1/k}{\sqrt{[1 - (\omega/\omega_n)^2]^2 + [2\zeta\omega/\omega_n]^2}}, \tag{8.56}$$

according to our previous discussion of frequency response. Plotting the magnitude of the transfer function of Equation 8.56 and the phase angle would determine the frequency response of the system to the input of a sinusoidal wave of radian frequency ω. Obviously, as ω is varied over a range of interest, the transfer function gives the frequency response to any sinusoid in that frequency range. Also, the product

$$Y(i\omega) = H(i\omega)F(i\omega)$$

determines the effect of the system on the frequency spectrum of the input function $f(t)$.

As a specific example, consider the differential equation

$$\frac{d^2y(t)}{dt^2} + 3\frac{dy(t)}{dt} + 2y(t) = P(t), \qquad (8.57)$$

where $P(t)$ is the pulse

$$P(t) = \begin{cases} A, & 0 \leq t \leq 1, \\ 0, & \text{elsewhere.} \end{cases}$$

Using our previous results, the system parameters are

$$\begin{aligned} 2\zeta\omega_n &= 3, \\ \omega_n{}^2 &= 2, \\ m &= 1, \end{aligned}$$

so that $k = m\omega_n{}^2 = 2$.

Using the result of Equation 8.55, the transform of the output is

$$|Y(i\omega)| = \frac{1/2}{\sqrt{[1 - (\omega/\sqrt{2})^2]^2 + [2\zeta\omega/\sqrt{2}]^2}} |P(i\omega)|,$$

where $P(i\omega)$ is the transform of the pulse.

The Fourier transform of the pulse in this example is computed as the Fourier transform of the even pulse as given in Equation 8.51. The input pulse has amplitude $A = 1$, pulse width $\tau = 1$ second, and is shifted to the right by $t_0 = 1/2$ second. Using the shift theorem,

$$P(i\omega) = \frac{\sin(\omega/2)}{\omega/2}e^{-i\omega/2}.$$

The result for the magnitude of the spectrum of the pulse is

$$|P(i\omega)| = \left| \frac{\sin(\omega/2)}{\omega/2} \right|,$$

because the magnitude of the phase shift $\exp(-i\omega/2)\exp(i\omega/2) = 1$.

Combining these results, the output signal has a frequency spectrum

$$|Y(i\omega)| = \frac{1/2}{\sqrt{[1 - (\omega/\sqrt{2})^2]^2 + [2\zeta\omega/\sqrt{2}]^2}} \left| \frac{\sin(\omega/2)}{\omega/2} \right|.$$

Setting $\zeta = 3/(2\omega_n)$, the response can be simplified to the form

$$|Y(i\omega)| = \frac{1}{\sqrt{\omega^4 + 5\omega^2 + 4}} \left| \frac{\sin(\omega/2)}{\omega/2} \right|,$$

which is suitable for plotting.

The results from the accompanying MATLAB script are shown in Figure 8.12. It is evident that the output response of the system has a much narrower frequency spread than that of the input pulse. The system cannot respond to the higher frequencies in the input signal, and these are attenuated in the output.

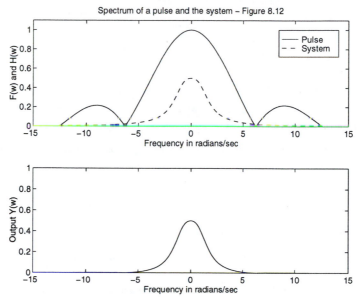

FIGURE 8.12 *Frequency response of the system of Example 8.14*

MATLAB Script _____

Example 8.14
```
% EX8_14.M Plot frequency response of a system with pulse input
%  y'' +3 y' +2 y=P(t);  Pulse width is tau = 1 second.
w=[-4*pi:.1:4*pi];            % Frequency range
w=w + eps;                    % Avoid a divide by zero
Y=zeros(size(w));
P=zeros(size(w));
H=zeros(size(w));
%
A=1;                          % Pulse amplitude
tau =1;                       % Pulse width in seconds
alp=w*tau/2;
P=(A*tau)*abs((sin(alp))./alp); % Transform of pulse
H=1./(sqrt(w.^4+w.^2+4));     % Transform of system
Y=P.*H;                       % Transform of output
%
clf
subplot(2,1,1),plot(w,P,'-',w,H,'--')
axis([-15 15 0 1.1])
title('Spectrum of a pulse and the system - Figure 8.12')
```

```
xlabel('Frequency in radians/sec')
ylabel('F(w) and H(w)')
legend('Pulse','System')
subplot(2,1,2), plot(w,Y);
axis([-15 15 0 1])
xlabel('Frequency in radians/sec')
ylabel('Output Y(w)')
```

☐

REINFORCEMENT EXERCISES AND EXPLORATION PROBLEMS

In these problems, do the computations by hand unless otherwise indicated and then check those
that yield numerical or symbolic results with MATLAB.

REINFORCEMENT EXERCISES

P8.1. **Fourier series of odd function** Compute the Fourier series for the odd function

$$f(x) = Ax, \quad -\lambda/2 < x < \lambda/2.$$

P8.2. **Fourier series of clock signal** Consider the computer clock signal shown in
Figure 8.13, with a pulse rate of 8 million pulses per second ($f_c = 8$ Megahertz) and amplitude of 4
volts and a pulse width of 0.05 microseconds. Find the Fourier series.

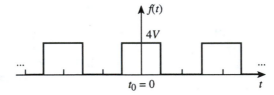

FIGURE 8.13 *Clock signal*

P8.3. **Fourier series** Compute the complex Fourier series for the following periodic function:

$$g(t) = \begin{cases} 1, & 0 \le t \le \pi, \\ -1, & -\pi \le t < 0. \end{cases} \tag{8.58}$$

P8.4. **Fourier series** Compute the Fourier series for the function

$$y(t) = \sin^5 t.$$

Hint: Expand in terms of exponentials.

Chapter 8 ■ FOURIER ANALYSIS

P8.5. **Fourier shift theorem**   The shifted pulse train of Figure 8.14 consists of a pulse of width d in each period of length T. Each pulse is shifted by t_0 from the center of the period. Using the Fourier time shift theorem, find the Fourier series for the pulse train.

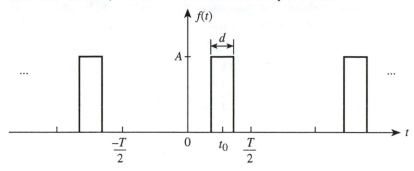

FIGURE 8.14 *Shifted periodic pulse train*

P8.6. **Power series sum**   Using the Fourier series of the function

$$f(t) = \begin{cases} 0, & -\pi < t < 0, \\ t, & 0 < t < \pi, \end{cases}$$

determine the sum of the series

$$\sum_{n=1}^{\infty} \frac{1}{(2n-1)^2}.$$

P8.7. **Power in Fourier series**   The average power dissipated in a resistor of R ohms is

$$\bar{P} = \frac{\bar{V}^2}{R} \quad \text{watts},$$

where $\bar{V}$ is the root-mean-square (rms) value of voltage in volts.

    a. Compute the average power that an odd square wave with amplitude A volts dissipates in a 1-ohm resistor by computing the rms value.

    b. Compute the power by summing the components of the Fourier series.

    c. Suppose the repetition rate of the pulses is 10,000 times a second, so that the fundamental frequency is 10 kilohertz and the signal is filtered by a system that passes only frequencies up to 60 kilohertz. Find the power in the 1-ohm resistor at the output of the system if the amplitude of the square wave is $A = 1$ volt.

P8.8. **Convergence of Fourier series**   Let $f(t)$ be a continuous, even function. Show that the Fourier series coefficients a_n decrease as $1/n^2$.

P8.9. **Fourier series solution of differential equations**   Consider the mass, spring, damper system governed by the following differential equation:

$$\ddot{y} + 4\dot{y} + 40y = P(t),$$

where $P(t)$ is a periodic function. Solve for $y(t)$ when the input function is defined as follows:

$$P(t) = \begin{cases} 9.81, & 0 \le t \le \dfrac{T}{2}, \\ -9.81, & \dfrac{T}{2} < t < T. \end{cases}$$

Let $y(0) = \dot{y}(0) = 0.0$ and $T = 1$ second. Express the solution as a Fourier series.

P8.10. Fourier transform and series Consider the period pulse train with a pulse of width τ centered in a period T with $T > \tau$. Compare the Fourier series coefficients of the complex series with the Fourier transform of a pulse of width τ centered at $t = 0$. Derive a relationship between the Fourier transform magnitude and the magnitude of the Fourier coefficients.

P8.11. Fourier transform of triangular pulse Compute the Fourier transform of the triangular pulse shown in Figure 8.15.

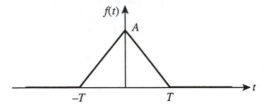

FIGURE 8.15 *Triangular pulse*

Express the answer as magnitude and phase functions and plot the result.

P8.12. Fourier transform and rise time Compute the Fourier transform of the pulse with a finite rise time shown in Figure 8.16.

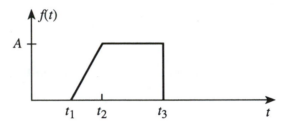

FIGURE 8.16 *Pulse with finite rise time*

P8.13. Fourier transform of double gate Compute the Fourier transform of the *double gate* function shown in Figure 8.17.

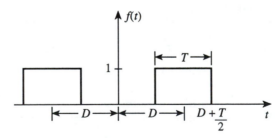

FIGURE 8.17 *Double-gate function*

The pulse train consists of two pulses, each of width T, separated by a distance $2D$.

P8.14. Energy spectrum Using the results of Example 8.10 for the signal $f(t) = \exp(-t)$, $t \geq 0$ and zero elsewhere, compute the following:

    a. The energy in the signal using $f(t)$ and $F(i\omega)$;

b. The magnitude of the component of the signal at 1 hertz.

P8.15. Differential equation Solve the differential equation of Equation 8.57 and show how the input pulse is changed in time by the system.

P8.16. MATLAB Fourier series Plot the Fourier spectrum of the clock pulse in Problem 8.2 up to 48 Megahertz.

P8.17. MATLAB Fourier series Plot the series for Problem 8.3 over the interval $(-\pi, \pi)$ for the following approximations:

    a. 1 term;

    b. 5 terms;

    c. 10 terms.

EXPLORATION PROBLEMS

P8.18. MATLAB Fourier Series Plot the input function $P(t)$ and the response $y(t)$ on the same graph versus time for the equation of Problem 8.9.

P8.19. MATLAB Fourier transform Plot the Fourier transform for positive frequencies for the pulse of width τ centered at the origin. On the same plot, compare the transforms for $\tau = 4, 8, 16$ seconds.

P8.20. Physical systems Draw the mechanical and electrical systems that are equivalent to the system described in Example 8.14 and explain Figure 8.12 in terms of the response of the systems.

P8.21. MATLAB Bode plot The *Signals and Systems Toolbox* in the student edition of MATLAB has a function (**bode**) that plots the magnitude of the frequency response and phase for a transfer function. The MATLAB function to plot the frequency response requires that the equation be written in state-space form.

The state-space equation for Equation 8.54 is

$$\dot{x}_1(t) = x_2(t),$$

$$\dot{x}_2(t) = -\omega_n{}^2 x_1(t) - 2\zeta\omega_n x_2(t) + \frac{f(t)}{m},$$

as explained in Chapter 5. The MATLAB command **bode** will plot the frequency response of the system

$$\dot{\mathbf{x}} = A\mathbf{x} + B\mathbf{u},$$

$$\mathbf{y} = C\mathbf{x} + D\mathbf{u},$$

where $\mathbf{u}$ is a vector of input functions, $\mathbf{x}$ is the state vector for the system, and $\mathbf{y}$ is the vector of output functions. In our case, $B = [0\ 1]^T$, $u = f(t)/m$ and $y = x_1$ so $C = [1\ 0]^T$ and $D = 0$. The resulting plot represents the magnitude of the frequency response in decibels $[20\log_{10}(|H(i\omega)|)]$ versus radian frequency on a logarithmic scale. The phase is plotted in degrees. Make a Bode plot of the frequency response of the system treated in Example 8.14. The plot should use $\omega_n = \sqrt{2}$, $\zeta = 1.06$, and $m = 1$.

Comment: The Bode plot was developed by H. W. Bode to analyze the characteristics of amplifiers using feedback. Although it is primarily used by electrical engineers to analyze circuits and control systems, the Bode plot is convenient to define the frequency response of any system since it indicates the change in amplitude and phase of the frequency components of any input that can be represented in the frequency domain.

ANNOTATED BIBLIOGRAPHY

1. Bracewell, Ronald N., *The Fourier Transform and Its Applications*, McGraw-Hill Book Company, New York, 1986. *This text presents a variety of applications for the Fourier series, Fourier transform, and a variety of related transforms.*

2. Kaplan, Wilfred, *Operational Methods For Linear Systems*, Addison Wesley Publishing Company, Reading, MA, 1962. *A fairly rigorous treatment of linear systems in general, and Fourier methods in particular.*

3. Strum, Robert D., and Donald E. Kirk, *Contemporary Linear Systems*, PWS Publishing Company, Boston, MA, 1994. *Another in the BookWare Companion Series, this text covers Fourier methods with numerous MATLAB examples.*

ANSWERS

$\blacksquare$

P8.1. Fourier series of odd functions Since the function $f(x) = x$ centered at zero with period λ is an odd function, the coefficients of the sine series are

$$
\begin{aligned}
b_m &= \frac{2A}{\lambda} \int_{-\lambda/2}^{\lambda/2} x \sin mkx \, dx \\
&= \frac{2A}{\lambda} \left[\frac{\sin(mkx)}{(mk)^2} - x \frac{\cos(mkx)}{mk} \right]_{-\lambda/2}^{\lambda/2} \\
&= -\frac{2A}{mk} \cos m\pi,
\end{aligned}
$$

where $k = 2\pi/\lambda$. The series is

$$
f(x) = \frac{-2A}{k} \sum_{m=1}^{\infty} \frac{(-1)^m}{m} \sin mkx = \frac{A\lambda}{\pi} \left[\sin kx - \frac{1}{2} \sin 2kx + \cdots \right].
$$

P8.2. Fourier series of clock signal Following the derivation of Example 8.7, the Fourier series of an even periodic pulse is

$$
f(t) = \frac{A\tau}{T} + 2\frac{A\tau}{T} \sum_{n=1}^{\infty} \text{sinc}(n\omega_0 \tau/2) \cos(n\omega_0 t).
$$

In the case of the computer clock signal

$$
A = 4 \, \text{volts}, \quad \tau = 0.05 \, \text{microseconds}, \quad T = \frac{1}{f_c} = 0.125 \, \text{microseconds},
$$

with $\omega_0 = 2\pi/T = 2\pi(8\text{MHz})$.

P8.3.   Fourier series   Using the results of Example 8.5, the complex series is

$$f(t) = \frac{2A}{i\pi} \sum_{n=-\infty}^{\infty} \frac{e^{i(2n-1)\omega_0 t}}{(2n-1)}$$

with $\omega_0 = 1$ and $A = 1$.

P8.4.   Fourier series   This problem is straightforward

$$\sin^5 t \ = \ \left(\frac{e^{it} - e^{-it}}{2i} \right)^5$$

$$= \ \frac{1}{32i} \left(e^{i5t} - 5e^{i3t} + 10e^{it} - 10e^{-it} + 5e^{-i3t} - e^{-i5t} \right)$$

using the binomial theorem. Collecting terms leads to the result

$$\sin^5 t = \frac{5}{8} \sin t - \frac{5}{16} \sin 3t + \frac{1}{16} \sin 5t.$$

As a quick check, notice that at $t = \pi/2$, $\sin^5 t = 1$. The series expansion yields the same result.

P8.5.   Fourier shift theorem   Notice that if the pulse train of Figure 8.14 was shifted by t_0 to the left, the function would be even. Hence, the Fourier series can be written

$$f(t) = \frac{a_0}{2} + \sum_{n=1}^{\infty} a_n \cos[n\omega_0(t - t_0)]$$

with $\omega_0 = \dfrac{2\pi}{T}$.

The series coefficients are

$$a_n = \frac{2}{T} \int_{-d/2}^{d/2} A \cos(n\omega_0 t)\, dt, \qquad n = 0, 1, \ldots,$$

considering the pulse train as an even function.

The coefficients become

$$\frac{a_0}{2} \ = \ \frac{Ad}{T} = \frac{\text{area of pulse}}{\text{period}},$$

$$a_n \ = \ \frac{2Ad}{T} \text{sinc}(n\omega_0 d/2).$$

Thus, the complete series is

$$f(t) = \frac{Ad}{T} + \frac{2Ad}{T} \sum_{n=1}^{\infty} \text{sinc}(n\omega_0 d/2) \cos[n\omega_0(t - t_0)].$$

P8.6.   Power series sum   The Fourier series is

$$f(t) = \frac{\pi}{4} \ - \ \frac{2}{\pi} \cos(t) - \frac{2}{9\pi} \cos(3t) \cdots$$

$$+ \ \sin(t) - \frac{1}{2} \sin(2t) + \frac{1}{3} \sin(3t) + \cdots,$$

and at $t = \pi$ the function converges to $\pi/2$. Thus, evaluating the series at $t = \pi$ yields

$$f(\pi) = \frac{\pi}{2} = \frac{\pi}{4} - \frac{2}{\pi}(x),$$

where x is the sum of the odd series $1/(2n-1)^2$. Solving for x yields the result

$$\sum_{n=1}^{\infty} \frac{1}{(2n-1)^2} = \frac{\pi^2}{8}.$$

Also, the fact that $f(0) = 0$ could be used to find the result.

P8.7. Power in Fourier series

    a. The average power in a signal is

$$\bar{P} = \frac{1}{T} \int_{-T/2}^{T/2} |f(t)|^2 \, dt,$$

    so for the square wave

$$\bar{P} = \frac{1}{T} \int_{-T/2}^{T/2} A^2 dt/R = A^2 \text{ watts},$$

    with $R = 1$ ohm;

    b. The Fourier series of the square wave is

$$f(t) = \frac{4A}{\pi} \sum_{n=1}^{\infty} \frac{\sin[(2n-1)\omega_0 t]}{(2n-1)}.$$

Thus, the power dissipated by a 1-ohm resistor by Parseval's theorem is

$$\begin{aligned}
\bar{P} &= \left(\frac{a_0}{2}\right)^2 + \frac{1}{2}\sum_{n=1}^{\infty}(a_n^2 + b_n^2) \\
&= \frac{1}{2}\sum_{n=1}^{\infty} b_n^2 \\
&= \frac{1}{2}\left(\frac{4A}{\pi}\right)^2 \sum_{n=1}^{\infty} \frac{1}{(2n-1)^2} \quad \text{watts}.
\end{aligned}$$

Using the result of Problem 8.6, the sum of the odd series is $\pi^2/8$ so the result is A^2 watts, as in Part a.

    c. Since the system passes frequencies up to 60 kHz, only the first three terms of the Fourier series of frequencies 10, 30 and 50 kHz are present in the output. The average power dissipated in a 1-ohm resistor at the output is thus

$$\bar{P}_0 = \frac{1}{2}\left(\frac{4A}{\pi}\right)^2 \left[1 + \frac{1}{9} + \frac{1}{25}\right] = 0.933 \quad \text{watts}.$$

Thus, comparing this power with the result of Part a with $A = 1$, we see that the first, third, and fifth harmonic contain 93.3% of the average power associated with a square voltage wave.

P8.8.   Convergence of Fourier Series   The Fourier coefficients for a function $f(t)$ that is even are

$$\pi a_n = \int_{-\pi}^{\pi} f(t) \cos nt \, dt.$$

To integrate by parts, let

$$u = f(t), \quad dv = \cos nt \, dt,$$
$$du = f'(t), \quad v = \frac{\sin nt}{n},$$

so that $\int u \, dv = uv - \int v \, du$ becomes

$$\pi a_n = f(t) \frac{\sin nt}{n} \Big|_{-\pi}^{\pi} - \int_{-\pi}^{\pi} f'(t) \left(\frac{\sin nt}{n} \right) dt. \tag{8.59}$$

Consider the case if $f(t)$ is not continuous. The integration would have to be taken over intervals where the function is piecewise continuous. Thus, the first term may not be zero and the coefficients would decrease with n as $1/n$.

If $f(t)$ is continuous, the first term is zero since $\sin(\pm n\pi) = 0$ (and $f(t)$ is periodic). Thus, taking the second term in Equation 8.59 and integrating by parts again with $u = f'$ and $v = -\cos(nt)/n$ yields

$$
\begin{aligned}
\pi a_n &= -\frac{1}{n} \left[\int_{-\pi}^{\pi} f'(t) \sin nt \, dt \right] \\
&= -\frac{1}{n} \left[f'(t) \frac{\cos nt}{-n} \Big|_{-\pi}^{\pi} - \int_{-\pi}^{\pi} f''(t) \frac{-\cos nt}{n} \, dt \right] \\
&= \frac{1}{n^2} f'(t) \cos nt \Big|_{-\pi}^{\pi} - \frac{1}{n^2} \int_{-\pi}^{\pi} f''(t) \cos nt \, dt.
\end{aligned}
$$

Suppose f' is continuous also. Then, the first term is zero and the integral would lead to the conclusion that the coefficients decrease at least as $1/n^3$. We summarize the results as follows:

$$f(t) \text{ continuous, } f' \text{ continuous,} \qquad a_n \to \frac{1}{n^3},$$
$$f(t) \text{ continuous, } f' \text{ discontinuous,} \quad a_n \to \frac{1}{n^2},$$
$$f(t) \text{ discontinuous,} \qquad\qquad\qquad a_n \to \frac{1}{n}$$

P8.9.   Fourier series solution of differential equations   To find the steady-state forced motion of the system, it is necessary only to determine the change in amplitude and phase shift for each component of the input forcing function. If the input is written as a Fourier series, the response will be another Fourier series with different coefficients. The frequencies in the output series are identical to those in the input since the frequencies are unchanged by a linear system. Thus, multiplying the coefficients of the input series by the transfer function of the system evaluated at the appropriate frequency will yield the output coefficients.

The transfer function of the system is given by Equation 8.56 as

$$|H(i\omega)| = \frac{1/k}{\sqrt{[1 - (\omega/\omega_n)^2]^2 + [2\zeta\omega/\omega_n]^2}},$$

where

$$\zeta = \frac{\beta}{2\sqrt{km}} = \frac{4}{2\sqrt{40}} = \frac{1}{\sqrt{10}}.$$

for this particular system. This transfer function must be evaluated at $\omega = i(2n - 1)\omega_0$ as described in Example 8.8.

The Fourier series of the input square wave is

$$f(t) = \frac{4A}{\pi} \sum_{n=1}^{\infty} \frac{\sin[(2n - 1)\omega_0 t]}{(2n - 1)}$$

according to the results of Example 8.4. In this case, $A = 9.81$ and $\omega_0 = 2\pi/1$.

The output series can be written

$$y(t) = \sum_{n=1}^{\infty} |\rho_n| \frac{\sin[(2n - 1)\omega_0 t]}{(2n - 1)}.$$

Since the coefficients of the input series are $4A/\pi(2n - 1)$,

$$\rho_n = \frac{\dfrac{1}{k}\left(\dfrac{4A}{\pi(2n - 1)}\right)}{\sqrt{[1 - (\omega/\omega_n)^2]^2 + [2\zeta\omega/\omega_n]^2}} e^{i\phi},$$

and the phase angle is

$$\phi = -\tan^{-1} \frac{2\zeta\omega/\omega_n}{1 - (\omega/\omega_n)^2}.$$

P8.10. Fourier transform and series The Fourier series for the periodic pulse train has the coefficients

$$c_n = \frac{1}{T} \int_{-\tau/2}^{\tau/2} e^{-in\omega_0 t}\, dt = \frac{\tau}{T} \frac{\sin(n\omega_0\tau/2)}{n\omega_0\tau/2}.$$

The Fourier transform of a pulse of width τ centered at the origin is

$$F(\omega) = \int_{-\tau/2}^{\tau/2} e^{-i\omega t}\, dt = \tau \frac{\sin(\omega\tau/2)}{\omega\tau/2}.$$

Notice that plotting Tc_n gives the same envelope for the frequency spectrum and the values of the Fourier transform and Fourier series spectrum are the same at $\omega = n\omega_0$.

P8.11. Fourier transform of triangular pulse The analytic expression describing the triangular pulse is

$$P_T(t) = \begin{cases} A\left(1 - \dfrac{|t|}{T}\right), & t < |T|, \\ 0, & \text{otherwise.} \end{cases}$$

Then, the Fourier transform is

$$F(\omega) = \int_{-T}^{0} A\left(1 + \frac{t}{T}\right) e^{-i\omega t}\, dt + \int_{0}^{T} A\left(1 - \frac{t}{T}\right) e^{-i\omega t}\, dt.$$

After integrating, a little algebra shows that

$$F(\omega) = AT\text{sinc}^2\left(\frac{\omega T}{2}\right).$$

P8.12. Fourier transform and rise time Since the Fourier transform is linear, the transform can be written as the sum of the Fourier transform of a sawtooth pulse $f_1(t)$ from t_1 to t_2 and a rectangular pulse $f_2(t)$ of duration $t_2 \leq t < t_3$. Thus, we have

$$F(\omega) = \int_{t_1}^{t_2} f_1(t)e^{-i\omega t}\, dt + \int_{t_2}^{t_3} f_2(t)e^{-i\omega t}\, dt.$$

Defining the variables

$$
\begin{aligned}
T_1 &= t_2 - t_1, \\
T_c &= \frac{t_3 + t_2}{2}, \\
T_2 &= t_3 - t_2,
\end{aligned}
$$

the sawtooth can be shifted by t_1 so that it starts at the origin and the rectangular pulse can be shifted by T_c. The width of the rectangular portion is T_2. Then,

$$
\begin{aligned}
F(\omega) &= F_1(\omega) + F_2(\omega) \\
&= e^{-i\omega t_1} \int_0^{T_1} At e^{-i\omega t}\, dt + e^{-i\omega T_c}\left[AT_2 \operatorname{sinc}\left(\frac{\omega T_2}{2}\right)\right].
\end{aligned}
$$

Computing $F_1(\omega)$ leads to the result

$$
\begin{aligned}
F(\omega) &= \frac{A}{\omega^2 T_1}e^{-i\omega t_1}\left[e^{-i\omega T_1}(1 + i\omega T_1) - 1\right] \\
&\quad + e^{-i\omega T_c}\left[AT_2 \operatorname{sinc}\left(\frac{\omega T_2}{2}\right)\right].
\end{aligned}
$$

P8.13. Fourier transform of double gate The double gate function is simply two pulses of width T. One is shifted by $t = D$ and the other, by $-D$. Thus, the Fourier transform is

$$
\begin{aligned}
F(\omega) &= e^{-i\omega D} AT \operatorname{sinc}\left(\frac{\omega T}{2}\right) + e^{i\omega D} AT \operatorname{sinc}\left(\frac{\omega T}{2}\right) \\
&= 2AT \cos(\omega D) \operatorname{sinc}\left(\frac{\omega T}{2}\right).
\end{aligned}
$$

P8.14. Energy Spectrum

a. For the function $f(t) = \exp(-t)$,

$$E = \int_{-\infty}^{\infty} |\exp(-t)|^2\, dt = \frac{1}{2}$$

and

$$
\begin{aligned}
E &= \frac{1}{2\pi} \int_{-\infty}^{\infty} \frac{d\omega}{1 + \omega^2} \\
&= \frac{1}{2\pi} \tan^{-1}(\omega)\Big|_{-\infty}^{\infty} = \frac{1}{2\pi}\left[\frac{\pi}{2} - \frac{-\pi}{2}\right] = \frac{1}{2},
\end{aligned}
$$

as expected by Parseval's Theorem.

Answers

b. At any frequency, the component is given by evaluating the frequency response $|F(i\omega)|$ at the frequency of interest. In this case, for $f = 1$ hertz, $\omega = 2\pi$. Thus,

$$|F(2\pi)| = \frac{1}{\sqrt{1+(2\pi)^2}} \approx 0.1572.$$

P8.15. Differential equation For the equation

$$\frac{d^2y(t)}{dt^2} + 3\frac{dy(t)}{dt} + 2y(t) = P(t)$$

the solution is

$$y(t) = \begin{cases} \dfrac{1}{2} - e^{-t} + \dfrac{1}{2}e^{-2t}, & 0 \le t \le 1, \\[2mm] e^{-t}(e-1) + \dfrac{1}{2}e^{-2t}(1-e^2), & t > 1. \end{cases}$$

assuming $y(0) = 0$ and $\dot{y}(0) = 0$. The solution to the pulse input is thus a rising exponential until $t = 1$ and then a decaying exponential for $t > 0$. In terms of frequency, the frequency spectrum of the input pulse with sharp changes in amplitude has been filtered by the system so that higher frequencies are attenuated. The resulting output is a much smoother function of time.

Comment: MATLAB solutions are on the disk accompanying this textbook.

9

THE DISCRETE FOURIER TRANSFORM AND THE FFT

PREVIEW_____

Classical numerical analysis techniques depend largely on polynomial approximation of functions for differentiation, integration, interpolation, and solution of differential equations. Fourier techniques, as presented in Chapter 8, use sinusoids and exponential functions to describe a function. Moreover, the Fourier techniques lead to the possibility of understanding physical phenomena in terms of the frequency components associated with a system or signal.

In Chapter 8, the Fourier series and Fourier transform were applied to signals and systems described by functions that are continuous functions of time. For computer analysis, which is in fact the modern approach to analyzing signals and systems, a signal or system is described by *samples* of the continuous function associated with the signal or system. Fourier techniques are ideally suited to studying the effects of sampling such continuous functions.

This chapter introduces the discrete Fourier transform (DFT) and an important algorithm to compute the DFT, called the fast Fourier transform (FFT). Our emphasis is on the practical use of the FFT and the errors that can arise from sampling a signal and then computing its DFT.

FREQUENCY ANALYSIS OF SIGNALS

One of the most important uses of Fourier analysis, although not the only one by any means, is to analyze the frequency components of a *signal* derived from measurements of a physical variable of interest. These signals can be functions that represent quantities changing with time, such as voltage, force, or temperature. Light from a star can be analyzed to determine the spectrum of the light from the star. Similarly, the Fourier transform of radar echoes or audio (speech) signals is used to analyze the characteristics of the source of the signals. The study of the frequency content of such signals is called *spectral analysis*, and the Fourier transform is the primary method to compute the spectrum analytically.

Before attempting to apply Fourier analysis by computer to physical signals, it is necessary to thoroughly understand the frequency properties of such signals. This is one area where the motto, "The purpose of computing is insight, not numbers," mentioned in the preface, must be taken very seriously.[1]

First, suppose the physical signal of interest is a continuous function of time or another continuous variable. For computer analysis it is necessary to represent the signal by a finite number of values of the signal, usually called *samples* of the signal. Second, the analytical form of the Fourier methods, as presented in Chapter 8, assumes that the signal being analyzed is *infinitely* long in time. Obviously, any computer representation of the signal must be finite in length. Thus, representing a continuous signal for computer analysis leads to a finite-length vector of sampled points with the possibility of various errors in representation. Later in the chapter, we shall see that the spacing of the samples of a time-dependent signal and the length of time of observation of the signal will determine the information that can be derived from frequency analysis of the signal.

As with any physical problem, some knowledge of the characteristics of the system being analyzed must be introduced into the Fourier analysis

---

[1]The motto is from Richard Hamming's book, *Numerical Methods for Scientists and Engineers*.

of the signals created by the system. For example, in human speech processing, the frequencies of interest generally range from 0 hertz (dc) to about 5000 hertz. White light, by contrast, has a frequency range from about 0.4×10^{15} hertz to 0.7×10^{15} hertz. The range of frequencies computed as the maximum frequency in the signal minus the minimum frequency is called the *bandwidth* of the signal. Thus, a typical speech signal would have a bandwidth of 5000 hertz. The bandwidth for white light is about 0.3×10^{15} hertz.

☐ EXAMPLE 9.1   *Frequency Spectrum*

Comparing the frequency spectrum of single pulses gives important insight into the effect of changes in the characteristics of the pulse on the spectrum. Consider the pulse

$$P(t) = \begin{cases} A, & |t| < T/2, \\ 0, & \text{otherwise}. \end{cases} \qquad (9.1)$$

According to the analysis in Chapter 8, the frequency spectrum of the pulse is determined by the Fourier transform of the pulse as

$$\mathcal{F}[P(t)] = F(f) = AT \frac{\sin(2\pi f T/2)}{(2\pi f T/2)}. \qquad (9.2)$$

The effect on the spectrum $F(f)$ of changing the pulse width is shown clearly in Figure 9.1. The first zero of the 2-second pulse occurs at $f = 1/2$ hertz.

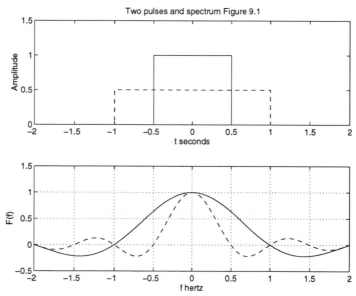

FIGURE 9.1 *Rectangular pulses and their spectra*

From Equation 9.2, the zeros of the frequency spectrum occur at the points where $2\pi f T/2 = n\pi$ and n is an integer. Solving for f leads to the zero points as $f = n/T$, in which T is the total width of the pulse.

As the pulse width is decreased, the first zero crossing of the frequency axis moves up in frequency. The spectrum of the narrower 1-second pulse has its first zero at $f = 1$ hertz. If we assume that the frequencies of interest are primarily contained in the region defined from 0 hertz to the first zero, the bandwidth of the pulse for analysis could be considered to be approximately $1/T$ hertz. However, studying the analytical solution of Equation 9.2 shows that the actual bandwidth is infinite.

The magnitude of the frequency spectrum for a pulse of width T seconds is

$$|F(f)| = AT \left| \frac{\sin(2\pi fT/2)}{(2\pi fT/2)} \right|,$$

which has zeros at $f = n/T$ and maxima where

$$AT \frac{d}{df} \left| \frac{\sin(2\pi fT/2)}{(2\pi fT/2)} \right| = 0.$$

Differentiating leads to the equation

$$AT \left\{ -\frac{1}{f^2} \sin\left[2\pi f \left(\frac{T}{2} \right) \right] + \frac{2\pi T/2}{f} \cos\left[2\pi f \left(\frac{T}{2} \right) \right] \right\} = 0.$$

After dividing by the cosine term, the equation for the maxima is

$$\tan\left[2\pi f \left(\frac{T}{2} \right) \right] = \left[2\pi f \left(\frac{T}{2} \right) \right],$$

which becomes $x = \tan x$ by substituting $x = 2\pi fT/2$. The solutions are tabulated in the *Handbook of Mathematical Functions* by Abramowitz and Stegun listed in the Annotated Bibliography for this chapter. For a pulse with amplitude $A = 1$ and width $T = 1$ second, Table 9.1 shows the first four frequencies at which there are relative maxima of the magnitude of the spectrum. The values at these points show that the magnitudes of the frequency components diminish rapidly with increasing frequency. For practical purposes, the spectrum can be assumed to be zero above some frequency.

TABLE 9.1 *Maxima of $F(f)$*

f hertz	$|F(f)|$
0.0000	1.0000
1.4303	0.2172
2.4590	0.1284
3.4709	0.0913

The ideal pulses just discussed cannot exist in nature since every physical pulse has a finite rise time. The *rise time* is usually defined as the time for a signal to change from 10% to 90% of its maximum value. A triangular pulse, as shown in Figure 9.2, will be used to demonstrate the effect of the rise time on its frequency spectrum.

Triangular Pulse. The triangular pulse of width T is defined by the equation

$$P_T(t) = \begin{cases} A\left(1 - 2\dfrac{|t|}{T}\right), & |t| < \dfrac{T}{2}, \\ 0, & \text{otherwise.} \end{cases} \tag{9.3}$$

This pulse has the Fourier transform spectrum

$$F(f) = \frac{AT}{2}\frac{\sin^2(2\pi fT/4)}{(2\pi fT/4)^2}, \tag{9.4}$$

with zeros at $f = 2n/T$. Comparing the spectra in Figure 9.1 and Figure 9.2 shows the frequency spectrum to the first zero for the triangular pulse is wider than that of a square pulse of the same width. However, the higher-frequency components in the spectrum of the triangular pulse go to zero more rapidly than those of a pulse with a zero rise time.

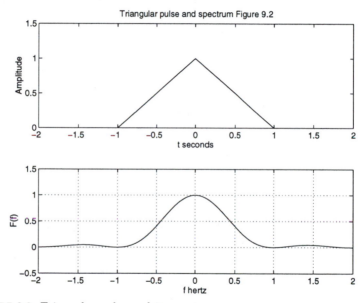

FIGURE 9.2 *Triangular pulse and its spectrum*

For the triangular pulse, the time taken for the signal to change from 0 to A is $T/2$ seconds, and the slope of the rising edge of the triangular pulse is $dP_T/dt = 2A/T$. Rewriting Equation 9.4 as

$$F(f) = \left(\frac{2A}{T}\right)\frac{\sin^2(2\pi fT/4)}{(\pi f)^2}$$

shows that the magnitude of the frequency components is proportional to the slope of the rising edge of the signal. In the case of typical pulses used in modern electronic equipment, the signal could change from 0 to 5 volts in 10 nanoseconds (10^{-8} seconds) or less. Thus, significant frequencies greater than $1/10^{-8}$ hertz, or 100 megahertz, could be present in the spectrum of the pulse.

$\square$

DISCRETE AND FAST FOURIER TRANSFORMS

In this section, we treat the approximation of the exponential Fourier series and the integral Fourier transform of real-valued signals by sums of finite lengths and then present an algorithm for efficient computation. The discrete sum is called the *discrete Fourier transform*, or DFT, and the algorithm is the *fast Fourier transform*, or FFT. Table 9.2 summarizes the relationship between the time function and various Fourier techniques. In the table, $f(t)$ and $f(t_i)$ are functions of time. The parameter $\omega = 2\pi f$ radians per second is angular frequency, where f is frequency in hertz.

TABLE 9.2 *Table of Fourier techniques*

Name	Characteristics	Typical use
Fourier series	$f(t)$ continuous $F(\omega_i)$ discrete	Analysis of periodic functions and signals
Fourier transform	$f(t)$ continuous $F(i\omega)$ continuous	Frequency analysis of signals and systems
Discrete Fourier transform (DFT)	$f(t_i)$ discrete $F(\omega_i)$ discrete	Computation of other transforms Analysis of sampled signals
Fast Fourier transform (FFT)	$f(t_i)$ discrete $F(\omega_i)$ discrete	Algorithm to compute the DFT

Both the DFT and FFT deal with discrete functions in time and frequency. Thus, the DFT and the FFT transform a series of values $f(t_i)$, $i = 0, \ldots, N-1$, into a series of components

$$F(f_i), \quad i = 0, \ldots, N-1,$$

where each f_i is a discrete frequency in the spectrum computed by the DFT.

Consider the portion of a continuous signal $f(t)$ shown in Figure 9.3. To determine the Fourier transform of $f(t)$ by computer analysis requires that the signal be sampled at a finite number of points. Typically, signals are sampled at equally spaced points in time. In Figure 9.3, assume the signal is sampled at intervals $\Delta t = T_s$ seconds to create a set of N points. The length in time of the sampled signal is $T = (N-1)T_s$ seconds. Thus, the original continuous signal is *sampled* and *truncated* for computer processing.

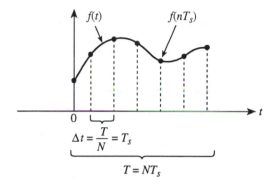

FIGURE 9.3 *Approximation of a signal by sampling*

Definition of DFT Assume that a function $f(t)$ is defined at a set of N points, $f(nT_s)$ for $n = 0, \ldots, N-1$ values, as shown in Figure 9.3. The DFT yields the frequency spectrum at N points by the formula

$$F\left(\frac{k}{NT_s}\right) = \sum_{n=0}^{N-1} f(nT_s)e^{-i2\pi nk/N} \tag{9.5}$$

for $k = 0, \ldots, N-1$. Thus, N sample points of the signal in time lead to N frequency components in the discrete spectrum spaced at intervals $f_s = 1/(NT_s)$.

Real f(t) The DFT yields N frequency components with a spacing of $f_s = 1/T$, where $T = NT_s$ is the period of the sampled time signal. The period of the frequency spectrum is $F_p = Nf_s = N/T = 1/T_s$. However, for a real function of time sampled at an even number of points N, the transform produces symmetry about the point $N/2$ because the real part of the transform is even and the imaginary part of the transform is odd. Table 9.3 summarizes the properties of the DFT applied to a real function of time. The DFT is assumed to have the form

$$F(f) = F_r(f) + iF_i(f),$$

where $F_r(f)$ is the real part of the transform and $F_i(f)$ is the imaginary part. A real and even signal has a transform that is a real and even function. Similarly, a real and odd signal produces an imaginary and odd function for the transform, as described in Table 9.3.

TABLE 9.3 *Properties of the DFT of $f(t)$*

$f(t)$	$F_r(f)$ and $F_i(f)$
Real	$F_r(f)$ even; $F_i(f)$ odd
Real and even	$F(f) = F_r(f)$; even
Real and odd	$F(f) = F_i(f)$; odd

In this chapter, we consider only functions $f(t)$ that are real functions of time.

Frequency Range The DFT frequencies described by Equation 9.5 appear to range from $f = 0$ to $f = (N - 1)f_s$. However, this is not correct because of the symmetry of the transform results. Summarizing the relationship between sampling in time and the frequency components in the spectrum leads to the following conclusions:

1. The sample spacing T_s determines the highest frequency as

$$F_{\max} = \frac{1}{2T_s} \; ;$$

2. The period of the time signal determines the frequency spacing as

$$f_s = \frac{1}{T} = \frac{1}{NT_s}.$$

We stress these conclusions in discussing the DFT since it is necessary to understand that the frequency components computed by DFT analysis of a signal are determined by the choice of sample spacing and number of samples, *not* necessarily by the characteristics of the signal being analyzed. An important part of the analysis is to define the sampling parameters correctly so that the DFT represents the spectrum of the signal accurately. We shall reinforce these results in the chapter by numerous examples, discussions, and problems at the end of the chapter.

Figure 9.4 presents a comparison of the Fourier series, Fourier transform, and the DFT. The spectrum of a periodic signal computed as a Fourier series consists of a sequence of frequency components spaced at $\omega = 2\pi f_s = 2\pi/T$ hertz apart, where T is the period of the signal in seconds. The Fourier transform is used to determine the spectrum of an infinite length signal which is continuous in time. The spectrum of the

Fourier transform and Fourier series is defined on the frequency interval $(-\infty, \infty)$.

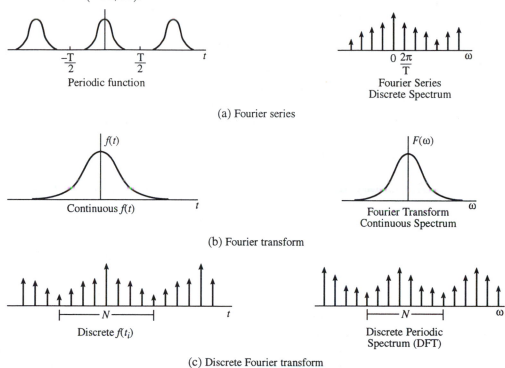

(a) Fourier series

(b) Fourier transform

(c) Discrete Fourier transform

FIGURE 9.4 *Comparison of Fourier techniques*

Notice that use of the discrete Fourier transform implies *periodicity* in both the time and frequency domain. This property of the DFT is proven in several of the texts listed in the Annotated Bibliography at the end of this chapter. In Problem 9.1, you are asked to show that the DFT as defined in Equation 9.5 is periodic.

The discrete signal and its transform have the periodic properties

$$F(kf_s) = F([k+N]f_s) = F(kf_s + F_p),$$
$$f(nT_s) = f([n+N]T_s) = f(nT_s + T),$$

where $F_p = 1/T_s$ is the period of the spectrum computed by the DFT. Thus, it is easily shown by direct substitution in the DFT that

$$F(N) = F(0).$$

The FFT algorithm presented later computes only the frequency components from $k = 0$ to $k = N - 1$ as defined in Equation 9.5. The sampled function of time also becomes periodic with period T seconds. This periodicity in time and frequency is characteristic of the DFT regardless of the true characteristics of the signal being analyzed.

APPROXIMATION OF FOURIER TRANSFORMS

The DFT can be used to approximate the continuous Fourier transform. As defined in Chapter 8, the continuous Fourier transform is

$$\mathcal{F}[f(t)] = F(f) = \int_{-\infty}^{\infty} f(t)e^{-i2\pi ft} \, dt. \tag{9.6}$$

The frequency f in hertz is used as the parameter in this integral. The function $F(i\omega)$, where $\omega = 2\pi f$ is the frequency in radians per second, could be calculated as well. For the transform of a physical signal, we assume that $f(t) = 0$ for $t < 0$. Such signals are called *causal*. If the signal samples need to be shifted in time to meet this restriction, only the phase of the Fourier transform changes according to the shifting theorem presented in Chapter 8.

Using the sampled $f(t)$ with $t = nT_s$ and replacing f by the discrete frequencies $f_s = k/(NT_s)$ leads to the approximation of the Fourier transform as

$$F\left(\frac{k}{NT_s}\right) = T_s \sum_{n=0}^{N-1} f(nT_s)e^{-i2\pi nk/N} \tag{9.7}$$

for $k = 0, \ldots, N - 1$. The factor $\Delta t = T_s$ replaced dt in the integral and is used as a multiplier of the DFT defined by Equation 9.5 in order to approximate the continuous Fourier transform. Problem 9.2 presents another derivation of the DFT approximation to the Fourier transform.

Various computer algorithms, called collectively *fast Fourier transforms*, have been developed to compute the DFT. Few of them take the sampling time T_s in seconds into account since the typical FFT result is simply a function of the index k. Thus, the spectrum must be interpreted and scaled properly if the spectral components are to be displayed versus frequency in hertz.

FAST FOURIER TRANSFORM (FFT)

There are many algorithms that are generally called fast Fourier transform algorithms. In fact, developing techniques to improve the efficiency of calculation for the Fourier transform has been an active research area for many years. Several of the textbooks and articles listed in the Annotated Bibliography for this chapter describe these algorithms and give many of the details not included here. The FFT algorithm discussed here has the properties of the MATLAB function **fft**. The number of samples N in time is usually taken to be a power of 2, so that $N = 2^m$, where m is an integer. This is done for efficiency in many algorithms. However, choosing N as a power of 2 is not required by the DFT definition previously given.

The FFT algorithms take advantage of the symmetry in the exponential functions $\exp(-i2\pi nk/N)$ to reduce the number of computations while computing the DFT. For example, a direct calculation of the DFT requires N^2 multiplications. The basic FFT requires approximately

$N \log_2 N$ multiplications. If $N = 4096$ points, the FFT reduces the number of multiplications from more than 16 million to less than 50,000.

Figure 9.5 shows the important parameters for the DFT and the common FFT algorithms when applied to physical signals. The two important parameters in the time domain are the sampling interval T_s and the number of sample points N. Other parameters can be derived in terms of these. For example, the period of the sampled wave form is $T = NT_s$. One caution is that the Nth point is the periodic extension of the first point. Counting from zero, only the points $N = 0, 1, \ldots, N - 1$ are used by the algorithm.

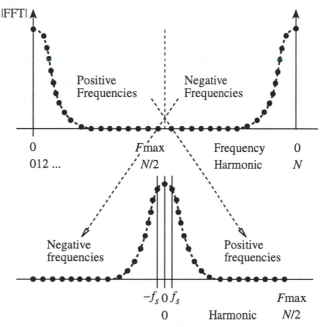

FIGURE 9.5 *Discrete Fourier transform spectrum*

In the frequency domain, the spectral lines are spaced a distance f_s hertz or $\omega_s = 2\pi f_s$ radians/second apart. The frequency *resolution* is thus $f_s = 1/(NT_s)$ hertz. However, as previously discussed, the real part of the transform is even whenever the signal is real. Thus, the results for $k > N/2$ are simply negative frequency results.

For example, $F[(N - 1)/NT_s]$ corresponds to the frequency component $F(-1/NT_s)$. The result is that the output of most FFT algorithms is folded in frequency, as shown in the figure. The component at $N/2$ could be used as the maximum positive frequency or the minimum negative frequency since $F(-N/2) = F(N/2)$ for the DFT. Thus, negative frequency values are the mirror image of the positive values around the frequency $F_p/2$, which is called the *folding* frequency.

Folding Frequency The theory of the folding shown in Figure 9.5 is that N points in time produce N points in the frequency domain. However, for a real signal, there are N complex numbers in the transform with N real parts and N imaginary parts. The real part is even and the imaginary part is odd around the folding point as indicated in Table 9.3. Thus, including the component at $f = 0$, there are really only $N/2 + 1$ unique points in frequency that yield the magnitude and phase of the signal.

The frequencies of the components in hertz are thus

$$f = -\frac{N f_s}{2}, \ldots, 0, f_s, 2f_s, \ldots, \left(\frac{N}{2} - 1\right) f_s,$$

and the spectrum has a period of $F_p = 1/T_s = N f_s$ hertz.

Table 9.4 summarizes the DFT or FFT parameters when a real signal is sampled every T_s seconds for $(N - 1)T_s$ seconds. Be careful to not confuse the period of the signal $T = NT_s$ with the total time interval for the time signal. Also, the maximum frequency in the spectrum is $(N/2)f_s = 1/(2T_s)$ hertz, but the spectrum is periodic with period $N f_s = F_p$ hertz.

TABLE 9.4 *DFT parameters*

Parameter	Notation
Time domain:	
Sample interval	T_s (s)
Sample size	N points
Length	$(N - 1)T_s$ (s)
Period	$T = NT_s$ (s)
Frequency domain:	
Frequency Spacing	$f_s = \dfrac{1}{T} = \dfrac{1}{NT_s}$ (Hz)
Spectrum size	N components
Maximum frequency	$\dfrac{N}{2}f_s = F_{\max}$ (Hz)
Frequency period	$F_p = N f_s = \dfrac{1}{T_s}$ (Hz)

The practical consequences of the characteristics of the DFT and FFT are presented later in the section discussing practical signal analysis. First, we present computer examples to demonstrate the use of MATLAB commands to compute the DFT.

MATLAB FOURIER COMMANDS

---■---

MATLAB contains a number of commands to compute, manipulate and plot the DFT of a function. These are listed in Table 9.5. Except for the command **fourier**, the commands are used for numerical computation. The symbolic command **fourier** is part of the *Symbolic Math Toolbox*. If a function can be defined symbolically, **fourier** computes the Fourier integral transform. The *Signals and Systems Toolbox* has additional commands for more advanced signal processing.

TABLE 9.5 *MATLAB commands for frequency analysis*

Command	Result
abs	Magnitude of FFT
angle	Phase angle of FFT in radians ($-\pi$ to π)
fft	FFT
fftshift	Moves zero frequency to center of spectrum
ifft	Inverse FFT
nextpow2	Returns next power of 2
unwrap	Unwraps phase angle beyond $-\pi$ to π
stem	Plots discrete sequence data
fourier	Symbolic Fourier transform

The MATLAB commands **fft** and **ifft** compute the DFT and inverse DFT, respectively, without regard for the actual sample spacing. The input to the functions is a given number of samples of the time function for **fft** or frequency function for **ifft**. Thus, the discrete functions that result from MATLAB Fourier computations functions must be scaled if an answer in physical units is desired.

The result of **fft** is generally an array of complex numbers. The commands **abs** and **angle** can be used to compute the magnitude and angle of the complex values, respectively. For plotting a two-sided transform, the command **fftshift** will rearrange the output of **fft** to move the zero component to the center of the spectrum.

INDEXING AND TRANSPOSE

MATLAB vectors are indexed from 1 to N, where N is the number of elements. However, the standard definition of the DFT, as in Equation 9.5, is indexed from 0 to $N-1$. Care must be taken in indexing when analyzing and plotting the results of MATLAB FFT calculations.

If the MATLAB transpose operator (') is used on a complex vector, the operation gives the complex conjugate transpose; that is, the sign of

---

the imaginary part is changed as part of the transpose operation. This is not generally desired if the transpose of the DFT results are to be used since the sign of the phase angle would be changed. The operator (.') transposes a complex array but does not conjugate it.

□ EXAMPLE 9.2   *Function to Compute DFT*

The accompanying MATLAB script presents the function clfftf, which computes the approximate Fourier transform, two-sided spectrum, and phase in degrees of the function ft, defined at N sample points. The variable Ts defines the sample interval in seconds. Notice that the MATLAB command **fft** is multiplied by the sampling interval to approximate the actual Fourier transform.

This function is used to compute the Fourier transform in other examples in this chapter. The file can also be modified to provide the spectrum in radians per second or to plot the results.

MATLAB Script _____

```
Example 9.2
function [FT,FTmag,FTang] = clfftf(ft,N,Ts)
% CALL: [FT,FTmag,FTang] = clfftf(ft,N,Ts) Compute the DFT
%  approximation of the Fourier Transform
% Inputs:
% ft Sampled function of time f(nTs)
% N Number of sample points
% Ts Sample interval in seconds
% Outputs:
% FT Approximate Fourier transform using DFT
% FTmag Magnitude of spectrum
% FTang Phase in degrees
% Determine the two-sided spectrum
FT1=Ts*(fft(ft,N));             % Scale to approximate FT
FT=fftshift(FT1);              % Shift 0 to center
%
% Compute the frequency values in hertz fs=1/(N*Ts); fmax=1/(2*Ts)
%
FTmag=abs(FT);                % Magnitude
FTang=(180/pi)*angle(FT);     % Phase in degrees
```

□

You will notice in this function and other scripts in this chapter that calculation of the FFT only requires a call to the MATLAB command **fft**. Most of the other executable statements scale or shift the result for plotting in physical units of frequency.

DFT OF THE EXPONENTIAL FUNCTION

The decaying exponential function is a good test function to use for exploring the accuracy and problems of the DFT applied to a continuous function.

FFT Computation of Spectrum

Consider the function

$$f(t) = \begin{cases} e^{-t}, & t \geq 0, \\ 0, & t < 0, \end{cases}$$

with the Fourier amplitude spectrum

$$F(\omega) = \frac{1}{\sqrt{1 + \omega^2}},$$

as computed in Chapter 8. The phase of the spectrum is

$$\theta(f) = -\tan^{-1}\left(\frac{2\pi f}{1}\right).$$

The accompanying MATLAB script is used to compute the DFT of the function. The magnitude and phase of the result are plotted in Figure 9.6. After the number of points is input and the various parameters are computed, the FFT of the sampled signal is computed using the function clfftf described in Example 9.2. Even though the computed spectrum is discrete, the function **plot**, which interpolates through the points, is used when a large number of points are to be plotted. For a small number of plotted points, the MATLAB plotting command **stem** could be used.

A comparison of the DFT and exact Fourier transform plotted in the figure shows a close agreement in the amplitude spectrum. Experimenting with the number of points or the length of the period (T) in this example will show that taking a shorter interval or fewer sampling points leads to decreased accuracy.

MATLAB Script _____

```
Example 9.3
% EX9_3.M  Compute and plot the DFT of f(t)=exp(-t)
%  Creates f(t) sampled each Ts seconds for T seconds
%  Input:  N   -number of points input
%          T   -Period of signal
%       t0=0   -start of time points
%  Calls clfftf to compute DFT
N=input('Number of points N= ')  % Sample N points
T=input('Period of signal T= ')
Ts=T/N;                          % Sampling interval
% Form the vector of time points and f(n*Ts)
t0=0;                            % Start of signal
ts=(t0:Ts:Ts*(N-1));            % Compute N points
ft=exp(-ts);
% Determine the spectrum
[Fft,Ffmag,Ffang]=clfftf(ft,N,Ts);
% Compute the frequency values in hertz fs=1/(N*Ts); fmax=1/(2*Ts)
%
fs=1/(N*Ts);                     % Frequency spacing
f=fs*linspace(-N/2,N/2-1,N);     % N points in frequency
% Plot Fexact and DFT result
w=2*pi*f;
Fexact=1./(sqrt(1+w.^2));        % Magnitude
```

```
Thetaex=-(180/pi)*atan(w);          % Angle in degrees
clf
subplot(2,1,1),plot(f,Fexact(1:N),'--',f,Ffmag(1:N));
Title(['FT and DFT of exp(-t), N=',num2str(N)])
xlabel('Frequency in hertz')
ylabel('FT and DFT')
legend('FT','DFT')
subplot(2,1,2),plot(f,Thetaex(1:N),'--',f,Ffang(1:N));
xlabel('Frequency in hertz')
ylabel('Phase FT and DFT')
legend('FT','DFT')
```

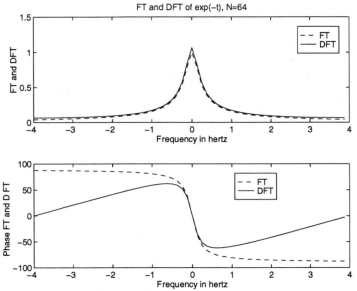

FIGURE 9.6 *Magnitude and phase of spectrum of e^{-t}*

□

DFT Phase Errors One potential problem with the DFT is that the signal being analyzed appears as a periodic function with period NT_s, where T_s is the sampling interval. Considering the decaying exponential in Example 9.3, the function is actually zero for all $t < 0$. However, applying the DFT to the time function can be viewed as taking the Fourier transform of a periodic waveform of exponential pulses.

The effect of periodicity of the analyzed signal was not obvious in computing the magnitude of the transform in Example 9.3. However, comparing the computed phase with the correct value ($-\arctan 2\pi f$) shows that the computed phase angle can be badly in error at the higher frequencies. The true phase changes from $\pi/2$ at negative frequencies to $-\pi/2$ at positive frequencies. One method for reducing the error is to increase the period of the function being analyzed by increasing the number of sample points N. If the signal is not defined beyond a certain point in time, zeros can be added to the signal sample to increase the period. This is called *padding* with zeros. Interested readers should try Problem 9.10 to explore the errors involved.

Caution. The function **angle** computes the arctangent of the ratio of the real to the imaginary part of a complex number. However, the result is very sensitive to the magnitude and sign of the elements. Values close to the minimum representable by the computer may cause unpredictable variations in the phase angle of the DFT when the command **angle** is used. These minimum values should be zero but are not due to roundoff errors. In some cases, it is best to plot the real and imaginary part of the DFT and set to zero any values that are below a predefined amount. A value such as $10 \times$ **eps** could be used as the minimum allowed value, where **eps** is the smallest MATLAB number for a particular computer.

PRACTICAL SIGNAL ANALYSIS

Assume that a physical signal is to be analyzed using the DFT to determine the spectral components. The signal will be assumed to be a *continuous-time* signal, which is often referred to as an *analog* signal. A system that converts the analog signal to a sequence of time samples suitable for computer processing is called a *data acquisition system*. Figure 9.7 presents a simplified diagram of the input stage of a data acquisition system.

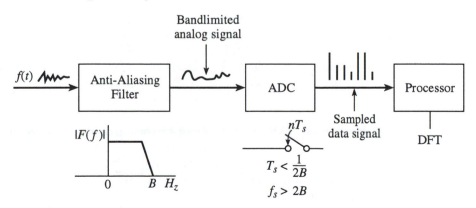

FIGURE 9.7 *Data acquisition system*

The first component is a filter that eliminates unwanted frequency components of a signal. In this case, the filter limits the frequencies in the signal to the range 0 to B. A signal filtered in this way is said to be *bandlimited* to B hertz. In a data acquisition system, the filter is sometimes called a *presampling filter*. Such a filter is also called an *anti-aliasing* filter because it is intended to prevent an error called aliasing, as explained later after the sampling theorem is presented. The analog-to-digital converter (ADC) samples the filtered signal each T_s seconds to create a *sampled-data* or *digital* signal, which is the discrete-time representation of the analog signal after sampling. Typically, the samples are stored in a computer's memory for processing.

SAMPLING

Two of the most important questions in the specification of a data acquisition system such as that shown in Figure 9.7 are the following:

1. How often should the analog signal be sampled?

2. How long should the signal be sampled?

If the highest frequency of interest in the signal is B hertz and the frequency spectrum of the signal is limited to B hertz by the anti-aliasing filter, the *sampling theorem* answers the question in Part 1. The theorem is the cornerstone of practical and theoretical studies in electronic communication.

Sampling Theorem Suppose the highest frequency contained in an analog signal $f(t)$ is $f_{\max} = B$ hertz. Then, if $f(t)$ is sampled periodically at a rate of

$$f_{\text{sample}} = 1/T_s > 2B \qquad (9.8)$$

samples per second, the signal can be exactly recovered from the sample values.[2]

Considering the sampling interval T_s, the sampling theorem states that the analog signal must be sampled so that more than two samples per cycle of the highest frequency in the signal are taken. Since a cycle of the sinusoid at B hertz is $1/B$ seconds in length, the sampling interval must be less than $1/(2B)$ seconds, as indicated in the sampling theorem. The frequency $2B$ is called the *Nyquist frequency*, and the corresponding sampling rate is the *Nyquist rate*. Thus, an analog signal must be sampled at a rate greater than the Nyquist rate if errors due to sampling at too low a rate are to be avoided.

For example, if the signal contains frequencies up to $B = 1000$ hertz, the Nyquist frequency is 2000 hertz and the sampling theorem requires sampling at a rate higher than 2000 samples per second. In terms of the parameters defined earlier for the DFT, the sampling interval corresponding to the Nyquist rate must be

$$T_s < \frac{1}{2B} = \frac{1}{2000} \quad \text{seconds,}$$

---

[2]The interpolation function to reconstruct the analog signal is discussed in most textbooks that cover digital signal processing. Several of these books are listed in the Annotated Bibliography at the end of this chapter.

or $T_s < 0.5$ *milliseconds, between samples. Another way to view this is to realize that a 1000-hertz sinusoid repeats every millisecond. Sampling more than two samples per cycle requires a sampling interval of less than* $1/(2 \times 10^3)$ *seconds. The length of time the signal is sampled would be*

$$T = (N - 1)T_s$$

seconds if N samples are taken.

Sampling and the DFT   *Suppose that an analog signal is sampled at a rate*

$$f_{\text{sample}} = \frac{1}{T_s} > 2B \quad samples/second.$$

The highest possible frequency in the DFT spectrum would be $F_{\max} = 1/(2T_s)$ hertz. If the signal is bandlimited to B hertz and sampled properly, the component at the DFT maximum frequency should be zero since $F_{\max} > B$ hertz.

Although the sampling theorem gives an elegant solution to the sampling problem, a number of practical considerations intervene to complicate the selection of a sampling rate. First, the reader should review the examples in this chapter and in Chapter 8 and notice that none of the ideal example signals have a bandlimited spectrum. In practice, this is not a problem because any physical signal is in fact bandlimited. The physical system that created the signal cannot oscillate above some finite frequency, and the energy content of the signal is negligible beyond a certain frequency. However, the highest frequency may not be known, so an anti-aliasing filter can be used in a data acquisition system to limit the maximum frequency to a known value, say, B hertz. An ideal filter would leave frequency components of the signal below B hertz unaltered and eliminate all components above B hertz. Sampling the signal at a rate greater than $2B$ samples per second would thus satisfy the sampling theorem.

Because the physical filter in a data acquisition system is not ideal, frequencies above the desired bandlimited frequency B can be present in the signal even after filtering. To compensate for the nonideal characteristics of the anti-aliasing filter, the filtered signal is often sampled at a much higher sampling rate than that required by the theorem. In practice, a signal may be sampled at 3 to 10 or more times the rate dictated by the minimum ideal rate of $2B$ samples per second. This oversampling reduces sampling errors but increases the number of samples that must be stored and processed.

Another problem cannot be completely overcome. For the sampling theorem to be correct, the signal must theoretically be sampled in time over the interval $(-\infty, \infty)$. This is often stated as "a bandlimited signal cannot be limited in time." The ideal pulses in previous examples are time limited and thus cannot have bandlimited spectra. However, the desired frequency resolution in the spectrum of a signal can be used to define the define the length of time necessary to sample the signal. Since the DFT resolution is

$$\Delta f = f_s = 1/T \quad \text{hertz,}$$

where T is the time of sampling the analog signal, if a signal is sampled for 2 seconds, it is possible to resolve frequencies as low as $f = 1/2 = 0.5$ hertz.

If the signal is not sampled at a high-enough rate to satisfy the sampling theorem, the errors cannot be undone once the sampling occurs. Also, if the signal is not sampled long enough, the resolution in frequency cannot be improved.

☐ EXAMPLE 9.4   *Sampling Example*

This example defines the relationship between sampling interval, frequency resolution, and number of samples for the DFT. In terms of previous notation, T_s is the sampling interval in seconds, f_s is the frequency resolution, and N is the number of sample points in time and in frequency.

Consider an analog signal with frequencies of interest up to 1200 hertz. The desired frequency resolution is 0.5 hertz. Thus, the signal should be filtered so that $B = 1200$ hertz. This filtering removes frequencies in the signal above 1200 hertz and *noise* above B hertz. The noise consists of unwanted signals added to the desired signal that are the result of environmental effects as the signal is transmitted to the data acquisition system.

By the sampling theorem, the sampling interval in time must be

$$T_s < \frac{1}{2B} = \frac{1}{2400} \quad \text{seconds,}$$

so that at least 2400 samples per second are needed. For a resolution of 0.5 hertz, $T = 1/0.5 = 2$ seconds. The total number of points required is thus

$$N = \frac{T}{T_s} = \frac{2}{(2400)^{-1}} = 4800.$$

If N is to be a power of 2 for the FFT algorithm, $2^{13} = 8192$ samples would be taken. The sampling rate could be increased to 4096 samples per second, which is sampling at a rate corresponding to about 3.4 times the highest frequency of interest.

☐

Spectral Analysis   The measurement and study of the frequency content of a signal is called *spectral analysis*. Many systems designed to measure the Fourier spectrum of an arbitrary signal $f(t)$ sample the signal and perform Fourier analysis using the DFT. If the sampled data points are stored in computer memory, the discrete signal can be analyzed using the DFT.

☐ **EXAMPLE 9.5**   *MATLAB Spectral Analysis*

The time signal shown in Figure 9.8 is stored in the file CLEX95.MAT on the disk accompanying this textbook. The signal was sampled with the parameters

$$N = 128 \quad \text{and} \quad Ts = 1/128.$$

The DFT analysis yields the spectrum of Figure 9.8, with a maximum frequency of 64 hertz and a resolution of 2 hertz. From the DFT analysis, the spectrum appears to have a strong component at 20 hertz. In fact, the MATLAB function that generated the signal was

```
ft=sin(2*pi*20*t) + randn(size(t))
```

where the MATLAB function **randn** generates random values (noise) superimposed on the sine wave. The MATLAB script analyzes the signal and plots the result.

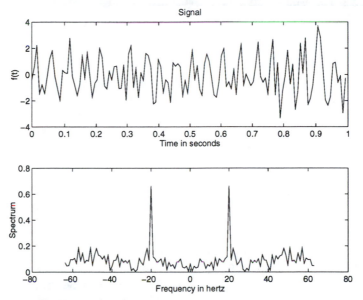

FIGURE 9.8 *Noisy signal and spectrum*

MATLAB Script _____

```
Example 9.5
% EX9_5.M Compute the spectrum of a signal saved
%   in file CLEX95.MAT. The data are:
%   N    samples of the |FFT| are plotted
%   Ts   sampling interval in seconds
%   t    time points
%   ft   function f(t)
% Calls clfftf to compute DFT
load clex95.mat                    % Load N,Ts,t,ft
fs=1/(N*Ts);                       % Frequency spacing
fhertz=fs*linspace(-N/2,N/2-1,N);  % Create frequency axis
```

```
[FT,FTmag,FTarg]=clfftf(ft,N,Ts);   % Compute DFT
% Plot f(t) and DFT
subplot(2,1,1), plot(t,ft)          % Time function
title('Signal')
xlabel('Time in seconds'), ylabel('f(t)')
subplot(2,1,2),plot(fhertz, FTmag) % Spectrum
xlabel('Frequency in hertz')
ylabel('Spectrum')
```

□

PRACTICAL SIGNAL SAMPLING AND DFT ERRORS

One way to approach an error analysis of the DFT is to compare the transform to the exponential (complex) Fourier series studied in Chapter 8. The Fourier series represents a periodic analog signal that exists for $-\infty < t < \infty$. The resulting line spectrum can cover the range $-\infty < f < \infty$ with a resolution of $1/T$ hertz, where T is the period of the analog signal in seconds. In contrast, the DFT analyzes the analog signal using a finite number of N samples in the range $0 \leq nT/N < T$. The useful frequency range is limited to $0 \leq k/T < N/2T$ and is periodic with a period of N frequency points. The limits imposed by the finite nature of the DFT lead to several possible errors that must be considered when the DFT is used to determine the spectrum of an analog signal.

ALIASING
ERROR

If a signal is sampled at a rate that is equal to or less than the Nyquist rate defined by the sampling theorem, *aliasing errors* can occur. The name implies that one signal can be "impersonated" by another signal. The impersonation caused by too-coarse sampling (undersampling) describes high-frequency components of the true spectrum appearing as low-frequency components in the DFT spectrum.

The effect can be seen in several practical situations. Spoked wagon wheels on a stagecoach going forward occasionally appear to rotate backward when western films are shown on TV because the image of the rotating wheel is being sampled. The continuous motion is represented by a series of pictures (samples) taken at a time interval that is too slow to capture the true motion of the wheel. The effect can also be seen when a stroboscope is used to capture the motion of rotating equipment. If the light from the stroboscope is flashed at times that are slightly less that the period of rotation, the equipment appears to be rotating a slower rate than the true rate. If the stroboscope flashes exactly at the period of rotation, the rotating portion of the machine appears to be stationary.

□ **EXAMPLE 9.6**   *Aliasing*

A dramatic example of aliasing in signal processing is shown by the function

$$f(t) = \cos(2\pi f t),$$

which should be sampled at a rate greater than $2f$ samples per second. Suppose we sample at a rate corresponding to one-half the Nyquist frequency, or only f samples per second. The sampling interval is thus $T_s = 1/f$, and the times of sampling are

$$0, \frac{1}{f}, \frac{2}{f}, \ldots, \frac{N-1}{f} \quad \text{seconds}$$

if N points are taken. Notice that $f(nT_s) = \cos(2\pi f n T_s) = 1$ for every point $0 \le n < N$. Now consider the constant function $f(t) = 1$ sampled at the same rate. Both the cosine and the constant yield the time series

$$f(nT_s) = [1, 1, 1 \ldots, 1],$$

so we conclude that the cosine sampled at one-half of the Nyquist rate impersonates a constant or dc value. The DFT of the cosine signal would indicate a spectral value at $f = 0$ but not at the frequency of the cosine, since the maximum frequency in the DFT spectrum is $f/2$. The only solution to eliminate the aliasing error is to sample at a rate that satisfies the sampling theorem. In practice, if the high-frequency components of the signal diminish rapidly, the effect of aliasing may not be significant. The anti-aliasing filter in Figure 9.7 assures that the signal will be bandlimited in a practical data acquisition system.

□

LEAKAGE ERROR

The DFT processes an N-sample periodic time series. An error called *leakage* occurs when the signal being sampled is not perfectly periodic over an N-sample interval. The term implies that energy "leaks" from one frequency to another in the DFT spectrum. Hence, the signal appears to have frequency components that may not be present in the spectrum of the analog signal.

Truncating the time signal at $t = T$ has the effect of introducing a discontinuity in the signal unless it is zero at T or the signal is periodic with period T. Increasing the DFT period of the signal by increasing N tends to reduce the leakage effect unless the analog signal is actually periodic. Problem 9.9 allows you to investigate the effect.

PICKET FENCE EFFECT

The fact that only discrete frequencies k/T appear in the DFT spectrum leads to the conclusion that actual signal components at other frequencies will not be present in the spectrum. This is sometimes colorfully referred to as the *picket fence effect* since the scene behind this fence can only be viewed between the pickets. The pickets in this case correspond to the frequencies that cannot be resolved in the spectrum. Increasing the overall sampling time T while holding the sampling interval constant leads to greater resolution in the spectrum. This is equivalent to increasing the number of samples N.

Table 9.6 summarizes the major DFT errors that may arise from sampling at too low a rate or not sampling long enough. For the parameters listed in the table,

$$T = NT_s,$$

so there are really only two parameters to manipulate. Several problems at the end of this chapter are intended to emphasize the effects on the DFT spectrum of changing the sampling interval or the number of samples.

TABLE 9.6 *DFT errors*

Condition	Cure
Aliasing	Increase maximum frequency (decrease T_s)
Leakage	Increase frequency resolution (increase N)
Picket fence	Increase frequency resolution (increase N)

There are also more sophisticated ways to improve the accuracy of the DFT spectrum. Such methods are treated in the books dealing with signal processing listed in the Annotated Bibliography at the end of the chapter.

REINFORCEMENT EXERCISES AND EXPLORATION PROBLEMS

In these problems, do the computations by hand unless otherwise indicated and then check those that yield numerical or symbolic results with MATLAB.

REINFORCEMENT EXERCISES

P9.1.   Periodicity of the DFT   Show that the DFT defined in Equation 9.5 is periodic. Thus, for an arbitrary integer r,

$$F\left(\frac{r+N}{NT_s}\right) = F\left(\frac{r}{NT_s}\right).$$

P9.2.   Approximation to Fourier transform   Consider the Fourier transform of a function

$$F(\omega) = \int_{-\infty}^{\infty} f(t)e^{-i\omega t}\, dt$$

applied to a function sampled at intervals $t = nT_s$. Assume that the function is defined only for $t > 0$ and further that $f(nT_s)$ is almost zero for $n \geq N$ when N is a sufficiently large number. By integrating over one sampling interval and summing the results, show that the discrete approximation to the Fourier transform is

$$F(k\Omega) \approx T_s \sum_{n=0}^{N-1} f(nT_s)e^{-ink\Omega T_s} \quad \Omega = \frac{2\pi}{NT_s},$$

which agrees with the definition of the DFT in this chapter.

Hint: Assume that the sampling interval is sufficiently small that exponentials not involving k or n can be expanded in a Taylor series with only two terms.

P9.3. DFT conditions Determine the functions $f(t)$ that have the properties that meet the criteria for exact DFT representation:

    a. $f(t)$ is periodic and the spectrum is bandlimited.

    b. $f(t)$ can be sampled at a rate that yields more than two samples per cycle of the highest frequency in $F(f)$.

    c. $f(t)$ can be sampled exactly over one period or an integral number of periods.

Hint: Consider the answer in terms of the Fourier series expansion of the function.

P9.4. Direct calculation of DFT A good way to understand the DFT is to calculate the DFT for a simple function by hand. Determine the DFT for the discrete function

$$f_n = \begin{cases} 1, & 0 \le n \le 4, \\ 0, & 5 \le n \le 9. \end{cases}$$

Plot the amplitude and phase of the transform and describe the symmetries in the transform.

P9.5. Aliasing. Given that the sampling rate is 10 samples per second for the following signals, are they aliased?

    a. $f(t) = \cos(2\pi t)$

    b. $f(t) = \cos(4\pi t)$

    c. $f(t) = \cos(10\pi t)$

    d. $f(t) = \cos(12\pi t)$

    e. $f(t) = \cos(14\pi t)$

P9.6. MATLAB discrete signal Create a 256-point function with a 25% duty cycle of ones (64 points) and compute the Fourier spectrum and plot it.

    a. Try using the MATLAB command **kron** to create the discrete pulse.

    b. Compute the maximum error when compared with the Fourier transform of a pulse.

P9.7. **MATLAB FFT** Compare the Fourier spectrum for the signal

$$f(t) = \begin{cases} \sin\left(\dfrac{\pi t}{4}\right), & \text{for } 0 \le t \le 4, \\ 0, & \text{for } t < 0 \text{ or } t > 4 \end{cases}$$

sampled every T_s seconds at N points in the following cases:

    a. $N = 16$ and $T_s = 1$ second;

    b. $N = 64$ and $T_s = 1$ second;

    c. $N = 128$ and $T_s = 0.5$ second.

P9.8. **MATLAB FFT resolution** Numerical techniques usually require some trial and error. For example, the proper number of sample points N and the sampling period of the FFT are not generally known for an arbitrary signal. Assume that a signal is given analytically and various values of sampling times and sampling period are to be tried.

One method is to choose a reasonable value for the number of samples N and compute the spectrum of the signal. Then, let $N1 = 2 * N$ and recompute. This is repeated until the spectrum of two subsequent calculations are very close. Start with 64 samples.

Assume the function

$$\sin(0.6\pi t) + 0.5\sin(0.64\pi t)$$

sampled every second for 512 points is to be analyzed.

    a. Compute the magnitude of the FFT for 64, 128, 256, and 512 points and plot the results.

    b. Analyze each plot and compare the frequency resolution for each.

    c. For the final plot, if there appear to be several distinct frequencies, pick off the peaks and print the frequencies that correspond.

Hint: Use MATLAB command **ginput** if you have a mouse.

P9.9. **MATLAB FFT leakage** Compare the FFT results for the functions:

$$f(t) = \sin(2\pi 20t) \text{ and } g(t) = \sin(2\pi 19t)$$

by sampling 64 points with a time resolution of $1/128$ second.

    a. Plot the magnitude of each FFT on a subplot to compare the results.

    b. Explain the extra frequency components in each result.

EXPLORATION PROBLEMS

P9.10. MATLAB FFT phase Given the function

$$f(t) = \begin{cases} e^{-t}, & t \ge 0, \\ 0, & t < 0, \end{cases}$$

as treated in Example 9.3, compute the phase of the function and compare the results with the actual value. Use the MATLAB command **fftshift** to plot the two-sided spectrum from $-F_{\max}$ to

F_{max}, where the frequency range will be determined by the sampling interval of $f(t)$ and the number of points taken.

Hint: The way to approach the Fourier transform result is to add zeros to increase the period of the time function as much as possible.

P9.11. MATLAB discrete spectrum plot Write an M-function to plot a discrete spectrum versus frequency in hertz or radians per second. The function call should be:

```
function plotdscf(f,F,xunit)
% CALL: plotdscf(f,F,xunit); plot a discrete spectrum [f F]
% Inputs to function are
%    f    - frequencies
%    F    - spectral values
%  xunit - units of frequency (Hz or rad/s)
```

P9.12. FFT Investigation (Optional) Consider the unit step function, $(U(t) = 1$ for $f > 0$, defined in Chapter 6. Its ordinary Fourier transform does not exist (because the signal would have infinite energy) but a *generalized transform* can be defined. The transform is written

$$\mathcal{F}[U(t)] = \pi\delta(\omega) + \frac{1}{i\omega},$$

where $\delta(\omega)$ is the *unit impulse function*. Attempt to compute the DFT of the unit step function and explain the result. Use the inverse FFT (**ifft**) on the result and reconstruct the time function.

ANNOTATED BIBLIOGRAPHY

1. Abramowitz, Milton, and Irene A. Stegun, *Handbook of Mathematical Functions*, Dover Publications, Inc., New York, 1972. *This handbook contains many formulas, graphs and tables of mathematical functions. In particular, the solutions to the equation $x = \tan x$ are tabulated.*

2. Bracewell, Ronald N., *The Fourier Transform and Its Applications*, 2nd edition, revised, McGraw-Hill Book Company, New York, 1986. *An excellent book that covers many applications of the Fourier transform as well as a number of other transforms.*

3. Brigham, E. Oran, *The Fast Fourier Transform and its Applications*, Prentice Hall, Englewood Cliffs, NJ, 1988. *A very readable treatment of Fourier techniques with emphasis on the FFT. There is also an extensive bibliography.*

4. Burrus, C. Sidney, J. H. McClellan, A. V. Oppenheim, T. W. Parks, R. W. Schafer, and H. W. Schuessler, *Computer-Based Exercises for Signal Processing Using MATLAB*, Prentice Hall, Englewood Cliffs, NJ, 1994. *The book contains a collection of computer exercises about signal processing using MATLAB.*

5. Cooley, James W., P. Lewis, and P. Welch, "Application of the Fast Fourier Transform to Computation on Fourier Integrals, Fourier Series, and Convolution Integrals," AU-15, No. 2, June 1967. *This paper discusses the use of the FFT and the errors involved in various approximations.*

6. Hamming, Richard W., *Numerical Methods for Scientists and Engineers,* McGraw-Hall Book Company, New York, 1973. *An interesting treatment of numerical techniques with a number of personal insights from the author. The book is particularly notable for dividing numerical analysis into classical analysis (polynomials) and modern analysis (Fourier methods).*

ANSWERS

P9.1. Periodicity of the DFT For an arbitrary integer r, the DFT component at index $r + N$ is

$$F\left(\frac{r+N}{NT_s}\right) = \sum_{n=0}^{N-1} f(nT_s)e^{-i2\pi k(r+N)/N}$$

for $k = 0, \ldots, N - 1$. Considering the exponential term, we find

$$e^{-i2\pi k(r+N)/N} = e^{-i2\pi kr/N}e^{-i2\pi k} = e^{-i2\pi kr/N}$$

since

$$e^{-i2\pi k} = \cos 2\pi k - i\sin 2\pi k = 1$$

for k an integer. Thus,

$$F\left(\frac{r+N}{NT_s}\right) = F\left(\frac{r}{NT_s}\right),$$

which shows that the DFT defined in Equation 9.5 is periodic with a period N.

P9.2. Approximation to Fourier transform Set

$$
\begin{aligned}
F(\omega) &= \sum_{n=0}^{\infty} \int_{nT_s}^{NT_s+T_s} f(nT_s)e^{-i\omega t}\, dt \\
&= \sum_{n=0}^{\infty} \left[\int_{nT_s}^{NT_s+T_s} e^{-i\omega t}\, dt\right] f(nT_s) \\
&= \frac{-e^{-i\omega T_s} + 1}{i\omega} \sum_{n=0}^{\infty} e^{-i\omega nT_s} f(nT_s).
\end{aligned}
$$

If $f(nT_s) = 0$ for $n \geq N$, the sum to infinity can be replaced by the sum between 0 and $N - 1$. Also, replacing ω in the exponent by $k\Omega$, where

$$k\Omega = \frac{2\pi k}{T} = \frac{2\pi k}{NT_s},$$

leads to the expression

$$F(k\Omega) = \frac{-e^{-i\omega T_s} + 1}{i\omega} \sum_{n=0}^{N-1} e^{-i2\pi nk/N} f(nT_s).$$

Expanding the exponential term that multiplies the sum in a Taylor series yields

$$\frac{-e^{-i\omega T_s} + 1}{i\omega} = \frac{1 - [1 - i\omega T_s + \cdots]}{i\omega} \approx \frac{i\omega T_s}{i\omega} = T_s.$$

Thus, ignoring higher order terms in the Taylor series leads to the desired result.

P9.3.   DFT conditions   The only functions $f(t)$ for which all the ideal DFT conditions hold are those functions with finite Fourier series.

P9.4.   Direct calculation of DFT   The frequencies and the computed discrete Fourier transform are as follows:

0	5.0000
1.0000	1.0000 + 3.0777i
2.0000	0
3.0000	1.0000 + 0.7265i
4.0000	0
5.0000	1.0000 - 0.0000i
6.0000	0
7.0000	1.0000 - 0.7265i
8.0000	0
9.0000	1.0000 - 3.0777i
10.0000	5.0000

The MATLAB file P9_4.M on the disk contains the program to compute the DFT.

P9.5.   Aliasing   Given that the sampling rate is 10 samples per second, the sampling theorem requires the maximum frequency in the signal to be less than $f_{\max} = 10/2 = 5$ hertz to avoid aliasing.

     a. $\cos(2\pi t)$, $f = 1$ Hz, not aliased;

     b. $\cos(4\pi t)$, $f = 2$ Hz, not aliased;

     c. $\cos(10\pi t)$, $f = 5$ Hz, aliased;

     d. $\cos(12\pi t)$, $f = 6$ Hz, aliased;

     e. $\cos(14\pi t)$, $f = 7$ Hz, aliased.

P9.6.   MATLAB discrete signal   See the disk file P9_6.M.

P9.7.   MATLAB FFT   See the disk file P9_7.M.

P9.8.   MATLAB FFT Resolution   See the disk file P9_8.M.

P9.9.   FFT Leakage   The 20-hertz sine wave is sampled over a number of periods with a DFT spectrum resolution of 2 hertz. If a sufficient number of sampling points are taken, the spectrum appears as a "spike" at 20 hertz (with a component at negative frequencies). The 19 hertz signal is truncated by sampling but not over an integral number of periods. The energy at 19 hertz "leaks" into adjacent frequencies, and the spectrum near 19 hertz appears broadened. Another way to view this is to recognize that there is no component of this DFT spectrum at 19 hertz. Energy at 19 hertz will appear at all the frequencies in the spectrum, with the largest values at 18 and 20 hertz. See the disk file P9_9.M.

Comment: Other MATLAB programs are contained on the disk.

10

ADVANCED CALCULUS

PREVIEW———————————————————————

This chapter introduces the differential calculus of functions of several real variables. As with functions of a single variable $f(x)$, functions of several variables, such as $g(x, y)$ in two dimensions or $F(x_1, \ldots, x_n)$ in n dimensions, are of interest in various problems. The chapter first defines the limit, continuity, and partial derivatives for functions of several variables. The discussion generalizes these concepts as presented in Chapter 6 for functions of a single variable.

By taking partial derivatives of a function of several variables, various properties of the function can be determined. Other useful concepts presented here are the differential, the Jacobian, and the Taylor's series for a function of several variables. A useful application of these operations is to create a linear function that approximates a more general function.

Functions of two and three variables can be displayed in various ways for visualization. Throughout this chapter, MATLAB is used to plot functions as well as perform other operations, such as differentiation and finding the minimum of a function. The chapter also introduces interpolation of two-dimensional functions using MATLAB.

The last section of this chapter introduces vector functions in rectangular and curvilinear coordinate systems. Several applications of vector functions are shown in cylindrical and spherical coordinate systems.

The topics in this chapter are typically included in the subject commonly called *advanced calculus*. The main purpose of this chapter and the following chapters is to study generalizations of derivatives and integrals to higher dimensions. Extensions of topics in this chapter lead to the study of vector differential operators and the integral calculus of vector functions, as treated in later chapters.

FUNCTIONS OF SEVERAL VARIABLES

Functions of two or more independent variables arise frequently in applications. Such functions are called *multivariate functions* or *functions of several variables*. One common example is a function of the form

$$z = f(x, y), \tag{10.1}$$

which represents a surface if plotted as z-values versus points in the xy-plane. As another example, the expression for the kinetic energy of an ideal gas may contain a large number of independent variables since the energy is

$$E = \frac{m}{2} \sum_{i=1}^{N} (u_i^2 + v_i^2 + w_i^2) \tag{10.2}$$

in terms of the velocity components of the individual molecules.

The theory of functions of several variables can be considered to be an extension of the theory of functions of a single variable. In one dimension, we have considered mainly real functions, such as

$$y = f(x), \tag{10.3}$$

where $x \in \mathbf{R}$ and $f(x) \in \mathbf{R}$. In this chapter, only real functions of several real variables will be considered.

For a function of a single variable, the function f assigns a point $f(x)$ to each point x for which f is defined. The set of such points is the *domain* of the function. The *range*, or *image*, is the set of values assumed by the function. For example, the domain of $f(x) = x^2$ could be taken to be the set of all real numbers so the range is the set of nonnegative real numbers.

The *graph* of $f(x)$ is a subset of the points in $\mathbf{R}^2$ and so consists of all the points $[x, f(x)]$ in the xy-plane. Thus, the single-variable function in Equation 10.3 represents a *mapping* or *transformation* from a point $x \in \mathbf{R}$ to another point $f(x) \in \mathbf{R}$. A notation for the function is also

$$\mathbf{R} \xrightarrow{f} \mathbf{R},$$

emphasizing the fact that the domain of the function is a subset of $\mathbf{R}$ and the range is also a subset of the real numbers. A more general transformation is written

$$\mathbf{R}^n \xrightarrow{f} \mathbf{R}^m.$$

In the special case that $m = 1$, the function is said to be a *scalar-valued function*. For example, the function $z = f(x, y)$ in Equation 10.1 represents a transformation from $\mathbf{R}^2$ to $\mathbf{R}$. The kinetic energy expression of Equation 10.2 is also a scalar-valued function.

Suppose f is a real-valued function defined at the point

$$\mathbf{x} = (x_1, x_2, \ldots, x_n)$$

in $\mathbf{R}^n$. Then, the function

$$u = f(x_1, x_2, \ldots, x_n) = f(\mathbf{x}) \tag{10.4}$$

assigns a value $u \in \mathbf{R}$ to each n-tuple of values $x_i, i = 1, \cdots, n$. When this multivariate function is designated as $f(\mathbf{x})$, $\mathbf{x}$ is considered to be a vector in $\mathbf{R}^n$. The transformation can be written

$$\mathbf{R}^n \xrightarrow{f} \mathbf{R},$$

since f is a scalar valued function. In general, the domain of the function will be a subset of $\mathbf{R}^n$ and the range will be a subset of $\mathbf{R}^m$. If $m > 1$, the function is called a *vector function*. Vector functions will be studied later in this chapter.

For a function of a single variable $f(x)$, the allowable values of x are usually defined by an interval such as $a \leq x \leq b$. For a multivariate function, the domain might be defined on a more complicated region of the space involved. Defining the domain and range of a multivariate function is sometimes difficult and requires special attention in operations such as integration. Careful specification of the domain and range is also necessary when the function is being plotted.

CONTOURS AND THREE-DIMENSIONAL PLOTS

There are a number of ways to represent a multivariate function. Analytically, the function can be represented by a functional relationship such as that of Equation 10.4. In some cases, it is possible to visualize the graph of the function. For a function of a single variable, the graph is a curve in $\mathbf{R}^2$. For a general scalar transformation from $\mathbf{R}^n$ to $\mathbf{R}$, the graph is a set of points $[\mathbf{x}, f(\mathbf{x})]$ in $\mathbf{R}^{n+1}$. Since we live in a three-dimensional world, it is difficult for us to visualize the graph of a function such that

$$\mathbf{R}^n \xrightarrow{f} \mathbf{R}$$

for $n \geq 3$. Even with $n = 3$, the graph is a set in $\mathbf{R}^4$. To visualize such functions, the concept of level sets is introduced. The *level set* is a subset of $\mathbf{R}^3$ on which the function is constant. In the case of functions of two variables, $f(x, y)$, the level sets are usually called level curves.

For display, a function of two variables can be represented graphically by a surface in 3D space. Another way to represent functions of two variables is by drawing *level curves*, or *contour lines*. The technique is to plot the graphs of the functions

$$f(x, y) = c_1, \quad f(x, y) = c_2, \ldots$$

for various values of the constants $c_1, c_2, \ldots$, assuming the function attains these values. Thus, the contour lines for the function $f(x, y) = x^2 + y^2$ would be circles of various radii in the xy-plane.

In three dimensions, the level surfaces for a function $f(x, y, z)$ could be plotted. The magnitude of the gravitational force surrounding a spherical mass according to Newton's law of gravitation is of the form

$$|\mathbf{f}| = \frac{k}{r^2},$$

where $r = \sqrt{x^2 + y^2 + z^2}$ is the radial distance from the center of the mass and k is a constant that depends on the units of measurement. The constant surfaces of this function are spheres surrounding the mass.

Another approach to visualizing a multivariate function is to assume that one of the variables is constant in the function $f(x, y, z)$ and to plot the level curves versus the other variables. A common example is the use of lines of constant pressure, called *isobars*, versus altitude on weather maps. Examples in this chapter show the plots of various multivariate functions. Frequently, a combination of techniques is used to aid our visualization of a surface in space.

MATLAB
PLOTTING

MATLAB has plotting commands to plot functions of one, two, or three variables as well as create axes, labels and annotation. Table 10.1 lists a few of the many commands for generating 2D graphs, contour plots, surface plots, and 3D plots. Most of these commands are used in this textbook in various examples.

Chapter 10 ■ ADVANCED CALCULUS

TABLE 10.1 *MATLAB commands for plotting*

Purpose	*Command*
2D Plots:	
Plot $y = f(x)$	**plot, fplot, ezplot**
Plot $f(r, \theta)$	**polar**
Variable spacing	**linspace, logspace**
Logarithmic plots	**loglog, semilogx, semilogx**
Miscellaneous	**axis, fill, grid, legend, zoom**
Contour Plots:	
Plot	**contour**
Label	**clabel**
3D Plots:	
Plots	**plot3, contour3**
Surfaces	**mesh, meshc, surf, surfc**
Miscellaneous	**fill3, slice, view**

The **ezplot** command is part of the *Symbolic Math Toolbox*. It is used to plot the graph of a function $f(x)$ that is defined as a symbolic expression. The other commands are to create plots of functions defined numerically.

There are many MATLAB commands to create and manipulate figures, as well as add color to plots. There are also commands to add titles, axes annotation, text, and a legend to a graph. The reader is encouraged to explore MATLAB's graphics capability by viewing the demonstration plots included with MATLAB. Also, please refer to the *MATLAB User's Guide* for a detailed description of these commands.

☐ **EXAMPLE 10.1**   *MATLAB Plots*

The accompanying MATLAB script and figures show a contour plot and two surface plots of the function

$$z = x^2 y + x^2 + 2y^2.$$

Figure 10.1 shows the contour plot created by commands **contour** and **clabel** for the contour values $(0.5, 1, 2)$ defined in the vector **v** in the script.

The command **meshgrid** converts the values (x, y) to a matrix of values for subsequent use in function evaluation. This is necessary for 3D plotting, as explained in more detail in Example 10.9.

Since this M-file creates two separate figures, the command **figure** is used so that another figure window will be created for the surface plots of Figure 10.2. The MATLAB commands **surf** and **plot3** are used to produce a surface plot and a 3D line plot, respectively. The results shown in Figure 10.2 were created by two **subplot** commands. Each **subplot** command creates a separate axis in one figure window on the computer screen.

```
Example 10.1
% EX10_1.M Plot a contour map and surface of the function
%   z=x^2y+x^2+2y^2
clear                    % Clear variables and figures
clf
x=[-2.0:.15:2.0];       % Define x,y points
y=[-2.0:0.15:2.0];
v=[0.5 1 2]; % Define level values
[X,Y]=meshgrid(x,y);    % Create matrix of points
Z=Y.*(X.^2)+X.^2+2*(Y.^2);
c=contour(x,y,Z,v);
clabel(c)               % Add height values to contours
title('Contour Plot of z=x^2y+x^2+2y^2')
pause
% Plot a surface and mesh
figure                  % Create a new figure window
subplot(2,1,1),surf(X,Y,Z)
xlabel('x')
ylabel('y')
subplot(2,1,2),plot3(X,Y,Z)
title('Surface and Mesh Plots')
xlabel('x')
ylabel('y')
```

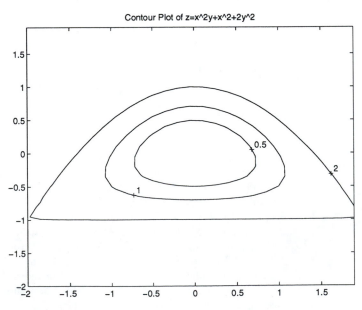

FIGURE 10.1 _Contour plot_

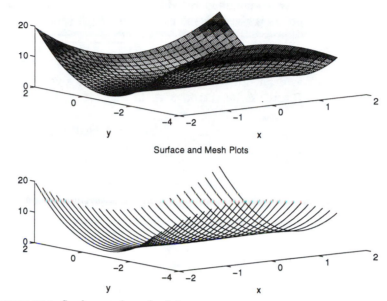

FIGURE 10.2 *Surface and mesh plots*

☐

WHAT IF?   Suppose you are not satisfied with the view of the surface shown in Figure 10.2. In this figure, your viewpoint is 30° of elevation above the xy-plane. Experiment with the **view** command to display views at other evaluations.

LIMITS AND CONTINUITY

The concepts of limits and continuity of a function of one variable are discussed in Chapter 6. This section extends the definitions of limit and continuity to multivariate functions.

Consider the limit

$$\lim_{x \to 0} \frac{\sin x}{x} = 1.$$

This expression means that $(\sin x)/x$ is arbitrarily close to 1 provided that x is sufficiently close to 0. Thus, for every positive number ϵ, there exists a positive number δ such that if

$$0 < |x - 0| = |x| < \delta,$$

then

$$\left| \frac{\sin x}{x} - 1 \right| < \epsilon.$$

Extending this idea of limits to $\mathbf{R}^n$ is straightforward.

The limit of a multivariate function is defined in terms of the distance between points in $\mathbf{R}^n$. For the vectors $\mathbf{x}$ and $\mathbf{y}$, the distance between the points is given by the Euclidean length (norm)

$$\| \mathbf{x} - \mathbf{y} \| = \sqrt{(x_1 - y_1)^2 + \cdots + (x_n - y_n)^2},$$

as defined in Chapter 2. For example, given any $\epsilon > 0$ and point $\mathbf{x}_0$ in $\mathbf{R}^3$, the set of vectors $\mathbf{x} \in \mathbf{R}^3$ that satisfy the inequality

$$\| \mathbf{x} - \mathbf{x}_0 \| < \epsilon$$

is a ball with radius ϵ and centered at $\mathbf{x}_0$.

Limit of a Multivariate Function To define the limit in $\mathbf{R}^n$, we say that $\mathbf{y}_0$ is the limit if, for any $\epsilon > 0$, there is a $\delta > 0$ such that

$$\| f(\mathbf{x}) - \mathbf{y}_0 \| < \epsilon$$

whenever $0 < \| \mathbf{x} - \mathbf{x}_0 \| < \delta$. The relation is written

$$\lim_{\mathbf{x} \to \mathbf{x}_0} f(\mathbf{x}) = \mathbf{y}_0. \tag{10.5}$$

Note. There is an important difference between the existence of the limit of Equation 10.5 and the equivalent formula when only one variable is involved. For a function of a single variable, the limit formula is

$$\lim_{x \to x_0} f(x) = f(x_0).$$

In this case, x_0 is a point in $\mathbf{R}$; if the limit exists, it is *unique*. For a function of several variables if the limit exists and is unique, the limit must have the same value regardless of the approach of $\mathbf{x}$ to $\mathbf{x}_0$. Problem 10.2 illustrates that the limit does not exist if two different approaches yield different values.

Continuity of a Multivariate Function The concept of continuity is fundamental to calculus. If a function does not change abruptly, we expect that if $\mathbf{x}$ is close to $\mathbf{x}_0$, then $f(\mathbf{x})$ must be close to $f(\mathbf{x}_0)$. Thus, continuity is defined in terms of the limit in the following way.

Continuity. A function of several variables is continuous at $\mathbf{x}_0$ if $\mathbf{x}_0$ is in the domain of f and

$$\lim_{\mathbf{x} \to \mathbf{x}_0} f(\mathbf{x}) = f(\mathbf{x}_0).$$

Many of the properties of continuous functions of a single variable can, with suitable modification, be extended to functions of two or more variables. The property that concerns us most here is the definition of derivatives for a continuous multivariate function.

DERIVATIVES OF A MULTIVARIATE FUNCTION

In Chapter 6 we considered derivatives of functions of a single variable. The extension to functions of several variables can proceed in several ways. The most common generalization is to introduce partial derivatives in which one variable is treated at a time, holding the others fixed. The partial derivative is introduced first in this section and applied to define the chain rule for multivariate functions and the Jacobian of transformations. Other generalizations of the derivative are considered later in this chapter and in the next chapter.

PARTIAL DERIVATIVES

Let $f(\mathbf{x})$ be a real-valued function with domain $\mathbf{R}^n$. The *partial derivative* of f with respect to the ith variable is denoted either by $\partial f/\partial x_i$ or f_{x_i}. The formula for the partial derivative is

$$\frac{\partial f(\mathbf{x})}{\partial x_i} = \lim_{h \to 0} \frac{f(x_1, \ldots, x_i + h, \ldots, x_n) - f(x_1, \ldots, x_i, \ldots, x_n)}{h}. \quad \textbf{(10.6)}$$

Thus, the partial derivative of f with respect to x_i is computed by holding $x_1, \ldots, x_{i-1}, x_{i+1}, \ldots, x_n$ constant so that for differentiation f is considered as a function of the ith variable only. When the partial derivative is viewed this way, many of the derivative formulas from the calculus of functions of one variable apply. As an example, let $f(x, y) = x^2 - y^2$. Then

$$\frac{\partial f(x, y)}{\partial x} = 2x, \quad \frac{\partial f(x, y)}{\partial y} = -2y$$

since y is considered constant in the first expression and x is considered constant in the second.

Higher Derivatives Each partial derivative of a function of several variables could in turn have partial derivatives, usually called *second partial derivatives*. For example, a function $f(x, y)$ of two variables has four possible second partial derivatives as follows:

$$f_{xx} = \frac{\partial}{\partial x} f_x(x, y) = \frac{\partial^2 f(x, y)}{\partial x^2}; \quad f_{yy} = \frac{\partial}{\partial y} f_y(x, y) = \frac{\partial^2 f(x, y)}{\partial y^2};$$

$$f_{xy} = \frac{\partial}{\partial y} f_x(x, y) = \frac{\partial^2 f(x, y)}{\partial y \partial x}; \quad f_{yx} = \frac{\partial}{\partial x} f_y(x, y) = \frac{\partial^2 f(x, y)}{\partial x \partial y}.$$

If the partial derivative functions are continuous in a rectangular region of the xy plane, $f_{xy}(x, y) = f_{yx}(x, y)$ within that region. Notice carefully the order of differentiation in the notation $f_{x_i x_j}$. For example, the notation f_{xy} means the partial derivative of $\partial f/\partial x$ with respect to y.

CHAIN RULE

From the calculus of functions of one variable, the derivative of the composite of two functions is

$$\frac{d}{dx} f[g(x)] = f'[g(x)] \, g'(x).$$

Suppose that $z = f(x, y)$ but x and y are, in turn, functions of other variables. For example, let

$$x = g(u, v), \qquad y = h(u, v),$$

so that z is a function of u and v. Then, the partial derivatives of z with respect to u and v are

$$\frac{\partial z}{\partial u} = \frac{\partial z}{\partial x} \frac{\partial x}{\partial u} + \frac{\partial z}{\partial y} \frac{\partial y}{\partial u}, \quad \frac{\partial z}{\partial v} = \frac{\partial z}{\partial x} \frac{\partial x}{\partial v} + \frac{\partial z}{\partial y} \frac{\partial y}{\partial v}. \tag{10.7}$$

This is an example of the *chain rule* for computing the derivatives of composite functions. The extension to functions of n variables is straightforward.

□ **EXAMPLE 10.2**   *Chain Rule*

The area of the rectangle in Figure 10.3 is

$$A = xy,$$

where x and y are the lengths of the sides. It is desired to determine the change of the area resulting from a change in the diagonal r. This change is determined as $\partial A / \partial r$.

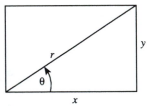

FIGURE 10.3 *Area of a rectangle*

To find the derivative of A with respect to r, we apply the chain rule of Equation 10.7 with $F(x, y) = A$ and define the functions of (r, θ) from the figure as

$$x = g(r, \theta) = r \cos \theta, \quad y = h(r, \theta) = r \sin \theta.$$

The required partial derivatives are

$$\frac{\partial A}{\partial x} = y = r \sin \theta, \quad \frac{\partial A}{\partial y} = x = r \cos \theta, \quad \frac{\partial x}{\partial r} = \cos \theta, \quad \frac{\partial y}{\partial r} = \sin \theta.$$

Then, using the chain rule,

$$\begin{aligned}
\frac{\partial A}{\partial r} &= \frac{\partial A}{\partial x} \frac{\partial x}{\partial r} + \frac{\partial A}{\partial y} \frac{\partial y}{\partial r} \\
&= r \sin \theta \cos \theta + r \cos \theta \sin \theta = 2r \sin \theta \cos \theta \\
&= r \sin 2\theta
\end{aligned}$$

In this example, $A = (r\cos\theta)(r\sin\theta) = (r^2\sin 2\theta)/2$ so that the derivative $\partial A/\partial r$ can be calculated directly. However, the chain rule will play an important role in several of our later examples.

□

COORDINATE TRANSFORMS

The functional relationships

$$x = g(u,v), \quad y = h(u,v) \tag{10.8}$$

can be regarded as pairing corresponding points in an xy-plane and a uv-plane. The curves

$$u = \text{constant}, \quad v = \text{constant}$$

in the uv-plane then could be considered to determine a system of coordinates in the xy-plane. In this case, the functions in Equation 10.8 are said to perform a transformation of coordinates.

Polar Coordinates   Many problems are simplified by a transformation of coordinate systems. A common example is the transformation that relates Cartesian and polar coordinates,

$$x = r\cos\theta, \quad y = r\sin\theta.$$

This transformation simplifies many problems involving concentric circles in the xy-plane.

Linear Transformation   A commonly used transformation is the general linear transformation in two dimensions,

$$\begin{aligned} x &= a_1 u + a_2 v &= g(u,v), \\ y &= b_1 u + b_2 v &= h(u,v). \end{aligned} \tag{10.9}$$

This can be written in matrix notation $\mathbf{x} = T\mathbf{u}$ as

$$\begin{bmatrix} x \\ y \end{bmatrix} = \begin{bmatrix} a_1 & a_2 \\ b_1 & b_2 \end{bmatrix} \begin{bmatrix} u \\ v \end{bmatrix}, \tag{10.10}$$

where $\mathbf{x} = [x,y]^T$, $\mathbf{u} = [u,v]^T$, and T is the matrix of coefficients. This transformation between two coordinate systems is thought of as a transformation *from* the uv-plane to the xy-plane. The values (u,v) are said to be *mapped* into points in the xy-plane. Also, the linear transformation described in Equation 10.9 *maps* lines into lines.

For the linear transformation, the equations can be solved explicitly for u and v if the matrix T can be inverted. The result would be *explicit equations* for u and v as functions of x and y

$$u = \phi(x,y), \quad v = \psi(x,y). \tag{10.11}$$

This transformation would be from the xy-plane to the uv-plane and it is called the *inverse transformation* of the system in Equation 10.9.

☐ **EXAMPLE 10.3**   *Transformations*

Consider the transformation from the uv-plane to the xy-plane represented by

$$\begin{aligned} x &= u + v &= g(u, v), \\ y &= 3u + 2v &= h(u, v). \end{aligned} \qquad (10.12)$$

The matrix formulation of this transformation is $\mathbf{x} = T\mathbf{u}$ if

$$\mathbf{x} = \begin{bmatrix} x \\ y \end{bmatrix}, \qquad \mathbf{u} = \begin{bmatrix} u \\ v \end{bmatrix},$$

and

$$T = \begin{bmatrix} 1 & 1 \\ 3 & 2 \end{bmatrix}.$$

The inverse transformation is then

$$\mathbf{u} = \begin{bmatrix} u \\ v \end{bmatrix} = T^{-1}\mathbf{x} = \begin{bmatrix} -2 & 1 \\ 3 & -1 \end{bmatrix} \begin{bmatrix} x \\ y \end{bmatrix} = \begin{bmatrix} -2x + y \\ 3x - y \end{bmatrix}. \qquad (10.13)$$

Suppose the derivatives of u or v in the transformation of Equation 10.12 are required for a certain application. For this transformation, relationships for $u(x, y)$ and $v(x, y)$ can be determined explicitly, as shown in Equation 10.13. Thus, calculation of derivatives such as

$$\frac{\partial u}{\partial x} = -2, \qquad \frac{\partial u}{\partial y} = 1$$

can be performed immediately. However, there is another approach to determining the partial derivatives of the inverse functions. After defining the Jacobian of a transformation, we will return to the transformation of Equation 10.12.

☐

JACOBIAN

It is often necessary to find various derivatives of a function $F(x, y)$, which is related to the function $G(u, v)$ by a coordinate transformation. These derivatives are required, for example, to determine the area or volume element for integration when coordinates are transformed, as will be shown in Chapter 12. Another example arises in robotics when it is necessary to relate the coordinates of the gripper and those of the robot joints. For precise control of the velocity of the robot, the rates of change of these coordinates with respect to each other must be calculated. The Jacobian of a transformation is often used to aid in computing derivatives of the functions in these applications.[1]

Transformations between coordinate systems and more general problems that involve change of variables can be simplified in some cases by the use of the *Jacobian determinant*, which is usually called simply the Jacobian. In the following discussion, the continuity of the functions and their partial derivatives is assumed.

---

[1] The case of robot control using the Jacobian is discussed in detail in Schilling's textbook, listed in the Annotated Bibliography at the end of the chapter.

We introduce the Jacobian by using it to determine derivatives using a transformation of coordinates, defined as

$$x = g(u, v) \quad \text{and} \quad y = h(u, v). \tag{10.14}$$

Suppose that we wish to calculate the derivative terms

$$\frac{\partial u}{\partial x}, \; \frac{\partial v}{\partial x}, \; \frac{\partial u}{\partial y}, \; \frac{\partial v}{\partial y}$$

without explicitly finding the relationships $u = \phi(x, y)$ and $v = \psi(x, y)$. Applying the chain rule to differentiate the relationships in Equation 10.14 with respect to x yields

$$1 = \frac{\partial g}{\partial u}\frac{\partial u}{\partial x} + \frac{\partial g}{\partial v}\frac{\partial v}{\partial x},$$

$$0 = \frac{\partial h}{\partial u}\frac{\partial u}{\partial x} + \frac{\partial h}{\partial v}\frac{\partial v}{\partial x}. \tag{10.15}$$

In matrix form $A\mathbf{x} = \mathbf{b}$, the equations become

$$\begin{bmatrix} \dfrac{\partial g}{\partial u} & \dfrac{\partial g}{\partial v} \\[2ex] \dfrac{\partial h}{\partial u} & \dfrac{\partial h}{\partial v} \end{bmatrix} \begin{bmatrix} \dfrac{\partial u}{\partial x} \\[2ex] \dfrac{\partial v}{\partial x} \end{bmatrix} = \begin{bmatrix} 1 \\[2ex] 0 \end{bmatrix}. \tag{10.16}$$

The partial derivatives in the matrix multiplying the vector of the desired derivatives are determined from the functions g and h. This matrix in Equation 10.16 is called the *Jacobian matrix* and its determinant, designated $J(u, v)$, is

$$J(u, v) = \begin{vmatrix} \dfrac{\partial g}{\partial u} & \dfrac{\partial g}{\partial v} \\[2ex] \dfrac{\partial h}{\partial u} & \dfrac{\partial h}{\partial v} \end{vmatrix}. \tag{10.17}$$

Following the usual practice, we will call this determinant the *Jacobian*. Other notations for this Jacobian are

$$J(u, v) = \frac{\partial(g, h)}{\partial(u, v)}, \quad \text{or} \quad \frac{\partial(x, y)}{\partial(u, v)}.$$

If the Jacobian is not zero, then it is possible to solve for the derivatives of u and v with respect to x in Equation 10.16.

It is convenient to apply a solution method called *Cramer's rule* to solve for the unknown partial derivatives in Equation 10.16. Cramer's rule can be used to solve a system of linear equations in the form $A\mathbf{x} = \mathbf{b}$ for the unknown $\mathbf{x}$ in terms of determinants. One purpose of this approach is to help us remember the position of the terms in the solutions.

If the determinant $J(u, v)$ is not zero, Cramer's rule leads to the solution of the system in Equation 10.16 as

$$\frac{\partial u}{\partial x} = \frac{J_1}{J(u, v)}, \quad \frac{\partial v}{\partial x} = \frac{J_2}{J(u, v)},$$

where $J(u, v)$ is defined as the Jacobian determinant and the $J_i, i = 1, 2$ are the determinants obtained by replacing the ith column of $J(u, v)$ with the constant vector. Thus, the derivatives sought are

$$\frac{\partial u}{\partial x} = \frac{\begin{vmatrix} 1 & \dfrac{\partial g}{\partial v} \\ 0 & \dfrac{\partial h}{\partial v} \end{vmatrix}}{J(u, v)}, \quad \frac{\partial v}{\partial x} = \frac{\begin{vmatrix} \dfrac{\partial g}{\partial u} & 1 \\ \dfrac{\partial h}{\partial u} & 0 \end{vmatrix}}{J(u, v)}. \tag{10.18}$$

Differentiating the relationships in Equation 10.14 with respect to y yields

$$\begin{bmatrix} \dfrac{\partial g}{\partial u} & \dfrac{\partial g}{\partial v} \\ \dfrac{\partial h}{\partial u} & \dfrac{\partial h}{\partial v} \end{bmatrix} \begin{bmatrix} \dfrac{\partial u}{\partial y} \\ \dfrac{\partial v}{\partial y} \end{bmatrix} = \begin{bmatrix} 0 \\ 1 \end{bmatrix}. \tag{10.19}$$

The derivatives with respect to y are

$$\frac{\partial u}{\partial y} = \frac{\begin{vmatrix} 0 & \dfrac{\partial g}{\partial v} \\ 1 & \dfrac{\partial h}{\partial v} \end{vmatrix}}{J(u, v)}, \quad \frac{\partial v}{\partial y} = \frac{\begin{vmatrix} \dfrac{\partial g}{\partial u} & 0 \\ \dfrac{\partial h}{\partial u} & 1 \end{vmatrix}}{J(u, v)}. \tag{10.20}$$

□ **EXAMPLE 10.4**   *Jacobians*

Assume that the relationship between the coordinates (x, y) and (u, v) is

$$\begin{aligned} x &= u + v &&= g(u, v), \\ y &= 3u + 2v &&= h(u, v), \end{aligned} \tag{10.21}$$

as defined in Example 10.3.

The elements of the Jacobian are

$$\frac{\partial g}{\partial u} = 1, \quad \frac{\partial g}{\partial v} = 1,$$

$$\frac{\partial h}{\partial u} = 3, \quad \frac{\partial h}{\partial v} = 2.$$

The Jacobian of the transformation is thus

$$J(u, v) = \begin{vmatrix} 1 & 1 \\ 3 & 2 \end{vmatrix} = -1.$$

Using the Jacobian determinants of Equation 10.18, the partial derivatives of u and v with respect to x have the values

$$\frac{\partial u}{\partial x} = \frac{\begin{vmatrix} 1 & 1 \\ 0 & 2 \end{vmatrix}}{J(u,v)} = -2, \qquad \frac{\partial v}{\partial x} = \frac{\begin{vmatrix} 1 & 1 \\ 3 & 0 \end{vmatrix}}{J(u,v)} = 3. \qquad (10.22)$$

From Equation 10.20, a similar calculation shows that

$$\frac{\partial u}{\partial y} = 1 \quad \text{and} \quad \frac{\partial v}{\partial y} = -1.$$

Now assume that a transformation is defined as

$$w(u,v) = w[u(x,y), v(x,y)] = uv,$$

where u and v are related to x and y as given in Equation 10.21. The partial derivatives of w with respect to x and y are

$$\frac{\partial w(u,v)}{\partial x} = v\frac{\partial u}{\partial x} + u\frac{\partial v}{\partial x},$$
$$\frac{\partial w(u,v)}{\partial y} = v\frac{\partial u}{\partial y} + u\frac{\partial v}{\partial y}. \qquad (10.23)$$

Using the results from Equation 10.22 for the partial derivatives of u and v with respect to x and y, the changes in w with respect to x and y can be calculated without explicitly writing w as a function of these variables. Substituting the values in the expressions of Equation 10.23 yields

$$\frac{\partial w(u,v)}{\partial x} = -2v + 3u,$$
$$\frac{\partial w(u,v)}{\partial y} = v - u. \qquad (10.24)$$

In this example, the partial derivatives of Equation 10.24 can be computed directly without resort to the Jacobian since the inverse relationship between the coordinates can be determined. Thus, according to the result of Example 10.3,

$$w = u(x,y)v(x,y) = (-2x + y)(3x - y) = -6x^2 + 5xy - y^2.$$

For example,

$$\frac{\partial w}{\partial x} = -12x + 5y.$$

Comparing the solution previously given by the Jacobian method shows that

$$\frac{\partial w(u,v)}{\partial x} = 3u - 2v = -12x + 5y,$$

as expected.

□

WHAT IF?   The MATLAB command **jacobian** from the *Symbolic Math Toolbox* will compute the Jacobian of a function defined symbolically. Determine the Jacobian determinant in Example 10.4 using MATLAB's symbolic capability.

DIFFERENTIALS AND LINEAR APPROXIMATION

The *differential* of a function $f(x)$ at the point x is defined by the formula

$$dy = f'(x)\,dx, \qquad (10.25)$$

where dx is an arbitrary increment of the independent variable x. As shown in Figure 10.4, dy can be viewed as the change in height of a point that moves along the tangent line at the point $[x, f(x)]$ rather than along the curve $f(x)$. If $\Delta x = dx$ is an increment in x, then

$$\Delta y = f(x + \Delta x) - f(x) \qquad (10.26)$$

according to the definitions of the variables in Figure 10.4.

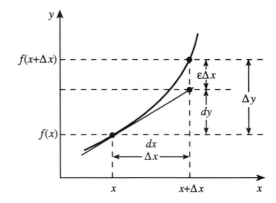

FIGURE 10.4 *The estimate of increment Δy by the differential dy*

Using the definition of the derivative,

$$f'(x) = \lim_{\Delta x \to 0} \frac{\Delta y}{\Delta x} = \lim_{\Delta x \to 0} \frac{f(x + \Delta x) - f(x)}{\Delta x},$$

and the properties of limits, this expression can be written in the form

$$\frac{\Delta y}{\Delta x} = f'(x) + \epsilon, \qquad (10.27)$$

where $\lim_{\Delta x \to 0} \epsilon = 0$. Multiplying by Δx yields the result

$$\Delta y = f'(x)\Delta x + \epsilon \Delta x. \qquad (10.28)$$

Substituting the expression for dy from Equation 10.25 with $\Delta x = dx$ leads to a relationship between the increment Δy and the differential dy,

$$\Delta y = dy + \epsilon \Delta x,$$

with $\lim \epsilon = 0$ as $\Delta x \to 0$. The conclusion is that, for small values of Δx, the increment Δy is a good approximation to the differential dy and $\Delta y \approx f'(x)\Delta x$. Considering the definition of Δy in Equation 10.26, the function f at the point $x + \Delta x$ can be approximated as

$$f(x + \Delta x) \approx f(x) + \Delta y = f(x) + f'(x)\Delta x. \qquad (10.29)$$

This is called the *linear approximation* to $f(x + \Delta x)$ because the change in f depends linearly on Δx.

The linear approximation to $f(x)$ near a point x_0 is usually written by replacing x with x_0 in Equation 10.29 and letting $\Delta x = x - x_0$ so that

$$f(x) \approx f(x_0) + f'(x_0)(x - x_0). \qquad (10.30)$$

This linear expression represents the first two terms of the Taylor series expansion of function near the point x_0, as treated in Chapter 6. The approximation will also be discussed in more detail later. First, we wish to define the differential for a function of many variables.

TOTAL DIFFERENTIAL

When forming partial derivatives, one variable is changed while the others are held constant. We now consider the effect of changing several variables together. For a function of two variables $f(x, y)$, let Δf be the change in $f(x, y)$ as the independent variables are changed to $(x + \Delta x, y + \Delta y)$. Then,

$$\Delta f = f(x + \Delta x, y + \Delta y) - f(x, y).$$

The *differential* of $f(x, y)$ is

$$df = \frac{\partial f}{\partial x}dx + \frac{\partial f}{\partial y}dy. \qquad (10.31)$$

It can be shown that if $f(x, y)$ has continuous first partial derivatives in a region, then df is a good approximation to Δf in the sense that

$$\Delta f = \frac{\partial f}{\partial x}\Delta x + \frac{\partial f}{\partial y}\Delta y + \epsilon_1\Delta x + \epsilon_2\Delta y,$$

where ϵ_1 and ϵ_2 approach zero as Δx and Δy approach zero. If Δx can be approximated by dx and Δy can be approximated by dy, then df is a close approximation to Δf.

Generalizing the result of Equation 10.31, the differential of a function of n variables $f(x_1, x_2, \ldots, x_n)$ is

$$df = \frac{\partial f}{\partial x_1}dx_1 + \frac{\partial f}{\partial x_2}dx_2 + \cdots + \frac{\partial f}{\partial x_n}dx_n. \qquad (10.32)$$

The differential of f in Equation 10.32 is often called the *total differential* of f.

A theorem of multivariate calculus states that if f has a differential at $\mathbf{x}$, then f is continuous at $\mathbf{x}$. In contrast to the one-dimensional case, for which existence of the derivative at x implies the continuity of $f(x)$, a multivariate function may have partial derivatives at a point and yet not be continuous at the point. The stronger condition that the function have *continuous* partial derivatives is sufficient to assure the existence of the differential and hence continuity of the functions.[2]

Significance. The result in Equation 10.32 has important practical applications both mathematically and physically. In order to compute partial derivatives, we can first compute the differential using the rules of ordinary differential calculus and then immediately obtain the partial derivatives. The physical significance corresponds to the principle of superposition.

In Equation 10.32 for the differential of a function, each small change dx_i can take place simultaneously. Since the relationship is a linear function in these increments, the total change is the sum of the effects due to the independent changes.

☐ **EXAMPLE 10.5**   *Derivatives and Differentials*

a. Consider the ideal gas law $pV = cT$, in which p is pressure, V is the volume, T is temperature, and c is a constant. Then, the volume can be written in the form

$$V = V(p, T) = \frac{cT}{p},$$

and the differential of V is

$$dV = \frac{\partial V}{\partial p}\, dp + \frac{\partial V}{\partial T}\, dT.$$

Calculating the partial derivatives yields

$$dV = -\frac{cT}{p^2} dp + \frac{c}{p} dT.$$

From this formula, we can calculate the change in volume due to "small" changes in pressure or temperature or both.

b. Let $z = f(x, y) = x^2 y$. Then, the incremental change in z is

$$
\begin{aligned}
\Delta z &= f(x + \Delta x, y + \Delta y) - f(x, y) \\
&= (x + \Delta x)^2 (y + \Delta y) - x^2 y \\
&= 2xy\Delta x + x^2 \Delta y + \left[(\Delta x)^2 y + 2x\Delta x\Delta y + (\Delta x)^2 \Delta y \right].
\end{aligned}
$$

The term in brackets becomes zero as Δx and Δy go to zero and thus the linear terms in Δx and Δy become the differential of z. The differential is

$$dz = 2xy\, dx + x^2\, dy = \frac{\partial z}{\partial x}\, dx + \frac{\partial z}{\partial y}\, dy.$$

A numerical comparison of Δz and dz is made in Problem 10.4.

☐

---

[2] The theorems are proven in Tom M. Apostol's excellent textbook *Mathematical Analysis*, Addison Wesley, Reading, MA, 1974.

LINEARIZATION   Linearization of functions is one of the most fundamental techniques in engineering, mathematics, and physics. In the mathematics we have studied so far, many of the techniques involved treating a set of simultaneous *linear equations*. If an equation or a system of equations involves a non-linear dependence on the variables, it is often possible to analyze the problem using linear approximations.

According to Equation 10.30, a real-valued function that is differentiable at $x = x_0$ theoretically can be represented as closely as desired by the expression

$$f^*(x) = f(x_0) + f'(x_0)(x - x_0)$$

if x is close enough to x_0. Figure 10.5 shows the approximation graphically as the tangent line to $f(x)$ at the point $[x_0, f(x_0)]$.

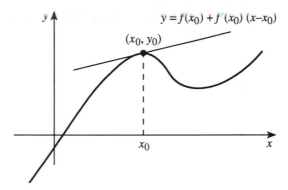

FIGURE 10.5 *Affine approximation of a function*

A function of the form

$$f^*(x) = \alpha + \beta x$$

is called an *affine* function. The discussion of linear functions in Chapter 3 indicates that this function is not linear unless $\alpha = 0$.

One useful approach to studying a complicated and nonlinear function is to determine its behavior near a given point. If the first partial derivatives are known at the point, the total differential previously discussed is one method of obtaining a linear approximation to the function subject to small changes in its arguments. This is equivalent to expanding the function in a Taylor series and retaining only the constant and linear term so that an affine function replaces the function being approximated.

When studying a physical system, we seek to describe the system by means of the simplest possible equations. In general, a physical system would be expected to be nonlinear unless the variables involved are restricted to a limited range. In fact, when dealing with electrical and mechanical systems, most of the governing equations are linear approximations if a specific element is being described.

A typical example of a linearized equation is Ohm's approximation describing the voltage E across the terminals of a resistor due to a current I flowing through the resistor. For a constant current,

$$E = RI,$$

where R is the resistance. Mathematically, the voltage is proportional to current, with R as the proportionality constant. In fact, the resistance of resistors made of carbon and similar substances varies with temperature. The temperature of the resistor is, in turn, a function of the current flowing through it since the power involved is $P = I^2 R$ and the temperature increases with the power. Fortunately, the dependence of resistance on current is slight in most electronic devices as long as the current does not cause an excessive temperature rise in the resistor. Therefore, a linear relationship between E and I is justified in many cases.

Even though Ohm's approximation is usually called Ohm's law, such approximations are *not* fundamental in physics, as are Newton's laws. Newton's laws are a general statement about nature. Ohm's law is an approximation to the behavior of resistors subject to a limited range of voltage and current.[3]

Hooke's law is a fundamental approximation in mechanics. A spring with a restoring force governed by the equation $F = kx$ is called a *linear spring* since the displacement is proportional to the force involved. An actual spring force may be governed by a more complicated equation, such as

$$F(x) \approx kx + k_1 x^3,$$

where a cubic nonlinear term is used since we expect that $F(x) = -F(-x)$ if the spring is symmetric. Notice that the motion of the spring will be approximately linear if $k_1 x^3 << kx$. If either k_1 or the displacement x is small enough, the second term can be neglected compared with the linear term. The expression $dF = k\,dx + 3k_1 x^2\,dx$ shows the dependence of F on displacement for small displacements dx.

☐ EXAMPLE 10.6   *Linearization*

Suppose the acceleration of an automobile can be described by the equation of motion

$$M \frac{dv(t)}{dt} = cu(t) - \alpha v^2(t), \tag{10.33}$$

where the first term represents the acceleration caused by the engine at a throttle setting u and the second term is the drag caused by air resistance. Since this force is proportional to the square of the speed, the equation is nonlinear. Solving this by numerical techniques would not be difficult if the constants

---

[3]The purist may comment that Newton's laws are also approximations. This is so, but the range of applicability is enormous. Only at velocities approaching the speed of light (relativity), in the quantum world (quantum mechanics), or in other very special cases would we abandon Newton's laws and substitute a more complicated theory.

were known. However, we can analyze the motion of the car for small changes in throttle position by linearizing the equation.

Take (U_0, V_0) to be the "operating point," so that the car travels at a constant speed V_0 for a constant throttle position U_0. Inserting this condition into Equation 10.33 yields

$$M \frac{dV_0}{dt} = cU_0 - \alpha V_0{}^2(t) = 0,$$

or $V_0 = \sqrt{cU_0/\alpha}$. Let us assume that a small change in the throttle position leads to a small change in speed. Thus, we set

$$u(t) = U_0 + \Delta u, \quad v(t) = V_0 + \Delta v$$

and substitute again into the equation of motion. The result is

$$M \frac{d}{dt}[V_0 + \Delta v] = c[U_0 + \Delta u] - \alpha[V_0 + \Delta v]^2.$$

Using the result that $cU_0 = \alpha V_0^2$ and expanding the terms but neglecting the second-order term $-\alpha \Delta v^2$ leads to the approximation

$$M \frac{d}{dt}[\Delta v] \approx c[\Delta u] - 2\alpha V_0 \Delta v.$$

Writing $\Delta v = v_a(t)$ and $\Delta u = u_a(t)$ to indicate the linearized variables, the original differential equation of motion described by Equation 10.33 becomes the linear differential equation

$$M \frac{dv_a(t)}{dt} + 2\alpha V_0 v_a(t) = c u_a(t). \tag{10.34}$$

This is a linear differential equation with constant coefficients, which can be solved by the techniques presented in Chapter 5.

$\square$

TWO-DIMENSIONAL TAYLOR SERIES

The notion of sequences and series of functions of a single variable as described in Chapter 6 can be extended to functions of several variables. For example, the power series expansion for a function of two variables $F(x, y)$ is

$$F(x, y) = \sum_{n=0}^{\infty} f_n(x, y) = f_0(x, y) + f_1(x, y) + \cdots + f_n(x, y) + \cdots, \tag{10.35}$$

with the terms

$$f_n(x, y) = c_{n,0} \, x^n + c_{n,1} \, x^{n-1}y + \cdots + c_{n,n-1} \, xy^{n-1} + c_{n,n} \, y^n.$$

Expanding about the origin, the series would be

$$F(x, y) = c_{0,0} + (c_{1,0}\, x + c_{1,1}\, y) + (c_{2,0}\, x^2 + c_{2,1}\, xy + c_{2,2}\, y^2) + \cdots. \quad \textbf{(10.36)}$$

Assuming that the function can be differentiated at the origin, the first six coefficients of the 2D Taylor series expansion become

$$F(0,0) = c_{0,0}, \qquad \frac{\partial F(0,0)}{\partial x} = c_{1,0}, \qquad \frac{\partial F(0,0)}{\partial y} = c_{1,1},$$

$$\frac{1}{2!}\frac{\partial^2 F(0,0)}{\partial x^2} = c_{2,0}, \qquad \frac{2}{2!}\frac{\partial^2 F(0,0)}{\partial x \partial y} = c_{2,1}, \qquad \frac{1}{2!}\frac{\partial^2 F(0,0)}{\partial y^2} = c_{2,2}, \quad \textbf{(10.37)}$$

where the notation $F(0,0)$ here means that the partial derivatives of $F(x,y)$ are to be evaluated at the point $(0,0)$.

A way to remember the Taylor series for $f(x,y)$ is to set

$$h = x - x_0, \qquad k = y - y_0,$$

and expand the function near (x_0, y_0) as

$$F(x_0 + h, y_0 + k) \;=\; F(x_0, y_0) + \left(h\frac{\partial}{\partial x} + k\frac{\partial}{\partial y} \right) F(x_0, y_0) + \cdots$$

$$+ \frac{1}{n!}\left(h\frac{\partial}{\partial x} + k\frac{\partial}{\partial y} \right)^n F(x_0, y_0) + R_n, \quad \textbf{(10.38)}$$

where R_n is the remainder after n terms. The operator terms for the partial derivative

$$\left(h\frac{\partial}{\partial x} + k\frac{\partial}{\partial y} \right)^n, \qquad n = 0, 1, \ldots,$$

can be expanded by the binomial theorem. Then, the Taylor series of Equation 10.38 can be written

$$F(x_0 + h, y_0 + k) \;=\; F(x_0, y_0) + h\frac{\partial F(x_0, y_0)}{\partial x} + k\frac{\partial F(x_0, y_0)}{\partial y}$$

$$+ \frac{h^2}{2}F_{xx}(x_0, y_0) + hk F_{xy}(x_0, y_0)$$

$$+ \frac{k^2}{2}F_{yy}(x_0, y_0) + \cdots$$

$$+ \sum_{r=0}^{n} C_r^n \frac{\partial^n F(x_0, y_0)}{\partial x^r \partial y^{n-r}}(x - x_0)^r (y - y_0)^{n-r}$$

$$+ \cdots, \quad \textbf{(10.39)}$$

where C_r^n are the binomial coefficients

$$C_r^n = \frac{n(n-1)\cdots(n-r+1)}{r!} = \frac{n!}{r!(n-r)!}. \quad \textbf{(10.40)}$$

The general term in Equation 10.39 is the nth *differential* $d^n F$ of the function $F(x,y)$.

☐ **EXAMPLE 10.7**   *Taylor Series*

The function $f(x,y) = \sqrt{1 + x^2 + y^2}$ can be expanded in a 2D Taylor series about the point $(0,0)$ using the coefficients from Equation 10.37. Computing the terms up to the second derivative in x and y leads to the results

$$
\begin{aligned}
f(0,0) &= c_{0,0} = 1, \\
f_x(0,0) &= c_{1,0} = 0, \quad f_y(0,0) = c_{1,1} = 0, \\
f_{xy}(0,0) &= c_{2,1} = 0, \\
f_{xx}(0,0) &= c_{2,0} = 1, \quad f_{yy}(0,0) = c_{2,2} = 1.
\end{aligned}
$$

Thus, only the coefficients $c_{0,0}$, $c_{2,0}$, and $c_{2,2}$ are nonzero. Substituting these coefficients in Equation 10.36, the approximation is thus

$$
f(x,y) \approx 1 + \frac{1}{2}(x^2 + y^2).
$$

The accompanying MATLAB script calculates and plots the values for

$$
z = f(x,y) = \sqrt{1 + x^2 + y^2},
$$

and the Taylor series approximation for a fixed value of y as input. Figure 10.6 shows the results for $y = 0$. Here we determined the shape of the graphs by the *method of sections*. The section of $f(x,y)$ shown is in the vertical xz-plane, defined by $y = 0$.

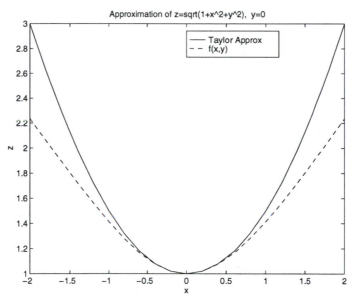

FIGURE 10.6 *Taylor approximation*

Example 10.7

```
% EX10_7.M Plot Taylor series approximation and
%   function f(x,y)= sqrt(1+x^2+y^2); Input y value.
clear               % Clear variables
clf                 %  and figures
x=[-2:0.2:2];
y=input('Input y value= ')
z1=x.^2+y.^2;       % Quadratic terms
Z=sqrt(1+z1);       % Function
ztay=1+z1/2;        % Taylor approximation
% Plot the two graphs in one figure
plot(x,ztay,'-')    % Dotted lines
hold                % Plot on same axis
plot(x,Z,'--')
legend('Taylor Approx','f(x,y)')
title(['Approximation of z=sqrt(1+x^2+y^2), ',...
' y=',num2str(y)])
xlabel('x')
ylabel('z')
zoom                % Change resolution with mouse
```

When x and y are near zero, the approximation is good, but the Taylor series increases much more rapidly than $f(x, y)$ with increasing x or y. The command **zoom** allows us to zoom in or out on portions of the graph using the computer's mouse. The y-value input was displayed in the title because the **num2str** command converted the numerical value to a text string for use by the **title** command.

□

TAYLOR SERIES LINEARIZATION   The 2D Taylor series representation of $F(x, y)$ about the point (x_0, y_0) can be written

$$F(x, y) \quad = \quad F(x_0, y_0) + \frac{\partial F(x_0, y_0)}{\partial x}(x - x_0) + \frac{\partial F(x_0, y_0)}{\partial y}(y - y_0)$$
$$+ \quad R(x - x_0, y - y_0),$$

where $R(x - x_0, y - y_0)$ contains only powers of $(x - x_0)$ and $(y - y_0)$ higher than the first power. For (x, y) close enough to (x_0, y_0), the constant and linear terms can be used and the higher terms neglected. The result is the expression for the first differential of $F(x, y)$,

$$dF(x_0, y_0; dx, dy) = \frac{\partial F}{\partial x} dx + \frac{\partial F}{\partial y} dy = dF,$$

if $x - x_0 = dx$ and $y - y_0 = dy$.

☐ **EXAMPLE 10.8** **2D Taylor Series**

Consider the exponential function of n variables,

$$f(\mathbf{x}) = e^{(x_1 + \cdots + x_n)}.$$

Ignoring higher-order terms that are nonlinear in the variables, the Taylor series approximation near $\mathbf{x} = \mathbf{0}$ is

$$
\begin{aligned}
f(\mathbf{x}) &\approx 1 + \frac{1}{1!}\left(x_1\frac{\partial}{\partial x_1} + \cdots + x_n\frac{\partial}{\partial x_n}\right)f(\mathbf{0}) \\
&= 1 + \frac{1}{1!}(x_1 + \cdots + x_n).
\end{aligned}
$$

From this approximation, it follows that

$$e^{x+y} \approx 1 + (x + y).$$

The remainder is of the form $(x + y)^2/2!$.

☐

MATLAB TWO-DIMENSIONAL INTERPOLATION

Many engineering problems involve the tabulation of a function of two or more variables. When the values arise as a result of measurements, the number of data points is usually limited by practical considerations, such as the amount of physical space available to place sensors or the number of sensors and data channels available. For example, a number of temperature measurements at different positions and depths could be made to collect data about the temperature profile of a region of a body of water. The resulting function might be of the form $T(x, y, z)$, where x, y are positions on a rectangular grid defined in terms of the position on the surface and z is the depth at which a measurement is made.

Consider the temperature data for the function $T(x, y, z)$ at a fixed depth z. At each depth, a table can be constructed with the measurements $Tz(x, y)$. Table 10.2 shows the results for measurements from the points $(x_1, y_1), \ldots, (x_M, y_N)$.

TABLE 10.2 *Function of two variables*

x	$y = y_1$	$y = y_2$	$\cdots$	$y = y_N$
x_1	$Tz(x_1, y_1)$	$Tz(x_1, y_2)$	$\cdots$	$Tz(x_1, y_N)$
x_2	$Tz(x_2, y_1)$	$Tz(x_2, y_2)$	$\cdots$	$Tz(x_2, y_N)$
$\vdots$	$\vdots$	$\vdots$	$\vdots$	$\vdots$
x_M	$Tz(x_M, y_1)$	$Tz(x_M, y_2)$	$\cdots$	$Tz(x_M, y_N)$

The notation $Tz(x, y)$ indicates that z is fixed as x and y vary. If there are M points in x and N points in y, a table of size $M \times N$ results.

Assuming that the intervals in the tabulated data points are too large to approximate the function to the desired degree of accuracy, interpolation between the points is possible. For a table with two variables, one approach is to interpolate twice using one-dimensional interpolation for each of the variables. The interpolated values are then themselves used to perform a *surface* interpolation. Other methods for surface interpolation are discussed in references in the Annotated Bibliography for this chapter.

The MATLAB command **interp2** performs a 2D interpolation using various schemes. Linear interpolation is used if other methods are not specified. Cubic interpolation can be selected, but the data points must be evenly spaced in x and y. In either case, an important use of 2D interpolation is to define a grid of closely spaced points for surface plotting using the commands **mesh** or **surf**. Data values need not be evenly spaced for 2D linear interpolation.

☐ **EXAMPLE 10.9**   *MATLAB 2D Interpolation*

This example shows the effects of two interpolation resolutions to approximate a function using the MATLAB command **interp2**. The results show that care must be taken when using MATLAB commands for interpolation. Also, the command **interp2** has a number of variations that are explained in the MATLAB *User's Guide*.

The accompanying script shows the use of **interp2** to interpolate the functional values of the function

$$z = (x - 2)^2 + (y - 2)^2, \qquad 0 \le x \le 4, \ 0 \le y \le 4.$$

The original increment in x and y is 1 in the script, so that five points in each variable are used to define the grid over which z will be considered. The values (x, y) are converted to a matrix of values by the command **meshgrid** for subsequent use in function evaluation or 3D plotting.

Thus, the MATLAB statement that computes Z in the script yields a 5×5 table (matrix) of values. Sizes of the MATLAB variables involved are shown in the script as a summary from the command **whos**. This summary may help you to visualize the results of the various MATLAB commands.

The surface plot and contour lines are shown in the first plot in Figure 10.7. Notice that the surface points are connected by straight lines and the contour lines have sharp corners. This is the effect of taking a rather course grid for the original (x, y)-values.

The representation can be improved by 2D interpolation of the function Z. The interpolating points are defined as xi and yi with a resolution of 0.2. After Zi is computed by **interp2**, a 21×21 array results which is plotted in the second part of Figure 10.7.

The command **surfc** plots a surface and draws a contour plot below the surface. Evidently, after interpolation the surface is much smoother and the contour lines are nearly circles, as expected.

Surface without interpolation

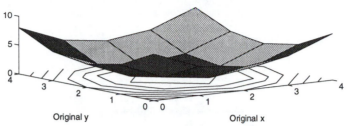

Surface (x–2)^2+(y–2)^2 from 2D Interpolation

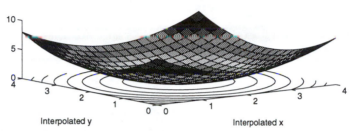

FIGURE 10.7 *MATLAB two-dimensional interpolation results*

MATLAB Script _____

```
Example 10.9
% CLINT2D.M Create data, plot original data and then
%  perform 2D interpolation for Example 10.9.
%
x=[0:1:4];              % Original x points
y=[0:1:4];              % Original y points
[X,Y]=meshgrid(x,y);    % Rows of X are x; columns of Y are y
Z=(X-2).^2+(Y-2).^2;    % A matrix Z(X,Y)
%
xi=[0:.2:4];            % x points to interpolate
yi=[0:.2:4]';           % y points to interpolate
Zi=interp2(X,Y,Z,xi,yi,'cubic');
% Plot surface and contours to compare plots
clf                     % Clear graphics window
subplot(2,1,1), surfc(x,y,Z)
xlabel('Original x ')
ylabel('Original y ')
title('Surface from z=f(x,y)')
subplot(2,1,2), surfc(xi,yi,Zi)
xlabel('Interpolated x ')
ylabel('Interpolated y ')
title('Surface from 2D Interpolation')
% ------------------------------------------------------------
% Size of Variables
>>whos
```

	Name	Size	Elements	Bytes
	X	5 by 5	25	200
	Y	5 by 5	25	200
	Z	5 by 5	25	200
Interp Z	Zi	21 by 21	441	3528
Original x	x	1 by 5	5	40
Interp x	xi	1 by 21	21	168
	y	1 by 5	5	40
	yi	21 by 1	21	168

□

MATLAB DIFFERENTIATION

Table 10.3 presents MATLAB commands for numerical and symbolic differentiation. These commands will be discussed in this section.

TABLE 10.3 *MATLAB commands for differentiation*

Command	Purpose
$y = f(x)$:	
diff	Computes differences
polyder	Derivative of a polynomial
Symbolic:	
diff	Differentiation

DIFFERENCE
CALCULUS

For continuous functions, we can consider the derivative operator d/dx operating on a function $f(x)$ to yield the derivative of the function. In a similar manner, we can define an operator that computes the difference between elements of a sequence $(f_0, f_1, \ldots)$. This difference operator can be applied to compute an approximate derivative of a function.

If a sequence contains values of a function $f(x)$ defined at equally spaced points $x_n = x_0 + nh$ for $n = 0, 1, \ldots$, then the function values can be designated

$$f_n = f(x_n) = f(x_0 + nh).$$

The *forward difference operator*, designated Δ, is defined as computing the value

$$\Delta f = f_{n+1} - f_n.$$

The difference operator applied to a function $f(x)$ tabulated at points in x spaced $\Delta x = h$ apart yields

$$\Delta f(x) = f(x + h) - f(x). \tag{10.41}$$

Since the first derivative of $f(x)$ is defined as

$$\frac{df(x)}{dx} = \lim_{h \to 0} \frac{f(x + h) - f(x)}{h},$$

the expression can be written in terms of the difference operator as

$$\frac{df(x)}{dx} = \lim_{h \to 0} \frac{\Delta f(x)}{h}.$$

If the step size h is sufficiently small, an accurate representation of the derivative of $f(x)$ may be computed using the approximation

$$\frac{df(x)}{dx} \approx \frac{\Delta f(x)}{\Delta x}. \tag{10.42}$$

The error involved in approximating derivatives using differences depends greatly on the step size. The effect of step size on various approximations was discussed in Chapter 6.

The second derivative can be approximated as

$$\frac{d^2 f(x)}{dx^2} \approx \frac{\Delta^2 f(x)}{\Delta x^2}, \tag{10.43}$$

where $\Delta^2 f(x) \equiv \Delta[\Delta f(x)]$.

☐ **EXAMPLE 10.10**   *Differences*

For a second-degree polynomial $f(x) = ax^2 + bx + c$, the first difference is

$$
\begin{aligned}
\Delta f(x) &= \Delta(ax^2 + bx + c) \\
&= a[(x + h)^2 - x^2] + b[(x + h) - x] + c[1 - 1] \\
&= 2axh + ah^2 + bh.
\end{aligned}
$$

According to Equation 10.42 with $h = \Delta x$, the approximate derivative is thus

$$\frac{\Delta f(x)}{h} = \frac{\Delta f(x)}{\Delta x} = 2ax + b + ah.$$

Letting $h \to 0$ shows that the ratio of differences does indeed yield the derivative of the polynomial when h becomes so small that the term ah becomes insignificant compared to the other terms.

☐

POLYNOMIAL INTERPOLA- TION

The derivative of a polynomial is easily computed if the coefficients are known. For numerical problems, the MATLAB command **polyder** will compute the coefficients of the derivative polynomial given the coefficients of the original polynomial.

Interpolation can also be used if a function to be differentiated is tabulated, as may be the case when the function is to be defined from experimental data. However, even if the interpolating polynomial adequately represents the function being analyzed, its derivatives may be quite different from those of the original functions. Higher derivatives of the interpolating polynomial may not be related at all.

☐ **EXAMPLE 10.11**   *MATLAB Derivatives*
For the polynomial

$$y(x) = 2x^4 - 7x^3 + 5x^2 - 1,$$

the derivative is $y'(x) = 8x^3 - 21x^2 + 10x$. The accompanying MATLAB script computes the derivative in three ways. First, the command **polyder** is used to compute the coefficients of the derivative polynomial. Then, the **diff** command approximates the numerical derivative using two different step sizes. The three results are plotted for comparison in Figure 10.8

Solution by **polyder**. First in the script, the coefficients of the polynomial are defined. Then, the command **polyder** computes the coefficients of the derivative function from the coefficients of the polynomial. These coefficients are the elements in the vector pd listed following the M-file statements.

The command **polyval** evaluates the derivative polynomial at the 100 points generated by the command **linspace**. The vector yder is created for plotting.

Solution by **diff**. In the script, **diff** commands are used to approximate the derivative as

```
diff(y)./diff(x)
```

The expression that is calculated is

$$\frac{y_{i+1} - y_i}{x_{i+1} - x_i}, \quad i = 1, 2, 3, \ldots, n - 1,$$

which contains one less point than the array of y values. The accuracy of the approximation depends on the step size $\Delta x = x_{i+1} - x_i$ and the number of points in the interval of interest.

The values of the derivative are first approximated using the **diff** commands on a coarse interval in x ($\Delta x = 0.5$) with only seven points. The evaluation points are listed as xd after the program.

To improve the accuracy of the approximation, a closer spacing in x can be taken. Thus, a more accurate numerical approximation is attempted by evaluating the function at 100 points and using **diff** again. The broken line plot of Figure 10.8 shows the graph of the derivative function evaluated at 100 points.

In Figure 10.8, the difference between the graph of the approximate derivative using **diff** with 99 points ($\Delta x = 0.0303$) and the assumed correct values computed by **polyder** and **polyval** is almost indistinguishable. The **zoom** command can be used to compare the results more closely. The result calculated with the **diff** commands using ($\Delta x = 0.5$) noticeably deviates from the actual derivative with increasing x, as shown by the points plotted in the figure.

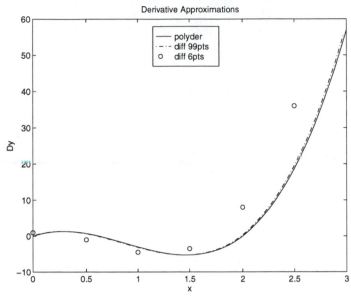

FIGURE 10.8 *MATLAB derivatives*

MATLAB Script ⎯⎯⎯⎯⎯⎯⎯⎯⎯⎯⎯⎯⎯⎯⎯⎯⎯⎯⎯⎯⎯⎯⎯⎯⎯⎯⎯⎯

```
Example 10.11
% CLPOLYDV.M  Compute derivatives of
% f(x)=2x^4-7x^3+5x^2-1
% Compare diff and polyder results
p=[2 -7 5 0 -1];            % Coefficients
pd=polyder(p)               % Polynomial derivative
xi=linspace(0,3,100);       % 0-3 for 100 points
yder=polyval(pd,xi);        % Evaluate at xi
%
% Derivative using diff
%
x=[0:.5:3];                 % Coarse interval
y=2*x.^4-7*x.^3+5*x.^2-1;
% Using diff with 6 points
dely=diff(y)./diff(x);
xd=x(1:length(x)-1)
%
% More accurate diff using 100 points
yder99=2*xi.^4-7*xi.^3+5*xi.^2-1;
dely1=diff(yder99)./diff(xi);
xd1=xi(1:length(xi)-1)
%
```

```
clf                          % Clear any figures
plot(xi,yder,'-'),hold on    % Ployder
plot(xd1,dely1,'-.')         % diff 99 points
plot(xd,dely,'o')            % diff  6 points
title('Derivative Approximations')
xlabel('x'),ylabel('Dy')
legend('polyder','diff 99pts','diff 6pts')
hold off
% -------------------------------------------

% Results (From edited diary file)
pd =
    8   -21    10     0
xd =
    0    0.5000    1.0000    1.5000    2.0000    2.5000
```

□

APPROXIMATE
PARTIAL
DERIVATIVES

There are a number of approximations for the partial derivatives of a function. The approach presented here uses *central differences* to approximate the derivatives. Central difference expressions contain the same number of function values ahead of the reference location as backward from it.

Let the notation $f_{i,j}$ mean to evaluate the function $f(x,y)$ at the point (x_i, y_j), as shown in Figure 10.9.

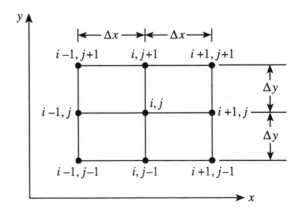

FIGURE 10.9 *Grid for partial derivatives*

Then, the first partial derivatives can be approximated as

$$\frac{\partial f}{\partial x} \approx \frac{1}{2\Delta x}(f_{i+1,j} - f_{i-1,j}),$$
$$\frac{\partial f}{\partial y} \approx \frac{1}{2\Delta y}(f_{i,j+1} - f_{i,j-1}), \tag{10.44}$$

where the 2 in the denominator arises from the fact that the derivative is taken in two steps in x or y. The second derivative with respect to x can be approximated as

$$\frac{\partial^2 f}{\partial x^2} \approx \frac{1}{\Delta x^2}(f_{i-1,j} - 2f_{i,j} + f_{i+1,j}), \qquad \textbf{(10.45)}$$

with a similar expression for the second derivative with respect to y.

Problem 10.10 asks you to find the expression for the mixed partial derivative

$$\frac{\partial^2 f}{\partial x \partial y}$$

for a function $f(x, y)$

NUMERICAL DIFFERENTIA- TION

Numerical differentiation may cause severe numerical errors, since small changes in the shape of a function can lead to large changes in the derivative. This is because the derivative is a function that depends on the shape of the function at each point. In contrast, numerical integration is not as sensitive to changes in the function since the integral generally depends on the functional values over a relatively wide interval where the function is defined.

There are a large number of formulas for numerical differentiation, and many are described in the references listed in the Annotated Bibliography at the end of this chapter. We have presented only some techniques that are conveniently programmed with MATLAB.

EXTREMA OF REAL-VALUED FUNCTIONS

The most basic geometric features of the graph of a function are its *extreme* points. At such points, the function attains its greatest and least values. Finding the location of the maxima and minima (extrema) is based on the observation from one-variable calculus that each extremum within the domain of definition of the function is a *critical point*. Fortunately, the technique of finding extrema of functions of several variables is easily generalized from the calculus of single-variable functions.

For a function $y = f(x)$, extreme values can occur at points where $df/dx = 0$, or at the ends of the interval where f is defined. In higher dimensions, partial derivatives are computed to find critical points, as described in this section. Also, this section presents MATLAB commands to find extrema of functions.

EXTREMA OF A
FUNCTION OF
ONE VARIABLE

In the calculus of functions of one variable $f(x)$, the domain of the function is an interval on the real axis. A maximum or minimum may exist within the interval over which the function is defined. For other functions, the extreme values may occur at the ends of an interval in x.

In general, the behavior of the function at the endpoints requires a separate investigation from that required to find the extrema at interior points of the domain. Because of this, we must distinguish between the open and closed interval over which $f(x)$ is defined. The *open* interval (a, b) is the set of all real numbers x such that $a < x < b$. The *closed* interval $[a, b]$ contains all the points $a \leq x \leq b$.

As an example, consider the continuous function

$$f(x) = x$$

defined only for $0 \leq x < 1$. The domain is thus a half-open interval. Drawing a simple graph of the function shows that f attains its minimum value of 0 at $x = 0$. There is no maximum value in the domain of f. If the function were defined on the closed interval $[0, 1]$, the maximum would occur at $x = 1$. It can be shown that a continuous function defined on a closed interval attains its minimum and maximum at points of the interval.

Local Extrema For a differentiable function $y = f(x)$ defined in an *open interval*, a local or relative minimum or maximum occurs only at a point where the derivative is zero so that

$$f'(x) = 0.$$

Such points are called *critical points* of the function. At such a point, the line tangent to the graph of the function is horizontal. The term *local extremum* is used because the function value is a minimum or maximum relative only to nearby points.

However, the converse statement regarding critical points of the functions is not true. In some cases, the function may have a zero derivative at a point but neither a maximum or minimum. Such points are called *inflection points*. To show the possible cases, the graph of $f(x)$ in Figure 10.10 has a maximum, an inflection point, and a minimum in the interval shown.

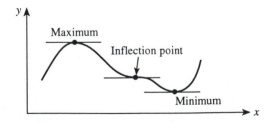

FIGURE 10.10 *Maximum, inflection point, and minimum for a function*

Second-Derivative Test From elementary calculus, we know that the sign of the first derivative indicates whether the graph of the function $f(x)$ is rising or falling. The sign of the second derivative indicates the way the curve is bending, upward or downward.

Let $f(x)$ be defined in some *open* interval in which f' exists. The *second-derivative test* for a maximum or a minimum of a function of one variable $y = f(x)$ in some open interval containing x_0 is as follows:

1. *If $f'(x_0) = 0$ and $f''(x_0) > 0$, then $f(x)$ has a relative minimum at x_0.*

2. *If $f'(x_0) = 0$ and $f''(x_0) < 0$, then $f(x)$ has a relative maximum at x_0.*

From the previous discussion, it is clear that $f(x) = 0$ is a necessary condition for the existence of an extremum of f in an open interval. The second derivative test provides both the necessary and sufficient conditions for a local extremum.

EXTREMA OF A MULTIVARIATE FUNCTION

Consider a region in which a scalar function of two or more variables is defined such that

$$\mathbf{R}^n \xrightarrow{f} \mathbf{R}.$$

Such a multivariate function may have a local maximum or minimum at points $\mathbf{x}_0 = [x_1, \ldots, x_n]$ in the domain for which

$$f'(\mathbf{x}_0) = 0. \tag{10.46}$$

These are the *critical points* of the multivariate function. The function is said to have a *local* minimum (or maximum) at $\mathbf{x} = \mathbf{x}_0$ if all the function values near this point are greater (or less) than $f(\mathbf{x}_0)$.

The function has an *absolute maximum value* at $\mathbf{x} = \mathbf{x}_0$ if, for all $\mathbf{x}$ in the domain of f,

$$f(\mathbf{x}) \leq f(\mathbf{x}_0).$$

Similarly, the function has an *absolute minimum value* at $\mathbf{x}_0$ if

$$f(\mathbf{x}_0) \leq f(\mathbf{x})$$

for all $\mathbf{x}$ in the domain.

Necessary Conditions for Extrema The test to find critical points at *interior points* of the region on which $f(\mathbf{x})$ is defined involves calculating the partial derivatives of the function. To apply the condition of Equation 10.46, we form the row matrix

$$f'(\mathbf{x}_0) = \begin{bmatrix} \dfrac{\partial f}{\partial x_1} & \dfrac{\partial f}{\partial x_2} & \cdots & \dfrac{\partial f}{\partial x_n} \end{bmatrix}, \tag{10.47}$$

where it is understood that each partial derivative is to be evaluated at $\mathbf{x}_0$. If $f(\mathbf{x})$ attains a local maximum or a local minimum at $\mathbf{x} = \mathbf{x}_0$, it is necessary that

$$\frac{\partial f}{\partial x_1} = \frac{\partial f}{\partial x_2} = \cdots = \frac{\partial f}{\partial x_n} = 0 \qquad (10.48)$$

at the point $\mathbf{x} = \mathbf{x}_0$.

In general, equating the partial derivatives to zero leads to a set of simultaneous equations in the n variables. Thus, there could be n points in $\mathbf{R}^n$ where the equations are satisfied. However, the vanishing of the partial derivatives is *not* a sufficient condition to guarantee a local extremum. Tests of the higher derivatives may be necessary, as is true for a function of a single variable.

□ EXAMPLE 10.12   **Extreme values**

Consider the function

$$z = f(x, y) = (x - 2)^2 + (y - 2)^2$$

defined in $\mathbf{R}^2$, which was studied in Example 10.9. Setting the partial derivatives to zero to find the critical points yields

$$\frac{\partial f}{\partial x} = 2(x - 2) = 0, \qquad \frac{\partial f}{\partial y} = 2(y - 2) = 0.$$

Thus, the only finite extreme value can occur at $(x, y) = (2, 2)$. The function achieves an absolute minimum at this point since $f(x, y) > f(2, 2)$ everywhere $f(x, y)$ is defined. Because of the simplicity of this function, there is no need to check further to verify our conclusions.

□

Second-Derivative Tests   For a function of n variables, a criterion can be developed to test a critical point as a relative extremum. The sufficient condition for extrema depends on the second derivative, as it does for functions of a single variable. For $n = 1$, the criterion must reduce to the familiar condition that $f'' > 0$ for a minimum and $f'' < 0$ for a maximum. Although functions in higher dimension can be treated, we restrict our discussion to functions $f(x, y)$ defined in $\mathbf{R}^2$.

Functions $f(x, y)$ The second partial derivatives of a function $f(x, y)$ must be computed to determine the sufficient conditions for local extrema. A theorem that is similar to the second derivative test for a function of one variable yields the necessary conditions.

■ THEOREM 10.1   **Sufficient Conditions for Local Extrema**

Let $f(x, y)$ have continuous first-order and second-order partial derivatives in an open region containing the point (x_0, y_0). If

$$\partial f(x_0, y_0)/\partial x = 0, \quad \partial f(x_0, y_0)/\partial y = 0,$$

and the product

$$D \equiv f_{xx}(x_0, y_0)f_{yy}(x_0, y_0) - f_{xy}^2(x_0, y_0) > 0,$$

a. $f(x, y)$ *has local minimum at* (x_0, y_0) *if* $f_{xx}(x_0, y_0) > 0$;

b. $f(x, y)$ *has local maximum at* (x_0, y_0) *if* $f_{xx}(x_0, y_0) < 0$.

──────────────────────────────■

If the expression $D < 0$, then $f(x, y)$ has neither a minimum or maximum at (x_0, y_0) and the point (x_0, y_0) is called a *saddle point* of $f(x, y)$. If $D = 0$, there is no information about the critical points of $f(x, y)$. In such cases, derivatives higher than the second must be investigated.

□ **EXAMPLE 10.13**   *Extremum of a function*

To find the critical points of the function

$$z = x^2 + xy + y^2 + 3x - 3y + 1,$$

we compute the partial derivatives as indicated by Equation 10.48 and find the zero points as

$$\frac{\partial z}{\partial x} = 2x + y + 3 = 0,$$

$$\frac{\partial z}{\partial y} = x + 2y - 3 = 0.$$

The solution to this set of equations yields $(x, y) = (-3, 3)$ as a critical point. This function is tested for extrema at the critical point by forming the second partial derivatives evaluated at $(-3, 3)$. The values are

$$f_{xx} = 2,$$
$$f_{yy} = 2,$$
$$f_{xy} = 1,$$

and the product

$$D = f_{xx}f_{yy} - f_{xy}^2 = 3.$$

The results indicate a minimum since $D > 0$ and $f_{xx} > 0$ according to Theorem 10.1. The functional value at the critical point is $f(-3, 3) = -8$.

□

W H A T   I F ?   For all interior points $\mathbf{x}$ of the domain of $f(\mathbf{x})$, we seek extreme points only among those for which $f'(\mathbf{x}) = 0$. However, suppose the function is restricted to a domain that is a subset of $\mathbf{R}^n$. Give examples of functions such that $f'(\mathbf{x}) = 0$, but $\mathbf{x}$ is not an extreme point for f. Also, think of functions that have extreme points without having $f'(\mathbf{x}) = 0$. See Problem 10.22 for an example function.

MATLAB COMMANDS FOR FINDING EXTREMA AND ZEROS

Table 10.4 lists the MATLAB commands for finding the minimum of a function and the zero crossings of a function.

TABLE 10.4 *MATLAB commands for minimum and zeros*

Command	Purpose
$y = f(x)$	
fmin	Finds minimum of a function
find	Finds specified values
fzero	Finds zero crossings
Multivariate functions:	
fmins	Finds minimum of a function

The MATLAB command **fmin** finds the location of a minimum for a function of one variable. For functions of several variables, the command **fmins** is used. The negative of the function can be used if the maxima are to be found. The command **fzero** finds points where a function of one variable $f(x)$ crosses the x axis.

Numerical methods to find a minimum of a function involve *searching* for the minimum by evaluating the function repeatedly. The computation time and perhaps even the success of a particular method depends on the search range defined. Thus, the MATLAB commands to find a minimum require an initial "guess" of a point reasonably near a critical point of the function. The initial point can often be found by plotting functions of one or two variables over regions of interest.

☐ EXAMPLE 10.14 **_MATLAB Minimum_**

Consider finding the minimum of the function

$$z = f(x, y) = x^2 + xy + y^2 + 3x - 3y + 1$$

from Example 10.13 using the **fmins** command. Assuming for the moment that the analytic solution is unknown, it is necessary to define an initial point (x_0, y_0) before using **fmins**. It is clear that as x and y become large in the positive direction, the function is monotonically increasing as $x^2 + y^2$, so the search can be limited to fairly small values of x and y.

Plotting the function using our M-file CLPLOT2.M yields the results of Figure 10.11. When CLPLOT2.M is executed, the function must be entered as a string using matrix notation for the operators. The period (.) operator causes element-by-element operations, as explained in Chapter 1. This operation is necessary to define the function in 3D for plotting. Results from the diary file are listed after the M-file. The figure shows that the minimum is reached in the quadrant with $(-x, y)$, so a guess for the starting point for **fmins** might be $(-1, 1)$.

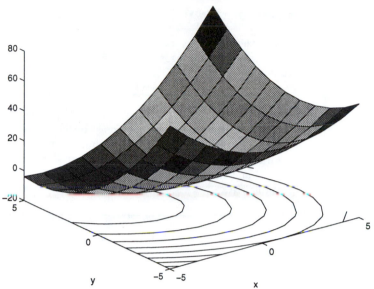

FIGURE 10.11 *Plot of* $f(x,y) = x^2 + xy + y^2 + 3x - 3y + 1$

MATLAB Script _____

Example 10.14

```
% CLPLOT2.M Plot an input as 'fn(x.,y.)'. Function must be input as
%   a string with matrix variables. Input plot limits in x and y.
%   Example:  'x.^2+y.^2+x.*y+2x+1'
clear                   % Clear variables and
clf                     %   figures
fn=input('Function to minimize -a string using matrices f(x.,y.)=')
% Plot the function using the limits input
x0=input('Input the plot limits [xmim xmax ymim ymax]= ')
x=[x0(1):.1:x0(2)];    % Create x and y points
y=[x0(3):.1:x0(4)];
%
% Use a mesh
X=x0(1):1:x0(2);
Y=x0(3):1:x0(4);
[x,y]=meshgrid(X,Y);    % Form matrices for x and y
fgrid=eval(fn);         % Form a grid of function values
surfc(x,y,fgrid)        % Plot surface
xlabel('x')
ylabel('y')
% -------------------------------------
% Inputs of function and limits (Diary)
>>clplot2      % Execute the M-file

Function to minimize as a string using matrices
```

```
f(x.,y.)='x.^2+x.*y+y.^2+3*x-3*y+1'
fn =
x.^2+x.*y+y.^2+3*x-3*y+1

Input the plot limits
[xmim xmax ymim ymax]= [-5 5 -5 5]
x0 =
  -5    5    -5    5
```

The M-file CLOPT.M uses **fmins** to find the minimum point of the function. For this M-file, the function must be defined as a scalar function of a vector variable for **fmins**. The accompanying script shows the inputs to CLOPT.M from the diary file and the results. As in Example 10.13, the minimum is at $(-3, 3)$ and the minimum value is $(-3, 3) = -8$, as indicated by the MATLAB variable **zmin**.

MATLAB Script _____
Example 10.14
```
% CLOPT.M Find the minimum of a function input as 'fn(x(1),x(2))'
%  Input function and initial guess.
clear
fn=input('Function to minimize as a string f(x(1),x(2))=')
%
% Input the limits for minimization
xguess=input('Input the starting vector [x(1) x(2)]= ')
%
xmin=fmins(fn,xguess) % Find minimum
% Display minimum value of function
x(1)=xmin(1);
x(2)=xmin(2);
zmin=eval(fn)
% Results from diary file
>>clopt      % Execute M-file
Function to minimize as a string

f(x(1),x(2))='x(1)^2+x(1)*x(2)+x(2)^2+3*x(1)-3*x(2)+1'
fn =
x(1)^2+x(1)*x(2)+x(2)^2+3*x(1)-3*x(2)+1

Input the starting vector [x(1) x(2)]= [-1 1]
xguess =
    -1     1
% Results
xmin =
   -3.0000    3.0000
zmin =
   -8.0000
>>quit
```

□

VECTOR FUNCTIONS AND CURVILINEAR COORDINATES

We now turn to vector functions defining transformations between the vector spaces

$$\mathbf{R}^n \longrightarrow \mathbf{R}^m.$$

We will designate vector functions by boldface notation, so that

$$\mathbf{F}(\mathbf{x}) = (\, F_1(\mathbf{x}), \ldots, F_m(\mathbf{x})\,)$$

is the vector function $\mathbf{F}(\mathbf{x})$ with scalar coordinate functions $F_1, \ldots, F_m$. For many applications, the properties of the vector function can be defined in terms of the coordinate functions. As an example, the following theorem defines the limit of a vector function in terms of the coordinate functions.

■ THEOREM 10.2   ***Limit of a Vector Function***

Given a function

$$\mathbf{R}^n \xrightarrow{\ \mathbf{F}\ } \mathbf{R}^m$$

with coordinate functions $F_1, \ldots, F_m$ and a point

$$\mathbf{y}_0 = (y_1, \ldots, y_m) \in \mathbf{R}^m,$$

then

$$\lim_{\mathbf{x} \to \mathbf{x}_0} \mathbf{F}(\mathbf{x}) = \mathbf{y}_0$$

if and only if

$$\lim_{\mathbf{x} \to \mathbf{x}_0} F_i(\mathbf{x}) = y_i, \quad i = 1, \ldots, m.$$

A consequence of this theorem is that a vector function is continuous at a point if and only if its coordinate functions are continuous there. The derivative of a vector function can also be defined in terms of the coordinate functions. For example, the partial derivative of $\mathbf{F}$ with respect to x_j would be

$$\frac{\partial \mathbf{F}(\mathbf{x})}{\partial x_j} = \left[\, \frac{\partial F_1(\mathbf{x})}{\partial x_j} \quad \frac{\partial F_2(\mathbf{x})}{\partial x_j} \quad \cdots \quad \frac{\partial F_m(\mathbf{x})}{\partial x_j} \,\right].$$

There are several special cases of vector functions that are of particular importance in this textbook. For example, the function

$$\mathbf{R}(t) = x(t)\,\mathbf{i} + y(t)\,\mathbf{j} + z(t)\,\mathbf{k}$$

represents a mapping from an interval in $\mathbf{R}$ into $\mathbf{R}^3$. As t varies, the vector function traces out a three-dimensional curve in space. Other vector functions to be studied in this section and in later chapters are mostly vectors defined on $\mathbf{R}^2$ or $\mathbf{R}^3$. In $\mathbf{R}^3$, the vector function would have the form

$$\mathbf{F}(x_1, x_2, x_3) = [\, F_1(x_1, x_2, x_3), F_2(x_1, x_2, x_3), F_3(x_1, x_2, x_3) \,]$$

written in terms of the components. As a vector in rectangular coordinates, $\mathbf{F}$ could be written

$$\mathbf{F}(x, y, z) = F_1(x, y, z)\, \mathbf{i} + F_2(x, y, z)\, \mathbf{j} + F_3(x, y, z)\, \mathbf{k},$$

using the unit vectors $\mathbf{i}$, $\mathbf{j}$, $\mathbf{k}$ defined in Chapter 2. Such vector functions will now be studied in the cylindrical and spherical coordinate systems.

CURVILINEAR COORDINATES

Many formulas in engineering and physics can be simplified by choosing the most convenient system of coordinates. Mathematically, the coordinates of the formula may originally be designated as $(x_1, \ldots, x_n)$ and a transformation is sought to assign new coordinates $(u_1, \ldots, u_n)$. Although there are many possible coordinate systems, this section discusses only the cylindrical and spherical systems shown in Figure 10.12. These systems are of primary interest in mathematical physics.

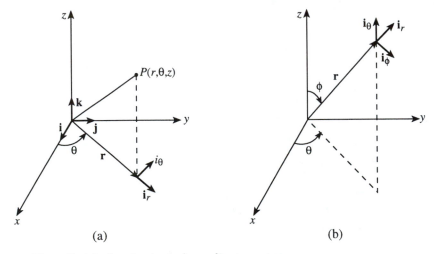

(a)   (b)

FIGURE 10.12 *The cylindrical and spherical coordinate systems*

Figure 10.12 shows the cylindrical and spherical coordinate systems and the unit vectors associated with each system. These two systems are examples of coordinate systems called *curvilinear* because if all but one of the nonrectangular coordinates are held fixed and the remaining one is varied, the coordinate transformation describes a curve in space. An important consequence of this is that the unit vectors defining a point in the curvilinear coordinate system may not be constant as the point moves in space since their direction changes as the point changes position. In

Figure 10.12, this is true for all of the unit vectors, except for $\mathbf{i}$, $\mathbf{j}$, $\mathbf{k}$ in rectangular coordinates and $\mathbf{i}_z = \mathbf{k}$ in cylindrical coordinates.

Notation. In this section, the vector $\mathbf{R}$ will be used as the vector from the origin to a point in space. The vector $\mathbf{r}$ is used as in Figure 10.12 to indicate the vector in cylindrical coordinates from the z-axis to the projection of the point of interest in the xy-plane. Some textbooks make no distinction and others use ρ in place of the vector $\mathbf{r}$. In any case, care should be taken not to confuse the two position vectors in cylindrical coordinates.

The dot notation,

$$\dot{f}(t) \equiv \frac{df(t)}{dt},$$

is used to indicate the derivative with respect to t for any function of time. The second derivative with respect to time is often designated $\ddot{f}(t)$.

Consider the position vector of a point moving in rectangular coordinates,

$$\mathbf{R}(t) = x(t)\,\mathbf{i} + y(t)\,\mathbf{j} + z(t)\,\mathbf{k}, \qquad (10.49)$$

where $x(t)$, $y(t)$, and $y(t)$ define the positions of the point at each moment along the x, y, and z axes, respectively. The time derivative of $\mathbf{R}$ represents the velocity of the point and is computed as

$$
\begin{aligned}
\frac{d\mathbf{R}}{dt} &= \dot{x}(t)\,\mathbf{i} + \dot{y}(t)\,\mathbf{j} + \dot{z}(t)\,\mathbf{k} + x(t)\frac{d\mathbf{i}}{dt} + y(t)\frac{d\mathbf{j}}{dt} + z(t)\frac{d\mathbf{k}}{dt} \\
&= \dot{x}(t)\,\mathbf{i} + \dot{y}(t)\,\mathbf{j} + \dot{z}(t)\,\mathbf{k} \qquad (10.50)
\end{aligned}
$$

since the rectangular unit vectors are constants.

CYLINDRICAL COORDINATES

Referring to Figure 10.12, the cylindrical coordinates (r, θ, z) are related to rectangular coordinates by

$$x = r\cos\theta, \qquad y = r\sin\theta, \qquad z = z. \qquad (10.51)$$

Writing the unit vectors in cylindrical coordinates in terms of the rectangular set yields

$$
\begin{aligned}
\mathbf{i}_r &= \cos\theta\,\mathbf{i} + \sin\theta\,\mathbf{j}, \\
\mathbf{i}_\theta &= -\sin\theta\,\mathbf{i} + \cos\theta\,\mathbf{j}, \\
\mathbf{i}_z &= \mathbf{k}.
\end{aligned}
\qquad (10.52)
$$

Both the scalar transformation of Equation 10.51 and the transformations for the unit vectors of Equation 10.52 can be solved to define the inverse transformation. Table 10.5 indicates both transformations.

TABLE 10.5 *Cylindrical coordinates*

Cylindrical $r\theta z$	Rectangular xyz
$r = \sqrt{x^2 + y^2}$	$x = r\cos\theta$
$\theta = \arctan y/x$	$y = r\sin\theta$
$z = z$	$z = z$
$\mathbf{i}_r = \cos\theta\,\mathbf{i} + \sin\theta\,\mathbf{j}$	$\mathbf{i} = \cos\theta\,\mathbf{i}_r - \sin\theta\,\mathbf{i}_\theta$
$\mathbf{i}_\theta = -\sin\theta\,\mathbf{i} + \cos\theta\,\mathbf{j}$	$\mathbf{j} = \sin\theta\,\mathbf{i}_r + \cos\theta\,\mathbf{i}_\theta$
$\mathbf{i}_z = \mathbf{k}$	$\mathbf{k} = \mathbf{i}_z$

DISTANCE FORMULA

The distance between endpoints of two vectors in curvilinear coordinates can be computed by transforming to rectangular coordinates. Thus, the distance between two points in cylindrical coordinates $P_1(r_1, \theta_1, z_1)$ and $P_2(r_2, \theta_2, z_2)$ is

$$d = \left[(r_1\cos\theta_1 - r_2\cos\theta_2)^2 + (r_1\sin\theta_1 - r_2\sin\theta_2)^2 + (z_1 - z_2)^2 \right]^{1/2}.$$

Expanding and collecting terms results in the distance formula

$$d = \left[r_1^2 + r_2^2 - 2r_1 r_2 \cos(\theta_1 - \theta_2) + (z_1 - z_2)^2 \right]^{1/2}, \quad \textbf{(10.53)}$$

for the distance between two points in cylindrical coordinates.

☐ **EXAMPLE 10.15**

Cylindrical Coordinates

Suppose the position of a particle is defined by the vector

$$\mathbf{R}(t) = x(t)\,\mathbf{i} + y(t)\,\mathbf{j} + z(t)\,\mathbf{k}$$

in rectangular coordinates, and it is desired to determine the velocity of the particle in cylindrical coordinates. The position vector in cylindrical coordinates is easily determined by writing $\mathbf{R}(t)$ with x and y converted to cylindrical coordinates and using the conversions of the unit vectors in Table 10.5. The result is

$$\begin{aligned} \mathbf{R}(t) &= r\cos\theta\,\mathbf{i} + r\sin\theta\,\mathbf{j} + z\,\mathbf{k} \\ &= r\,\mathbf{i}_r + z\,\mathbf{i}_z, \end{aligned}$$

in which r, θ, z, and the unit vector $\mathbf{i}_r$ are functions of time.

The velocity of the particle is found by differentiating the position vector with respect to time, including the unit vector $\mathbf{i}_r$, to yield

$$\mathbf{v} = \frac{d\mathbf{R}(t)}{dt} = \frac{dr}{dt}\,\mathbf{i}_r + r\frac{d\mathbf{i}_r}{dt} + \frac{dz}{dt}\,\mathbf{i}_z.$$

Differentiating the unit vector i_r with respect to time leads to the expression

$$\frac{di_r}{dt} = \frac{d}{dt}(\cos\theta\,\mathbf{i} + \sin\theta\,\mathbf{j})$$

$$= \frac{d\theta}{dt}[-\sin\theta\,\mathbf{i} + \cos\theta\,\mathbf{j}] = \frac{d\theta}{dt}\mathbf{i}_\theta = \dot\theta\,\mathbf{i}_\theta,$$

so that the expression for $d\mathbf{R}(t)/dt$ becomes

$$\mathbf{v} = \dot r\,\mathbf{i}_r + r\dot\theta\,\mathbf{i}_\theta + \dot z\,\mathbf{i}_z.$$

The speed is determined by combining the radial velocity, angular velocity, and the velocity in the z-direction as

$$|\mathbf{v}| = \sqrt{\dot r^2 + r^2\dot\theta^2 + \dot z^2}.$$

$\square$

SPHERICAL COORDINATES

Referring to Figure 10.12, the spherical coordinates (r, θ, ϕ) are related to rectangular coordinates by the equations

$$x = r\cos\theta\sin\phi, \qquad y = r\sin\theta\sin\phi, \qquad z = r\cos\phi. \qquad \textbf{(10.54)}$$

Notice that if $\phi = 90°$, the spherical relationships become those for cylindrical coordinates in the xy-plane. If $\phi = 0°$, $r = z$. The transformations between rectangular and spherical coordinates, including the unit vectors, are listed in Table 10.6.

TABLE 10.6 *Spherical coordinates*

Rectangular xyz	*Spherical r$\theta\phi$*
$x = r\cos\theta\sin\phi$	$r = \sqrt{x^2 + y^2 + z^2}$
$y = r\sin\theta\sin\phi$	$\theta = \arctan y/x$
$z = r\cos\phi$	$\phi = \arctan\sqrt{x^2 + y^2}/z$

Spherical unit vectors:
$\mathbf{i}_r = \cos\theta\sin\phi\,\mathbf{i} + \sin\theta\sin\phi\,\mathbf{j} + \cos\phi\,\mathbf{k}$
$\mathbf{i}_\theta = -\sin\theta\,\mathbf{i} + \cos\theta\,\mathbf{j}$
$\mathbf{i}_\phi = \cos\theta\cos\phi\,\mathbf{i} + \sin\theta\cos\phi\,\mathbf{j} - \sin\phi\,\mathbf{k}$

Rectangular unit vectors:
$\mathbf{i} = \cos\theta\sin\phi\,\mathbf{i}_r - \sin\theta\,\mathbf{i}_\theta + \cos\theta\cos\phi\,\mathbf{i}_\phi$
$\mathbf{j} = \sin\theta\sin\phi\,\mathbf{i}_r + \cos\theta\,\mathbf{i}_\theta + \sin\theta\cos\phi\,\mathbf{i}_\phi$
$\mathbf{k} = \cos\phi\,\mathbf{i}_r - \sin\phi\,\mathbf{i}_\phi$

□ EXAMPLE 10.16　　*Electric Dipole Example*

In many problems, it is convenient to use several coordinate systems to simplify the results. Such cases frequently arise when the solution has symmetry in one coordinate system, but the problem is posed in another. Consider the charge configuration and the vectors in Figure 10.13.

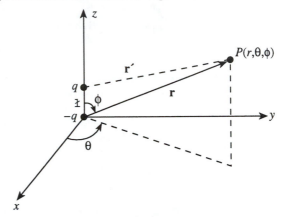

FIGURE 10.13 *Coordinate system for electric dipole*

An important result of electromagnetic field theory is that the *potential* caused by a charge at a point in space is

$$\Phi(r, \theta, \phi) = \frac{q}{4\pi\epsilon_0 r},$$

where r is the magnitude of the vector from the origin of the charge q to the field point Φ. The unit of Φ is volts and $\epsilon_0 = 8.85 \times 10^{-12}$ coulombs^2/newton·meter^2 in SI units when charge is measured in coulombs.[4]

Using superposition, the potential of two charges of opposite sign displaced by l meters along the z-axis, as shown in the figure, becomes

$$\Phi(P) = \frac{q}{4\pi\epsilon_0}\left[\frac{-1}{r} + \frac{1}{r'}\right] = \Phi(P)_- + \Phi(P)_+. \qquad \textbf{(10.55)}$$

By the law of cosines, $r' = (r^2 + l^2 - 2rl\cos\phi)^{1/2}$, so the potential expression is quite complicated in terms of the spherical coordinates. For field points far from the charges, the expression can be simplified, as described in Problem 10.14.

Using a Taylor expansion to compute the potential of the positive charge $+q$ with the charge displaced by a distance $-l$ yields the result

$$\Phi(P)_+ \;=\; \frac{q}{4\pi\epsilon_0}\frac{1}{r} + \frac{\partial}{\partial z}\left(\frac{q}{4\pi\epsilon_0 r}\right)(-l)$$
$$+ \text{ higher-order terms,}$$

where r is evaluated at the point P. Since $r = \sqrt{x^2 + y^2 + z^2}$,

$$\frac{\partial}{\partial z}\left(\frac{1}{r}\right) = \frac{-z}{r^3}.$$

---

[4]The SI system is the metric system and is often called the *mks* system, since the primary units are meters, kilograms, and seconds for length, mass, and time, respectively. SI is an abbreviation for the French designation *Système International d'Unités*.

Combining the expression for the potential of the negative charge and the approximation for the positive charge results in the approximation

$$\Phi(P) \approx \frac{q}{4\pi\epsilon_0}\left[-\frac{1}{r}+\frac{1}{r}+\frac{-z}{r^3}(-l)\right]$$

$$\approx \frac{qzl}{4\pi\epsilon_0 r^3}$$

as the total potential at the point P. Noticing that

$$\cos\phi = \frac{z}{r}$$

leads to the result

$$\Phi(P) = \frac{ql\cos\phi}{4\pi\epsilon_0 r^2} \tag{10.56}$$

for the potential far from the location of the charges. The charge configuration is called a *dipole*, and the product ql is the *dipole moment*, which is a measure of the strength of the dipole field at the point P.

$\square$

MATLAB CURVILINEAR COORDINATES

Table 10.7 lists MATLAB commands that are useful to convert between coordinate systems.

TABLE 10.7 *MATLAB commands for coordinate conversion*

Command	Purpose
Conversion:	
cart2pol	Cartesian to polar
pol2cart	Polar to Cartesian conversion
cart2sph	Cartesian to spherical
sph2cart	Spherical to Cartesian
Plots:	
polar	Polar coordinate plot
cylinder	Generate a cylinder
sphere	Generate a sphere

Typing the command **help specfun** will list some of the specialized functions of MATLAB. Use the **help** command with the specific command name to determine the format for these coordinate conversion commands.

REINFORCEMENT EXERCISES AND EXPLORATION PROBLEMS

———————————————■———————————————

In these problems, do the computations by hand unless otherwise indicated and then check those that yield numerical or symbolic results with MATLAB.

REINFORCEMENT EXERCISES

P10.1. Domains Find and draw the domain for the function

$$f(x, y) = \sqrt{6 - (2x + 3y)}$$

if the range is real.

P10.2. Limits Let

$$f(x, y) = \frac{x - y}{x + y},$$

and determine if

$$\lim_{x \to 0} \left(\lim_{y \to 0} \frac{x - y}{x + y} \right) \overset{?}{=} \lim_{y \to 0} \left(\lim_{x \to 0} \frac{x - y}{x + y} \right).$$

P10.3. Derivatives Let

$$f(x, y, z) = xe^{yz} + yze^x$$

and find $\partial f/\partial y$, f_{yx}, f_{yxz} and $\partial^3 f/\partial x^2 \partial y$.

P10.4. Differentials Compare Δz and the total differential dz for $z = x^2 y$ in Example 10.5, given the values

$$x = 4, \ y = 3, \quad \Delta x = -0.01, \quad \Delta y = 0.02.$$

P10.5. Chain rule Given that $w = f(x, y)$ with

$$x = r \cos \theta, \qquad y = r \sin \theta,$$

find $\partial^2 w/\partial r^2$.

P10.6. Chain rule Compute $\partial A/\partial \theta$ from $A = xy$ in Example 10.2.

P10.7. Transformations Consider the transformation

$$
\begin{aligned}
u &= x + y, \\
v &= x - 2y,
\end{aligned}
$$

which maps points in the xy-plane to the uv-plane. What is the transformation of the line $y = -x + 1$ in the uv-plane?

P10.8. Jacobian Compute the Jacobian $\partial(x, y, z)/\partial(r, \theta, z)$ for the transformation

$$x = r \cos \theta, \qquad y = r \sin \theta, \qquad z = z.$$

P10.9. Difference calculus For the polynomial of Example 10.10,

$$f(x) = ax^2 + bx + c,$$

find the second ($\Delta^2 f(x)$) and third ($\Delta^3 f(x)$) differences. Then, approximate the second and third derivative.

P10.10. Approximate partial derivatives   Using the grid of Figure 10.9 and the approximations for the partial derivatives given in Equation 10.44, find an expression for the mixed partial derivative

$$\frac{\partial^2 f}{\partial x \partial y}$$

for a function $f(x, y)$.

P10.11. Extrema   Find the critical points of the function

$$f(x, y) = x^3 - 12xy + 8y^3,$$

determine the type of extrema (if any), and compute the function values at the extreme points.

P10.12. Optimization   Generally speaking, an *optimization* problem is formulated to determine the best design given a mathematical model of a physical system. Mathematically, the word optimization denotes the minimization or maximization of a function or set of functions of n variables, such as $f(x_1, \cdots, x_n)$. The techniques have been applied to a wide variety of problems, including problems in economics. For example, we may seek to find the values of the n variables of a system that minimize the cost of producing a product or alternatively maximize the profit. In optimization problems, some of the equations may represent *constraints* placed on the variables. The following problem is a simple example.

Determine the maximum rectangular area that can be enclosed with a 200 foot length of fence. Would another geometry encompass a greater area?

P10.13. Acceleration in cylindrical coordinates   Using the position vector from Example 10.15,

$$\mathbf{R}(t) = r\,\mathbf{i}_r + z\,\mathbf{i}_z,$$

find the acceleration and explain the components in physical terms.

P10.14. Taylor expansion   Approximate the potential of Equation 10.55

$$\Phi(P) = \frac{q}{4\pi\epsilon_0}\left[\frac{-1}{r} + \frac{1}{r'}\right]$$

by expanding r' as a Taylor series assuming that $r \gg l$, and show that the approximation for the dipole potential of Equation 10.56 is correct for the potential far from the dipole charges.

P10.15. MATLAB derivatives   Experiment with the effect of the step size on the accuracy of the derivative for the function $y(x) = e^x$ at $x = 1$. Using the approximation

$$D_k \approx \frac{f(x + h_k) - f(x)}{h_k}$$

with $h_k = 10^{-k}, k = 1, 2, \ldots 10$, compute the difference between the exact derivative and the approximation and tabulate the error.

P10.16. MATLAB partial derivatives   Let $f(x, y) = x^3 y^3$. Using the approximation of Equation 10.44, find $\partial f / \partial x$ at the point $(1, 1)$. Try various values of Δx such as $\Delta x = 0.1, 0.01, \ldots$, and compare the errors using the exact value.

P10.17. MATLAB extrema   Given the function

$$f(x, y) = x^3 - 12xy + 8y^3$$

from Problem 10.11, determine the extrema (if any) of the function and plot the curve near the extrema.

P10.18. MATLAB navigation Suppose a submarine has a super gyroscope that acts as an accelerometer. Assume that the vector acceleration is measured accurately and that the initial position $\mathbf{r}(0)$ and velocity $\mathbf{v}(0)$ are known. As a test case, let

$$\begin{aligned}
\mathbf{r}(0) &= 2\,\mathbf{i}, \\
\mathbf{v}(0) &= \mathbf{i} - \mathbf{j}, \\
\mathbf{a}(t) &= 2\,\mathbf{i} + 6t\,\mathbf{j}.
\end{aligned}$$

Determine

    a. The position vector $\mathbf{r}(t)$ for the submarine;

    b. The course plotted on an xy-grid for $t = 0$ to $t = 3$.

EXPLORATION PROBLEMS

P10.19. Partial derivatives Let $f(x, y, z) = e^{xy} \cos z$ with the variables

$$x = tu, \qquad y = \sin tu, \qquad z = u^2.$$

Find $\partial f / \partial u$ in terms of t and u.

P10.20. Transformation Let a transformation be defined as

$$x = r \cos \theta, \qquad y = r \sin \theta, \qquad z = r.$$

If $f(x, y, z) = \sqrt{x^2 + y^2 + z^2}$, find

$$\frac{\partial f}{\partial r} \quad \text{and} \quad \frac{\partial f}{\partial \theta}$$

by first using the chain rule and then checking the result by direct substitution.

P10.21. Taylor expansion Let $f(x, y) = \ln(1 + x + 2y)$. Compute

    a. The 2D Taylor expansion up to degree 2;

    b. Using the fact that $\ln(1 + u) \approx u - u^2/2$, find the 2D expansion of $f(x, y)$ without computing the Taylor series.

P10.22. Extrema and critical points Consider the function

$$f(x, y, z) = xyz$$

in the region defined by $|x| \le 1$, $|y| \le 1$, $|z| \le 1$. Find the critical points and the extrema for the function. How are the critical points and the extrema related?

P10.23. MATLAB extrema Given the function

$$\begin{aligned}
f(x, y) &= 0.02 \sin x \sin y - 0.03 \sin 2x \sin y \\
&\quad + 0.04 \sin x \sin 2y + 0.08 \sin 2x \sin 2y,
\end{aligned}$$

plot the function and the contour plot. Also, find the numerical values of the extrema. Restrict the domain of the function to the square $0 \le x \le \pi$, $0 \le y \le \pi$.

P10.24. MATLAB great circle calculation Consider the two vectors in spherical coordinates

$$\begin{aligned}
\mathbf{R}_1 &= (3960 \text{ miles}, 286^\circ, 49.25^\circ), \\
\mathbf{R}_2 &= (3960 \text{ miles}, 0.0^\circ, 38.5^\circ).
\end{aligned}$$

Create an M-file to compute the angle and distance between two points defined in spherical coordinates. Test the program with the given values and display the following:

Chapter 10 ■ ADVANCED CALCULUS

a. The angle between the vectors;

b. The distance between the end points along the arc of a circle connecting them. (This is the great-circle distance between New York and London.)

P10.25. MATLAB dipole field Plot the potential field of the dipole described by Equation 10.56. Assume that $q = 1.6 \times 10^{-19}$ coulombs.

ANNOTATED BIBLIOGRAPHY

1. Forsythe, George E., M. A. Malcolm, and C. B. Moler, *Computer Methods for Mathematical Computations*, Prentice Hall, Englewood Cliffs, NJ, 1977. *This text describes many numerical techniques used in MATLAB routines, including optimization.*

2. Kunz, Kaiser S., *Numerical Analysis*, McGraw-Hill, New York, 1957. *This useful textbook covers practical numerical techniques, including interpolation in tables of two or more variables.*

3. Mathews, John M., *Numerical Methods for Mathematics, Science, and Engineering*, Prentice Hall, Englewood Cliffs, NJ, 1992. *This text explains various numerical algorithms for differentiation and optimization. There is a supplement that presents MATLAB implementation of many of these algorithms.*

4. Schilling, Robert J., *Fundamentals of Robotics*, Prentice Hall, Englewood Cliffs, NJ, 1990. *The use of Jacobians in robotic control is discussed in detail.*

ANSWERS

■

P10.1. Domain Since the function must remain real, the domain is defined by the line

$$2x + 3y = 6,$$

which cuts the axes at $(0, 2)$ and $(3, 0)$. Values of (x, y) below the line are valid.

P10.2. Limit One limit yields 1 and the other yields -1, so the limit does not exist.

P10.3. Derivatives Given $f(x, y, z) = xe^{yz} + yze^x$,

$$\partial f / \partial y = xze^{yz} + ze^x,$$

$$f_{yx} = \frac{\partial}{\partial x}\left(\frac{\partial f}{\partial y}\right) = ze^{yz} + ze^x,$$

$$f_{yxz} = \frac{\partial}{\partial z} f_{yx} = e^{yz} + yze^{yz} + e^x,$$

$$\frac{\partial^3 f}{\partial x^2 \partial y} = \frac{\partial}{\partial x} f_{yx} = ze^x.$$

P10.4. Differentials In this case Δx and Δy are relatively small and $dz \approx \Delta z$.

P10.5. Chain rule After some algebra, the result is

$$\frac{\partial^2 w}{\partial r^2} = \frac{\partial^2 w}{\partial x^2} \cos^2\theta + 2\frac{\partial^2 w}{\partial x \partial y} \cos\theta \sin\theta + \frac{\partial^2 w}{\partial y^2} \sin^2\theta.$$

P10.6. Chain rule The easiest method for finding the partial derivative is to form

$$\frac{\partial A}{\partial \theta} = \frac{\partial}{\partial \theta}\left(\frac{1}{2} r^2 \sin 2\theta\right) = r^2 \cos 2\theta.$$

P10.7. Transformations The result is the line $u = 1$.

P10.8. Jacobian The result is

$$\frac{\partial(x, y, z)}{\partial(r, \theta, z)} = \begin{vmatrix} \cos\theta & -r\sin\theta & 0 \\ \sin\theta & r\cos\theta & 0 \\ 0 & 0 & 1 \end{vmatrix} = r.$$

P10.9. Difference calculus For the function $f(x) = ax^2 + bx + c$, the second difference is

$$\Delta^2 f(x) = 2ah(x + h) - 2ah = 2ah^2.$$

Thus, the second derivative is

$$\frac{\Delta^2 f(x)}{\Delta x^2} = 2a.$$

The third difference is zero.

P10.10. Approximate partial derivatives Writing

$$\frac{\partial^2 f}{\partial x \partial y} = \frac{\partial}{\partial x}\left(\frac{\partial f}{\partial y}\right) \approx \frac{\partial}{\partial x}\left[\frac{1}{2\Delta y}(f_{i,j+1} - f_{i,j-1})\right]$$

leads to the relationship

$$\frac{\partial^2 f}{\partial x \partial y} \approx \frac{1}{2\Delta y}\left(\frac{\partial f_{i,j+1}}{\partial x} - \frac{\partial f_{i,j-1}}{\partial x}\right) \tag{10.57}$$

because Δy is constant and can be factored out.

Considering $f_{i,j+1}$ and $f_{i,j-1}$ as function of x, the partials with respect to x can be calculated. For example,

$$\frac{\partial f_{i,j+1}}{\partial x} = \frac{1}{2\Delta x}(f_{i+1,j+1} - f_{i-1,j+1}).$$

Substituting this expression and the equation for $\partial f_{i,j-1}/\partial x$ into Equation 10.57, leads to the result

$$\frac{\partial^2 f}{\partial x \partial y} \approx \frac{1}{4\Delta x \Delta y}(f_{i-1,j-1} - f_{i+1,j-1} - f_{i,j+1} + f_{i,j+1}).$$

P10.11. Extrema The function

$$f(x, y) = x^3 - 12xy + 8y^3$$

has critical points at $(0, 0)$ and $(2, 1)$. Applying the second derivative test shows that $f(0, 0)$ is not an extremum and $f(2, 1) = -8$ is a minimum.

P10.12. Optimization Let x and y represent the length and width of the enclosed area, respectively. If A is the area enclosed, the defining equations are

$$2x + 2y \quad = \quad 200 \quad \text{feet},$$
$$A \quad = \quad xy.$$

This sort of problem is called a *constrained optimization* problem. In this particular case, the problem can be transformed into an unconstrained problem so that the methods of ordinary calculus can be used to find the maximum area. Writing $y = (200 - 2x)/2$ and substituting in the area equation yields

$$\frac{\partial A}{\partial x} = 100 - 2x = 0$$

for the condition of a maximum or minimum. The solution is $x = 50$ feet. The second derivative of A is -2, indicating a maximum. The largest rectangular area enclosed is thus a square of 50 feet on each side. A little investigation would show that a circular fence of length 200 feet would enclose more area.

P10.13. Acceleration in cylindrical coordinates Writing

$$\mathbf{a} = \frac{d\mathbf{v}}{dt} = \left(\ddot{r}\,\mathbf{i}_r + \dot{r}\frac{d\mathbf{i}_r}{dt} \right) + \left(\dot{r}\dot{\theta}\,\mathbf{i}_\theta + r\ddot{\theta}\,\mathbf{i}_\theta + r\dot{\theta}\frac{d\mathbf{i}_\theta}{dt} \right) + \ddot{z}\,\mathbf{i}_z,$$

and substituting the relations

$$\frac{d\mathbf{i}_r}{dt} = \dot{\theta}\,\mathbf{i}_\theta \qquad \frac{d\mathbf{i}_\theta}{dt} = -\dot{\theta}\,\mathbf{i}_r$$

yields the acceleration.

Collecting terms, the acceleration becomes

$$\mathbf{a} = (\ddot{r} - r\dot{\theta}^2)\mathbf{i}_r + (2\dot{r}\dot{\theta} + r\ddot{\theta})\mathbf{i}_\theta + z\,\mathbf{i}_z.$$

To analyze the physical situation, consider an object of mass m attached to a mechanism such as a robot arm that can rotate about the origin and also extend in the r direction. The $\ddot{z}$ term is simply linear acceleration in the z direction. We can let r, θ, or z be constant and explain the acceleration in the direction of each unit vector.

For example, fixing r and z, the acceleration is $-r\dot{\theta}^2$ toward the origin and is called *centripetal* acceleration. The term $r\ddot{\theta}$ is the tangential acceleration. For uniform motion in a circle, $\dot{\theta} = \omega$, where ω is a constant. Then, $v = r\dot{\theta}$ is the linear velocity in a direction tangent to the circle, but there is no tangential acceleration. If r changes with θ constant, the acceleration is linear in the r direction. The force on the object is $m\ddot{r}$, as expected.

The term $2\dot{r}\dot{\theta}$ in the θ direction may not have been expected. Consider a constant rate of change in the r direction so that $\dot{r} = k$ and a fixed angular velocity $\dot{\theta} = \omega$. This is equivalent to a particle, such as a bead, sliding on a wire fixed to a disk rotating at constant angular velocity. As before, the acceleration in the r-direction is the centripetal term with the form $-kt\dot{\theta}^2 = -kt\omega^2$ in this case because $r(t) = kt$ for constant velocity $\dot{r} = k$. The force in the θ direction is called the *Coriolis force*. The bead experiences a force at right angles to the wire, as well as along the wire. Most books on advanced mechanics discuss the Coriolis force in some detail.

P10.14. Taylor expansion In the expression for potential

$$\Phi = \frac{q}{4\pi\epsilon_0} \left[\frac{-1}{r} + \frac{1}{r'} \right], \tag{10.58}$$

$r' = (r^2 + l^2 - 2rl\cos\phi)^{1/2}$ by the law of cosines.

Using a Taylor expansion in powers of l/r and neglecting terms higher than the linear term, the function

$$\frac{1}{r'} = \frac{1}{r}\left(1 - \frac{2l\cos\phi}{r} + \frac{l^2}{r^2}\right)^{-1/2}$$

becomes

$$\frac{1}{r'} \approx \frac{1}{r}\left(1 + \frac{l\cos\phi}{r}\right).$$

Substituting the approximation in Equation 10.58 yields the result

$$\Phi \approx \frac{q}{4\pi\epsilon_0}\frac{l\cos\phi}{r^2},$$

which agrees with the result of Example 10.16.

Comment: MATLAB programs are included on the disk accompanying this textbook.

11

VECTOR
DIFFERENTIAL
OPERATORS

PREVIEW_____

In applications, scalar and vector functions often are used to represent physical values that vary with position in space. The temperature at each point in a room, for example, defines a scalar function. A vector function would be needed to represent the instantaneous velocity of the surface water of a river. The values of a scalar function or a vector function associated with each point in space are said to constitute a *field* of the physical quantity involved. Determining the properties of these fields is the subject of *field theory*. In this chapter, fields are studied in terms of the spatial rate of change of the fields.

   This chapter presents the mathematical operators known as the gradient, the divergence, the curl, and the Laplacian. Mathematically, these operators are defined in terms of partial derivatives, and they are called *vector differential operators*. In physics and engineering, formulating the physical laws of nature using these operators often allows considerable simplifications.

Scalar and vector fields are defined first in this chapter. Then, the directional derivative and the gradient are presented. These operations determine the spatial rates of change of scalar functions in specific directions. The divergence and curl of a vector function are studied next. Combining the divergence of a vector field and the gradient of a scalar field leads to the operator known as the Laplacian. Table 11.1 lists the operators, their notation, and the type of function that results from each operation.

TABLE 11.1 *Table of vector operators*

Type	Notation	Resulting Function
Gradient	∇f	Vector
Divergence	$\nabla \cdot \mathbf{F}$	Scalar
Curl	$\nabla \times \mathbf{A}$	Vector
Laplacian	$\nabla^2 \Phi$	Scalar

These differential operators yield information about fields encountered in applications, such as electromagnetic fields and the fields associated with fluid flow. In our presentation, the physical significance of the mathematical operations applied to scalar and vector fields will be stressed throughout the chapter. For example, the Laplacian operator defines Laplace's equation in the form

$$\nabla^2 \Phi = \frac{\partial^2 \Phi}{\partial x^2} + \frac{\partial^2 \Phi}{\partial y^2} + \frac{\partial^2 \Phi}{\partial z^2} = 0,$$

where $\Phi(x, y, z)$ describes steady distributions of quantities such as temperature. This equation will play a central role in our study of partial differential equations in Chapter 13.

After the operators are defined, a section of this chapter presents a summary of various applications using the operators and Laplace's equation. The last section presents the vector differential operators in curvilinear coordinates.

VECTOR AND SCALAR FIELDS

The measurement of certain physical quantities results in a single real number, called a *scalar*, defining the quantity. The volume and mass of a physical object are examples of scalars. Other physical quantities require a *vector* to describe them. Examples of vector quantities are displacement, acceleration, and force. In this chapter, it will be assumed that the scalar or vector values describing a measurable quantity are defined at all the points in a region of space. In this case, a scalar or vector *field* is said to exist in the region of space. Each type of field will be considered separately in this section.

SCALAR FIELD
Consider a physical quantity that varies with position in space. The function that specifies the scalar values at each point in space is said to define a *scalar field*. As an example of a scalar field, the temperature in a room can be defined by a function of the form $T = f(x, y, z)$ in rectangular coordinates. Such a representation assumes a steady-state temperature at various points in the room, since no variation with time is specified. Fields with no time variation are called *stationary*.

Notice that the temperature field $T = f(x, y, z)$ involves four variables, including the three coordinates and the temperature. The temperature field can be represented graphically by plotting contour lines or surfaces showing lines or surfaces of equal temperature. Points at the same temperature are connected to form *isothermal* lines or surfaces. Other scalar fields studied in this chapter represent electrostatic potential, magnitude of a force, and fields describing fluid flow. All these scalar fields are represented by scalar-valued functions ($\mathbf{R}^3 \to \mathbf{R}$), various examples of which were defined and plotted in Chapter 10.

WHAT IF? The mass of an object is unchanged if the object is moved around in space. However, other scalar quantities do vary with position. List various variables that might describe an object and show that certain properties are not invariant under a change of position if the object is subject to external forces.

VECTOR FIELD
A *vector field* is said to exist at points in space at which a vector function is defined. In general, the vector function can vary in magnitude and direction at each point. In mathematical terms, a vector field defined on $\mathbf{R}^n$ is a transformation that assigns to each point $\mathbf{x}$ in its domain a vector $\mathbf{F}(\mathbf{x})$. A physical vector field usually manifests itself by the action on a particle placed in the field. For example, the electric and magnetic fields studied in electromagnetics are vector fields. The fields act on charged particles placed in the field. The gravitational force exerted on a body by

another body also can be described by a vector field. The gravitational field is considered to exert an attractive force on any particle in the field.

Other vector fields describe the flow of some material. In fluid flow, if $\mathbf{v}(x, y, z)$ is the fluid velocity and ρ is the density at point (x, y, z), then $\rho\mathbf{v}$ is a vector representing the flow of mass per unit area at the point. The flow field can be defined by specifying a vector at each point in space where the fluid flows. Similarly, the electric current density $\mathbf{J}(x, y, z)$ can be viewed as a current flow field that represents the flow of charge.

It is often useful to be able to visualize a vector field in $\mathbf{R}^2$. One way is to draw a collection of typical vectors $\mathbf{F}(x, y)$, each represented by an arrow of length $\|\mathbf{F}(x, y)\|$ and placed so that (x, y) is its initial point. Figure 11.1 of Example 11.2 shows a MATLAB plot of a typical vector field.

Notation. In Cartesian coordinates, the vector $\mathbf{F}$ can be represented as

$$\mathbf{F}(x, y, z) = F_x(x, y, z)\mathbf{i} + F_y(x, y, z)\mathbf{j} + F_z(x, y, z)\mathbf{k}.$$

Such vector fields have three scalar fields as components so that

$$\mathbf{F}(x, y, z) = [F_x(x, y, z), F_y(x, y, z), F_z(x, y, z)].$$

The notation F_x, F_y, and F_z is sometimes used to represent the partial derivatives of $\mathbf{F}$ with respect to x, y, and z, respectively. We prefer to designate the rectangular components of $\mathbf{F}$ by this notation.

In general, we use the terms vector function and vector field interchangeably. In the application of mathematics to physics and engineering problems, we emphasize the notion of a vector field since the field is often given physical significance. Thus, the term *vector field* should be taken to mean a vector field on $\mathbf{R}^2$ or $\mathbf{R}^3$ unless otherwise stated.

☐ EXAMPLE 11.1   *Vector Field*

The position vector in $\mathbf{R}^3$ describes a simple vector field

$$\mathbf{R}(x, y, z) = x\mathbf{i} + y\mathbf{j} + z\mathbf{k}.$$

At each point (x, y, z), the vector points directly away from the origin. The length of this vector is

$$\|\mathbf{R}(x, y, z)\| = \sqrt{x^2 + y^2 + z^2},$$

which is equal to the distance from the origin to (x, y, z). Notice that three scalar fields

$$F_x = x, \qquad F_y = y, \qquad F_z = z$$

are associated with this vector field in $\mathbf{R}^3$.

☐

Newton's Gravitational Law   Newton's gravitational law defines the force of attraction between bodies. The force acting on a body of mass

m at a distance R from the origin due to a particle with mass M fixed at the origin is

$$\begin{aligned} \mathbf{F}(x,y,z) &= -\frac{GMm}{R^3}\,\mathbf{R}(x,y,z) \\ &= -\frac{GMm}{R^2}\,\mathbf{a}_R, \end{aligned} \qquad (11.1)$$

where $\mathbf{R}$ is a vector from one particle to the other and $R = \|\mathbf{R}\|$. The vector $\mathbf{a}_R = \mathbf{R}/R$ is a unit vector in the same direction, and G is a universal constant called the *gravitational constant*.

Since the particle with mass M is located at the origin, the position vector $\mathbf{R}$ points outward from the origin toward the particle with mass m. The negative sign in the equation indicates that the force acts opposite to the direction of the vector $\mathbf{R}$ so that the particle with mass m is *attracted* to the particle at the origin. According to another of Newton's laws, a force of the same magnitude acts on the particle of mass M.

In terms of Cartesian coordinates, the force of Equation 11.1 is

$$\mathbf{F}(x,y,z) = \left(-\frac{GMm}{R^3}\,x, \ -\frac{GMm}{R^3}\,y, \ -\frac{GMm}{R^3}\,z \right).$$

This formulation of Newton's gravitational law assumes that the particles have no extent in space. Since it is assumed that the force between any two particles is not modified by the presence of other particles, the principle of superposition applies. If another particle with mass is introduced, the force on each mass is the vector sum of the forces caused by each other mass treated separately. When the bodies involved have spatial extent so that their mass is distributed, integral relationships described in Chapter 12 must be used to determine the forces.

A body with mass can be considered to generate a *gravitational force field* that attracts any other particle with mass. The attractive field of force due to the particle of mass M is defined by the equation

$$\mathbf{F} = -\frac{GM}{R^3}\,\mathbf{R} = -\frac{GM}{R^2}\,\mathbf{a}_R. \qquad (11.2)$$

At each point a distance $R = \|\mathbf{R}\|$ from the particle, a force field of attraction exists that is directed toward the particle. If another particle of mass m is subject to the effect of the field, the force on the particle is determined by Newton's gravitational law of Equation 11.1.

Newton's Second Law In the formulation of Newton's gravitational law, any motion of the particles caused by the forces is neglected. With these assumptions, the forces involved can be described by the differential operators to be introduced in this chapter. Differential equations of motion must be applied if the particles are subject to acceleration due to Newton's second law of motion $\mathbf{F} = m\mathbf{a}$.

Coulomb's Law   Coulomb's law in electrostatics,

$$
\begin{aligned}
\mathbf{F}(x, y, z) &= \frac{Qq}{4\pi\epsilon_0 R^3}\,\mathbf{R}(x, y, z) \\
&= \frac{Qq}{4\pi\epsilon_0 R^2}\,\mathbf{a}_R,
\end{aligned}
\tag{11.3}
$$

describes the force between two charged particles in a vacuum (free space) using SI (mks) units. In the equation, ϵ_0 is the permittivity of free space. If the charges are both positive or negative, the force tends to separate the particles. Charges of opposite sign attract.

A charged particle q is considered to create a force field that acts on other charged particles. The field that a charge creates is normally defined by introducing the electric field $\mathbf{E}(x, y, z)$ and deriving the force by the equation

$$
\mathbf{F}(x, y, z) = q\mathbf{E}(x, y, z).
\tag{11.4}
$$

One reason for introducing the electric field is that the effect on charges placed in the field can be calculated without considering explicitly the distribution of charges that created the electric field. An electric field associated with fixed electric charges is called an *electrostatic field* because there is no time variation in the field.

In electromagnetic field theory, the electric field plays a fundamental role. It is considered to define the field of force that acts on charged particles by Equation 11.4. The electric field is studied in more detail later in this chapter and in Chapter 12.

Inverse-Square Forces   Coulomb's law has the same form as Newton's gravitational law with respect to the dependence on $\mathbf{R}$. A force field with a variation in distance as $1/R^2$ is called an *inverse-square* force field. In the relationships of Equation 11.2 and Equation 11.3, $\|\mathbf{F}\| \to \infty$ as $\mathbf{R} \to 0$. Away from the origin, the gravitational field and the electrostatic field are continuous and decrease smoothly with distance.

Conditions for the Fields   The fields we consider in most examples will be *continuous* and *single-valued*. A *single-valued* field has a unique value of the scalar or vector at each point. These conditions basically mean that the fields we study vary smoothly from point to point. Also, the fields are assumed to be continuously differentiable, which requires that the partial derivatives of the function defining the field exist and are continuous.

In certain cases, discontinuities in the field will be allowed at various points or along a line or a surface. These special cases will be treated as they arise in our discussions.

MATLAB COMMANDS FOR VECTOR DIFFERENTIAL CALCULUS

MATLAB has a number of commands that are useful for computing and plotting functions encountered in vector differential calculus. Some of these were listed and used in Chapter 10. Commands such as **contour**, **mesh**, **plot3**, and **surf** are examples of plotting commands that were used previously. Those additional MATLAB commands of particular interest in this chapter are listed in Table 11.2.

TABLE 11.2 *MATLAB commands for vector calculus*

Command	Purpose
Operations:	
del2	Laplacian
diff	Numerical difference or approximate derivative
gradient	Gradient
jacobian	Symbolic gradient
Plots:	
quiver	Quiver plot (vectors)
surfnorm	3D surface normals

□ EXAMPLE 11.2   ***MATLAB Field Plot***

The accompanying MATLAB script uses the command **quiver** to plot a vector field. The field is the 2D projection of an inverse-square force field, such as the gravitational force field in Equation 11.1 with the constants set to 1. The result is the field

$$\mathbf{F}(x,y) = -\frac{1}{R^3}\,\mathbf{R}(x,y)$$

on the xy-plane, where $z = 0$. Notice the use of the MATLAB variable eps to avoid a division by zero. This technique was previously described in Chapter 2.

Since the gravitational field of Equation 11.1 is spherically symmetrical with respect to the origin, the 2D field plotted in the xy-plane represents the projection of the 3D field on any plane through the origin. Figure 11.1 shows the result. The command **zoom** is used to allow closer inspection of the plot using the computer's mouse.

MATLAB Script _____

Example 11.2

```
% EX11_2.M Plots field of inverse square field using
%  quiver command
clear                          % Clear variables
clf                            %  and figures
[x,y]=meshgrid(-1:.2:1);       % Define a grid
R=sqrt(x.^(2)+y.^(2)) + eps;   % Avoid divide by zero
fx=-x./R.^(3);
fy=-y./R.^(3);
quiver(x,y,fx,fy);             % Plot vectors
title('Plot of Inverse-square Force Field')
xlabel('x')
ylabel('y')
zoom                           % Change resolution with mouse
```

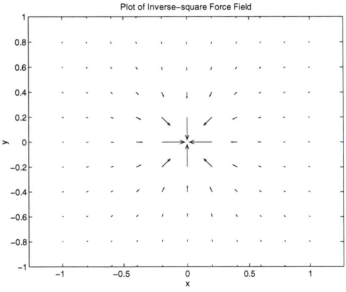

Plot of Inverse-square Force Field

FIGURE 11.1 *MATLAB field plot*

□

DIRECTIONAL DERIVATIVES AND THE GRADIENT

In this section, the directional derivative is first introduced as the operation that allows the study of the rate of change of a function in an arbitrary direction. Then, the gradient is shown to be the vector function that indicates the direction along which a function is increasing the fastest.

DIRECTIONAL DERIVATIVE

Suppose $f(x, y, z)$ is a given function defined in a region of space. The partial derivatives of f with respect to x, y, or z compute the ratio of the change Δf in the function along the x-, y-, or z-directions, respectively. For example, the partial derivative with respect to x at the point (x, y, z) is

$$\frac{\partial f(x, y, z)}{\partial x} = \lim_{\Delta x \to 0} \frac{f(x + \Delta x, y, z) - f(x, y, z)}{\Delta x}.$$

This partial derivative considers the values of f along a line parallel to the x-axis. Similarly, $\partial f / \partial y$ and $\partial f / \partial z$ determine how f changes along lines parallel to the y- and z-axes, respectively.

The *directional derivative* is used to determine the rate of change of f in an arbitrary direction as the limit of the ratio

$$\frac{\Delta f}{\Delta s}, \tag{11.5}$$

which yields the ratio of the change in f to the distance Δs in the given direction as Δs approaches zero. To be specific, suppose that the change in $f(x, y, z)$ from the point $P(x, y, z)$ to the point $Q(x + \Delta x, y + \Delta y, z + \Delta z)$ is defined as

$$\Delta f = f(Q) - f(P). \tag{11.6}$$

According to the discussion of total differentials in Chapter 10, the increment in Equation 11.6 can be approximated by the linear relationship

$$\Delta f \approx \frac{\partial f}{\partial x} \Delta x + \frac{\partial f}{\partial y} \Delta y + \frac{\partial f}{\partial z} \Delta z. \tag{11.7}$$

Further, let $\mathbf{v} = \overrightarrow{PQ} = (\Delta x, \Delta y, \Delta z)$ be the *displacement vector* from P to Q so that $\Delta s = \|\mathbf{v}\|$. Then, the directional derivative in Equation 11.5 can be written in terms of the components of $\mathbf{v}$ as

$$\frac{\Delta f}{\Delta s} \approx \frac{\partial f}{\partial x} \frac{v_x}{\|\mathbf{v}\|} + \frac{\partial f}{\partial y} \frac{v_y}{\|\mathbf{v}\|} + \frac{\partial f}{\partial z} \frac{v_x}{\|\mathbf{v}\|}. \tag{11.8}$$

Notice that the terms $(v_x/\|\mathbf{v}\|, v_y/\|\mathbf{v}\|, v_x/\|\mathbf{v}\|)$ are the components of a unit vector in the direction of $\mathbf{v}$. Designating this unit vector $\mathbf{u}$, the rate of change of f with respect to the distance between P and Q becomes

$$\frac{df}{ds} = \lim_{\Delta s \to 0} \frac{\Delta f}{\Delta s} = \frac{\partial f}{\partial x} u_x + \frac{\partial f}{\partial y} u_y + \frac{\partial f}{\partial z} u_z,$$

which can be written as a dot product

$$\frac{df}{ds} = \left(\frac{\partial f}{\partial x} \mathbf{i} + \frac{\partial f}{\partial y} \mathbf{j} + \frac{\partial f}{\partial z} \mathbf{k} \right) \cdot (u_x \mathbf{i} + u_y \mathbf{j} + u_z \mathbf{k}). \tag{11.9}$$

GRADIENT

The first vector in the scalar (dot) product of Equation 11.9 is called the *gradient* of $f(x, y, z)$. Thus, defining the gradient as

$$\nabla f = \left(\frac{\partial f}{\partial x}, \frac{\partial f}{\partial y}, \frac{\partial f}{\partial z} \right)$$

$$= \frac{\partial f}{\partial x}\mathbf{i} + \frac{\partial f}{\partial y}\mathbf{j} + \frac{\partial f}{\partial z}\mathbf{k}, \tag{11.10}$$

the directional derivative of f in the direction $\mathbf{u}$ is the dot product of the gradient of f and $\mathbf{u}$, indicated as

$$\nabla_u f(x, y, z) = \nabla f \cdot \mathbf{u}. \tag{11.11}$$

The gradient is thus useful for computing directional derivatives. However, this is only one of its many uses. In operator notation, the gradient operator that maps a scalar function $f(x, y, z)$ to a vector function in $\mathbf{R}^3$ is

$$\nabla \equiv \frac{\partial}{\partial x}\mathbf{i} + \frac{\partial}{\partial y}\mathbf{j} + \frac{\partial}{\partial z}\mathbf{k}. \tag{11.12}$$

The operator is called *del* or sometimes *nabla*, from the Greek word meaning "harp."

☐ EXAMPLE 11.3   *Gradient*

The distance from the origin to a point $P(x, y, z)$ is

$$f(x, y, z) = \sqrt{x^2 + y^2 + z^2}.$$

The gradient of f is

$$\nabla f = \left(\frac{\partial f}{\partial x}, \frac{\partial f}{\partial y}, \frac{\partial f}{\partial z} \right)$$

$$= \left(\frac{x}{\sqrt{x^2 + y^2 + z^2}}, \frac{y}{\sqrt{x^2 + y^2 + z^2}}, \frac{z}{\sqrt{x^2 + y^2 + z^2}} \right).$$

Notice that the result can be written as $\mathbf{R}/R$, where $\mathbf{R} = (x, y, z)$ is a vector from the origin to the point. Thus, the gradient in this case is a unit vector in the direction of P. It is easily shown that

$$\nabla \left(\frac{1}{R} \right) = \nabla \left(\frac{1}{\sqrt{x^2 + y^2 + z^2}} \right) = -\frac{\mathbf{R}}{R^3} = -\frac{1}{R^2}\mathbf{a}_R. \tag{11.13}$$

☐

MEANING OF THE GRADIENT

The gradient has useful geometric significance. First, consider a *level surface* defined as a surface for which

$$f(x, y, z) = \text{constant}.$$

If $\hat{\mathbf{u}}$ is a vector tangent to the level surface at a point,

$$\nabla_{\hat{\mathbf{u}}} f(x, y, z) = \nabla f(x, y, z) \cdot \hat{\mathbf{u}} = 0$$

since the function does not change along the direction $\hat{\mathbf{u}}$. Assuming that neither $\nabla f(x, y, z)$ nor $\hat{\mathbf{u}}$ is the zero vector, we conclude that the vector $\nabla f(x, y, z)$ is perpendicular to the level surface. In two dimensions, the gradient is perpendicular to the level curves of a function $f(x, y)$.

Now we explore the magnitude and direction of the gradient with respect to the spatial rate of change of a function. Considering the directional derivative in the direction $\mathbf{u}$ in Equation 11.11, the directional derivative can be interpreted as the component of ∇f in the direction of $\mathbf{u}$ if $\mathbf{u}$ is a unit vector $\mathbf{v}/\|\mathbf{v}\|$.

The relationship is shown in Figure 11.2, using a two-dimensional function for simplicity. The gradient, the directional derivative in direction $\mathbf{v}$, and a level curve of the function $f(x, y) = $ constant are shown. From the relationship

$$\nabla_{\mathbf{u}} f(x, y) = \nabla f \cdot \mathbf{u} = \nabla f \cdot \frac{\mathbf{v}}{\|\mathbf{v}\|},$$

as shown in the figure, it is evident that the directional derivative at a given point is a maximum when it is in the direction $\nabla f(x, y)$. This conclusion can be stated as a theorem.

■ THEOREM 11.1

Gradient

The gradient vector $\nabla f(x, y, z)$ points in the direction in which f increases most rapidly, and its length

$$\|\nabla f(x, y, z)\| = \sqrt{\left(\frac{\partial f}{\partial x}\right)^2 + \left(\frac{\partial f}{\partial y}\right)^2 + \left(\frac{\partial f}{\partial z}\right)^2}$$

is the rate of increase in that direction.

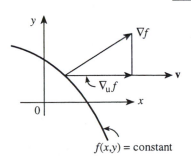

FIGURE 11.2 *Gradient of $f(x, y)$ and directional derivative*

□ EXAMPLE 11.4   *MATLAB Numerical Gradient*

Figure 11.3 shows the contour lines and the gradient field of the function

$$f(x,y) = \sqrt{x^2 + y^2}.$$

The MATLAB script computes the gradient of the scalar function using the **gradient** command. The contour lines are plotted by the command **contour**, and the command **quiver** plots the gradient field.

MATLAB Script _____

```
Example 11.4
% EX11_4.M Plot the gradient of f(x,y)=[x^(2)+y^(2)]^(1/2)
%
clear                    % Clear variables
clf                      %  and figures
[x,y]=meshgrid(-1:.1:1); % Create a grid of points
R=sqrt(x.^(2)+y.^(2));
[dx,dy]=gradient(R,.1,.1); % Compute gradient
contour(x,y,R);          % Draw contours
axis('square')           % Force radial symmetry
hold on
quiver(x,y,dx,dy);       % Plot vectors
title('Plot of Gradient Field-f(x,y)=R')
xlabel('x')
ylabel('y')
zoom            % Change resolution with mouse
```

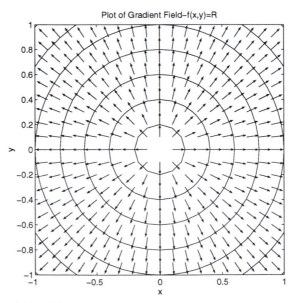

FIGURE 11.3 *MATLAB gradient plot*

From the figure, it is clear that the gradient vectors are perpendicular to the contour lines of $f(x, y)$. The axes are made "square" so the field will appear radially symmetric.

□

MATLAB Symbolic Gradient   The command **jacobian** listed in Table 11.2 can compute the gradient of a function defined symbolically. When the argument of the command is a scalar function, **jacobian** returns the gradient of the function.

Tangent Plane   The gradient can be used to analyze the level sets of a real-valued function. As discussed in Chapter 10, the level set of a function f is the set of points satisfying

$$f(\mathbf{x}) - k$$

for some constant k. For the transformation

$$\mathbf{R}^2 \xrightarrow{f} \mathbf{R},$$

the set usually defines a curve. For $\mathbf{R}^3 \xrightarrow{f} \mathbf{R}$, the sets are most often surfaces.

Assuming the gradient is not zero, the gradient is perpendicular to the level surfaces of f. The *tangent plane* at a point $\mathbf{x}_0$ is the set of points $\mathbf{x}$ satisfying

$$\nabla f(\mathbf{x}_0) \cdot (\mathbf{x} - \mathbf{x}_0) = 0. \qquad (11.14)$$

GRADIENT OF A POTENTIAL FUNCTION   The gradient computes a vector field from a scalar field. In many problems in physics, the gradient of a scalar field can be used to derive the vector field of interest. When used in this way, the scalar field is called the *scalar potential*. The scalar potential is often introduced because the scalar fields are usually simpler to manipulate mathematically than the vector fields that are derivable from them. Thus, the scalar field may be a fictitious field used to simplify the mathematics.

In some applications, the scalar potential leads to physical insights that are vital to our understanding of the physical phenomenon involved. In these cases, the scalar potential is usually related to the potential energy. In other problems, the scalar potential represents the solution to the differential equation describing the physical system. In summary, the vector character of a field in many problems can be determined from the scalar field associated with a physical quantity. After defining the scalar potential, several examples will be presented.

Scalar Potential   A scalar function $\Phi(\mathbf{x})$ whose gradient $\nabla \Phi(\mathbf{x})$ is the vector field

$$\mathbf{F}(\mathbf{x}) = \nabla \Phi(\mathbf{x}) \qquad (11.15)$$

is called the *scalar potential function* of the vector field $\mathbf{F}(\mathbf{x})$.

The mathematical conditions for the existence of a scalar potential will be stated after the divergence and curl are defined. In physics, potentials are associated with conservative vector fields, as discussed in Example 11.11. The fields associated with heat flow, gravitation, and electrostatics can be derived from potential functions.

☐ EXAMPLE 11.5   ***Potential Functions***

In discussing abstract fields, the sign of the gradient in Equation 11.15 is typically taken as positive. However, in many applications the vector field is defined as the negative of the gradient of the scalar potential. This is usually determined by the physics of the problem. For example, experiments indicate that the flow of heat in a body with a temperature distribution $T(x, y, z)$ is described by *Fourier's law of heat conduction*,

$$\mathbf{q} = -K\nabla T,$$

where $K > 0$ is the heat conductivity of the body. We know from thermodynamics that the heat flows from a region of higher temperature to a region of lower temperature. Hence, the minus sign is required in Fourier's law to indicate that the heat flow is opposite to that of the positive temperature gradient that points from a colder region to a hotter region.

Gravitational potential. Newton's gravitational law in Equation 11.1 can be derived from a scalar potential. Letting

$$V(R) = -\frac{GMm}{R}$$

be the gravitational potential, the gravitational force can be computed as the negative gradient of V since

$$\mathbf{F}(\mathbf{R}) = -\nabla V(R) = \nabla\left(\frac{GMm}{R}\right),$$

using the fact that $\nabla(1/R) = -\mathbf{a}_R/R^2$ according to the result of Example 11.3.

Coulomb's law. Suppose a charge Q is located at the origin of the coordinate system. Coulomb's force acting on another charge q at a position defined by the position vector $\mathbf{R}$ can be derived from the electric field associated with Q. The electric field due to Q is the negative of the gradient of the electrostatic potential

$$\Phi(R) = \frac{Q}{4\pi\epsilon_0 R}.$$

Thus, the electric field derived from $-\nabla\Phi(R)$ is

$$\mathbf{E}(\mathbf{R}) = \frac{Q}{4\pi\epsilon_0 R^2}\mathbf{a}_R.$$

The force on the charge q is $\mathbf{F} = q\mathbf{E}$ according to Equation 11.4. Comparing this to Coulomb's law in Equation 11.3 shows that the results are the same. Thus, a free positively charged particle would move in the direction of the electric field and in the opposite direction to the gradient of the potential.

☐

If f is a differentiable real-valued function, $\mathbf{R}^n \xrightarrow{f} \mathbf{R}$, then the gradient ∇f is defined as

$$\nabla f(\mathbf{x}) = [\,\partial f(\mathbf{x})/\partial x_1, \ldots, \partial f(\mathbf{x})/\partial x_n\,].$$

Thus, the gradient is a function from $\mathbf{R}^n$ to $\mathbf{R}^n$ that defines a vector field in $\mathbf{R}^n$.

WHAT IF? Problems in economics, transportation, and production often involve dozens or even hundreds of variables. Certain problems, called optimization or minimization problems, require finding the minimum values of the multivariate function describing the system.

If you wish to solve such problems, investigate the use of the n-dimensional gradient operator as it is applied to optimization problems involving many variables. Problem 11.22 defines the project.

THE DIVERGENCE

The gradient operation defines the direction and magnitude of the rate of change of a scalar field in the direction of its maximum spatial rate of change. For a vector field that changes from point to point, the mathematical description of its spatial rate of change must include changes in direction as well as in magnitude. To fully specify a vector field, both the divergence and the curl must be computed at each point where the field is defined. The divergence is presented in this section and the curl is treated in the next section.

Consider the vector function in $\mathbf{R}^3$ that can be represented as

$$\mathbf{F}(x, y, z) = F_x(x, y, z)\mathbf{i} + F_y(x, y, z)\mathbf{j} + F_z(x, y, z)\mathbf{k}.$$

The definition of the *divergence* for vector fields in $\mathbf{R}^3$ is

$$\operatorname{div} \mathbf{F}(x, y, z) = \frac{\partial F_x}{\partial x} + \frac{\partial F_y}{\partial y} + \frac{\partial F_z}{\partial z}. \tag{11.16}$$

Using operator notation

$$
\begin{aligned}
\nabla \cdot \mathbf{F} &= \left(\frac{\partial}{\partial x}\mathbf{i} + \frac{\partial}{\partial y}\mathbf{j} + \frac{\partial}{\partial z}\mathbf{k}\right) \cdot (F_x\mathbf{i} + F_y\mathbf{j} + F_z\mathbf{k}) \\
&= \frac{\partial F_x}{\partial x} + \frac{\partial F_y}{\partial y} + \frac{\partial F_z}{\partial z} = \operatorname{div} \mathbf{F}.
\end{aligned}
$$

Since the unit vectors in rectangular coordinates are constant, the partial derivative operators yield zero when applied to them. This is not necessarily the case in other coordinate systems, as explained in Example 11.14.

The divergence measures the expansion, or divergence, of a field at a point. In terms of the flow of a fluid in space, the divergence measures the rate of expansion (or contraction) per unit volume of the fluid. For example, suppose $\mathbf{F}(x, y, z) = x\,\mathbf{i} + y\,\mathbf{j} + z\,\mathbf{k}$ is the velocity field for an expanding gas. Since $\nabla \cdot \mathbf{F} = 3$, the gas is expanding at the rate of 3 cubic units per unit of volume per unit of time. This interpretation of the divergence is explored further when physical applications are discussed.

☐ **EXAMPLE 11.6** *Divergence*

Given the function $\mathbf{F} = x^2\,\mathbf{i} - xy\,\mathbf{j} + xyz\,\mathbf{k}$, the divergence is

$$\nabla \cdot \mathbf{F} = 2x - x + xy = x(1 + y).$$

Inverse-square forces. The Newtonian gravitational force and Coulomb's law both have an inverse-square form,

$$\mathbf{F} = \frac{k}{R^2}\mathbf{a}_R.$$

You are asked in Problem 11.6 to show that the divergence of such forces is zero. Except at the point where $R = 0$, the divergence is zero since the divergence exists everywhere and the derivatives of the force are continuous. In this case, the origin is considered to be the location of the source of the field. The source would be a body with mass in the gravitational case or an electric charge in the electrostatic problem. When the origin is not included, the divergence of an inverse-square force field is zero at the points of empty space. A vector field in a region of space where the divergence is zero has special properties that will be explored later.

☐

SOLENOIDAL FIELDS

If $\mathbf{F}$ is such that $\nabla \cdot \mathbf{F} = 0$, the field is said to be *solenoidal*. Solenoidal fields and various other properties of the divergence applied to a vector field will be considered again after the curl is introduced in the next section.

GENERALIZED DIVERGENCE

The divergence can be applied to a vector in $\mathbf{R}^n$ so that the divergence is a transformation $\mathbf{R}^n \to \mathbf{R}$. If $\mathbf{F} = (F_1, \ldots, F_n)$ is a vector function in $\mathbf{R}^n$, the divergence is

$$\nabla \cdot \mathbf{F}(x_1, \ldots x_n) = \frac{\partial F_1}{\partial x_1} + \frac{\partial F_2}{\partial x_2} + \cdots + \frac{\partial F_n}{\partial x_n},$$

assuming the partial derivatives on the right exist.

THE CURL

———————————————— ■ ————————————————

The *curl* is a vector operator that associates a vector field with $\mathbf{F}(x, y, z)$ in $\mathbf{R}^3$. The curl is usually written as $\nabla \times \mathbf{F}$ and has the definition

$$\nabla \times \mathbf{F} = \left(\frac{\partial F_z}{\partial y} - \frac{\partial F_y}{\partial z} \right) \mathbf{i} + \left(\frac{\partial F_x}{\partial z} - \frac{\partial F_z}{\partial x} \right) \mathbf{j} + \left(\frac{\partial F_y}{\partial x} - \frac{\partial F_x}{\partial y} \right) \mathbf{k}. \quad (11.17)$$

The formula for the curl can be remembered by viewing the curl operation as taking the cross product of the ∇ operator with the vector $\mathbf{F}$ in the expression

$$\nabla \times \mathbf{F} = \begin{vmatrix} \mathbf{i} & \mathbf{j} & \mathbf{k} \\ \frac{\partial}{\partial x} & \frac{\partial}{\partial y} & \frac{\partial}{\partial z} \\ F_x & F_y & F_z \end{vmatrix}. \quad (11.18)$$

□ **EXAMPLE 11.7**   *Curl*

Given the vector

$$\mathbf{A} = (x + 2y + 4z)\mathbf{i} + (2x - 3y - z)\mathbf{j} + (4x - y + 2z)\mathbf{k},$$

the curl as defined in Equation 11.18 is

$$\nabla \times \mathbf{A} = \begin{vmatrix} \mathbf{i} & \mathbf{j} & \mathbf{k} \\ \frac{\partial}{\partial x} & \frac{\partial}{\partial y} & \frac{\partial}{\partial z} \\ x + 2y + 4z & 2x - 3y - z & 4x - y + 2z \end{vmatrix}.$$

Expanding this expression leads to the result

$$\nabla \times \mathbf{A} = (-1 + 1)\mathbf{i} + (4 - 4)\mathbf{j} + (2 - 2)\mathbf{k} = 0.$$

In this case, $\nabla \times \mathbf{A} = \mathbf{0}$ everywhere in space. Fields with zero curl in a region of space are fundamental fields in certain applications, such as electrostatics.

□

IRROTATIONAL FIELD

Suppose that $\mathbf{F}$ is a continuously differentiable vector function specified in a region R of space. If $\nabla \times \mathbf{F} = \mathbf{0}$ at every point of R, the vector field is called *irrotational*. We shall see later that under certain conditions such fields can be derived from a scalar function as $\mathbf{F} = \nabla \Phi$.

PHYSICAL IN-
TERPRETATION
OF THE CURL

For a body or fluid rotating at a constant angular velocity, the curl of the linear velocity of points on the body or in the fluid defines a vector field. This field is directed parallel to the axis of rotation with magnitude twice the angular velocity, as you are asked to show in Problem 11.7.

A nonzero curl is often associated with fluid flow along a curved path, such as seen in whirlpools. However, even in straight-line flow, torque could be exerted on a small paddle wheel placed in the fluid if the flow is not uniform, as demonstrated in Problem 11.8. In other problems where there is no flow of material, it is perhaps better not to try to relate the curl to actual rotation or circulation of the field.

☐ EXAMPLE 11.8

Curl of Vector Field

Consider the vector field $\mathbf{F} = y\,\mathbf{i} - x\,\mathbf{j}$. The curl of $\mathbf{F}$ is

$$\nabla \times \mathbf{F} = \begin{vmatrix} \mathbf{i} & \mathbf{j} & \mathbf{k} \\ \dfrac{\partial}{\partial x} & \dfrac{\partial}{\partial y} & \dfrac{\partial}{\partial z} \\ y & -x & 0 \end{vmatrix} = (-1 - 1)\mathbf{k} = -2\mathbf{k} \neq 0.$$

From the nonzero result for the curl, we conclude that the field is not irrotational, and thus the field cannot be derived from a scalar potential.

☐

WHAT IF? You should plot the vector function $\mathbf{F}$ in Example 11.8. Although the curl of $\mathbf{F}$ does not vanish, if this function could be derived from the gradient of a scalar function Φ, the equations to be satisfied are

$$F_x = y = \frac{\partial \Phi}{\partial x}, \quad F_y = -x = \frac{\partial \Phi}{\partial y}, \quad F_z = 0 = \frac{\partial \Phi}{\partial z}.$$

Convince yourself there is no function Φ that satisfies these three equations.

THE LAPLACIAN AND LAPLACE'S EQUATION

It is possible to form scalar and vector products in which the operator ∇ is used more than once. For example, since the gradient is a vector, it is possible to take the divergence of the gradient of a scalar function. The result is the Laplacian of the scalar function. In this section, the Laplacian operator and Laplace's differential equation are presented.

LAPLACIAN

Let $\Phi(x, y, z)$ define a function with continuous first and second partial derivatives. Taking the divergence of the gradient of Φ leads to the relationship

$$\nabla \cdot \nabla \Phi = \left(\frac{\partial}{\partial x}\mathbf{i} + \frac{\partial}{\partial y}\mathbf{j} + \frac{\partial}{\partial z}\mathbf{k} \right) \cdot \left(\frac{\partial \Phi}{\partial x}\mathbf{i} + \frac{\partial \Phi}{\partial y}\mathbf{j} + \frac{\partial \Phi}{\partial z}\mathbf{k} \right)$$

$$= \frac{\partial^2 \Phi}{\partial x^2} + \frac{\partial^2 \Phi}{\partial y^2} + \frac{\partial^2 \Phi}{\partial z^2}, \qquad (11.19)$$

which is the *Laplacian* of Φ.

The *Laplacian operator* is also called *del-squared*, and it is denoted ∇^2, motivated by the operator notation

$$\nabla^2 = \nabla \cdot \nabla$$

$$= \left(\frac{\partial}{\partial x}\mathbf{i} + \frac{\partial}{\partial y}\mathbf{j} + \frac{\partial}{\partial z}\mathbf{k} \right) \cdot \left(\frac{\partial}{\partial x}\mathbf{i} + \frac{\partial}{\partial y}\mathbf{j} + \frac{\partial}{\partial z}\mathbf{k} \right)$$

$$= \frac{\partial^2}{\partial x^2} + \frac{\partial^2}{\partial y^2} + \frac{\partial^2}{\partial z^2}. \qquad (11.20)$$

For a vector field $\mathbf{F} = F_x\mathbf{i} + F_y\mathbf{j} + F_z\mathbf{k}$, the Laplacian is defined as

$$\nabla^2 \mathbf{F} = \nabla^2 F_x\mathbf{i} + \nabla^2 F_y\mathbf{j} + \nabla^2 F_z\mathbf{k}. \qquad (11.21)$$

LAPLACE'S EQUATION

The scalar function $\Phi(x, y, z)$ is said to satisfy *Laplace's equation* when

$$\nabla^2 \Phi = \frac{\partial^2 \Phi}{\partial x^2} + \frac{\partial^2 \Phi}{\partial y^2} + \frac{\partial^2 \Phi}{\partial z^2} = 0. \qquad (11.22)$$

This partial differential equation arises in many applications that treat steady state phenomena.

□ **EXAMPLE 11.9**   *Laplace's Equation*

Suppose that a vector field $\mathbf{F}$ is both *solenoidal* and *irrotational* in a region of space, so that $\nabla \cdot \mathbf{F} = 0$ and $\mathbf{F} = \nabla\Phi$. The divergence of the gradient of Φ for this vector field is zero, so that

$$\mathrm{div}(\nabla\Phi) = \nabla \cdot (\nabla\Phi) = \nabla^2 \Phi = 0.$$

In applications, Laplace's equation represents a partial differential equation to be solved for the potential function Φ. The equation describes steady-state temperature distributions, the gravitational potential, and the electrostatic potential.

□

VECTOR FIELD THEORY

—————————— ■ ——————————

We now consider some of the results from vector field theory. The fundamental types of vector fields are solenoidal fields that have zero divergence everywhere and irrotational fields that have zero curl everywhere. However, the most general vector field will have both a nonzero divergence and a nonzero curl. A powerful theorem of vector field theory states that any continuous vector field with continuous derivatives in a region can be resolved into a sum of solenoidal and irrotational fields. This result is presented here as Helmholtz's theorem. Then, a selection of the properties of vector differential operators is presented, and a number of useful vector identities are listed.

IRROTATIONAL FIELDS

Suppose that $\mathbf{F} = (F_x, F_y, F_z)$ is an irrotational vector field and that the first partial derivatives are continuous in an open region of $\mathbf{R}^3$. A theorem of vector field theory states that $\mathbf{F}$ is the gradient of a potential $f(x, y, z)$ if and only if curl$\mathbf{F} = \mathbf{0}$. Thus,

$$\text{if } \nabla \times \mathbf{F} = 0, \quad \text{then} \quad \mathbf{F} = \nabla f$$

for some f. The proof of this assertion will be discussed in Chapter 12.

The converse statement that a vector function derived from a scalar potential will have zero curl is proven by substituting the gradient components in the definition of the curl in Equation 11.17. The result is

$$\nabla \times \nabla f = \mathbf{0}$$

because of the symmetry property of the mixed partial derivatives that result. Thus, the curl of the gradient of a scalar function is zero if the appropriate derivatives exist.

☐ **EXAMPLE 11.10**

Vector Field from a Scalar Potential

For the vector function

$$\mathbf{F} = y^2\,\mathbf{i} + 2xy\,\mathbf{j} - z^2\,\mathbf{k},$$

the curl is $\nabla \times \mathbf{F} = 0\mathbf{i} + 0\mathbf{j} + (2y - 2y)\mathbf{k} = \mathbf{0}$, so $\mathbf{F}$ can be derived from the gradient of a scalar function. Thus, $\mathbf{F} = \nabla\Phi$ becomes

$$\left(\frac{\partial\Phi}{\partial x}\mathbf{i} + \frac{\partial\Phi}{\partial y}\mathbf{j} + \frac{\partial\Phi}{\partial z}\mathbf{k} \right) = y^2\mathbf{i} + 2xy\mathbf{j} - z^2\mathbf{k}.$$

Equating each component leads to the differential equations

$$\frac{\partial\Phi}{\partial x} = y^2, \qquad \frac{\partial\Phi}{\partial y} = 2xy, \qquad \frac{\partial\Phi}{\partial z} = -z^2, \tag{11.23}$$

which must be simultaneously satisfied for $\mathbf{F}$ to be derived from the gradient of Φ.

Integrating the equations with respect to x, y, and z, respectively, yields

$$
\begin{aligned}
\Phi &= xy^2 + f(y,z), \\
\Phi &= xy^2 + g(x,z), \\
\Phi &= -\frac{z^3}{3} + h(x,y),
\end{aligned}
$$

where $f(y,z)$, $g(x,z)$, and $h(x,y)$ are arbitrary functions. Notice that in solving partial differential equations, the constant of integration that occurs in ordinary differential equations becomes a function of the variables that are considered constant in the equation.

Using the first equation for Φ and substituting into the second of the relations of Equation 11.23 lead to the conclusion that

$$
\frac{\partial \Phi}{\partial y} = 2xy + \frac{\partial f}{\partial y} = 2xy,
$$

so that $\partial f / \partial y = 0$. Thus, f can be viewed as a function of z only. Substituting into the third equation for z in Equation 11.23 yields another equation for f as

$$
\frac{\partial \Phi}{\partial z} = \frac{\partial f}{\partial z} = \frac{df}{dz} = -z^2,
$$

which is an ordinary differential equation. This can be integrated, with the result

$$
f(z) = -\frac{z^3}{3} + C,
$$

where C is a constant of integration. The final result is that the scalar potential function is

$$
\Phi(x,y,z) = xy^2 - \frac{z^3}{3} + C.
$$

It is easily verified that $\nabla \Phi = \mathbf{F}$, as desired.

$\square$

SOLENOIDAL FIELDS

Some vector fields can be derived from the curl of a vector field, called the *vector potential*. For solenoidal fields, $\nabla \cdot \mathbf{F} = 0$, and $\mathbf{F}$ can be written as

$$
\nabla \times \mathbf{F} = \mathbf{J}(x,y,z).
$$

Maxwell's equations, treated in Chapter 12, show that the magnetic field is derivable from a vector potential. In many cases, the vector potential may not have a physical realization, but it is introduced to simplify the mathematical operations on the field of interest.

Helmholtz's theorem states that the most general vector field will have both a nonzero divergence and a nonzero curl. Thus, a general vector field $\mathbf{F}(x, y, z)$ can be derived as

$$\mathbf{F}(x, y, z) = \nabla \Phi(x, y, z) + \nabla \times \mathbf{A}(x, y, z). \tag{11.24}$$

In a specific area of study, the scalar potential Φ and vector potential $\mathbf{A}$ are determined by the physical sources of the vector field. For example, in electrostatics, the scalar sources are charge distributions and the vector sources are currents.

Summary Table 11.3 summarizes the relationships for vector fields that are the consequence of Helmholtz's theorem.

TABLE 11.3 *Properties of vector fields*

Type	Condition	Source
Irrotational	$\nabla \times \mathbf{F} = 0$	$\mathbf{F} = \nabla \Phi$
Solenoidal	$\nabla \cdot \mathbf{F} = 0$	$\mathbf{F} = \nabla \times \mathbf{A}$
General		$\mathbf{F} = \nabla \Phi + \nabla \times \mathbf{A}$

As we shall see, many of the properties of vector fields arise directly from the solenoidal or irrotational characteristics of the field. The significance of the statements in Table 11.3 can be better understood when vector integral operations for vector fields are presented in Chapter 12. Usually, the integral relationships are used to actually determine the source functions Φ or $\mathbf{A}$ for the fields.

Let $\mathbf{F}$ and $\mathbf{G}$ be differentiable vector functions and f and g be differentiable scalar functions. The operators gradient, divergence, and curl are linear since

$$\begin{aligned}
\nabla(f + g) &= \nabla f + \nabla g, \\
\nabla \cdot (\mathbf{F} + \mathbf{G}) &= \nabla \cdot \mathbf{F} + \nabla \cdot \mathbf{G}, \\
\nabla \times (\mathbf{F} + \mathbf{G}) &= \nabla \times \mathbf{F} + \nabla \times \mathbf{G}.
\end{aligned} \tag{11.25}$$

Applications often involve vector differential operations on products of scalar functions and products of a scalar function and a vector function. A few identities are as follows:

$$\begin{aligned}
\nabla(fg) &= f \nabla g + g \nabla f, \\
\nabla \cdot (f\mathbf{F}) &= f\nabla \cdot \mathbf{F} + \mathbf{F} \cdot \nabla f, \\
\nabla \times (f\mathbf{F}) &= \nabla f \times \mathbf{F} + f\nabla \times \mathbf{F}.
\end{aligned} \tag{11.26}$$

The four operators, gradient, curl, divergence, and Laplacian, are related by the identity

$$\nabla \times \nabla \times \mathbf{F} = \nabla(\nabla \cdot \mathbf{F}) - \nabla^2 \mathbf{F}, \qquad (11.27)$$

which you are asked to prove in Problem 11.23. This identity and the others given here can be proven by writing out each expression in terms of coordinates.

PHYSICAL APPLICATION AND INTERPRETATION

In this section we consider three applications of vector field theory. First, the significance of a conservative field is studied. Then, field theory is applied to fluid flow to derive the continuity equation. Finally, the properties of the electrostatic field are studied.

POTENTIAL ENERGY

In physics, the law of conservation of energy is one of the most important results to emerge from Newtonian mechanics. The concept is also extended to thermal and electromagnetic energy, as well as various other forms of energy. Using an example, we study the relationship between force and potential energy in the special case that the force is conservative.

□ EXAMPLE 11.11   *Potential Energy*

The vector field $\mathbf{F}$ defined in a region of space is called *conservative* if $\mathbf{F}$ is derived from a scalar potential function as

$$\mathbf{F} = -\nabla\Phi. \qquad (11.28)$$

As shown in Problem 11.9, the curl of the Newtonian gravitational force and the Coulomb force is zero. Thus, these forces can be derived from a scalar potential because the force field is irrotational. The scalar potential in these cases can be associated with the potential energy of a particle in the field. We will show this for the case of a particle subject to a conservative force.

The kinetic energy T of a particle with mass m moving with a speed v is

$$T = \frac{1}{2}m\dot{x}^2 + \frac{1}{2}m\dot{y}^2 + \frac{1}{2}m\dot{z}^2,$$

where $v = (\dot{x}^2 + \dot{y}^2 + \dot{z}^2)^{1/2}$. We can designate the path of the particle as

$$\mathbf{x}(t) = [x(t), y(t), z(t)],$$

so that the kinetic energy becomes $\frac{1}{2}m\|\dot{\mathbf{x}}(t)\|^2$. Let the potential energy of the particle be denoted by V. A mechanical system for which the total energy,

$$E = T + V,$$

is constant is called a *conservative* system. Systems with friction, electrical resistance, or air resistance, are not conservative.

Applying Newton's second law to a particle with mass m moving along a path $\mathbf{x}(t)$ leads to the equation

$$m\ddot{\mathbf{x}}(t) = \mathbf{F}[\mathbf{x}(t)].$$

Assuming that $\mathbf{F}$ can be derived from $V[\mathbf{x}(t)]$, the expression

$$\mathbf{F}[\mathbf{x}(t)] = -\nabla V[\mathbf{x}(t)]$$

can be substituted in Newton's law to yield

$$m\ddot{\mathbf{x}}(t) + \nabla V[\mathbf{x}(t)] = 0. \qquad (11.29)$$

A system described by this equation can be shown to be a conservative system by showing that the time derivative of the energy is zero. Taking the dot product of $\dot{\mathbf{x}}(t)$ with both terms yields

$$m\ddot{\mathbf{x}}(t) \cdot \dot{\mathbf{x}}(t) + \nabla V[\mathbf{x}(t)] \cdot \dot{\mathbf{x}}(t) = 0. \qquad (11.30)$$

Equation 11.30 can be transformed to the desired form by rewriting the derivative terms. Since

$$\frac{d}{dt}(\dot{\mathbf{x}})^2 = 2\dot{\mathbf{x}} \cdot \ddot{\mathbf{x}},$$

the first term is m times one-half of the time derivative of $||\dot{\mathbf{x}}||^2$. This term represents the *kinetic energy* of the particle. Applying the chain rule to the potential $V[\mathbf{x}(t)] = V[x(t), y(t), z(t)]$ results in the expression

$$
\begin{aligned}
\frac{dV[\mathbf{x}(t)]}{dt} &= \frac{\partial V}{\partial x}\frac{\partial x(t)}{\partial t} + \frac{\partial V}{\partial y}\frac{\partial y(t)}{\partial t} + \frac{\partial V}{\partial z}\frac{\partial z(t)}{\partial t} \\
&= \frac{\partial V}{\partial x}\dot{x}(t) + \frac{\partial V}{\partial y}\dot{y}(t) + \frac{\partial V}{\partial z}\dot{z}(t) \\
&= \nabla V[\mathbf{x}(t)] \cdot \dot{\mathbf{x}}.
\end{aligned}
$$

With these substitutions, Equation 11.30 becomes

$$\frac{d}{dt}\left[\frac{1}{2}m||\dot{\mathbf{x}}||^2 + V[\mathbf{x}(t)]\right] = 0,$$

which leads to the conclusion that

$$\frac{1}{2}m||\dot{\mathbf{x}}||^2 + V[\mathbf{x}(t)] = \text{constant}.$$

Thus, the scalar potential function V is the potential energy, and the total energy is constant when the force field is derived from a scalar potential. Such a system and the force field are said to be conservative.

In physics, energy is identified with the ability to do work. This aspect of energy is explored in the next chapter.

□

In fluid dynamics, the divergence can measure the rate of change of density at any point of the fluid. Let $\mathbf{v}(x, y, z, t)$ be the velocity vector of the fluid particles at a point. If the amount of fluid flowing through a region does not change so that the density of the fluid is constant, the fluid is said to be *incompressible* and

$$\nabla \cdot \mathbf{v} = 0 \qquad (11.31)$$

Since electric current is the flow of charge, a steady current also has a zero divergence. The divergence equation is also called a *continuity equation* because it indicates a conservation of mass in the case of fluid flow or charge in current flow when the divergence is zero.

Suppose the density of a fluid is changing with time so that $\rho(x, y, z, t)$ defines the density in mass units per unit volume. In this case, the vector $\mathbf{u} = \rho\mathbf{v}$ can be used to describe the fluid flow. It can be shown that when the density of the fluid varies with time and space, the *continuity equation* becomes

$$\frac{\partial \rho}{\partial t} + \nabla \cdot (\rho\mathbf{v}) = 0. \qquad (11.32)$$

This equation expresses the law of conservation of matter as will be shown when the equation is derived using vector integrals in the next chapter.

□ EXAMPLE 11.12   ***Equation of Continuity***

Let $\nabla \cdot \mathbf{u} = -\dfrac{\partial \rho}{\partial t}$ with $\mathbf{u} = \rho\mathbf{v}$. Then, taking one component of the divergence yields

$$(\nabla \cdot \mathbf{u})_x = (\nabla \cdot \rho\mathbf{v})_x = \frac{\partial \rho}{\partial x}v_x + \rho\frac{\partial \rho}{\partial x},$$

with similar results for the y- and z-components. Noting that

$$\mathbf{v} = v_x\mathbf{i} + v_y\mathbf{j} + v_z\mathbf{k},$$
$$\nabla\rho = \frac{\partial \rho}{\partial x}\mathbf{i} + \frac{\partial \rho}{\partial y}\mathbf{j} + \frac{\partial \rho}{\partial z}\mathbf{z}$$

leads to the result

$$\frac{\partial \rho}{\partial t} + \nabla\rho \cdot \mathbf{v} + \rho\nabla \cdot \mathbf{v} = 0,$$

as stated in Equation 11.32.

□

ELECTROMAGNETIC THEORY

Problems in an area of application called *classical electromagnetic theory* attempt to relate the electric field $\mathbf{E}(x, y, z)$ and magnetic field $\mathbf{B}(x, y, z)$ to the charges and currents that produce the fields. The term *classical* refers to the theory developed by James Clerk Maxwell in the last century that relates the fields to their sources through a set of equations now known as *Maxwell's equations*.

We present here a very restricted application that treats the *electrostatic field* in vacuum. In this case, the fields do not vary with time, and there are no sources for the fields in the region of space being considered. In such a region

$$\nabla \times \mathbf{E}(x, y, z) = 0, \qquad (11.33)$$

so that the electric field is irrotational and can be derived from a scalar potential. By convention, the negative gradient of the potential is taken, so that

$$\mathbf{E}(x, y, z) = -\nabla \Phi(x, y, z). \qquad (11.34)$$

The potential in the electrostatic case is measured in volts. Another of Maxwell's equations states that

$$\nabla \cdot \mathbf{E}(x, y, z) = 0 \qquad (11.35)$$

for the field in a vacuum. Thus, the potential satisfies Laplace's equation,

$$\nabla \cdot (-\nabla \Phi) = -\nabla^2 \Phi(x, y, x) = 0, \qquad (11.36)$$

in a source-free region of space.

Although these results are limited to a special case of a stationary electric field, the equations for the electrostatic field have many applications. The magnetic field is more complicated because it derives from a vector potential $(\nabla \cdot \mathbf{B} = 0)$, and we defer the study of the magnetic field until the next chapter.

□ EXAMPLE 11.13   *Electrostatic Field*

Consider the two conducting plates in Figure 11.4 with the lower plate at $z = 0$ held at zero potential and the top plate at $z = d$ at V_0 volts. Such a configuration acts as a capacitor with capacitance

$$C = \frac{Q}{V_0} \quad \text{farads}, \qquad (11.37)$$

where Q is the magnitude of the total charge in coulombs on each plate with negative charge on the lower plate.

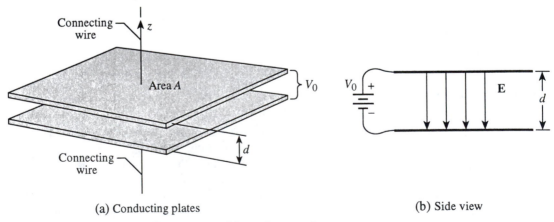

(a) Conducting plates   (b) Side view

FIGURE 11.4 *Conducting plates separated by a distance d*

Since a potential difference exists between the plates, an electric field is present inside the capacitor that is directed from the upper plate to the lower plate, as shown in the side view.

For our analysis, the plates are taken to be infinite in the x- and y-directions so that the field varies with z only. This avoids consideration of the electric field at the edges of the plates, which is not uniform. However, the total charge is also infinite in this case, so the capacitance formula of Equation 11.37 could not be used in the form shown. One way to avoid these complications is to assume the plates are finite in x and y but to consider only the nearly uniform electric field near the center of the plates.[1]

Since the interior of the capacitor has no sources, Laplace's equation determines the potential Φ between the plates, as defined in Equation 11.36. The constant potential of each plate yields the boundary conditions that are necessary to solve the equation. Because the variation in potential is only in the z-direction, Laplace's equation becomes

$$\frac{d^2\Phi(z)}{d^2z} = 0 \qquad 0 < z < d, \tag{11.38}$$

with the boundary conditions

$$\Phi(0) = 0 \quad \text{and} \quad \Phi(d) = V_0.$$

Laplace's equation is now an ordinary differential equation, which can be integrated twice to yield the result

$$\Phi(z) = c_1 z + c_2.$$

Applying the boundary conditions at $z = 0$ and $z = d$ shows that $c_2 = 0$ and $c_1 = V_0/d$.

The potential between the plates is

$$\Phi(z) = \frac{V_0}{d} z,$$

and the electric field according to Equation 11.34 becomes

$$\mathbf{E}(x, y, z) = -\nabla\Phi = -\frac{V_0}{d}\mathbf{k},$$

measured in volts per meter in SI (mks) units.

Symmetry. One additional point is worth noting. The solution of Laplace's equation was greatly simplified due to the symmetry of the problem. The lack of variation of potential in the x- and y-directions reduced the partial differential equation in three dimensions to an ordinary differential equation in one dimension to which we can apply the techniques discussed in Chapter 5.

$\square$

---

[1]Capacitors are discussed in every textbook on electromagnetic theory. The effect at the edges of the plates is called *fringing* of the field and requires special techniques to solve for the field.

Table 11.4 summarizes the main results of vector field theory for the specific applications treated in this chapter. Each of the equations describes a vector or scalar field at a point. These applications will be discussed further in the next chapter, where integral relationships for the field quantities will be derived. Such relationships will allow us to compute the fields from distributed sources given the source distributions.

TABLE 11.4 *Summary of physical applications*

Description	Relationship
Force:	
Potential energy	$V(x, y, z)$
Conservative force	$\mathbf{F} = -\nabla V$
Fluid flow:	
Velocity	$\mathbf{v}(x, y, z)$
Density	$\rho(x, y, z)$
Velocity potential	$\Phi(x, y, z)$
Continuity	$\dfrac{\partial \rho}{\partial t} + \nabla \cdot (\rho \mathbf{v}) = 0$
Incompressible	$\nabla \cdot \mathbf{v} = 0$
Irrotational	$\nabla \times \mathbf{v} = 0$
Irrotational and incompressible	$\nabla^2 \Phi = 0$
Electrostatics in vacuum:	
Potential	$\nabla^2 \Phi = 0$
Electric field	$\mathbf{E} = -\nabla \Phi$
	$\nabla \times \mathbf{E} = 0$

CURVILINEAR COORDINATES

One advantage of formulating relations among geometrical and physical quantities in vector form is that the relations are valid in all coordinate systems. For certain applications, particularly when numerical computations are required, it is desirable to translate the vector equations into a specific coordinate system to simplify the problem being solved.

In this section, we consider curvilinear coordinate systems in three dimensions that have orthonormal basis vectors. Figure 11.5 shows the coordinates for an arbitrary orthogonal curvilinear coordinate system.

The coordinate lines designated u_1, u_2, u_3 are, in general, curved. Unit vectors $\mathbf{a}_1, \mathbf{a}_2, \mathbf{a}_3$ can be defined that are tangent to the coordinate lines at a point. These unit vectors are assumed to be mutually perpendicular for any orthogonal coordinate system.

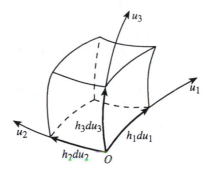

FIGURE 11.5 *Curvilinear coordinate system*

Thus, a vector field in these curvilinear coordinates is written as

$$\mathbf{F}(\mathbf{x}) = F_1(\mathbf{x})\,\mathbf{a}_1 + F_2(\mathbf{x})\,\mathbf{a}_2 + F_1(\mathbf{x})\,\mathbf{a}_3,$$

where $\mathbf{x} = (u_1, u_2, u_3)$. A scalar field has the form $\Phi(u_1, u_2, u_3)$.

The scale factors h_1, h_2, and h_3 in Figure 11.5 are used to define the length elements along the coordinate axes. These scale factors can be a function of the coordinates u_1, u_2, and u_3. Table 11.5 summarizes the gradient, divergence, curl, and Laplacian operators.

TABLE 11.5 *Vector operations in curvilinear coordinates*

Operation	Equation
Gradient	$\nabla\Phi = \dfrac{\mathbf{a}_1}{h_1}\dfrac{\partial\Phi}{\partial u_1} + \dfrac{\mathbf{a}_2}{h_2}\dfrac{\partial\Phi}{\partial u_2} + \dfrac{\mathbf{a}_3}{h_3}\dfrac{\partial\Phi}{\partial u_3}$
Divergence	$\nabla\cdot\mathbf{F} = \dfrac{1}{h_1 h_2 h_3}\left[\dfrac{\partial}{\partial u_1}(h_2 h_3 F_1) + \dfrac{\partial}{\partial u_2}(h_3 h_1 F_2) + \dfrac{\partial}{\partial u_3}(h_1 h_2 F_3)\right]$
Curl	$\nabla\times\mathbf{F} = \dfrac{1}{h_1 h_2 h_3}\begin{vmatrix} h_1\mathbf{a}_1 & h_2\mathbf{a}_2 & h_3\mathbf{a}_3 \\ \dfrac{\partial}{\partial u_1} & \dfrac{\partial}{\partial u_2} & \dfrac{\partial}{\partial u_3} \\ h_1 F_1 & h_2 F_2 & h_3 F_3 \end{vmatrix}$
Laplacian	$\nabla^2\Phi = \dfrac{1}{h_1 h_2 h_3}\left[\dfrac{\partial}{\partial u_1}\left(\dfrac{h_2 h_3}{h_1}\dfrac{\partial\Phi}{\partial u_1}\right) + \dfrac{\partial}{\partial u_2}\left(\dfrac{h_3 h_1}{h_2}\dfrac{\partial\Phi}{\partial u_2}\right) + \dfrac{\partial}{\partial u_3}\left(\dfrac{h_1 h_2}{h_3}\dfrac{\partial\Phi}{\partial u_3}\right)\right]$

Coordinate parameters for rectangular, cylindrical, and spherical coordinates will be presented in the next subsection. In rectangular coordinates, the scale factors are 1 and the coordinates are $u_1 = x$, $u_2 = y$ and $u_3 = z$.

However, in the cylindrical coordinate system introduced in Chapter 10, $u_1 = r, u_2 = \theta, u_3 = z$. Notice that $du_2 = d\theta$ is not a length, so the infinitesimal distance caused by a change $d\theta$ is $r\,d\theta$. Thus, the scale factors serve to convert angular motion to linear distance when necessary.

RECTANGULAR, CYLINDRICAL, AND SPHERICAL COORDINATES

The primary coordinate systems for engineering and physics are the rectangular, cylindrical, and spherical coordinate systems, as shown in Figure 11.6. Table 11.6 lists the coordinates, volume elements, and scale factors for these coordinate systems.

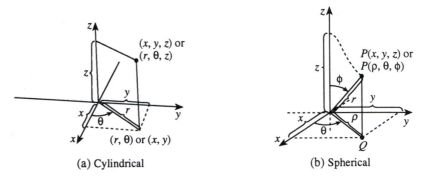

(a) Cylindrical   (b) Spherical

FIGURE 11.6 *Coordinate systems: (a) cylindrical (b) spherical*

The parameters for the three primary coordinate systems are used in operators such as the gradient, as listed in Table 11.5.

TABLE 11.6 *Coordinate parameters*

Parameter	*Rectangular*	*Cylindrical*	*Spherical*
Coordinates	(x, y, z)	(r, θ, z)	(r, θ, ϕ)
Volume element	$dx\,dy\,dz$	$r\,dr\,d\theta\,dz$	$r^2 \sin \phi \, dr \, d\theta \, d\phi$
Scale factors	$h_1 = 1$	$h_1 = 1$	$h_1 = 1$
	$h_2 = 1$	$h_2 = r$	$h_2 = r \sin \phi$
	$h_3 = 1$	$h_3 = 1$	$h_3 = r$

As defined in Chapter 10, the polar angle ϕ for spherical coordinates is measured down from the z-axis and the azimuth angle θ is measured from the x-axis in the xy-plane. Other authors may reverse the notation by using θ as the polar angle.

□ EXAMPLE 11.14   *Operators in Cylindrical Coordinates*

Combining the expression for an operator in Table 11.5 in curvilinear coordinates with the coordinate parameters yields the expression in a particular coordinate system. For example, $h_1 = 1, h_2 = r, h_3 = 1$ for cylindrical coordinates (r, θ, z) yields the formula for the gradient of the scalar function $\Phi(r, \theta, z)$ in cylindrical coordinates as

$$\nabla \Phi = \frac{\partial \Phi}{\partial r} \mathbf{i}_r + \frac{1}{r} \frac{\partial \Phi}{\partial \theta} \mathbf{i}_\theta + \frac{\partial \Phi}{\partial z} \mathbf{i}_z. \tag{11.39}$$

The divergence of $\mathbf{F}$ can be computed as the dot product of the gradient and the vector as

$$\nabla \cdot \mathbf{F} = \left(\frac{\partial}{\partial r} \mathbf{i}_r + \frac{1}{r} \frac{\partial}{\partial \theta} \mathbf{i}_\theta + \frac{\partial}{\partial z} \mathbf{i}_z \right) \cdot (F_r \mathbf{i}_r + F_\theta \mathbf{i}_\theta + F_z \mathbf{i}_z). \tag{11.40}$$

When the divergence is computed in this manner, the change of $\mathbf{i}_r$ and $\mathbf{i}_\theta$ as θ changes must be considered. Thus, the quantities $\partial \mathbf{i}_r / \partial \theta$ and $\partial \mathbf{i}_\theta / \partial \theta$ are not zero. Problem 11.11 asks you to expand the expression and compute the divergence.

Table 11.5 yields the divergence in cylindrical coordinates as

$$\nabla \cdot \mathbf{F} = \frac{1}{r} \frac{\partial (r F_r)}{\partial r} + \frac{1}{r} \frac{\partial (F_\theta)}{\partial \theta} + \frac{\partial F_z}{\partial z}. \tag{11.41}$$

□

MATLAB AND CURVILINEAR COORDINATES   MATLAB has a number of commands to convert values between coordinate systems and to plot results in various coordinate systems, such as polar coordinates. These commands were presented in Chapter 10 as part of the discussion of curvilinear coordinates.

REINFORCEMENT EXERCISES AND EXPLORATION PROBLEMS

■

In these problems, do the computations by hand unless otherwise indicated and then check those that yield numerical or symbolic results with MATLAB.

REINFORCEMENT EXERCISES

P11.1. Fields Give examples of common scalar and vector quantities that constitute fields.

P11.2. Directional derivative For a temperature distribution of the form $f(x, y) = x^3 y^2$ compare the rate of change of the temperature at the point $(-1, 2)$ in the directions

      a. $\mathbf{v} = 4\mathbf{i} - 3\mathbf{j}$;

      b. Parallel to the x-axis;

      c. Parallel to the y-axis.

P11.3. Temperature gradient Let the temperature of a surface be

$$T = x^2 + xy + yz.$$

At the point $(2, 1, 4)$ determine the following:

      a. A unit vector in the direction of the maximum rate of change of temperature;

      b. The rate of change of temperature along the vector $\mathbf{i} - 2\mathbf{j} + 2\mathbf{k}$.

P11.4. Gradient Find the gradient of the function $F = x^2 + y^2 + z^2$. Where does the gradient point?

P11.5. Gradient and surfaces Find a unit vector normal to the surface

$$2x^2 + 4yz - 5z^2 = -10$$

at the point $P(3, -1, 2)$. Is there more than one?

P11.6. Divergence of inverse-square forces The Newtonian gravitational force and the Coulomb force in a source free region of space obey an inverse-square force such as

$$\mathbf{F} = -\frac{k}{R^3}\,\mathbf{R} = -\frac{k}{R^2}\,\mathbf{a}_R,$$

where k is a constant. Show that the force field is solenoidal.

P11.7. Curl of steady rotation Consider a fluid rotating with constant angular velocity ω in a counterclockwise direction around an origin at $(0, 0, 0)$. If the rotation is in the xy-plane, describe the velocity vector field and determine the divergence and curl of the field.

P11.8. Curl of straight-line flow Consider two vector fields

      a. $\mathbf{v} = v_0 e^{-y^2}\mathbf{j}$;

      b. $\mathbf{v} = v_0 e^{-x^2}\mathbf{j}$.

Draw the field vectors for several values of x and y. Are these fields irrotational?. If not, why not if the fields are considered to be fluid flow?

P11.9. Curl of inverse-square forces Given a force field that varies with R as $\mathbf{R}/R^3$ in a source free region of space, show that this force is irrotational since

$$\nabla \times \frac{\mathbf{R}}{R^3} = \mathbf{0}.$$

Considering the result of Problem 11.6, what conclusion do you draw about inverse-square forces?

P11.10. Derivatives Let $\Phi = 2x^2 y - xz^3$ and find

      a. $\nabla\Phi$;

                                          Chapter 11 ■ VECTOR DIFFERENTIAL OPERATORS

b. $\nabla^2 \Phi$.

P11.11. Divergence in cylindrical coordinates   Compute the divergence $\nabla \cdot \mathbf{u}$ in cylindrical coordinates by taking the dot product of the gradient and $\mathbf{u}$.

P11.12. Laplace's equation in curvilinear coordinates   Write Laplace's equation

$$\nabla^2 \Phi = 0$$

in rectangular, cylindrical, and spherical coordinates.

P11.13. Laplace's equation   The Newtonian gravitational potential is

$$\Phi = -\frac{GMm}{R}.$$

Show that this potential and any scalar function $\Phi = \alpha/R$ satisfies Laplace's equation, where α is a constant.

P11.14. Fields   Given the field

$$\mathbf{F} = e^x \sin y \, \mathbf{i} + e^x \cos y \, \mathbf{j},$$

find the scalar potential.

P11.15. Fluid flow   Can the field

$$\mathbf{v} = 3x^2 \, \mathbf{i} + 5xy^2 \, \mathbf{j} + xyz^3 \, \mathbf{k}$$

represent incompressible flow? Is the flow irrotational?

P11.16. Cylindrical and spherical capacitors   Consider a cylindrical and a spherical capacitor made of concentric conducting surfaces with the inner surface held at V_0 volts and the outer surface at zero volts. In each case, the inner radius is R_1 and the outer radius is R_2. The cylindrical capacitor is assumed to be infinitely long in the z-direction. Solve for the potential in between the surfaces in each case.

P11.17. MATLAB surface normal   Plot the surface normals for the function

$$f(x, y, z) = \sqrt{x^2 + y^2 + z^2},$$

as described in Example 11.3.

P11.18. MATLAB field plots   Plot the vectors defined by the fields in Problem 11.8.

EXPLORATION PROBLEMS

P11.19. Directional derivative   Given the function

$$f(x, y, z) = xyz + e^{2x+y},$$

find the directional derivative in the direction of

$$\mathbf{u} = \left[\frac{1}{2}, \frac{1}{2}, \frac{1}{\sqrt{2}} \right].$$

P11.20. Tangent plane Given a level surface or curve S of $f(\mathbf{x})$ passing through $\mathbf{x}_0$, the tangent plane to S at $\mathbf{x}_0$ is the set of points

$$\nabla f(\mathbf{x}_0)\cdot(\mathbf{x} - \mathbf{x}_0) = 0,$$

if the gradient is not zero at the point. The function

$$f(x,y,z) = x^2 + y^2 - z^2 = 0$$

is a cone which is a level surface of f. Find the equation of the tangent plane to the cone at the point $\mathbf{x}_0 = (1, 1, \sqrt{2})$. Plot the function and show the plane.

P11.21. Potential function Suppose that $f(x,y)$ and $g(x,y)$ are differentiable functions. If the partial derivatives of f and g exist and are continuous and if

$$\frac{\partial f}{\partial y} = \frac{\partial g}{\partial x},$$

then the vector field $\mathbf{F}[f(x,y), g(x,y)]$ has a potential function. Find the potential function for the vector field

$$\mathbf{F}(x,y) = 2xy\,\mathbf{i} + (x^2 + 3y^2)\,\mathbf{j},$$

if $\mathbf{F}$ has a potential function.

P11.22. Report on optimization techniques Write a report on minimization of multivariate functions. Include the *method of steepest descent* and *conjugate gradient methods*.

P11.23. Curl-curl operation Prove the following identity

$$\nabla \times \nabla \times \mathbf{F} = \nabla(\nabla \cdot \mathbf{F}) - \nabla^2\mathbf{F},$$

presented in Equation 11.27.

ANNOTATED BIBLIOGRAPHY

1. Kaplan, Wilfred, *Advanced Calculus, fourth edition*, Addison Wesley, Reading, MA, 1991. *An excellent treatment of advanced calculus.*

2. Kellogg, Oliver D., *Foundations of Potential Theory*, Dover Publications, New York, 1953. *A classical treatment of potential and vector fields. The material is presented rigorously but the text is very readable.*

ANSWERS

P11.1. Fields Scalar quantities that vary with position can be described by a scalar field. Example scalar values that vary with position are temperature, humidity, and air pressure in a region of space. The population density of a country can also be represented as a scalar field. Notice that most of these quantities do in fact vary with time so the scalar field is an instantaneous representation of the values at a specific time.

Quantities such as the velocity of the surface water in a flowing river or the velocity of raindrops on a windy day at various points can be represented as a vector field since the velocity includes both the speed and direction of the water particles.

P11.2. Directional derivative The gradient of the function $f = x^3 y^2$ is $\nabla f = 3x^2 y^2 \mathbf{i} + 2x^3 y \mathbf{j}$. The unit vector in the direction of $\mathbf{v}$ is

$$\mathbf{u} = \frac{\mathbf{v}}{||\mathbf{v}||} = \frac{1}{5}(4\mathbf{i} - 3\mathbf{j}).$$

Then, the directional derivative in the direction of $\mathbf{v}$ is

$$\nabla_u f(x, y) = \nabla f \cdot \mathbf{u} = 12$$

when evaluated at the point $(-1, 2)$. Thus, the temperature is increasing at $12°$ per unit change in distance. The partial derivatives with respect to x and y evaluated at $(-1, 2)$ yield 12 and -4, respectively.

P11.3. Temperature gradient Since $T = x^2 + xy + yz$ at the point $(2, 1, 4)$,

$$\nabla T = (2x + y)\mathbf{i} + (x + z)\mathbf{j} + y\mathbf{k}|_{(2,1,4)} = 5\mathbf{i} + 6\mathbf{j} + \mathbf{k}$$

in units of degrees per unit distance.

    a. The unit vector in the direction of the gradient is

$$\frac{\nabla T}{||\nabla T||} = \frac{5\mathbf{i} + 6\mathbf{j} + \mathbf{k}}{(25 + 36 + 1)^{1/2}}.$$

    b. The directional derivative in the direction $v = \mathbf{i} - 2\mathbf{j} + 2\mathbf{k}$ is

$$\nabla_u f = (\nabla T) \cdot \frac{\mathbf{v}}{||\mathbf{v}||} = (5\mathbf{i} + 6\mathbf{j} + \mathbf{k}) \cdot \left(\frac{\mathbf{i} - 2\mathbf{j} + 2\mathbf{k}}{3}\right) = \frac{-5}{3}$$

P11.4. Gradient The gradient of the function $F = x^2 + y^2 + z^2$ is

$$\nabla F = 2x\mathbf{i} + 2y\mathbf{j} + 2z\mathbf{i},$$

which is $2\mathbf{r}$, where $\mathbf{r}$ is a position vector from the origin.

P11.5. Gradient and surfaces The gradient at a point on the surface is normal to the surface. At the point $P(3, -1, 2)$, the gradient for $f(x, y, z) = 2x^2 + 4yz - 5z^2$ is

$$\nabla f = (4x)\mathbf{i} + (4z)\mathbf{j} + (4y - 10z)\mathbf{k}|_{(3,-1,2)} = 12\mathbf{i} + 8\mathbf{j} - 24\mathbf{k}$$

and the unit normal is

$$\mathbf{u} = \frac{12\mathbf{i} + 8\mathbf{j} - 24\mathbf{k}}{\sqrt{(12)^2 + (8)^2 + (-24)^2}} = \frac{3\mathbf{i} + 2\mathbf{j} - 6\mathbf{k}}{7}.$$

Note that $-\mathbf{u}$ is also perpendicular to the surface.

P11.6. Divergence of inverse-square force   Consider each component of the force as

$$\mathbf{F}_{x_i} = -\frac{kx_i}{R^3},$$

in which $x_i = x, y,$ or z. Differentiating with respect to x_i yields

$$\frac{\partial \mathbf{F}_{x_i}}{\partial x_i} = -k\left(\frac{1}{R^3} - \frac{3x_i^2}{R^5}\right).$$

Summing the three terms of the divergence yields the result that $\nabla \cdot \mathbf{F} = 0$ so the field is solenoidal.

P11.7. Curl of steady rotation   Let the vector describing the rotation be $\omega\mathbf{k}$. The linear velocity and the angular velocity are related as

$$\mathbf{v} = \omega\mathbf{k} \times \mathbf{r}.$$

Using the fact that $||\mathbf{v}|| = \omega||\mathbf{r}||$, the field is

$$\mathbf{v}(x, y) = \omega(-y\mathbf{i} + x\mathbf{j}).$$

The results are that

$$\nabla \cdot \mathbf{v} = 0, \qquad \nabla \times \mathbf{v} = 2\omega\mathbf{k},$$

so the field is solenoidal and the curl is twice the angular velocity, directed normal to the circulation of the fluid.

P11.8. Curl of straight-line flow

     a. $\nabla \times (v_0 e^{-y^2}\mathbf{j}) = 0$;

     b. $\nabla \times (v_0 e^{-x^2}\mathbf{j}) = -v_0 2x e^{-x^2}\mathbf{k}$.

In the case of the field varying with y there is no variation with x. In case b, there is a variation with x so there would be a net torque on a small paddle wheel placed in the field.

P11.9. Curl of inverse-square forces   Since

$$\nabla\left(\frac{1}{R}\right) = -\frac{\mathbf{R}}{R^3},$$

the function is irrotational so that

$$\nabla \times \frac{\mathbf{R}}{R^3} = \mathbf{0}.$$

P11.10. Derivatives   For $\Phi = 2x^2 y - xz^3$,

     a. $\nabla\Phi = (4xy - z^3)\mathbf{i} + 2x^2\mathbf{j} - 3xz^2\mathbf{k}$;

     b. $\nabla^2\Phi = \nabla \cdot \nabla\Phi = 4y - 6xz$.

P11.11. Divergence in cylindrical coordinates   In cylindrical coordinates

$$\nabla \cdot \mathbf{u} = \left(\frac{\partial}{\partial r}\mathbf{i}_r + \frac{1}{r}\frac{\partial}{\partial \theta}\mathbf{i}_\theta + \frac{\partial}{\partial z}\mathbf{i}_z\right) \cdot (u_r\mathbf{i}_r + u_\theta\mathbf{i}_\theta + u_z\mathbf{i}_z).$$

Note that each partial derivative operates on both the component of $\mathbf{u}$ and the associated unit vector. Because the unit vectors are orthonormal and constant with respect to changes in r and z, the terms from the partial derivatives of r and z yield

$$\frac{\partial u_r}{\partial r} + \frac{\partial u_z}{\partial z}.$$

The unit vector in the z direction does not change with θ. However, $\mathbf{i}_r$ and $\mathbf{i}_\theta$ are not constant as θ changes. The remaining terms arising from the partial derivative with respect to θ are

$$\left(\frac{1}{r}\frac{\partial}{\partial\theta}\,\mathbf{i}_\theta\right)\cdot(u_r\,\mathbf{i}_r + u_\theta\,\mathbf{i}_\theta).$$

Expanding the first term yields

$$\left(\frac{1}{r}\frac{\partial}{\partial\theta}\,\mathbf{i}_\theta\right)\cdot(u_r\,\mathbf{i}_r) = \frac{1}{r}\frac{\partial u_r}{\partial\theta}\,\mathbf{i}_r\cdot\mathbf{i}_\theta + \frac{u_r}{r}\mathbf{i}_\theta\cdot\frac{\partial\mathbf{i}_r}{\partial\theta}.$$

Because $\mathbf{i}_\theta\cdot\mathbf{i}_r = 0$, the first term is zero. Then, since

$$\frac{\partial\mathbf{i}_r}{\partial\theta} = \mathbf{i}_\theta,$$

the second term becomes u_r/r. Returning to Equation 11.11 and expanding the second term yields

$$\left(\frac{1}{r}\frac{\partial}{\partial\theta}\,\mathbf{i}_\theta\right)\cdot(u_\theta\,\mathbf{i}_\theta) = \frac{1}{r}\frac{\partial u_\theta}{\partial\theta} + \frac{u_\theta}{r}\mathbf{i}_\theta\cdot\frac{\partial\mathbf{i}_\theta}{\partial\theta}.$$

Recognizing that $\partial\mathbf{i}_\theta/\partial\theta = -\mathbf{i}_r$, the second term becomes zero, and the result after collecting terms is

$$\frac{\partial u_r}{\partial r} + \frac{u_r}{r} + \frac{1}{r}\frac{\partial u_\theta}{\partial\theta} + \frac{\partial u_z}{\partial z}.$$

The terms containing u_r are typically written

$$\frac{1}{r}\frac{\partial}{\partial r}(r u_r).$$

P11.12. Laplace's equation in curvilinear coordinates Writing the Laplacian $\nabla^2\Phi$ in the three coordinate systems using du_1, du_2, du_3 shown yields

    a. Rectangular: (dx, dy, dz)

$$\nabla^2\Phi(x,y,x) = \frac{\partial^2\Phi}{\partial x^2} + \frac{\partial^2\Phi}{\partial y^2} + \frac{\partial^2\Phi}{\partial z^2};$$

    b. Cylindrical: $(dr, r d\theta, dz)$

$$\nabla^2\Phi(r,\theta,z) = \frac{1}{r}\frac{\partial}{\partial r}\left(r\frac{\partial\phi}{\partial r}\right) + \frac{1}{r^2}\frac{\partial^2\Phi}{\partial\theta^2} + \frac{\partial^2\Phi}{\partial z^2};$$

    c. Spherical: $(dr, r\sin\phi\, d\theta, r d\phi)$

$$\nabla^2\Phi(r,\theta,\phi) = \frac{1}{r^2}\frac{\partial}{\partial r}\left(r^2\frac{\partial\phi}{\partial r}\right) + \frac{1}{r^2\sin^2\phi}\frac{\partial^2\Phi}{\partial\theta^2} + \frac{1}{r^2\sin\phi}\frac{\partial}{\partial\phi}\left(\sin\phi\frac{\partial\Phi}{\partial\phi}\right).$$

P11.13. Laplace's equation For a function of r only, $\Phi = \Phi(r)$, and Laplace's operator in spherical coordinates becomes

$$\nabla^2\Phi = \frac{1}{r^2}\frac{\partial}{\partial r}\left(r^2\frac{\partial\Phi}{\partial r}\right).$$

For any function of the form $\Phi = \alpha/r$, where α is a constant,

$$\nabla^2\left(\frac{\alpha}{r}\right) = \alpha\frac{1}{r^2}\frac{\partial}{\partial r}\left[r^2\frac{\partial}{\partial r}\left(\frac{1}{r}\right)\right]$$

$$= \alpha\frac{1}{r^2}\frac{\partial}{\partial r}\left[r^2\left(\frac{-1}{r^2}\right)\right] = 0.$$

Thus, Laplace's equation is satisfied.

P11.14. Fields Given the field
$$\mathbf{F} = e^x \sin y \, \mathbf{i} + e^x \cos y \, \mathbf{j},$$
the partial derivative with respect to x is
$$\frac{\partial \Phi}{\partial x} = e^x \sin y,$$
and
$$\Phi = e^x \sin y + f(y, z).$$
Thus,
$$\frac{\partial \Phi}{\partial y} = e^x \cos y = \frac{\partial}{\partial y} \left[e^x \sin y + f(y, z) \right]$$
$$= e^x \cos y + \frac{\partial f}{\partial y}.$$

So we conclude that $\partial f / \partial y = 0$. Also, $\partial f / \partial z = 0$. Thus,
$$\Phi = e^x \sin y + c,$$
where c is a constant.

P11.15. Fluid flow The field $\mathbf{v} = 3x^2 \mathbf{i} + 5xy^2 \mathbf{j} + xyz^3 \mathbf{k}$ cannot represent incompressible flow and is not irrotational since
$$\nabla \cdot \mathbf{v} = 6x + 10xy + 3xyz^2 \neq 0,$$
$$\nabla \times \mathbf{v} = xz^3 \mathbf{i} - yz^3 \mathbf{j} + 5y^2 \mathbf{k} \neq 0.$$

P11.16. Cylindrical and spherical capacitors In each case, Laplace's equation must be solved with the conditions
$$\Phi(R_1) = V_0 \quad \text{and} \quad \Phi(R_2) = 0,$$
and the variation is only in the r-direction.

In cylindrical coordinates
$$\nabla^2 \Phi(r) = \frac{1}{r} \frac{\partial}{\partial r} \left(r \frac{\partial \Phi}{\partial r} \right) = 0,$$
which after integrating twice and solving for the constants leads to the result
$$\Phi(r) = c_1 \ln r + c_2 = \frac{V_0}{\ln(R_1/R_2)} \ln \left(\frac{r}{R_2} \right).$$

In spherical coordinates
$$\nabla^2 \Phi(r) = \frac{1}{r^2} \frac{\partial}{\partial r} \left(r^2 \frac{\partial \Phi}{\partial r} \right) = 0,$$
which after integrating twice and solving for the constants leads to the result
$$\Phi(r) = \frac{c_1}{r} + c2 = V_0 \frac{R_1 R_2}{R_2 - R_1} \left(\frac{1}{r} - \frac{1}{R_2} \right).$$

Comment: MATLAB solutions are on the disk accompanying this textbook.

12

VECTOR INTEGRAL CALCULUS

PREVIEW————————————————————————

This chapter presents multiple integrals, line integrals, and surface integrals as well as various theorems of vector integral calculus that are important in applications. The chapter begins with a brief discussion of definite integrals, including multiple integrals and their applications. Then, line integrals are introduced as a generalization of definite integrals and surface integrals as generalizations of double integrals.

The powerful formulas of Green, Stokes, and Gauss are presented to describe transformations between various types of integrals. Green's theorem relates line integrals and double integrals. Stokes' theorem defines transformations between surface integrals and line integrals. Gauss' divergence theorem shows that triple integrals can be transformed into surface integrals. In addition to providing integral transformations, these theorems lead to a better understanding of the physical meaning of the divergence and curl of a vector function.

Finally, a number of physical applications are considered. Vector integral calculus plays an important role in practical studies of gravity, fluid flow, and electromagnetism.

INTEGRATION

The key concept of integral calculus is *integration*, a procedure that can be performed by computing the limit of sums that are called the *definite integral*. We shall discover that various integrals can be applied to determine useful quantities such as area, distance, volume, mass, work, and flux. These quantities can be approximated by sums and then obtained as accurately as desired by taking appropriate limits of the sums.

This section and the sections to follow that describe integrals begin with a definition of the integral to be discussed in terms of a sum. Then, properties of the integral and its applications are presented. We start with the so-called Riemann integral of a function of one variable. This integral forms the basis for the generalization of integration to multiple integrals, line integrals, and surface integrals.

RIEMANN INTEGRATION

Consider a continuous function $f(x)$ that is defined on a closed interval $a \leq x \leq b$. In elementary calculus, the *definite integral* $\int_a^b f(x)\,dx$ is defined as the limit of Riemann sums. Figure 12.1 illustrates the Riemann sum as the sum of areas of rectangles.

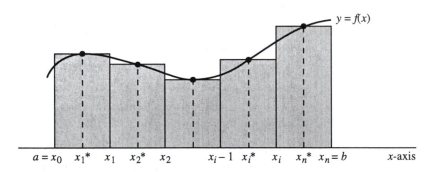

FIGURE 12.1 *The definite integral as a Riemann sum*

The integral can be computed by forming a partition of the interval $[a, b]$ as

$$a = x_0 < x_1 < \cdots < x_n = b,$$

and choosing points $x_i^* \in [x_i, x_{i+1}]$ to form the Riemann sum. The definite integral of a function $f(x)$ on the interval $a \leq x \leq b$ is defined by the limiting process

$$\int_a^b f(x)\,dx = \lim_{\substack{n \to \infty \\ \max(\Delta x_i \to 0)}} \sum_{i=0}^{n-1} f(x_i^*)\,\Delta x_i, \qquad (12.1)$$

where $\Delta x_i = x_{i+1} - x_i$. The point x_i^* in the sum can be any point of the ith subinterval. For example, to form the Riemann sum using the midpoints of the intervals,

$$x_i^* = \frac{x_{i-1} + x_i}{2}.$$

As discussed later in connection with numerical methods for evaluating integrals, Equation 12.1 can be used to compute an approximate numerical value of the definite integral by computing the sum for a finite number of terms. Also, since $f(x)$ is positive in Figure 12.1, the definite integral is seen to approximate the area bounded by the graph of $y = f(x)$ from $x = a$ to $x = b$. This is one of the possible applications of the definite integral.

FUNDAMENTAL THEOREM OF CALCULUS

The fundamental theorem of calculus shows the intimate relationship between the derivative and the integral. This theorem allows the evaluation of integrals without computing $\int_a^b f(x)\,dx$ from its definition as the limit of a sum. To apply the theorem, the *antiderivative* of $f(x)$ is defined as a function $F(x)$ such that $F'(x) = f(x)$.

■ **THEOREM 12.1**

Fundamental Theorem of Calculus

If $f(x)$ is a continuous function in the closed interval $[a, b]$, then

$$\int_a^b f(x)\,dx = F(b) - F(a),$$

where F is an antiderivative of f.

─────────────────────────────── ■

The theorem shows that the definite integral can be computed by finding an antiderivative F on the interval $[a, b]$ and evaluating it at the limits of integration a and b. The theorem does not say *how* to find the antiderivative, nor it does not imply that an antiderivative F *exists*. Another theorem of integral calculus states that a continuous function defined on an open interval has an antiderivative on that interval.

For elementary functions, the fundamental theorem can be applied directly since the derivatives are easily found. For example, if

$$F(x) = \frac{x^{n+1}}{n+1} + C,$$

where C is a constant, $F'(x) = x^n$, so that F is an·antiderivative of $f(x) = x^n$. Therefore, the integral of x^n is given by

$$\int_a^b x^n\,dx = \frac{x^{n+1}}{n+1}\bigg|_a^b, \quad n \neq -1. \tag{12.2}$$

The *indefinite* integral $\int f(x)\,dx$ is the collection of *all* antiderivatives of $f(x)$. It has the form

$$\int f(x)\,dx = F(x) + C,$$

where C is a constant and $F(x)$ is one particular antiderivative of $f(x)$. The antiderivatives of $f(x)$ thus can only differ by a constant.

SOME TECHNIQUES OF INTEGRATION

When $f(x)$ is a power of x, a trigonometric function such as $\sin x$, an exponential function e^x, or a simple combination of these functions, the integral of $f(x)$ is usually easily evaluated. Problem 12.1 presents a number of such integrals. The answer to that problem serves as a short *Table of integrals* by listing a few basic integral formulas. For certain other functions, integration by parts or substitution to change variables leads to a simple integral, as demonstrated in this section. However, for many functions of practical importance, numerical methods must be used to evaluate the integrals. Various numerical methods will be discussed later in the chapter.

Integration by Parts Consider the formula for the derivative of a product of functions,

$$\frac{d}{dx}[f(x)g(x)] = f(x)g'(x) + g(x)f'(x).$$

Integrating both sides of the product equation yields

$$f(x)g(x) = \int f(x)g'(x)\,dx + \int g(x)f'(x)\,dx,$$

or

$$\int f(x)g'(x)\,dx = f(x)g(x) - \int g(x)f'(x)\,dx. \tag{12.3}$$

This is the formula for *integration by parts*. It is usually written in condensed form by letting

$$u = f(x), \qquad du = f'(x)\,dx,$$
$$v = g(x), \qquad dv = g'(x)\,dx,$$

so that the formula for integration by parts becomes

$$\int u\,dv = uv - \int v\,du. \tag{12.4}$$

Change of Variables Suppose that the function $g(x)$ has a continuous derivative on $[a, b]$ and that $f(x) = F'(x)$ is continuous on the set $g([a, b])$. Applying the chain rule to the function $F[g(x)]$ yields the formula

$$\frac{d}{dx}F[g(x)] = F'[g(x)]\,g'(x)$$

and therefore,

$$\int_a^b F'[g(x)]\,g'(x)\,dx = F[g(x)]\Big|_a^b. \tag{12.5}$$

Since

$$F[g(x)]\Big|_a^b = F[g(b)] - F[g(a)] = F(u)\Big|_{g(a)}^{g(b)}$$

and

$$\int_{g(a)}^{g(b)} F'(u)\,du = F(u)\Big|_{g(a)}^{g(b)}.$$

Equation 12.5 may be written

$$\int_{g(a)}^{g(b)} F'[g(x)]g'(x)\,dx = \int_{y(u)}^{g(b)} F'(u)\,du.$$

Replacing $F'(x)$ with $f(x)$ and letting

$$\begin{aligned} u &= g(x) \\ du &= g'(x)\,dx \end{aligned}$$

yields the useful *change-of-variable* formula

$$\int_a^b f[g(x)]\,g'(x)\,dx = \int_{g(a)}^{g(b)} f(u)\,du. \tag{12.6}$$

Equation 12.6 emphasizes the fact that the limits of integration a to b for x must be changed to $g(a)$ and $g(b)$ for the u-limits. Alternatively, after the integral is computed as a function of u, the function $g(x)$ can be resubstituted and the original limits used to evaluate the definite integral. The technique of changing variables will be applied to multiple integrals in a later section of this chapter.

☐ EXAMPLE 12.1   *Integral Examples*

a. Since integration is a linear operation, the result of Equation 12.2 can be used repeatedly to yield the integral of a sum of terms, such as

$$\begin{aligned} \int (x^4 - 3x^3 + x^2)\,dx &= \int x^4\,dx - 3\int x^3\,dx + \int x^2\,dx \\ &= \frac{x^5}{5} - \frac{3x^4}{4} + \frac{x^3}{3} + C. \end{aligned}$$

b. To evaluate the integral

$$\int_0^\pi x \sin x\,dx,$$

we apply the integration by parts formula of Equation 12.4. There are two possible choices for the variables as

$$\begin{aligned} u &= x, & dv &= \sin x\,dx \\ \text{or} \quad u &= \sin x, & dv &= x\,dx. \end{aligned}$$

The first choice simplifies the integral to be evaluated by parts, with the result

$$\int_0^\pi x \sin x \, dx = -x \cos x \Big|_0^\pi - \int_0^\pi (-\cos x) \, dx$$

$$= -\pi \cos \pi + 0 \cos 0 + \sin \pi - \sin 0 = \pi.$$

c. To use the change-of-variable formula of Equation 12.6 to evaluate

$$\int_0^{\pi/4} \sin^3 2x \cos 2x \, dx,$$

notice that if $u = \sin 2x$ then the term $\cos 2x \, dx$ is one-half the value of du. Thus, since $u = 0$ when $x = 0$ and $u = 1$ when $x = \pi/4$,

$$\int_0^{\pi/4} \sin^3 2x \cos 2x \, dx = \frac{1}{2} \int_0^1 u^3 \, du = \frac{1}{2} \left[\frac{u^4}{4} \right]_0^1 = \frac{1}{8}.$$

To avoid changing the limits of integration, the result can be obtained by substituting $\sin 2x$ for u after integration to yield

$$\frac{1}{2} \int u^3 \, du = \frac{1}{2} \left[\frac{u^4}{4} \right] = \frac{1}{2} \left[\frac{\sin^4 2x}{4} \right]_0^{\pi/4} = \frac{1}{8}.$$

$\square$

APPLICATIONS OF SINGLE INTEGRALS

Integration finds wide application in physical problems. For example, an important use of integrals is to determine geometric properties such as the length of curves, area of a surface, or volume of solid objects. In these cases, the function being studied often has the form $y = f(x)$ or $x = g(y)$, so that one coordinate variable is defined explicitly in terms of the other. Another representation of a curve in a plane or in space is in terms of a parameter. For example, the motion of a point in a plane can be described by giving its position $[x(t), y(t)]$ at time t. The equation

$$\mathbf{R} = [x(t), y(t)] = x(t)\mathbf{i} + y(t)\mathbf{j}$$

is the *parametric representation* of the curve traced by the point as it moves in the plane. In cases in which a mass is associated with the point, it is possible to determine the acceleration, velocity, or the position of the point at a specified time.

AREA

Referring to Figure 12.1, the definite integral of a continuous positive-valued function $f(x)$ defined on a closed interval $[a, b]$ represents the area that lies below the curve and above the x-axis in the interval. Thus,

$$A = \int_a^b f(x)\, dx \quad \text{if } f(x) \geq 0,$$

where A is the area bounded by $f(x)$, the x-axis, and the vertical lines $x = a$ and $x = b$. If the units of the x-axis and y-axis are length, the area has units such as square feet or square meters.

In some cases, the area of a region may represent a physical quantity of interest. For example, the thermodynamic work done by expansion of a gas is given by the area enclosed by the curve of pressure versus volume.[1] Analytically, the work W is

$$W = \int_{V_i}^{V_f} P\, dV,$$

where V_i is the initial volume, V_f is the final volume after expansion, and P is the pressure.

Length of a Curve   The length of a curve can be computed by integration. The results are just stated here since the equations for lengths are usually derived in detail in introductory calculus textbooks. We consider only continuous curves such as the one shown in Figure 12.2. The quantity ds is the hypotenuse of the differential right triangle at a point along the curve.

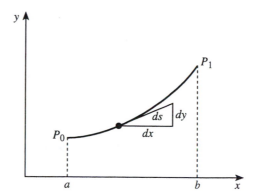

FIGURE 12.2 *Curve from P_0 to P_1.*

Assuming that a curve from points P_0 to P_1 is described by a smooth function $y = y(x)$ with continuous derivative, the length of the curve from

---

[1] If the work done during expansion is considered reversible, the work can be expressed in terms of the thermodynamic variables pressure and volume only. Hence, such effects as friction or any other external forces are neglected.

$x = a$ to $x = b$ is

$$s = \int_a^b ds = \int_a^b \sqrt{1 + \left(\frac{dy}{dx}\right)^2}\, dx. \qquad (12.7)$$

Parametric Equations Let a curve be defined in parametric form as

$$x = f(t), \quad y = g(t), \quad t_1 \le t \le t_2.$$

One interpretation of the parametric equations is that t is time and the length s of the curve from t_1 to t_2 is the distance the point (x, y) has moved in this time interval. To compute the distance traveled, we can substitute

$$dx = \frac{dx}{dt}\, dt, \quad dy = \frac{dy}{dt}\, dt$$

in the equation $ds^2 = dx^2 + dy^2$ for the differential triangle in Figure 12.2. Then, ds/dt in terms of the parametric equations becomes

$$\frac{ds}{dt} = \sqrt{\left(\frac{dx}{dt}\right)^2 + \left(\frac{dy}{dt}\right)^2} = \sqrt{[f'(t)]^2 + [g'(t)]^2}. \qquad (12.8)$$

This equation represents the *speed* of the moving point, and the terms dx/dt and dy/dt can be interpreted as the x and y components of the velocity vector of (x, y). The total distance traveled is then

$$s = \int_{t_1}^{t_2} \sqrt{\left(\frac{dx}{dt}\right)^2 + \left(\frac{dy}{dt}\right)^2}\, dt. \qquad (12.9)$$

In three dimensions, the space curve can be described by the equations

$$x = x(t), \quad y = y(t), \quad z = z(t)$$

for $t_1 \le t \le t_2$, and the length formula becomes

$$s = \int_{t_1}^{t_2} \sqrt{\left(\frac{dx}{dt}\right)^2 + \left(\frac{dy}{dt}\right)^2 + \left(\frac{dz}{dt}\right)^2}\, dt. \qquad (12.10)$$

Vector Representation If a curve in three dimensions is represented parametrically by a vector function

$$\mathbf{R} = [x(t), y(t), z(t)] = x(t)\mathbf{i} + y(t)\mathbf{j} + z(t)\mathbf{k}, \qquad (12.11)$$

the *velocity vector* is

$$\mathbf{v} = \frac{d\mathbf{R}}{dt} = \dot{\mathbf{R}} = \dot{x}(t)\mathbf{i} + \dot{y}(t)\mathbf{j} + \dot{z}(t)\mathbf{k}. \qquad (12.12)$$

Considering the formula for curve length in Equation 12.10, the length in terms of the vector $\mathbf{R}$ can be written

$$s = \int_{t_1}^{t_2} \sqrt{\frac{d\mathbf{R}}{dt} \cdot \frac{d\mathbf{R}}{dt}}\, dt = \int_{t_1}^{t_2} (\mathbf{v} \cdot \mathbf{v})^{1/2}\, dt. \qquad (12.13)$$

Parametric Arc Length

Consider the parametric curve for a helix

$$x = r \cos t, \quad y = r \sin t, \quad z = at.$$

If $a = 0$, the curve describes a circle of radius r in the xy-plane, as shown by eliminating the parameter using the equation

$$x^2 + y^2 = (r \cos t)^2 + (r \sin t)^2 = r^2.$$

When $a > 0$, the curve is a helix above the xy-plane if $t > 0$. As t varies from 0 to 2π, the curve traced by the equations increases in the z-direction by $2\pi a$. According to Equation 12.10, the length of the curve when t changes from 0 to angle θ is

$$
\begin{aligned}
L &= \int_0^\theta \sqrt{(-r \sin t)^2 + (r \cos t)^2 + a^2}\, dt \\
&= \int_0^\theta \sqrt{r^2 + a^2}\, dt = \sqrt{r^2 + a^2}\,\theta. \qquad \text{(12.14)}
\end{aligned}
$$

The length of the curve for one revolution is thus $2\pi\sqrt{r^2 + a^2}$.

☐

SUMMARY OF INTEGRAL APPLICATIONS

Table 12.1 presents a summary of the applications presented previously in this section.

TABLE 12.1 *Applications of integrals*

Type	Integral formula
Area defined by $y = f(x)$	$A = \displaystyle\int_a^b f(x)\, dx, \quad f(x) \geq 0$
Length of $y = f(x)$	$s = \displaystyle\int_a^b \sqrt{1 + y'^2}\, dx$
Length of parametric curve	$s = \displaystyle\int_{t_1}^{t_2} \sqrt{\left(\dfrac{dx}{dt}\right)^2 + \left(\dfrac{dy}{dt}\right)^2 + \left(\dfrac{dz}{dt}\right)^2}\, dt$

The area formula computes the area of a region between the continuous positive function $f(x)$ and the x-axis ($y = 0$) in the interval between $x = a$ and $x = b$. The length formula is for a curve $y = f(x)$ with continuous derivative $y' = dy/dx$ over the interval.

DOUBLE AND TRIPLE INTEGRALS

This section is devoted to a brief study of double and triple integrals. The double integral is first defined in terms of a limit of rectangular areas in the plane. Iterated integrals are then introduced because they are useful for numerical computations, and they are easily related to the definite integral for a function of a single variable. Triple integrals are also defined in this section.

For double and triple integrals, a transformation is often possible that simplifies the integrand or the region of integration. It will be shown that the Jacobian determinant of the transformation plays a central role in the simplification. Finally, this section presents examples of multiple integrals in various coordinate systems.

DOUBLE INTEGRALS

The *double integral* of $f(x, y)$ can be written

$$\int \int_R f(x, y)\, dx dy. \tag{12.15}$$

In this section, it will be assumed that the function $f(x, y)$ is defined over a closed bounded region R of the xy-plane. Saying that the region R is bounded means that R can be enclosed by a circle of sufficient radius.[2]

We define the double integral as the limit of a sequence of sums in a manner similar to that used for a function of one variable in terms of a Riemann sum. Figure 12.3 shows the region of integration R subdivided into rectangles.

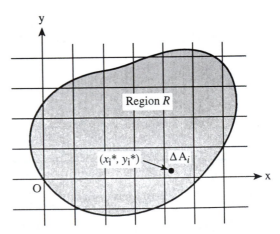

FIGURE 12.3 *The double integral*

---

[2]If the region is not bounded, the integral is called *improper*. The limits usually include ∞ in such cases.

Chapter 12 ■ VECTOR INTEGRAL CALCULUS

If the area of the ith rectangle is designated ΔA_i, the area of the region is approximately

$$\sum_{i=1}^{n} f(x^*, y^*)\, \Delta A_i, \qquad (12.16)$$

where (x^*, y^*) is an arbitrary point in the ith rectangle.

Suppose the sum in Equation 12.16 approaches a limit as the number of rectangles n becomes infinite and the maximum of the diagonals of the rectangles approaches zero. The double integral is then defined as that limit, so that

$$\int\int_R f(x, y)\, dx dy = \lim \sum_{i=1}^{n} f(x_i^*, y_i^*)\Delta A_i. \qquad (12.17)$$

In the following discussions, it is assumed that $f(x, y)$ is continuous on a bounded region in the xy-plane to assure that the double integral exists. Double integrals over rectangular regions R are studied first. Then, more general regions of integration are considered.

ITERATED INTEGRALS

Consider a continuous function $f(x, y)$ of two variables whose domain is a rectangle R with sides parallel to the coordinate axes, as shown in Figure 12.4.

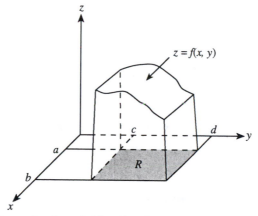

FIGURE 12.4 *The region bounded by $f(x, y)$, the rectangle R, and four vertical sides.*

Assume that $f(x, y) \geq 0$ on R, so the region in 3D space is bounded by the surface $f(x, y)$, the rectangle R, and the four planes

$$x = a, \quad x = b, \quad y = c, \quad y = d.$$

This is the boundary of a volume V in space, and it can be proven that the volume of the region above R and under the graph of $f(x, y)$ is computed by the double integral of $f(x, y)$ over R. This double integral of $f(x, y)$ can be designated as

$$V = \int\int_R f(x, y)\, dx\, dy.$$ (12.18)

For any general $f(x, y)$, not necessarily positive in the region of integration, the double integral has applications without recourse to the notion of volume. Such applications are presented later in the chapter.

We state a useful theorem that allows the computation of the double integral in terms of two ordinary definite integrals. The theorem that is known as *Fubini's theorem* states that the double integral can be calculated as an iterated integral.

■ THEOREM 12.2   *Iterated Integrals*

If the function $f(x, y)$ of two variables is continuous in the rectangular region bounded by the lines

$$x = a, \quad x = b, \qquad y = c, \quad y = d,$$

then

$$\int\int_R f(x, y)\, dx\, dy = \int_a^b \left[\int_c^d f(x, y)\, dy \right] dx = \int_c^d \left[\int_a^b f(x, y)\, dx \right] dy.$$
(12.19)

────────────────────────────────■

These integrals to compute the double integral of a function of two variables are called *repeated*, or *iterated*, integrals. The first iterated integral is computed by integrating with respect to y, holding x constant and then integrating the result with respect to x. The second iterated integral reverses the roles of x and y.

A useful consequence of the theorem for iterated integrals is that if $g(x)$ is continuous on $[a, b]$ and $h(y)$ is continuous on $[c, d]$, functions of the form $f(x, y) = g(x)h(y)$ can be separated into the product of two integrals

$$\int\int_R [g(x)h(y)]\, dx\, dy = \left[\int_a^b g(x)\, dx \right] \left[\int_c^d h(y)\, dy \right].$$ (12.20)

Nonrectangular Regions of Integration   The double integral of a continuous function over a region S other than a rectangle in the xy-plane may also be evaluated by a repeated integral. Two possibilities are shown in Figure 12.5.

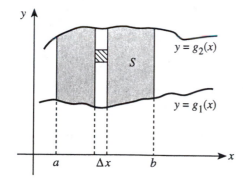 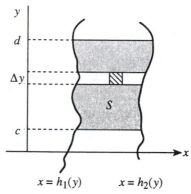

(a) Nonconstant y limits;
constant x limits

(b) Nonconstant x limits;
constant y limits

FIGURE 12.5 *Various regions of integration*

In Part a, S is bounded by the curves $y = g_1(x)$ and $y = g_2(x)$ and the lines $x = a$ and $x = b$. The double integral of $f(x,y)$ over a strip of width Δx is approximated as

$$\Delta x \int_{g_1(x)}^{g_2(x)} f(x,y)\,dy.$$

Letting Δx approach 0 and summing the areas of the vertical strips, the double integrals over the region becomes

$$\int \int_S f(x,y)\,dx\,dy = \int_a^b dx \int_{g_1(x)}^{g_2(x)} f(x,y)\,dy. \qquad (12.21)$$

If the region S is bounded by the curves $x = h_1(y)$ and $x = h_2(y)$ and the lines $y = c$ and $y = d$, as shown in Figure 12.5 b, the double integral is

$$\int \int_S f(x,y)\,dx\,dy = \int_c^d dy \int_{h_1(x)}^{h_2(x)} f(x,y)\,dx. \qquad (12.22)$$

WHAT IF?   Plotting the region of integration is usually helpful when determining the limits of integration. Also, the choice of the order of integration for iterated integrals may well determine the difficulty in evaluating the result. The solution to Problem 12.2 shows the results of selecting different limits to evaluate an integral over a given region.

Triangular Region of Integration. Consider the triangular region S in Figure 12.6 to be used in evaluating the double integral

$$\int \int_S f(x,y)\,dx\,dy.$$

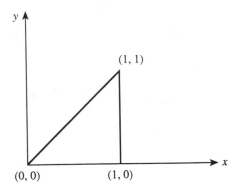

FIGURE 12.6 *Triangular region of integration*

To perform the x-integration first, the triangular region can be considered bounded by

$$y \leq x \leq 1, \quad 0 \leq y \leq 1.$$

This represents an integration in the vertical direction ($y = 0$ to $y = 1$) over strips of width dy in the horizontal direction. The integrals become

$$\int\int_S f(x,y)\,dx\,dy = \int_0^1 \left[\int_y^1 f(x,y)dx\right]dy.$$

If the y-integration is to be performed first, the region S is defined as

$$0 \leq y \leq x, \quad 0 \leq x \leq 1,$$

and the integration is in the horizontal direction ($x = 0$ to $x = 1$) over strips of width dx in the vertical direction. In this case, the integrals are

$$\int\int_S f(x,y)\,dx\,dy = \int_0^1 \left[\int_0^x f(x,y)\,dy\right]dx.$$

☐ **EXAMPLE 12.3**   *Double Integrals*

This example presents several approaches for evaluating double integrals. Problem 12.3 asks you to evaluate similar integrals.

a. To evaluate

$$\int\int_R x \cos y\,dx\,dy$$

over the rectangle $0 \leq x \leq 1, 0 \leq y \leq \pi/2$, the integrals can be separated according to Equation 12.20, with the result

$$\int_0^1 \int_0^{\pi/2} x \cos y\,dx\,dy = \int_0^1 x\,dx \int_0^{\pi/2} \cos y\,dy$$

$$= \left.\frac{x^2}{2}\right|_0^1 \times \left.\sin y\right|_0^{\pi/2} = \frac{1}{2}.$$

b. The integral

$$\iint xy\,dxdy$$

is to be integrated over the region in the first quadrant bounded by the two curves $y_1(x) = \sqrt{x}$ and $y_2(x) = x^3$. By analyzing the region enclosed by the curves, it is evident that $y_2(x) < y_1(x)$ for $0 \le x < 1$ and that the curves intersect at the points $(0,0)$ and $(1,1)$. Integrating using a vertical segment corresponds to integrating first with respect to y, so the integration is

$$\int_0^1 \int_{x^3}^{\sqrt{x}} xy\,dydx = \int_0^1 \left[\frac{1}{2}xy^2\right]_{y=x^3}^{y=\sqrt{x}} dx = \frac{5}{48}.$$

c. Let S be the quarter-circle $0 \le y \le \sqrt{1-x^2}$, $0 \le x \le 1$, as the region of integration for the function $f(x,y) = x^2 + y^2$. The double integral can be evaluated as

$$\int_0^1 \int_0^{\sqrt{1-x^2}} (x^2 + y^2)\,dydx = \int_0^1 \left(x^2\sqrt{1-x^2} + \frac{1}{3}(1-x^2)^{3/2}\right) dx.$$

One method of evaluating the integral is to make the trigonometric substitution

$$x = \sin\theta, \quad dx = \cos\theta\,d\theta,$$

which leads to the integral

$$\int_0^{\pi/2} \left(\sin^2\theta\cos^2\theta + \frac{1}{3}\cos^4\theta\right)d\theta = \frac{\pi}{8},$$

as you are asked to show in Problem 12.5. The value of the integral represents the volume in the first quadrant between the disk of radius 1 in the xy-plane and the graph of $f(x,y) = x^2 + y^2$.

$\square$

TRIPLE AND MULTIPLE INTEGRALS

The term multiple integrals *includes double integrals as well as integrals of three, four, or more variables*

$$\iiint_R f(x,y,z)\,dxdydz, \quad \iiiint_R f(x,y,z,w)\,dxdydzdw, \ldots,$$

called triple, quadruple,... integrals. *Only the triple integral is defined and discussed here.*

Suppose that $f(x,y,z)$ is continuous in a closed, bounded region R of 3D space. The region can be divided into rectangular parallelepipeds by planes parallel to the coordinate planes. Let there be n such parallelepipeds inside R, and denote the ith volume by ΔV_i. The triple integral of the function over the region can be written as the limit of a sum

$$\iiint_R f(x,y,z)\,dxdydz = \lim_{n\to\infty} \sum f(x^*,y^*,z^*)\,\Delta V_i \qquad (12.23)$$

as n approaches ∞, while the maximum diagonal of the ΔV_i approaches zero.

When the boundaries of the region R have a simple form, such as

$$x_1 \leq x \leq x_2, \quad y_1(x) \leq y \leq y_2(x), \quad z_1(x,y) \leq z \leq z_2(x,y),$$

a triple integral can be computed by means of iterated integrals. In the region of integration, there are six possible orders of integration. For example, we can integrate with respect to z, then y, then x or perform the z integration first, then x, then y, etc. The order of integration will not change the result.

Consider a region in space R that can be described by the inequalities

$$z_1(x,y) \leq z \leq z_2(x,y), \qquad (x,y) \in S,$$

where S is the projection of R onto the xy-plane. Further suppose that the plane region S is defined as

$$y_1(x) \leq y \leq y_2(x), \qquad x_1 \leq x \leq x_2.$$

With these descriptions of the limits of integration, the integration of Equation 12.23 can be written as

$$\int\int\int_R f(x,y,z)\,dx\,dy\,dz = \int_{x_1}^{x_2} \int_{y_1(x)}^{y_2(x)} \int_{z_1(x,y)}^{z_2(x,y)} f(x,y,z)\,dz\,dy\,dx.$$

$$(12.24)$$

CHANGE OF VARIABLES IN DOUBLE INTEGRALS

It is often convenient to change variables to perform an integration. Consider the change of variables formula in a definite integral

$$\int_a^b f(x)\,dx = \int_a^b f[x(u)]\,\frac{dx}{du}\,du = \int_{u_1}^{u_2} f(u)\,du, \qquad (12.25)$$

using $x = x(u)$. In the formula, $f(x)$ is assumed continuous in the interval $[a,b]$, and $x(u)$ has a continuous derivative and $f[x(u)]$ is continuous in the interval $[u_1, u_2]$. Notice that the factor $dx(u)/du$ is introduced in the integral when the change-of-variables formula is used, as previously discussed in connection with Equation 12.6.

For example, letting $x(u) = \sin u$, the integral of $f(x) = \sqrt{1-x^2}$ over the interval $[0,1]$ becomes

$$\int_0^1 \sqrt{1-x^2}\,dx = \int_0^{\pi/2} \cos^2 u\,du = \frac{\pi}{4}.$$

In general, the function being integrated and the limits of integration also change with the change of variables.

CHANGE OF VARIABLES FORMULA

For double integrals, the formula analogous to Equation 12.25 is

$$\int\int_{R_{xy}} f(x,y)\,dx\,dy = \int\int_{R_{uv}} f[g(u,v), h(u,v)] \left| \frac{\partial(x,y)}{\partial(u,v)} \right| du\,dv.$$

$$(12.26)$$

In these integrals, the region of integration and the function being integrated can be transformed according to the change of variables

$$x = g(u,v) \quad \text{and} \quad y = h(u,v), \qquad (12.27)$$

where the functions are assumed to be defined and have continuous derivatives in a region R_{uv} of the uv plane. Consider the double integral

$$\int\int_{R_{xy}} f(x,y)\,dx\,dy$$

to which is applied the transformation of coordinates of Equation 12.27. The change of variables is determined by a *transformation T* from the uv-plane to the xy-plane. A point (x,y) is called the *image* of a corresponding point (u,v). When no two different points in the uv-plane have the same image point in the xy-plane, the transformation is said to be *one-to-one*.

In Equation 12.27, the equivalent of the derivative factor for the change-of-variable formula in one dimension is the Jacobian

$$J(u,v) = \frac{\partial(x,y)}{\partial(u,v)},$$

introduced in Chapter 10. This term determines how the transformation distorts the area of a region.

JACOBIAN

For the transformation from the uv-plane to the xy-plane

$$x = g(u,v), \qquad y = h(u,v),$$

the determinant

$$J(u,v) = \frac{\partial(x,y)}{\partial(u,v)} = \begin{vmatrix} \dfrac{\partial g}{\partial u} & \dfrac{\partial g}{\partial v} \\[2mm] \dfrac{\partial h}{\partial u} & \dfrac{\partial h}{\partial v} \end{vmatrix} \qquad (12.28)$$

is the *Jacobian* of the transformation.

It can be shown that the Jacobian defines the change in scale for surface area given by the transformations. A rigorous proof defining the conditions under which this relation holds is provided in Kaplan's textbook referenced in the Annotated Bibliography at the end of this chapter. We will assume the appropriate conditions exist for the change of variables applied to examples and problems in this chapter.

In two dimensions, a useful transformation is from rectangular (x, y) to polar (r, θ) coordinates. Also, linear transformations are used to transform regions that are parallelograms into rectangular regions to simplify the evaluation of double integrals.

Polar Coordinates The transformation T that transforms polar coordinates to Cartesian coordinates is given by

$$x = r \cos \theta, \qquad y = r \sin \theta, \qquad (12.29)$$

and its Jacobian determinant is

$$J(u, v) = \frac{\partial(x, y)}{\partial(u, v)} = \begin{vmatrix} \cos \theta & -r \sin \theta \\ \sin \theta & r \cos \theta \end{vmatrix} = r.$$

Associating u with r and v with θ in Equation 12.26, the double integral of a region in the xy-plane is related to the integration in the $r\theta$-plane as

$$\iint_{R_{xy}} f(x, y) \, dx dy = \iint_{R_{r\theta}} f(r \cos \theta, r \sin \theta) \, r dr d\theta \qquad (12.30)$$

where the region $R_{xy} = T(R_{r\theta})$.

Circles centered at the origin in the xy-plane are bounded by lines of constant radius $(x^2 + y^2 = r^2)$. Thus, they are mapped into rectangular regions in the $r\theta$-plane. This can be shown by drawing the regions or by computing the inverse transform, to yield

$$r = \sqrt{x^2 + y^2}, \qquad \theta = \tan^{-1}\left(\frac{y}{x}\right). \qquad (12.31)$$

The inverse relationship is valid at all points that the Jacobian is not zero. However, since $J(u, v) = r$, the Jacobian is zero when $r = 0$. At such points, the transformation of Equation 12.29 cannot be inverted. Thus, the transformation T is not strictly one-to-one because all the points with $r = 0$ are mapped to $(0, 0)$. This does not affect the value of the integral over a region when the point $r = 0$ can be considered an isolated point on the edge of the region of integration.

☐ EXAMPLE 12.4   ***Polar Coordinates***

The area of the circle $x^2 + y^2 = a^2$ is πa^2. In this example, the area is computed using a single integral, a double integral in rectangular coordinates, and a double integral in polar coordinates.

a. Letting $y = \sqrt{a^2 - x^2}$, the area can be computed with a single integral as

$$A = 2 \int_{-a}^{a} \sqrt{a^2 - x^2} \, dx. \qquad (12.32)$$

The factor 2 arises because the integration yields the area of a half circle in the xy-plane so the total area is twice the value of the integral. Using the trigonometric substitution $x = a \sin \theta$, the value of the integral becomes

$$A = a^2 \left[\theta + \frac{\sin 2\theta}{2} \right]_{-\pi/2}^{\pi/2} = \pi a^2.$$

b. The double integral to determine the area of the circle in rectangular coordinates is

$$A = \int_{-a}^{a} \int_{-\sqrt{a^2-x^2}}^{\sqrt{a^2-x^2}} dy\, dx = 2 \int_{-a}^{a} \left[\sqrt{a^2 - x^2} \right] dx = \pi a^2,$$

using the result previously derived from the one-dimensional integral of Equation 12.32. Although this integral is not difficult to evaluate, changing to polar coordinates simplifies the integration considerably because the limits in polar coordinates will be constant in that coordinate system.

c. In polar coordinates, if $r_1(\theta) \le r \le r_2(\theta)$ and $\alpha \le \theta \le \beta$, the area enclosed is

$$A = \int_{\alpha}^{\beta} \int_{r_1(\theta)}^{r_2(\theta)} r\, dr\, d\theta.$$

For the circle $x^2 + y^2 = a^2$, the limits of integration are defined by the rectangle $0 \le r \le a$, $0 \le \theta \le 2\pi$ according to the transformation defined by Equation 12.29. The area of the circle is thus

$$\int_{0}^{2\pi} \int_{0}^{a} r\, dr\, d\theta = \int_{0}^{2\pi} \left[\frac{r^2}{2} \right]_{0}^{a} d\theta = \frac{1}{2} a^2 \int_{0}^{2\pi} d\theta = \pi a^2.$$

$\square$

Linear Transformations   The linear transformation

$$\begin{aligned} x &= au + bv, \\ y &= cu + dv, \end{aligned} \qquad (12.33)$$

with appropriate choice of coefficients, is often used to aid evaluation of double integrals over a region bounded by straight lines. When the region of integration in the xy-plane is a parallelogram, a linear transformation can be chosen so that the integration region in the uv plane is a rectangle. The Jacobian of the transformation of Equation 12.33 is $J(u, v) = ad - bc$.

□ EXAMPLE 12.5   *Linear Transformations*
Consider the integral

$$V = \int \int_{R_{xy}} (x + y)\, dx\, dy \qquad (12.34)$$

to be evaluated over the region R_{xy} as shown in Figure 12.7(b).

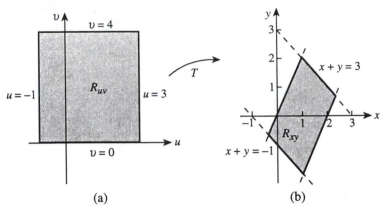

FIGURE 12.7 *Regions of integration:* (a). R_{uv} (b). R_{xy}

The region in the xy-plane is defined by the lines

$$
\begin{aligned}
x + y &= -1, \\
x + y &= 3, \\
2x - y &= 0, \\
2x - y &= 4.
\end{aligned}
\tag{12.35}
$$

Since the sides of the parallelogram defined by the region R_{xy} are the straight lines

$$
x + y = c_1, \qquad 2x - y = c_2,
$$

it is useful to introduce the coordinates

$$
\begin{aligned}
u &= x + y, \\
v &= 2x - y,
\end{aligned}
\tag{12.36}
$$

so that the region R_{xy} corresponds to a rectangle in the uv-plane. This is the inverse transformation with respect to the change-of-variables formula of Equation 12.26.

The limits for the uv variables are computed by substituting u and v in the relationships in Equation 12.35 for the straight lines in the xy-plane. The result is

$$
-1 \le u \le 3, \qquad 0 \le v \le 4,
$$

as shown in Figure 12.7(a).

In matrix notation, the transformation from the xy-plane to the uv-plane can be written

$$
\mathbf{u} = \begin{bmatrix} u \\ v \end{bmatrix} = \begin{bmatrix} 1 & 1 \\ 2 & -1 \end{bmatrix} \begin{bmatrix} x \\ y \end{bmatrix}.
$$

Thus, it is possible to solve for x and y explicitly, with the result

$$
\begin{aligned}
x &= \frac{u + v}{3}, \\
y &= \frac{2u - v}{3}.
\end{aligned}
\tag{12.37}
$$

The Jacobian of the transformation is

$$J(u,v) = \begin{vmatrix} \dfrac{\partial x}{\partial u} & \dfrac{\partial x}{\partial v} \\[2mm] \dfrac{\partial y}{\partial u} & \dfrac{\partial y}{\partial v} \end{vmatrix} = \begin{vmatrix} \dfrac{1}{3} & \dfrac{1}{3} \\[2mm] \dfrac{2}{3} & -\dfrac{1}{3} \end{vmatrix} = -\dfrac{1}{3}.$$

Applying the change-of-variables formula of Equation 12.26 to the original integral of Equation 12.34 yields

$$V = \iint_{R_{xy}} (x+y)\,dx\,dy \;=\; \iint u \left| -\frac{1}{3} \right| du\,dv$$

$$= \frac{1}{3} \int_{-1}^{3} u\,du \int_{0}^{4} dv = \frac{16}{3}.$$

□

WHAT IF? Show that the Jacobian of the inverse transformation in Equation 12.36 is -3 and thus that

$$\frac{\partial(u,v)}{\partial(x,y)} = \frac{1}{\dfrac{\partial(x,y)}{\partial(u,v)}}$$

for this example. Therefore, it is not necessary to solve for $x = g(u,v)$ and $y = h(u,v)$ explicitly as in Equation 12.37 to compute the Jacobian for the transformed double integral. This "chain rule" for Jacobians is true in general as long as the appropriate derivatives exist and the Jacobian is not zero.

CHANGE OF VARIABLES IN TRIPLE INTEGRALS

Let a three-dimensional function be defined by the equations

$$x = x(u,v,w), \quad y = y(u,v,w), \quad z = z(u,v,w).$$

The Jacobian of the transformation is

$$J(u,v,w) = \begin{vmatrix} \dfrac{\partial x}{\partial u} & \dfrac{\partial x}{\partial v} & \dfrac{\partial x}{\partial w} \\[2mm] \dfrac{\partial y}{\partial u} & \dfrac{\partial y}{\partial v} & \dfrac{\partial y}{\partial w} \\[2mm] \dfrac{\partial z}{\partial u} & \dfrac{\partial z}{\partial v} & \dfrac{\partial z}{\partial w} \end{vmatrix}.$$

For triple integrals, the change of variables formula takes the form

$$\int \int \int_{R_{xyz}} f(x, y, z) \, dx \, dy \, dz$$

$$= \int \int \int_{R_{uvw}} f[x(u, v, w), y(u, v, w), z(u, v, w)] \, |J(u, v, w)| \, du \, dv \, dw,$$

(12.38)

where R_{uvw} is the region of integration in uvw-space corresponding to R_{xyz} in xyz-space.

CYLINDRICAL AND SPHERICAL COORDINATES
In three dimensions, transformations from rectangular to cylindrical or spherical coordinates are frequently used to simplify certain triple integrals. Equation 12.38 can be used to determine the transformation.

Cylindrical Coordinates To transform from rectangular to cylindrical coordinates, the relationships are

$$x = r \cos \theta, \quad y = r \sin \theta, \quad z = z$$

so that the Jacobian is $J(r, \theta, z) = r$. Thus, the volume element in the $r\theta z$-space has the form

$$dV = r \, dr \, d\theta \, dz.$$

The triple integral in cylindrical coordinates is thus

$$\int \int \int_{R_{xyz}} f(x, y, z) \, dx \, dy \, dz = \int \int \int_{R_{r\theta z}} f(r \cos \theta, r \sin \theta, z) \, r \, dr \, d\theta \, dz.$$

(12.39)

Spherical Coordinates In spherical coordinates, the change of coordinate formulas are

$$x = r \cos \theta \sin \phi, \quad y = r \sin \theta \sin \phi, \quad z = r \cos \phi,$$

using the conventions introduced in Chapter 10 in which the angle ϕ is measured down from the z-axis and θ is measured in the xy-plane. To cover the entire space,

$$0 \le r \le \infty, \quad 0 \le \theta \le 2\pi, \quad 0 \le \phi \le \pi.$$

Using the results of Problem 12.10, the Jacobian of the transformation is

$$J(r, \theta, \phi) = -r^2 \sin \phi.$$

The triple integral in spherical coordinates is thus

$$\int \int \int_{R_{xyz}} f(x, y, z) \, dx \, dy \, dz$$

$$= \int \int \int_{R_{r\theta\phi}} f(r \cos \theta \sin \phi, r \sin \theta \sin \phi, r \cos \phi) \, r^2 \sin \phi \, dr \, d\theta \, d\phi.$$

(12.40)

APPLICATIONS OF MULTIPLE INTEGRALS

Multiple integrals arise in many branches of mathematics and engineering. They are used to compute volumes, mass, and moments of inertia in mechanics. Computing the gravitational potential from a distributed volume of mass or the electrostatic potential due to a volume charge distribution requires the evaluation of multiple integrals. In other applications, the integrals can be used to process images, determine diffraction patterns, and compute probabilities.

VOLUMES AND AREAS

Double and triple integrals can be used to compute areas and volumes, as shown in Table 12.2. These formulas apply to continuous functions defined in the region of interest.

TABLE 12.2 *Applications of multiple integrals to measurement*

Type	Integral Formula
Area of a region R	$A = \displaystyle\iint_R dx\,dy$
Volume between surface $f(x, y)$ and xy-plane	$V = \displaystyle\iint f(x, y)\,dx\,dy \quad z = f(x, y) \geq 0.$
Volume of a 3D region	$V = \displaystyle\iiint dx\,dy\,dz$
Mass of a solid object	$m = \displaystyle\iiint \rho(x, y, z)\,dx\,dy\,dz$

Various applications of the formulas for area and volume are presented in the problems for this chapter. In the equation for mass, $\rho(x, y, z)$ is the density in units of mass per unit volume.

☐ EXAMPLE 12.6   *Volume Integral with Change of Variables*

The formulas of Table 12.2 can sometimes be used with a change of variables to simplify the integration. As an example, suppose it is required to find the volume below the function

$$f(x, y) = e^{-(x^2 + y^2)}$$

and above the circle $x^2 + y^2 = R^2$ in the first quadrant of the xy-plane. The integral becomes

$$V = \int \int_{R_{xy}} e^{-(x^2+y^2)} \, dxdy, \qquad (12.41)$$

which is difficult to evaluate in rectangular coordinates. Changing to polar coordinates simplifies both the integration and the limits of integration.

Recalling the change-of-variables formula from rectangular to polar coordinates of Equation 12.30,

$$\int \int_{R_{xy}} f(x,y) \, dxdy = \int \int_{R_{r\theta}} f(r\cos\theta, r\sin\theta) \, rdrd\theta,$$

the function $f(x,y) = e^{-(x^2+y^2)}$ becomes

$$f(r\cos\theta, r\sin\theta) = \exp(-r^2\cos^2\theta - r^2\sin^2\theta) = e^{-r^2}.$$

The limits of integration in the first quadrant are

$$0 \leq r \leq R, \qquad 0 \leq \theta \leq \frac{\pi}{2}.$$

Thus, the original integral of Equation 12.41 becomes

$$V = \int_0^{\pi/2} \int_0^R re^{-r^2} \, drd\theta.$$

In polar coordinates, the function to be integrated separates and the limits of integration are constant, so V can be computed as iterated integrals, with the result

$$V = \int_0^{\pi/2} d\theta \int_0^R re^{-r^2} \, dr = \frac{\pi}{2}\left[-\frac{1}{2}e^{-r^2}\right]_0^R = \frac{\pi}{4}[1 - e^{-R^2}].$$

$\square$

TWO-DIMENSIONAL FOURIER TRANSFORMS

Applications of Fourier analysis to the analysis of 1D waveforms have been presented in previous chapters. Similar analysis of 2D images is possible with the 2D Fourier transform. Given the function $f(x,y)$, the Fourier transform can be defined as

$$F(u,v) = \mathcal{F}(f(x,y)) = \int_{-\infty}^{\infty} \int_{-\infty}^{\infty} f(x,y)e^{-i2\pi(ux+vy)} \, dxdy. \qquad (12.42)$$

A necessary condition for the existence of the 2D Fourier transform is the square-integrability condition

$$\int_{-\infty}^{\infty} \int_{-\infty}^{\infty} |f(x,y)|^2 \, dxdy < \infty.$$

When x and y in Equation 12.42 are coordinates in 2D space, the transform variables u and v are sometimes called *spatial frequencies*. Such transforms are of use in applications such as image processing and optics. In fact, there is an area of study called *Fourier optics* that is based on the 2D Fourier transform. Brigham's textbook, listed in the Annotated Bibliography at the end of this chapter, describes various applications of two-dimensional Fourier analysis, including the use of the 2D fast Fourier transform.

You are asked to compute a 2D Fourier transform in Problem 12.14 and to plot the result in Problem 12.44.

JOINT PROBABILITY

For a function of one variable, the *probability density function $f(x)$* defines the distribution of a random variable within a region of the x-axis. Assuming that the value x of a random variable lies somewhere on the x-axis, it is required that

$$\int_{-\infty}^{\infty} f(x)\, dx = 1,$$

and the probability that x is between a and b is defined as

$$P(a \leq x \leq b) = \int_{a}^{b} f(x)\, dx. \qquad (12.43)$$

The definition of probability density can be extended to two (or more) variables. For two variables x and y, the probability density $f(x,y)$ is called the *joint probability density*, and the probability that a point (x,y) lies in the region S of the xy-plane is defined as

$$P[(x,y) \in S] = \int\int_{S} f(x,y)\, dS. \qquad (12.44)$$

The conditions for $f(x,y)$ are that the function is nonnegative and

$$\int_{-\infty}^{\infty}\int_{-\infty}^{\infty} f(x,y)\, dx\, dy = 1. \qquad (12.45)$$

☐ **EXAMPLE 12.7**   *Joint Probability*

Suppose observations of several tasks lead to the probability distribution

$$f(x,y) = ke^{-x/10}e^{-y/20}$$

for the time to complete the tasks, where k is a constant. The interpretation is that x is the time to complete one task and y is the time to complete the second task. It is desired to find the probability that the two tasks will take no more than 30 units of time.

Since the variables represent time, we conclude that the valid range is

$$x \geq 0 \quad \text{and} \quad y \geq 0.$$

First, the probability density function must be normalized according to Equation 12.45, so that

$$k \int_0^\infty \int_0^\infty e^{-x/10} e^{-y/20} = 1.$$

Performing the integration and solving for the constant yields $k = 1/200$.

Computing the probability that the time of the tasks will take 30 time units or less means that $x + y \leq 30$ is the condition to be satisfied. Thus, in terms of the geometry of the region of interest, the probability must be calculated that a point (x, y) lies below the line $x + y = 30$ and above the coordinate axes in the first quadrant. Substituting the limits in Equation 12.44 for the joint probability yields the integrals

$$
\begin{aligned}
P[(x, y) \in S] &= \int \int_S f(x, y)\, dS = \frac{1}{200} \int_0^{30} \int_0^{30-x} e^{-x/10} e^{-y/20}\, dy dx \\
&= \frac{-20}{200} \int_0^{30} e^{-x/10} [e^{-(1/20)(30-x)} - 1]\, dx \\
&= \frac{-1}{10} \int_0^{30} [e^{-3/2} e^{-x/20} - e^{-x/10}]\, dx \\
&= e^{-3} - 2e^{-3/2} + 1 \approx 0.6035
\end{aligned}
$$

We conclude that the probability is about 60% that the tasks could be completed in less than 30 time units. The probability that the time required will be more than 30 units is

$$1 - 0.6035 = 0.3965.$$

$\square$

MATLAB COMMANDS FOR INTEGRATION

Numerical integration is used to obtain approximate values for definite integrals for which closed form evaluation using the fundamental theorem of calculus is difficult or impossible. The MATLAB functions **trapz**, **quad**, and **quad8** perform 1D numerical integration. Each algorithm approximates the definite integral of $f(x)$ over an interval $[a, b]$ by evaluating $f(x)$ at a finite number of sample points.

In MATLAB, the term *quadrature* is used to describe numerical approximation of integrals. The **quad** and **quad8** integration functions use techniques that are generally called *adaptive quadrature methods*. The basic adaptive routine automatically adjusts the subinterval lengths for integration to meet a desired accuracy requirement.

The particular MATLAB quadrature algorithms are described in detail in the book by Forsythe, Malcolm, and Moler listed in the Annotated Bibliography at the end of this chapter. The authors also analyze the errors and point out that "it is always possible to invent integrands $f(x)$ which fool the routines into producing a completely wrong result." Problem 12.43 gives you the opportunity to test the use of the MATLAB trapezoidal and quadrature routines.

MATLAB provides a number of commands for numerical and symbolic evaluation of integrals, as listed in Table 12.3. The table also lists the various special functions that are computed as numerical integrals by MATLAB. Problem 12.17 introduces several special functions that are not treated elsewhere in this book.

TABLE 12.3 *MATLAB commands for integration*

Command	Purpose
Quadrature:	
trapz	Trapezoidal numerical integration
quad	Simpson's rule integration
quad8	Newton-Cotes integration
Special functions:	
beta	Beta function
ellipj	Jacobian elliptic function
ellipke	Complete elliptic integrals
erf	Error function
expint	Exponential integral
fft	1D fast Fourier transform
fft2	2D fast Fourier transform
gamma	Gamma function
Symbolic:	
fourier	Fourier transform
funtool	Function calculator
int	Integration
laplace	Laplace transform
rsums	Riemann sum

TRAPEZOIDAL INTEGRATION

The command **trapz** performs trapezoidal numerical integration. This approximation evaluates the definite integral of $f(x)$ on an interval $[a, b]$, which is partitioned into n subintervals

$$a = x_1 < x_2 < \cdots < x_{n+1} = b.$$

The functional values are connected by straight-line segments with endpoints

$$[a, f(a)], [x_2, f(x_2)], \ldots, [b, f(b)].$$

For the MATLAB command **trapz**, the function can be evaluated at $n+1$ unequally spaced points $f(1) = f(a)$, $f(2)$, ..., $f(n+1) = f(b)$.

For equally spaced samples, with $x_1 = a$ and $x_{n+1} = b$, the interval between points is the constant value

$$h = x_{i+1} - x_i = \frac{b-a}{n}.$$

The *trapezoid rule* for integration uses the function values at the end points of each interval and approximates the integral as a sum of trapezoidal areas, so that

$$\int_a^b f(x)\,dx \approx \sum_{i=1}^n h\left(\frac{f_i + f_{i+1}}{2}\right) = \frac{h}{2}\left(f_1 + f_{n+1} + 2\sum_{i=2}^n f_i\right). \quad \textbf{(12.46)}$$

The sum of the trapezoidal areas over the interval of integration converges to the integral's value as the lengths of the subintervals decrease to zero. The advantage of the trapezoid integration formula is its simplicity. Problem 12.16 deals with the trapezoidal integral approximation. More sophisticated integration algorithms are used when it is necessary to control the accuracy of an integration routine automatically.

TWO-DIMENSIONAL INTEGRATION

Numerical computation of 2D and 3D integrals requires algorithms that are not available with MATLAB. The book *Numerical Recipes* and the textbook by Dahlquist and Björck listed in the Annotated Bibliography for this chapter discuss various algorithms to compute multiple integrals.

A MATLAB "Technical note" available from the MathWorks describes the function **dblquad**, which computes the double integral of a function between fixed limits of the variables. Such technical notes and other MATLAB functions can be downloaded from the MathWorks using the Internet.

The MATLAB symbolic math toolbox command **int** performs symbolic integration. Various forms of the command compute the symbolic form of the indefinite or definite integral. By calling **int** twice, 2D integrals can be evaluated. Example 12.8 shows an application of double integration.

☐ **EXAMPLE 12.8** *MATLAB Symbolic 2D Integration*

Consider evaluating the integral

$$I[f(x,y)] = \int_0^\pi \int_\pi^{2\pi} (x\sin y + y\cos x)\,dy\,dx.$$

This double integral is easily evaluated analytically, with the result

$$I = -\pi^2 = -9.8696$$

The accompanying MATLAB script shows the use of **int** to integrate this 2D function. First, **int** is called to evaluate the inner integral with respect to y, leaving x as a symbolic variable. Then, the outer integral is evaluated using the x limits. The results from the M-file (edited) are included following the script.

MATLAB Script _____

```
Example 12.8
% EX12_8.M Perform 2D symbolic integration of the function
%    x*sin(y)+y*cos(x) over fixed limits
%      pi < y < 2pi, 0 < x < pi
inty=int('x*sin(y)+y*cos(x)','y',pi,2*pi) % Inner integral
intx=int(inty,'x',0,pi)                   % Outer integral
twodint=numeric(intx)                % Convert to a number
% ------------------------------
% Results
inty = -2*x+3/2*cos(x)*pi^2

intx = -pi^2

twodint = -9.8696
```

_____

□

WHAT IF?   MATLAB's *Symbolic Math Toolbox* can solve only certain integrals symbolically. What does the program do if you try to integrate

$$y(x) = \int_0^x \exp(t^3) \sin t \, dt \quad ?$$

LINE INTEGRALS

The multiple integral was shown to be a generalization of the Riemann integral $\int_a^b f(x)dx$ in one dimension. Another way of extending the notion of integration is to replace the interval $[a, b]$ by a curve in $\mathbf{R}^n$ described by a vector-valued function. The result is called a *line integral*, or a *curvilinear integral*. In this section, we first define line integrals in terms of the limit of a sum. Then, line integrals of parametric curves and line integrals of vector fields are presented.

LINE INTEGRALS IN THE PLANE Let $f(x, y)$ be a continuous function of x and y, and let C be a continuous curve of finite length joining the points P_0 and P_n, as shown in Figure 12.8. It is assumed that $f(x, y)$ is a scalar function defined at every point of the curve between P_0 and P_n, but the function bears no relation to the equation for the curve.

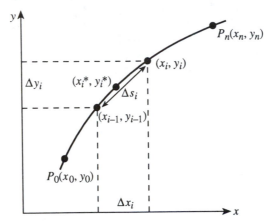

FIGURE 12.8 *Line integral*

If the curve is subdivided into n pieces of lengths $\Delta s_1, \ldots, \Delta s_n$, the projections on the x-axis and y-axis in the ith segment are Δx_i and Δy_i, respectively. The point (x_i^*, y_i^*) is an arbitrary point that lies in the ith subdivision.

To define the line integral over the curve C, $f(x, y)$ is evaluated at each of the points (x_i^*, y_i^*) and the products

$$f(x_i^*, y_i^*) \Delta x_i, \quad f(x_i^*, y_i^*) \Delta y_i, \quad f(x_i^*, y_i^*) \Delta s_i,$$

are formed. Summing over all the subdivisions of the curve results in the approximation to the line integrals as

$$\sum_{i=1}^{n} f(x_i^*, y_i^*) \Delta x_i, \quad \sum_{i=1}^{n} f(x_i^*, y_i^*) \Delta y_i, \quad \sum_{i=1}^{n} f(x_i^*, y_i^*) \Delta s_i.$$

Consider the sum along the curve. Taking the limit as n goes to infinity forces the maximum Δs_i to approach zero and leads to the expression for the line integral with respect to ds as

$$\int_C f(x, y) \, ds = \lim_{n \to \infty} \sum_{i=1}^{n} f(x_i^*, y_i^*) \Delta s_i. \tag{12.47}$$

Similarly, the line integrals with respect to x and y are

$$\int_C f(x, y) \, dx = \lim_{n \to \infty} \sum_{i=1}^{n} f(x_i^*, y_i^*) \Delta x_i,$$

$$\int_C f(x, y)\, dy \;=\; \lim_{n \to \infty} \sum_{i=1}^{n} f(x_i^*, y_i^*)\, \Delta y_i. \qquad (12.48)$$

The ordinary definite integrals of elementary calculus can also be viewed as line integrals, where the curve C is considered to be the x-axis. In fact, the evaluation of line integrals is often reduced to the evaluation of ordinary integrals to simplify the computation.

LINE INTEGRALS IN SPACE A slight extension of the discussion that led to Equation 12.47 and Equation 12.48 for the line integrals in the plane leads to the idea of line integrals in space. For example,

$$\int_C f(x, y, z)\, ds \;=\; \lim_{n \to \infty} \sum_{i=1}^{n} f(x_i^*, y_i^*, z_i^*)\, \Delta s_i \qquad (12.49)$$

if a curve in space is subdivided into n pieces of lengths $\Delta s_1, \ldots, \Delta s_n$ and the point (x_i^*, y_i^*, z_i^*) lies on the ith subdivision. The limit is taken as n goes to infinity. The line integrals with respect to the projections on the x-, y-, or z-axes are formed similarly.

☐ **EXAMPLE 12.9** *Mass of a Wire*
A practical application of Equation 12.49 is the determination of the mass of a wire with a variable mass density $f(x, y, z)$ at each point (x, y, z). The integral will be the total mass of the wire. If the density is constant, say ρ kilograms per meter, the mass is ρ times the length of the wire, as expected, since

$$M = \int_C f(x, y, z)\, ds = \rho \int_C ds \quad \text{kilograms.}$$

Considering this result, the method of computing the line integral with respect to ds is still to be determined. For a curve defined parametrically, the method has been given earlier in the chapter.
If the equation of the curve is defined in terms of parametric equations,

$$x = x(t), \quad y = y(t), \quad z = z(t),$$

Equation 12.10 is a line integral that defines length as

$$s = \int ds = \int_{t_1}^{t_2} \sqrt{\left(\frac{dx}{dt}\right)^2 + \left(\frac{dy}{dt}\right)^2 + \left(\frac{dz}{dt}\right)^2}\, dt. \qquad (12.50)$$

This formula shows that the line integral over the curve reduces to an ordinary definite integral when the curve is defined parametrically and the function $f(x, y, z)$ is constant. Other examples in this section consider more complicated forms of line integrals.

☐

Smooth Curves in Space   In this text, the curves of integration for line integrals are assumed to be either smooth or piecewise smooth. The

curve C is called a *smooth curve* if C has a parameterization or representation

$$\mathbf{R}(t) = [x(t), y(t), z(t)] = x(t)\mathbf{i} + y(t)\mathbf{j} + z(t)\mathbf{k} \qquad \textbf{(12.51)}$$

with a continuous derivative

$$\frac{d\mathbf{R}}{dt} = \frac{dx}{dt}\mathbf{i} + \frac{dy}{dt}\mathbf{j} + \frac{dz}{dt}\mathbf{k},$$

which is nowhere the zero vector. A *piecewise-smooth curve* consists of
a finite number of smooth curves joined at their endpoints. The value of
the line integral along such a curve is defined to be the sum of its values
along the smooth segments of the curve. We also deal only with *simple
curves*. A simple curve is a curve that does not intersect itself.

Line Integrals in Parametric Form Consider a function $f(x, y, z)$
that is continuous on a curve C, and let C be represented parametrically
in vector form by

$$\mathbf{R}(t) = x(t)\,\mathbf{i} + y(t)\,\mathbf{j} + z(t)\,\mathbf{k}.$$

Assume that the derivative $d\mathbf{R}/dt$ exists and is not zero for $a \leq t \leq b$.
Then, the line integrals with respect to x, y, and z can be converted to
ordinary Riemann integrals in the form

$$\int_C f(x, y, z)\, dx = \int_a^b f[x(t), y(t), z(t)] \frac{dx}{dt}\, dt$$

$$\int_C f(x, y, z)\, dy = \int_a^b f[x(t), y(t), z(t)] \frac{dy}{dt}\, dt$$

$$\int_C f(x, y, z)\, dz = \int_a^b f[x(t), y(t), z(t)] \frac{dz}{dt}\, dt \qquad \textbf{(12.52)}$$

Problem 12.18 presents an application of these equations.

**LINE INTEGRALS
OF VECTOR
FIELDS** Suppose a vector field in $\mathbf{R}^3$ is defined as

$$\mathbf{F}(x, y, z) = F_1(x, y, z)\,\mathbf{i} + F_2(x, y, z)\,\mathbf{j} + F_3(x, y, z)\,\mathbf{k},$$

with continuous component functions. The line integral of a vector field
over a smooth curve is denoted as

$$\int_C \mathbf{F} \cdot d\mathbf{R}. \qquad \textbf{(12.53)}$$

In terms of a parametric curve

$$x = x(t), \quad y = y(t), \quad z = z(t), \qquad \textbf{(12.54)}$$

with $a \leq t \leq b$, the line integral of $\mathbf{F}$ over C is defined to be

$$\int_C \mathbf{F} \cdot d\mathbf{R} = \int_a^b \mathbf{F}[x(t), y(t), z(t)] \cdot \frac{d\mathbf{R}}{dt} \, dt. \qquad (12.55)$$

To evaluate Equation 12.55, the following steps are required

1. Form the dot product $\mathbf{F} \cdot d\mathbf{R}/dt$.

2. In the dot product, replace x, y, z with the parametric representation of C given by $x(t)$, $y(t)$, and $z(t)$.

3. Integrate the resulting function of t from a to b.

Another Representation of Line Integrals   The line integral

$$\int_C \mathbf{F} \cdot d\mathbf{R}$$

can be written in terms of scalar functions with the substitution

$$d\mathbf{R}(t) = dx\mathbf{i} + dy\mathbf{j} + dz\mathbf{k}$$

from Equation 12.51 and

$$\mathbf{F} = F_1(x, y, z)\,\mathbf{i} + F_2(x, y, z)\,\mathbf{j} + F_3(x, y, z)\,\mathbf{k}.$$

Thus, the line integral becomes

$$\begin{aligned}
\int_C \mathbf{F} \cdot d\mathbf{R} &= \int_a^b (F_1\mathbf{i} + F_2\mathbf{j} + F_3\,\mathbf{k}) \cdot \left(\frac{dx}{dt}\,\mathbf{i} + \frac{dy}{dt}\,\mathbf{j} + \frac{dz}{dt}\,\mathbf{k} \right) dt \\
&= \int_a^b \left(F_1 \frac{dx}{dt} + F_2 \frac{dy}{dt} + F_3 \frac{dz}{dt} \right) dt \qquad (12.56)
\end{aligned}$$

Because of this equation, the integral $\int_C \mathbf{F} \cdot d\mathbf{R}$ is often written

$$\int_C F_1(x, y, z)\, dx + F_2(x, y, z)\, dy + F_3(x, y, z)\, dz. \qquad (12.57)$$

This expression can be evaluated by substituting the parametric functions $x(t)$, $y(t)$, and $z(t)$, as previously discussed. Other methods of evaluation of this integral are also considered in this chapter.

Line integrals in the form of Equation 12.56 are scalar functions. Another form of the line integral is

$$\int_C f(x, y, z)\, d\mathbf{r}$$

in which a vector sum of $f\,d\mathbf{r}$ along C is computed. The result is a vector that represents the integral of the scalar function $f(x, y, z)$ over the vector path. Problem 12.20 gives an example of this type of line integral.

Line Integral in the Plane

The line integral in the xy-plane

$$I = \int_C x^2 y \, dx - (x + y) \, dy$$

will be evaluated over two paths between the points $(0,0)$ and $(2,1)$, as shown in Figure 12.9.

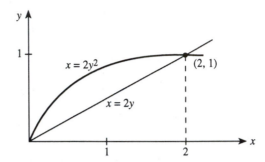

FIGURE 12.9 *Paths for Example 12.10*

One approach to computing the integral is to write the curves in terms of a parameter and apply Equation 12.55. For example, the representation

$$x(t) = 2t, \quad y(t) = t$$

for the straight line would serve as parametric equations. Problem 12.19 asks you to evaluate the line integral this way. In this example, the line integrals will be evaluated by converting the integrands to functions of x or y.

Along the straight line $y = x/2$, with $0 \le x \le 2$ and $dy = dx/2$, so that

$$I = \int_0^2 \left[x^2 \left(\frac{x}{2} \right) dx - \frac{1}{2} \left(x + \frac{x}{2} \right) dx \right] = \frac{1}{2}.$$

The other path shown in the figure is the segment of the parabola $x = 2y^2$ joining the two points. Substituting for x and $dx = 4y \, dy$ in the line integral yields the result

$$I = \int_0^1 \left[(2y^2)^2 y(4y) \, dy - (2y^2 + y) \, dy \right] = \frac{47}{42} = 1.1190.$$

In this example, the value of this line integral depends on the path. However, we shall see that for some integrands the value of a line integral is independent of the path.

□

W H A T   I F ?   The line integrals in Example 12.10 are easily evaluated by expressing the line integral along each path as an ordinary definite integral. Use the MATLAB symbolic command **int** to compute the integrals and verify the integrations in the example.

Line Integrals and Work   The line integral is associated with *work* in physics.[3] The basis is the formula

$$W = \mathbf{F} \cdot \mathbf{d}$$
$$= \text{(force)} \times \text{(displacement in direction of the force)}$$

when $\mathbf{F}$ is a constant force and $\mathbf{d}$ is a straight-line displacement. If the path is a continuous curve, the work W done by a continuous force field as an object moves along the curve is

$$W = \int_C \mathbf{F} \cdot d\mathbf{R}, \tag{12.58}$$

where $\mathbf{R}$ is the position vector for the object on the curve.

Consider the results of Example 12.10. In terms of work, it takes more energy to move an object along the parabolic path than along the straight line between the points in Figure 12.9. This seems reasonable since the parabolic path is longer. However, this is not true in all cases. In fact, integrals for which the result are independent of the path play an important role in applications. We shall return to the analysis of integrals that represent work in a later section of this chapter.

Direction of Integration   When C is a *closed curve*, the direction of integration must be indicated. By convention, the positive direction for a simple closed curve is defined as the counterclockwise direction. The line integral of $\mathbf{F} = F_1\, \mathbf{i} + F_2\, \mathbf{j} + F_3\, \mathbf{k}$ over a closed curve is often written

$$\oint \mathbf{F} \cdot d\mathbf{R} \quad \text{or} \quad \oint F_1\, dx + F_2\, dy + F_3\, dz,$$

using the result from Equation 12.57. For a curve C, the notation $-C$ denotes the curve with the direction of integration reversed. Changing direction on C changes the sign of the line integral, so that

$$\int_C \mathbf{F} \cdot d\mathbf{R} = -\int_{-C} \mathbf{F} \cdot d\mathbf{R}.$$

---

[3]The units of work are those of energy, which are joules in the SI (mks) system, and British thermal units, calories, or foot-pounds in engineering units.

SURFACE INTEGRALS

Let S be a smooth surface with continuous derivatives in the region of interest and let $f(x, y, z)$ be a scalar function defined and continuous on S. Then, the *surface integral* of f over S is defined as

$$I_S = \int \int_S f(x, y, z)\, dS = \lim_{n \to \infty} \sum_1^n f(x_i^*, y_i^*, z_i^*)\, \Delta S_i. \qquad \textbf{(12.59)}$$

The surface S is assumed to be cut into n pieces and ΔS_i denotes the area of the ith piece. In the limit, the ith piece shrinks to a point as $n \to \infty$. The surface integral is a generalization of the area of a surface, which corresponds to the integral over S with $f(x, y, z) = 1$. To evaluate Equation 12.59, it is often possible to reduce the surface integral to a double integral.

FORMULAS FOR SURFACE INTEGRALS

Consider the surface of Figure 12.10 with an angle γ between the normal to the surface and the positive z-axis. Let the projection of S on the xy-plane be R_{xy}. The differential areas ΔS and ΔA are related as $\Delta A = \Delta S \cos \gamma$.

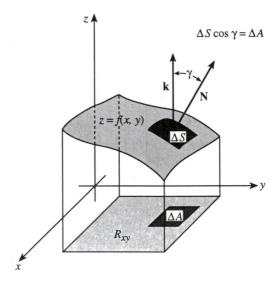

FIGURE 12.10 *Area for a surface integral*

The surface integral then becomes

$$\int \int_S f(x, y, z)\, dS = \int \int_{R_{xy}} f(x, y, z) |\sec \gamma|\, dA. \qquad \textbf{(12.60)}$$

The calculation of $\sec\gamma$ proceeds using the gradient to define the normal vector. Let the equation of the surface be given by $F(x, y, z) = 0$. A normal to the surface at (x, y, z) is

$$\mathbf{N} = \nabla F = \frac{\partial F}{\partial x}\mathbf{i} + \frac{\partial F}{\partial y}\mathbf{j} + \frac{\partial F}{\partial z}\mathbf{k}.$$

Then, $\nabla F \cdot \mathbf{k} = |\nabla F|\,|\mathbf{k}|\cos\gamma$, so that

$$\frac{\partial F}{\partial z} = \sqrt{\left(\frac{\partial F}{\partial x}\right)^2 + \left(\frac{\partial F}{\partial y}\right)^2 + \left(\frac{\partial F}{\partial z}\right)^2}\,\cos\gamma$$

from which

$$|\sec\gamma| = \frac{\sqrt{\left(\frac{\partial F}{\partial x}\right)^2 + \left(\frac{\partial F}{\partial y}\right)^2 + \left(\frac{\partial F}{\partial z}\right)^2}}{\left|\dfrac{\partial F}{\partial z}\right|}.$$

With this substitution, Equation 12.60 becomes

$$\int\int_S f(x, y, z)\frac{\sqrt{\left(\frac{\partial F}{\partial x}\right)^2 + \left(\frac{\partial F}{\partial y}\right)^2 + \left(\frac{\partial F}{\partial z}\right)^2}}{\left|\dfrac{\partial F}{\partial z}\right|}\,dx\,dy. \qquad (12.61)$$

If the surface does not have a projection in the xy-plane, the formula for the surface integral can be modified appropriately. The surface integral could be defined in terms of the projection on the yz-plane or on the xz-plane, if necessary.

☐ EXAMPLE 12.11   ***Surface Integral***
Consider the surface integral

$$\int\int_S z^2\,dS,$$

where S is the octant of a sphere of radius 1 centered at the origin. To apply Equation 12.61, let

$$F(x, y, z) = x^2 + y^2 + z^2 - 1 = 0,$$

so that

$$\frac{\partial F}{\partial x} = 2x, \quad \frac{\partial F}{\partial y} = 2y, \quad \frac{\partial F}{\partial z} = 2z.$$

The surface integral becomes

$$\int\int_S z^2\frac{2\sqrt{x^2 + y^2 + z^2}}{2z}\,dx\,dy = \int\int_S z\,dx\,dy$$

since the function inside the square root is unity. The integral is easily evaluated by converting to polar coordinates with the result

$$\int_0^{\pi/2} \int_0^1 (1-r^2)^{1/2}\, r\, dr\, d\theta = \frac{\pi}{6}.$$

□

SURFACE INTEGRALS OF VECTOR FIELDS

Many applications require that the normal component of a vector field be integrated over a surface. The integral can be written

$$\Phi_S = \int \int_S \mathbf{F} \cdot d\mathbf{S} = \int \int_S \mathbf{F} \cdot \mathbf{N}\, dS, \qquad (12.62)$$

where $\mathbf{N}$ is the outward unit normal to the surface S. Because $\mathbf{F} \cdot \mathbf{N}$ is a scalar function, this integral can be evaluated by the methods previously discussed.

In applications, the integral of Equation 12.62 is said to measure the *flux* Φ_S of a vector field $\mathbf{F}$ across S in the direction of $\mathbf{N}$, a unit normal vector field. In a problem of fluid flow, this integral can be interpreted as the net rate at which fluid is crossing the surface in the outward direction. A negative value would indicate inward flow. Surface integrals also play an important role in problems involving heat conduction and electromagnetics.

□ **EXAMPLE 12.12**   *Surface Integral as Heat Flux*

Suppose that the temperature of a region is controlled so that

$$T(x,y,z) = x^2 + y^2 + z^2.$$

Given this distribution, we wish to compute the heat flux across the unit sphere $x^2 + y^2 + z^2 = 1$. As described in Chapter 11, the heat flow field $\mathbf{F}$ can be calculated using Fourier's law as

$$\mathbf{F} = -K\, \nabla T(x,y,z) = K(-2x\,\mathbf{i} - 2y\,\mathbf{j} - 2z\,\mathbf{k}).$$

The outward normal on the sphere is $\mathbf{N} = x\,\mathbf{i} + y\,\mathbf{j} + z\,\mathbf{k}$ so the surface integral becomes

$$\begin{aligned}
\int \int_S \mathbf{F} \cdot d\mathbf{S} &= \int \int_S \mathbf{F} \cdot \mathbf{N}\, dS \\
&= K \int \int_S (-2x^2 - 2y^2 - 2z^2)\, dS = -2K(4\pi) = -8K\pi
\end{aligned}$$

because the integrand is -2 and the area of the sphere is $4\pi(1)^2$. The heat flow is directed toward the center.

□

THEOREMS OF VECTOR INTEGRAL CALCULUS

In Chapter 11, vector differential operators were studied. For fluid flow, the divergence of the associated vector field creates a scalar field which measures the rate at which fluid is leaving or entering the vicinity of each point. The curl creates another vector field that measures the tendency of the fluid to rotate at each point. Along with the gradient, these operators represent the basic differential operators used to study scalar and vector fields. In this chapter, the emphasis is on integrals that yield information about scalar and vector fields. This section presents theorems involving integral relationships that describe fields.

In studying vector fields, one way to simplify the analysis is to evaluate a function on a set of lower dimensions. For example, the fundamental theorem of calculus states that if $f' \equiv df(t)/dt$ is integrable on the interval $a \leq t \leq b$, then

$$\int_a^b f'(t)\, dt = f(b) - f(a).$$

We might say that the one-dimensional integral is reduced to a "zero-dimensional integral" since the end points have dimension zero and represent the boundary of the interval. Similarly, the line integral of a gradient $\nabla \Phi$ is

$$\int_{P_1}^{P_2} \nabla \Phi \cdot d\mathbf{R} = \Phi(P_2) - \Phi(P_1)$$

under suitable conditions for the function and the path.

It is often useful to express a surface integral in terms of a line integral or a volume integral in terms of a surface integral. The theorems of vector integral calculus not only relate such integrals but also lead to important relationships between integrals and the differential operators of gradient, curl, and divergence. This section first introduces line integrals that are independent of path. Then, Green's theorem in the plane, Stokes' theorem, and the divergence theorem are presented.

INDEPENDENCE OF PATH

In general, the value of a line integral over a path from point P_1 to point P_2 depends not only on the endpoints but also on the path between the points. Functions for which the value of the line integral does not depend on the path are of great practical importance. We define a line integral to be *independent of path* in a domain D in space if for every pair of endpoints P_1 and P_2 in D, the integral has the same value for *all* paths in D that start at P_1 and end at P_2. A useful criterion for path independence can be defined in terms of the gradient, as presented in this section. Another criterion derived from Green's theorem is presented later.

Consider the vector field $\mathbf{F}$ that is derived as $\mathbf{F} = \nabla\Phi$ for some real-valued function $\Phi(x, y, z)$. Thus,

$$\mathbf{F} = \frac{\partial\Phi}{\partial x}\mathbf{i} + \frac{\partial\Phi}{\partial y}\mathbf{j} + \frac{\partial\Phi}{\partial z}\mathbf{k}.$$

Considering the fundamental theorem of calculus,

$$\int_a^b g(x)\,dx = G(b) - G(a)$$

with $G' = g$, we ask if it is reasonable that the value of the line integral

$$\int_{P_1}^{P_2} \nabla\Phi \cdot d\mathbf{R}$$

is determined by the value of Φ at the endpoints. The next theorem shows this to be the case, with a form reminiscent of the fundamental theorem of calculus.

■ THEOREM 12.3   *Independence of Path*

Suppose that $\Phi(x, y, z)$ is a continuous scalar function and C is a continuously differentiable curve in space with endpoints P_1 and P_2. Then,

$$\int_{P_1}^{P_2} \nabla\Phi \cdot d\mathbf{R} = \Phi(P_2) - \Phi(P_1), \tag{12.63}$$

and we say that the integral is path-independent.

The proof of the path-independent theorem is straightforward using the fundamental theorem of calculus. This theorem shows that if the integrand is a gradient, then evaluation of the line integral reduces to two functional calculations. But, it is not always possible to find an "antiderivative" because a vector field need not be derivable from the gradient of a scalar field.

☐ EXAMPLE 12.13   *Central Force*

Consider the force field defined by

$$\mathbf{F} = -\frac{\alpha m}{R^2}\mathbf{a}_R, \qquad R \neq 0,$$

which could represent the force of attraction exerted on a particle of mass m by a particle of unit mass at the origin. Imagine that the particle is a distance R from the origin in the direction $\mathbf{a}_R = \mathbf{R}/R$. The constant α depends on the units to measure the force if the particle at the origin has unit mass. The work done displacing the particle along a path from P_1 to P_2, as shown in Figure 12.11, is

$$W = \int_{P_1}^{P_2} \mathbf{F} \cdot d\mathbf{R}.$$

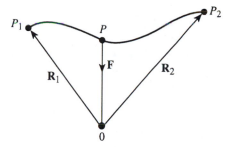

FIGURE 12.11 *Central force for Example 12.13*

Assuming that the path does not pass through the origin $\mathbf{R} = \mathbf{0}$, the work integral can be written as

$$W = \int_{P_1}^{P_2} -\frac{\alpha m}{R^3}\,\mathbf{R} \cdot d\mathbf{R},$$

and evaluated by noting that $\mathbf{R} \cdot d\mathbf{R}$ is the scalar function $R\,dR$ since the vectors point in the same direction. Thus, the work integral becomes

$$W = \int_{P_1}^{P_2} -\frac{\alpha m}{R^2}\,dR = \alpha m \int_{P_1}^{P_2} d\left(\frac{1}{R}\right) = \alpha m \left[\frac{1}{R}\right]_{P_1}^{P_2}. \tag{12.64}$$

Suppose the endpoints are defined by the vectors $\mathbf{R}_1$ and $\mathbf{R}_2$ with length R_1 and R_2, respectively. Then the work represented by Equation 12.64 becomes

$$W = \alpha m \left(\frac{1}{R_2} - \frac{1}{R_1}\right).$$

The scalar function $-\alpha m/R$ is known as the *gravitational potential*, and the force is derived from the gradient of this potential as

$$\mathbf{F} = -\nabla\left(-\frac{\alpha m}{R}\right) = -\frac{\alpha m}{R^3}\,\mathbf{R}.$$

Force fields derived as gradients of a scalar function will be discussed later as conservative force fields.

$\square$

CURVES, DOMAINS, AND SURFACES

In discussing surfaces, it is often useful to distinguish between the curve that defines the boundary of the surface and the surface itself without the boundary points. Some authors call the open set (not including boundary points) of points defining a surface the *domain D* for the surface. It is assumed that any two points in the domain can be connected by a piecewise continuous curve lying in D. The term *open region* is often used to define the domain in $\mathbf{R}^2$. The term *closed region* describes the domain of a surface and the boundary points. For example, the points

$x^2 + y^2 < r^2$ are the domain of the interior of a circle of radius r. The associated closed region consists of the points $x^2 + y^2 \leq r^2$.

The relation describing independence of path in Equation 12.63 is valid only when Φ is continuous and single-valued in a region of space that is called *simply connected*. Basically, a simply connected domain has no "holes," so that any simple closed curve within it can be continuously contracted into a point without having to leave the domain. The interiors of a circle and a rectangle are simply connected domains. Spaces that are not simply connected are said to be multiply connected. For example, the space between two infinitely long coaxial cylinders is multiply connected since it is impossible to shrink to a point any closed curve circling the inner cylinder. Vector theorems for multiply connected domains are discussed in the textbook by Kaplan listed in the Annotated Bibliography for this chapter

A *smooth surface* S in space is defined by a function, such as $f(x, y)$, that has continuous derivatives. If the surface is defined parametrically, then the parametric equations for a smooth surface have derivatives that are continuous. Intuitively, a smooth surface has no corners or breaks.

GREEN'S THEOREM IN THE PLANE

The two-dimensional analog of the fundamental theorem of integral calculus expresses a double integral over a surface S as a line integral taken along a closed path that represents the boundary of S. The theorem known as *Green's theorem in the plane* relates the line integral around the boundary of a region to an area integral over the interior of the region. Transforming between the integrals often helps to make the evaluation of an integral easier. The relationship is also useful in theoretical studies of fields.

■ **THEOREM 12.4**   *Green's Theorem in the Plane*

Let $P(x, y)$ and $Q(x, y)$ be continuous functions and have continuous first partial derivatives. Also, let S be a region of the plane bounded by the piecewise-smooth curve C that is simple and closed. Then,

$$\oint P(x, y)\, dx + Q(x, y)\, dy = \int\int_S \left(\frac{\partial Q}{\partial x} - \frac{\partial P}{\partial y} \right) dx dy. \qquad \textbf{(12.65)}$$

The positive direction (counterclockwise) is to be taken around the path C.

--- ■

Proving the theorem is straightforward for rectangular and other simple regions of the plane. For example, suppose that the domain of $P(x, y)$ is such that the region in the xy plane is defined as

$$a \leq x \leq b, \qquad f_1(x) \leq y \leq f_2(x).$$

Then, writing the double integral as an iterated integral

$$\int\int_S \frac{\partial P}{\partial y} dx dy = \int_a^b \left[\int_{f_1(x)}^{f_2(x)} \frac{\partial P}{\partial y} dy \right] dx$$

leads to the conclusions

$$\int_a^b \int_{f_1(x)}^{f_2(x)} \frac{\partial P}{\partial y} \, dy dx \;=\; \int_a^b \{P[x, f_2(x)] - P[x, f_1(x)]\} \, dx$$

$$= \; -\int_b^a P[x, f_2(x)] \, dx - \int_a^b P[x, f_1(x)] \, dx$$

$$= \; -\oint P(x, y) \, dx.$$

The same approach with the integral $\int \int_S \partial Q / \partial x \, dx dy$ over a region where the y limits are constant shows that

$$\int \int_S \frac{\partial Q}{\partial x} \, dx dy = \oint Q(x, y) \, dy.$$

Adding the two double integrals yields Green's theorem, as stated in Equation 12.65. The general proof of the theorem is contained in the book by Apostol listed in the Annotated Bibliography for this chapter.

Independence of Path Green's theorem leads to a number of interesting results, including a statement of the conditions for independence of path of two functions. With the definitions previously given for $P(x, y)$ and $Q(x, y)$ in a region of the plane R, the necessary and sufficient condition for

$$\oint P(x, y) \, dx + Q(x, y) \, dy = 0$$

around every closed path C in R is that $\partial P(x, y)/\partial y = \partial Q(x, y)/\partial x$ identically in R. The sufficiency condition follows immediately from Green's theorem in Equation 12.65, so that

$$\oint P(x, y) \, dx + Q(x, y) \, dy = \int \int_S \left(\frac{\partial Q}{\partial x} - \frac{\partial P}{\partial y} \right) dx dy = 0$$

when the partial derivatives are equal. If the line integral around every closed path C in R is zero, then the double integral must be zero.

☐ EXAMPLE 12.14   ***Area by Green's Theorem***
Letting $P = -y$ and $Q = x$ in Green's theorem leads to the result

$$\oint -y \, dx + x \, dy = \int \int_R (1 + 1) \, dx dy = 2 \times \text{area of R}.$$

Thus, the area bounded by a simple closed curve C is

$$\text{area} = \frac{1}{2} \oint x \, dy - y \, dx.$$

☐

Vector Form of Green's Theorem Suppose that a vector field in $\mathbf{R}^2$ is given as

$$\mathbf{F} = P(x, y)\,\mathbf{i} + Q(x, y)\,\mathbf{j}.$$

The vector form of Green's theorem is stated using the curl of $\mathbf{F}$:

$$\nabla \times \mathbf{F} = \begin{vmatrix} \mathbf{i} & \mathbf{j} & \mathbf{k} \\ \dfrac{\partial}{\partial x} & \dfrac{\partial}{\partial y} & \dfrac{\partial}{\partial z} \\ P(x, y) & Q(x, y) & 0 \end{vmatrix} = \left[\dfrac{\partial Q}{\partial x} - \dfrac{\partial P}{\partial y} \right] \mathbf{k}.$$

Applying this expression for $\nabla \times \mathbf{F}$ and letting $\mathbf{N} = \mathbf{k}$ be a unit normal to the region of integration in the xy-plane leads to the relationship

$$\oint_C \mathbf{F} \cdot d\mathbf{R} = \int\int (\nabla \times \mathbf{F}) \cdot \mathbf{k}\, dS. \qquad (12.66)$$

This form of Green's theorem is particularly suited for extension to surfaces in $\mathbf{R}^3$, as defined by Stoke's theorem, presented later. Another advantage of writing the formula in the vector form is that the curl is independent of the coordinate system and the curl has a physical significance.

Green's theorem can also be written in the form

$$\oint_C \mathbf{F} \cdot \mathbf{N}\, dR = \int\int \nabla \cdot \mathbf{F}\, dA, \qquad (12.67)$$

which relates the line integral of $\mathbf{F} \cdot \mathbf{N}$, the normal component of $\mathbf{F}$, to the divergence of $\mathbf{F}$. Equation 12.66 relates the tangential component of $\mathbf{F}$ to the curl of the vector field. The result of Equation 12.67 will be extended to vector fields of $\mathbf{R}^3$ in the form of the divergence theorem.

STOKES' THEOREM

For functions defined in space, *Stokes' theorem* transforms line integrals into surface integrals and conversely. Thus, this theorem generalizes Green's theorem in the plane.

■ THEOREM 12.5   ***Stokes' Theorem***

Let a surface S bounded by a simple closed curve C in $\mathbf{R}^3$, as shown in Figure 12.12. Let $\mathbf{F}(x, y, z)$ be a vector function that is continuously differentiable on S. Then,

$$\oint_C \mathbf{F}(x, y, z) \cdot d\mathbf{R} = \int\int (\nabla \times \mathbf{F}) \cdot \mathbf{N}\, dS, \qquad (12.68)$$

where $d\mathbf{R}$ is a directed line element of C and $\mathbf{N}$ is a unit vector normal to dS.

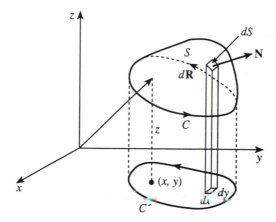

FIGURE 12.12 *Surface for Stokes' theorem*

Relationship to Green's Theorem   Let a vector function be

$$\mathbf{F} = F_1(x, y)\,\mathbf{i} + F_2(x, y)\,\mathbf{j}$$

in a domain in the xy-plane where $\mathbf{F}$ is continuously differentiable. Taking the z-component of the curl yields the result

$$(\nabla \times \mathbf{F}) \cdot \mathbf{k} = \frac{\partial F_1}{\partial x} - \frac{\partial F_2}{\partial y}.$$

Associating $F_1(x, y) = P(x, y)$ and $F_2(x, y) = Q(x, y)$, Stokes' theorem states

$$\int\int_S \left(\frac{\partial Q}{\partial x} - \frac{\partial P}{\partial y} \right) dx dy = \oint P(x, y)\,dx + Q(x, y)\,dy, \qquad \textbf{(12.69)}$$

which shows that Green's theorem as stated in Equation 12.65 is a special case of Stokes' theorem.

Conservative Fields and Stoke's Theorem   Stoke's theorem is often used to establish general properties of vector fields. For example, Stokes' theorem can be used to relate a conservative field to the curl of that field, as indicated by the following theorem.

■ THEOREM 12.6   ***Conservative Fields***

   If $\mathbf{F}$ and $\nabla \times \mathbf{F}$ are continuous in a simply connected region, then $\mathbf{F}$ is conservative if and only if $\nabla \times \mathbf{F} = 0$ in the region.

If $\mathbf{F}$ is conservative, $\mathbf{F} = \nabla f$ and

$$\nabla \times \mathbf{F} = \nabla \times \nabla(f) \equiv \mathbf{0},$$

as discussed in Chapter 11. Conversely, if $\nabla \times \mathbf{F} = 0$, let C be the boundary curve of the surface S. Then, Stokes' theorem gives the result

$$\oint_C \mathbf{F} \cdot d\mathbf{R} = \int\int (\nabla \times \mathbf{F}) \cdot \mathbf{N} \, dS = \int\int 0 \, dS = 0,$$

so the integral is independent of path, and hence $\mathbf{F}$ is a conservative vector field.

DIVERGENCE
THEOREM

A relationship known as both Gauss' theorem and the divergence theorem relates the volume integral of the divergence of a vector field to the normal component of the field over a surface enclosing the volume.[4]

■ THEOREM 12.7   *Divergence Theorem*
If $\mathbf{F}(x, y, z)$ is a continuous vector field with components that have continuous derivatives, then

$$\int\int_S \mathbf{F}(x, y, z) \cdot \mathbf{N} \, dS = \int\int\int_V \nabla \cdot \mathbf{F} \, dV, \qquad (12.70)$$

where $\mathbf{N}$ is the outward unit normal to the surface S enclosing the volume designated V.

━━━━━━━━━━━━━━━━━━━━━━━━━━━━■

For the vector field $\mathbf{F} = F_1 \mathbf{i} + F_2 \mathbf{j} + F_3 \mathbf{k}$, the divergence theorem can be written

$$\int\int_S F_1 \, dydz + F_2 \, dzdx + F_3 \, dxdy$$

$$= \int\int\int_V \left(\frac{\partial F_1}{\partial x} + \frac{\partial F_2}{\partial y} + \frac{\partial F_3}{\partial z} \right) dxdydz.$$

□ EXAMPLE 12.15   *Vector Integral Theorems*
    a. In this example, the line integral of the vector field $\mathbf{F} = 2y\,\mathbf{i} + x\,\mathbf{j}$ is to be computed by integrating counterclockwise around the curve defined by the sides of the triangle with vertices

$$(0, 0), \quad (2, 0), \quad (2, 4).$$

The line integral is

$$\int_C \mathbf{F} \cdot d\mathbf{R} = \int_C 2y \, dx + x \, dy + 0 \, dz.$$

Taking the sides of the triangle separately,

$$\int_C \mathbf{F} \cdot d\mathbf{R} = \int_{C_1} 2y \, dx + \int_{C_2} 2 \, dy + \int_{C_3} 2y \, dx + x \, dy.$$

---

[4] Karl Friedrich Gauss (1777-1855) made fundamental contributions to many areas of mathematics and physics that are now described by vector field theory. It is said that Gauss is the last mathematician to know everything in his subject. In any case, his name is associated with many mathematical theorems and physical laws. Gauss' flux theorem is an example presented later in this chapter.

On side C_1 from $(0,0)$ to $(2,0)$, $2y = 0$. On the side from $(2,0)$ to $(2,4)$, $x = 2$. On the third side, $2x = y$, so $2\,dx = dy$. The value of the line integral thus becomes

$$\Gamma = \int_0^4 2\,dy + \int_2^0 4x\,dx + 2x\,dx = -4.$$

Applying Stokes' theorem leads to the same result, since

$$
\begin{aligned}
\Gamma &= \int\int (\nabla \times \mathbf{F})\cdot \mathbf{N}\,dS \\
&= \int\int \left(\frac{\partial F_2}{\partial x} - \frac{\partial F_1}{\partial y}\mathbf{k}\right)\cdot \mathbf{k}\,dx\,dy\,dz \\
&= -\int\int dS = -4.
\end{aligned}
$$

b. Consider the vector field $\mathbf{F} = f(x, y, z)\,\mathbf{i}$ evaluated by the divergence theorem

$$\int\int\int_V \nabla \cdot (f\mathbf{i})\,dV = \int\int_S f\mathbf{i}\cdot \mathbf{N}\,dS.$$

Since the unit vector $\mathbf{i}$ is constant, $\nabla \cdot (f\mathbf{i}) = \mathbf{i}\cdot\nabla\Phi$, so that the integrals become

$$\mathbf{i}\cdot\left[\int\int\int_V \nabla f\,dV - \int\int_S f\mathbf{N}\,dS\right] = 0.$$

The vector $\mathbf{i}$ is never zero, and the quantity in brackets is not perpendicular to $\mathbf{i}$ in general. Consequently,

$$\int\int\int_V \nabla f\,dV = \int\int_S f\mathbf{N}\,dS. \tag{12.71}$$

This formula is another useful relationship of vector field theory.

□

SUMMARY

The main results discussed in this section are summarized in Table 12.4. The vector field in Green's formulas is a 2D function with the form

$$\mathbf{F} = P\,\mathbf{i} + Q\,\mathbf{j}.$$

In Stokes' theorem and the divergence theorem, $\mathbf{F}$ is a vector field in $\mathbf{R}^3$. The *divergence theorem* is also known as Gauss' theorem.

TABLE 12.4 *Theorems of vector field theory*

Name	Formula
Green's 2D	$\oint_C \mathbf{F}(x,y) \cdot d\mathbf{R} = \int\int \left(\dfrac{\partial P(x,y)}{\partial x} - \dfrac{\partial Q(x,y)}{\partial y} \right) dA$
	$\oint_C \mathbf{F}(x,y) \cdot d\mathbf{R} = \int\int (\nabla \times \mathbf{F}) \cdot \mathbf{N}\, dS$
	$\int_C \mathbf{F}(x,y) \cdot \mathbf{N}\, dR = \int\int \nabla \cdot \mathbf{F}\, dA$
Stokes	$\oint_C \mathbf{F}(x,y,z) \cdot d\mathbf{R} = \int\int (\nabla \times \mathbf{F}) \cdot \mathbf{N}\, dS$
Divergence	$\int\int \mathbf{F}(x,y,z) \cdot \mathbf{N}\, dS = \int\int\int \nabla \cdot \mathbf{F}\, dV$

Before applying these formulas, we must be clear about the conditions for the curves, surfaces, or volumes involved. These requirements were stated in this section. In the next section, these formulas and their variations are shown to have important physical significance for problems such as those involving fluid flow or electromagnetic fields.

APPLICATIONS OF VECTOR FIELD THEORY

We now turn to applications of the powerful theorems of vector integral calculus. Several specific application areas were introduced in the last section of Chapter 11. The approach to studying fields in Chapter 11 used vector differential operators. This section presents vector fields that arise in mechanics, fluid flow, and electromagnetics using the integral theorems to describe a physical situation.

CONSERVATION LAWS

One theme that runs through physics is the search for quantities that are constant. For example, if the forces acting on a particle are conservative, then the total energy of the particle is conserved. From the discussion in Chapter 11, the work in moving the particle in this case is equal to the change in the kinetic energy, but the sum of the kinetic and potential energies is constant. Notice that the two energies may change with time, but the total energy is constant.

Quantities, such as mass or charge, can also be conserved. Various equations called *continuity* equations have been derived to describe fluid flow, heat flow, and flow of electrical current. The integral theorems of vector field theory lead to these equations, and a few examples are presented in this section.

CONSERVATIVE FIELDS

The fields associated with electrostatics and magnetostatics, steady fluid flow, and gravitational force are conservative fields. The mathematical advantage of a conservative field is that it can be specified by its scalar potential function rather than its vector components. The other properties of a conservative field can be summarized in terms of the theorems and discussions given in previous sections.

Assume that a vector field exists in space as

$$\mathbf{F} = F_1(x, y, z)\,\mathbf{i} + F_2(x, y, z)\,\mathbf{j} + F_3(x, y, z)\,\mathbf{k},$$

and that the field possesses continuous first partial derivatives at all points of a simply connected domain. Then, the following statements are equivalent:

1. $\displaystyle\int_C \mathbf{F}\cdot d\mathbf{R} = \int_C F_1\,dx + F_2\,dy + F_3\,dz$ is independent of path.

2. $\displaystyle\oint_C \mathbf{F}\cdot d\mathbf{R} = \int_C F_1\,dx + F_2\,dy + F_3\,dz = 0$ around every closed curve.

3. $\mathbf{F}\cdot d\mathbf{R} = F_1\,dx + F_2\,dy + F_3\,dz$ is an exact differential.

4. $\mathbf{F}$ is the gradient of a scalar function $\Phi(x, y, z)$ such that

$$\frac{\partial \Phi}{\partial x} = F_1, \qquad \frac{\partial \Phi}{\partial y} = F_2, \qquad \frac{\partial \Phi}{\partial z} = F_3.$$

5. $\nabla \times \mathbf{F}$ vanishes identically.

□ **EXAMPLE 12.16**   *Nonconservative Forces*

The force of friction or of damping is not conservative. In systems with damping, the work depends on the force over the path of integration. As an example, consider the unforced, second-order mechanical system with damping described in Chapter 5 by the differential equation

$$m\ddot{x}(t) + b\dot{x}(t) + kx(t) = 0.$$

If $b = 0$, the total energy is $E = mv^2/2 + kx^2/2$, so that

$$\frac{dE}{dt} = (ma + kx)v,$$

where $a = \dot{v}(t)$ and $v = \dot{x}(t)$. Without damping, the term in parenthesis is zero, so $dE/dt = 0$ and mechanical energy is conserved. If damping is considered, $(ma + kx) = -bv$ and

$$\frac{dE}{dt} = -bv^2,$$

Applications of Vector Field Theory

579

so energy is being lost as the mass moves along any path. Thus, the force equation is not just $\mathbf{F} = -kx\mathbf{i} = \nabla(-kx^2/2)$, but it includes a term that is proportional to velocity, so the work done against the dissipation term is

$$dW_d = -\mathbf{F}_d \cdot d\mathbf{R} = -bv^2\,dt.$$

In many applications, the total force can be separated into conservative forces and nonconservative forces. In such cases, the effect of each type of force can be calculated separately and the results added. Problem 12.38 presents a situation that is conveniently analyzed by separating the forces involved and using superposition.

This example shows that the addition of a nonconservative force complicates the situation in several ways. First, the problem cannot be solved in terms of a scalar potential. Secondly, the properties of the materials involved must be considered to find a numerical solution. For example, if the damping is caused by sliding friction, the damping coefficient will depend on the roughness of the surfaces so the value is usually determined by experiment.

□

Potentials In mechanical and electrical systems described by ordinary differential equations, there are a finite number of components that are modeled as elements with no spatial extent. When the physical system covers a region of space, a mathematical description of the system often requires partial differential equations. In many problems, the solution can be expressed as a scalar or vector field.

As we have stated previously, the scalar fields associated with conservative forces are usually mathematically simpler than the vector fields that are derivable from them. Moreover, a scalar field is often called a *potential* because in physical applications the field is often associated with the potential energy of a system.

Many scalar fields vary with position as a power of the position vector, such as R^ν, where

$$R = (x^2 + y^2 + z^2)^{1/2}$$

is the magnitude of the position vector from the origin

$$\mathbf{R} = x\,\mathbf{i} + y\,\mathbf{j} + z\,\mathbf{k}.$$

Then, the vector field associated with R^ν is

$$\begin{aligned}
\nabla R^\nu &= \frac{\partial R^\nu}{\partial x}\mathbf{i} + \frac{\partial R^\nu}{\partial y}\mathbf{j} + \frac{\partial R^\nu}{\partial z}\mathbf{k} \\
&= \nu R^{\nu-1}\frac{\partial R}{\partial x}\mathbf{i} + \nu R^{\nu-1}\frac{\partial R}{\partial y}\mathbf{j} + \nu R^{\nu-1}\frac{\partial R}{\partial z}\mathbf{k} \\
&= \nu R^{\nu-2}\mathbf{R} = \nu R^{\nu-1}\mathbf{a}_R,
\end{aligned} \tag{12.72}$$

where $\mathbf{a}_R$ is a unit vector in the $\mathbf{R}$-direction. Notice that if $\nu = 1$, the gradient is just this unit vector. The scalar fields with $\nu = -1$ include the gravitational potential and Coulomb's potential for electric charge. The corresponding vector fields vary with $\mathbf{R}$ as $\mathbf{a}_R/R^2$ and can be derived from the gradient of the potential, as discussed in Chapter 11.

In previous examples, the potential is given and the vector field can be derived by taking the gradient of the potential. In many practical problems, the fields must be calculated from the source distributions. This generally involves integral relations. However, if a scalar potential exists, it is often simpler to derive a differential equation, which can be solved to determine the potential and hence the field. Both these approaches to compute a vector field are explored in the remainder of this chapter.

SOURCES OF THE FIELDS

From Chapter 11, we know that irrotational vector fields for which

$$\nabla \times \mathbf{F} = 0$$

can be derived from scalar potentials and solenoidal fields ($\nabla \cdot \mathbf{F} = 0$) are derivable from a vector potential. The question to be answered here is how the potentials are calculated from the *sources* of the fields.

Two types of vector field sources are considered here. We shall see that the divergence of a vector field defines the density of the scalar source that creates an irrotational field. The curl of a vector field specifies the density of the vector source which causes the solenoidal portion of a vector field.

Irrotational Fields Physically, a field must be created at points called *source* points or destroyed at *sink* points. The strength of the source (or sink) in a region is measured by computing the flux through a closed surface in the region of interest. If

$$\text{flux} = \int\int_S \mathbf{F}(x,y,z) \cdot \mathbf{N}\, dS$$

is nonzero, there must be sources or sinks within the region enclosed by S. However, the quantity of flux depends on the size of the surface so the flux defined in this way does not represent an intrinsic characteristic of the field. Assuming that the field and the surface approach zero smoothly as the volume enclosing the source or sink shrinks to zero, we define the *net flux per unit volume* diverging from or entering an infinitesimal neighborhood of a point as

$$\lim_{\Delta V \to 0} \frac{1}{\Delta V} \int\int_S \mathbf{F}(x,y,z) \cdot \mathbf{N}\, dS.$$

Considering the divergence theorem

$$\int\int_S \mathbf{F}(x,y,z) \cdot \mathbf{N}\, dS = \int\int\int_V \nabla \cdot \mathbf{F}\, dV,$$

the definition of the divergence is sometimes taken to be

$$\nabla \cdot F = \lim_{\Delta V \to 0} \frac{1}{\Delta V} \int \int_S \mathbf{F}(x, y, z) \cdot \mathbf{N} \, dS. \qquad (12.73)$$

This representation shows that the divergence of a vector field at a point measures the net normal component of the field passing out of an infinitesimal volume surrounding the point. If $\nabla \cdot F \neq 0$ at the point, a source or sink must be located there. For a vector field $\mathbf{F}$ at a point, the conclusions are as follows:

1. If $\nabla \cdot \mathbf{F} = 0$, the field is source (or sink) free at the point.

2. If $\nabla \cdot \mathbf{F} > 0$, the point is called a source.

3. If $\nabla \cdot \mathbf{F} < 0$, the point is called a sink.

☐ EXAMPLE 12.17   *Divergence of the Inverse-Square Field*
Consider an inverse-square force field

$$\mathbf{F} = \frac{\alpha}{R^2} \mathbf{a}_R = \frac{\alpha}{R^3} \mathbf{R},$$

where α is a constant and $\mathbf{R}$ is the position vector in $\mathbf{R}^3$. In Chapter 11, it was shown that both the Newtonian gravitational force and Coulomb's force vary with position in this way.

Computing the flux for the inverse-square field across a closed surface yields

$$\begin{aligned} \text{flux} \quad &= \quad \int \int_S \alpha \frac{\mathbf{R}}{R^3} \cdot \mathbf{N} \, dS \\ &= \quad \alpha \int \int \int_V \nabla \cdot \left(\frac{\mathbf{R}}{R^3} \right) \, dV. \qquad (12.74) \end{aligned}$$

The divergence in the volume integral becomes

$$\begin{aligned} \nabla \cdot \left(\frac{\mathbf{R}}{R^3} \right) \quad &= \quad \frac{1}{R^3} \nabla \cdot \mathbf{R} + \mathbf{R} \cdot \nabla \left(\frac{1}{R^3} \right) \\ &= \quad \frac{3}{R^3} - 3 \frac{\mathbf{R} \cdot \mathbf{R}}{R^5} = 0, \qquad (12.75) \end{aligned}$$

using Equation 12.72 to compute ∇R^{-3}. The conclusion from Equation 12.74 is that

$$\text{flux} = \int \int_S \alpha \frac{\mathbf{R}}{R^3} \cdot \mathbf{N} \, dS = 0. \qquad (12.76)$$

This result could be disturbing if you interpret this as stating that the gravitational and electrostatic fields are sourceless! The apparent contradiction will be resolved when Gauss' flux theorem is introduced later.

☐

W H A T   I F ?   Assuming that the mathematics that lead to Equation 12.76 is correct, can the conclusion be correct for all surfaces, including those that enclose mass or electric charge? Consider the field of a point charge as described in Chapter 11 and show that the result has been misinterpreted. What did we miss?

Potentials and Fields for Gravity and Electrostatics

The source of the gravitational field is mass and the source of the electrostatic field is electric charge. These fields were shown to be irrotational in Chapter 11. In each case, the potential varies as $1/r$ and the field varies as $1/r^2$, where $\mathbf{r}$ is the vector from the source to the field point. To compute the Newtonian potential due to a distribution of mass or the Coulomb potential from a distribution of charges, it is necessary to sum the potentials from point sources or integrate over distributed sources.

Figure 12.13 shows the vectors to determine the potential at the field point $P(x, y, z)$ due to an infinitesimal source at $P(x^*, y^*, z^*)$.

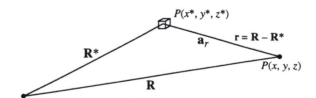

FIGURE 12.13 *Vectors for source and field points*

From the figure, the vector from the source to the field point is

$$\mathbf{r} = \mathbf{R} - \mathbf{R}^* = (x - x^*)\,\mathbf{i} + (y - y^*)\,\mathbf{j} + (z - z^*)\,\mathbf{k}$$

with magnitude $r = ||\mathbf{r}||$. For example, let the source have a continuous volume density $\rho(x^*, y^*, z^*)$ in a volume ΔV^*. The potential from this small distribution is given by

$$\Delta\Phi(x, y, z) = \alpha\,\frac{\rho(x^*, y^*, z^*)\Delta V^*}{r}.$$

Thus, the potential caused by the volume distribution is calculated as

$$\Phi(x, y, z) = \alpha \int\int\int \frac{\rho(x^*, y^*, z^*)}{r}\, dV^*. \qquad (12.77)$$

To determine Newton's force of attraction due to a distribution on mass, let $\alpha = G$ and take the gradient of the potential. To determine the electric field due to a charge distribution, let $\alpha = 1/4\pi\epsilon$ and $\mathbf{E} = -\nabla\Phi$.

Gauss' Flux Theorem

To determine the field in a general case, Equation 12.77 can be integrated over the distribution. In the case of extreme symmetry where the field is constant over a surface enclosing the distribution, Gauss' flux theorem can be applied to determine the vector field directly.

Consider the inverse-square force field

$$\mathbf{F} = \frac{\alpha}{R^2}\mathbf{a}_R = \frac{\alpha}{R^3}\mathbf{R},$$

where α is a constant and $\mathbf{R}$ is the position vector. Gauss' flux theorem states that

$$\int\int_S \alpha\frac{\mathbf{R}}{R^3}\cdot\mathbf{N}\,dS = \begin{cases} 4\alpha\pi, & \text{if } (0,0,0) \in V, \\ 0, & \text{if } (0,0,0) \notin V, \end{cases}$$

where V is the volume enclosed by S. This theorem is also called *Gauss' law*.

Now the question posed by Example 12.17 can be answered. The flux of the normal component of an inverse-square field is zero over a surface unless the origin is enclosed by the surface. When the origin is enclosed, the divergence theorem does not apply because the field does not have continuous first partial derivatives throughout the volume of integration. However, the surface integral can be computed by defining two surfaces enclosing the origin and applying the divergence theorem as described in Problem 12.27.

Gauss' flux theorem applied to an electrostatic problem asserts that the integral of the normal component of the electric flux density over any closed surface S is equal to the total electric charge enclosed by S. The mathematical statement is

$$\int\int_S \mathbf{E}\cdot d\mathbf{S} = \frac{1}{\epsilon_0}\int\int\int_V \rho\,dV, \tag{12.78}$$

where the permittivity $\epsilon_0 = (1/36\pi) \times 10^{-9}$ farad/meter is used if the charge is located in free space.[5] Applying the divergence theorem leads to a fundamental equation in electrostatics that defines the source density as

$$\nabla \cdot \mathbf{E} = \frac{\rho}{\epsilon_0}.$$

This is one of Maxwell's equations to be studied later.

Point Charge. Suppose a point charge of magnitude Q is located at the origin. The potential at any field point is

$$\Phi(R) = \frac{Q}{4\pi\epsilon_0 R}.$$

Thus, the electric field derived from $-\nabla\Phi(R)$ is

$$\mathbf{E}(\mathbf{R}) = \frac{Q}{4\pi\epsilon_0 R^2}\,\mathbf{a}_R. \tag{12.79}$$

Gauss' theorem of Equation 12.78 implies

$$\int\int_S \mathbf{E}\cdot d\mathbf{S} = \frac{Q}{\epsilon_0}.$$

---

[5]In electromagnetic theory, the term free space implies that there are no sources of the fields in the region being considered. In our applications, the terms free space and vacuum are used interchangeably.

Using spherical symmetry, we conclude that $\mathbf{E} = E_R\,\mathbf{a}_R$ is constant on any sphere surrounding the charge. For a sphere of radius R,

$$\int\int_S \mathbf{E}(\mathbf{R})\cdot d\mathbf{S} = E_R \int\int dS = 4\pi R^2 E_R = \frac{Q}{\epsilon_0}$$

and the field is as stated in Equation 12.79.

$\square$

WHAT IF? Suppose you wish to use Gauss' flux theorem to find the gravitational or electrical field. If fact, the theorem can be used only to define the vector field analytically in a few cases that have a great deal of symmetry. What are other configurations in which the theorem is useful to compute the field as in the case with spherical symmetry. Could numerical methods help in cases when the field is not constant on the surface?

Solenoidal Fields Suppose that $\nabla \cdot \mathbf{F} = 0$ for a continuously differentiable vector field. It can be shown that such a field $\mathbf{F}$ can be derived from the curl of a vector so that

$$\nabla \times \mathbf{F} = \mathbf{J},$$

as discussed in Chapter 11 in connection with Helmholtz's theorem. The vector $\mathbf{J}$ is called the vortex, or vector source, for the field $\mathbf{F}$. Using the vector identity $\nabla\cdot(\nabla \times \mathbf{A}) = 0$, the solenoidal field $\mathbf{F}$ can be derived as the curl of $\mathbf{A}$ so that

$$\mathbf{F} = \nabla \times \mathbf{A},$$

where $\mathbf{A}$ is called the vector potential.

In magnetostatics, the source for the magnetic field is a steady current. Thus, the equation for $\mathbf{B}$ can be written

$$\nabla \times \mathbf{B} = \mathbf{J},$$

in which $\mathbf{J}$ would have units such as amperes per unit area for a planar current or amperes per unit volume for a three-dimensional current. Given the current $\mathbf{J}$, it is possible to solve for the field $\mathbf{B}$. Several cases are described later.

FLUID FLOW   Hydrodynamics involves the study of the flow of fluids such as water or oil. Consider a fluid with density $\rho(x, y, z)$ that flows with velocity $\mathbf{v}$ in a region of space. If there are no sources or sinks in the region, the net outflow through a surface enclosing a volume of space must be equal to the amount of inflow. In terms of the integrals previously discussed in this chapter, the two flows are

$$\overbrace{-\int\int_S \rho\mathbf{v}\cdot\mathbf{N}\,dS}^{\text{OUTFLOW}} = \overbrace{\int\int\int_V \frac{\partial\rho}{\partial t}dV}^{\text{INFLOW}}\,.$$

Applying the divergence theorem leads to the result

$$\int\int\int_V \left[\nabla\cdot(\rho\mathbf{v}) + \frac{\partial\rho}{\partial t}\right]dV = 0,$$

where the equation must hold for any region V, even an infinitesimal volume. Because the integrand must be zero, the equation states a law of conservation of matter. The relationship is the *continuity equation* for fluids

$$\nabla\cdot(\rho\mathbf{v}) + \frac{\partial\rho}{\partial t} = 0. \tag{12.80}$$

In the special case of an incompressible fluid, $\partial\rho/\partial t = 0$, and the fluid flow is governed by the simple equation

$$\nabla\cdot(\rho\mathbf{v}) = 0.$$

☐ **EXAMPLE 12.19**   *Conservative Forces and Fluid Flow*

Consider the flow of an ideal fluid in 2D steady flow with velocity $\mathbf{v}$. For irrotational flow,

$$\mathbf{v} = \nabla u(x, y),$$

where $u(x, y)$ is the scalar potential.

As an example, let the potential be

$$u(x, y) = xy$$

so that

$$\mathbf{v} = y\mathbf{i} + x\mathbf{j}.$$

It is easily verified that the vorticity is zero since $\nabla \times \mathbf{v} = 0$.

The curves $u(x, y) = $ constant are called the *equipotential curves* and the gradient defining the velocity vector is everywhere perpendicular to these curves of constant potential. The tangent lines to the velocity curve have the equation

$$\frac{dy}{dx} = \frac{v_y}{v_x} = \frac{x}{y},$$

so the equation of the lines is

$$x^2 - y^2 = c.$$

These lines are called *streamlines*, or *flow lines*. They represent the lines along which the fluid is carried. Figure 12.14 shows the equipotentials and field lines for the problem. This flow field was plotted by the M-file EXFFLOW.M contained on the disk accompanying this textbook.

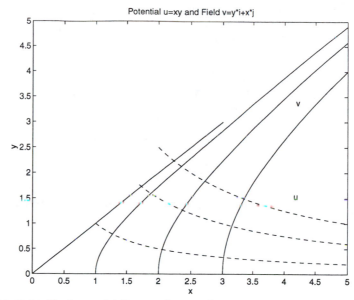

FIGURE 12.14 *Equipotential lines and streamlines*

☐

ELECTRICAL
AND MAGNETIC
FIELDS

One of the most elegant applications of vector field theory resulted in the statement of Maxwell's equations in vector terms. These equations are presented in both differential and integral form in Table 12.5.[6]

TABLE 12.5 *Maxwell's equations*

Differential Form	Integral Form
$\nabla \times \mathbf{E} = -\dfrac{\partial \mathbf{B}}{\partial t}$	$\oint_C \mathbf{E} \cdot d\mathbf{l} = \dfrac{\partial}{\partial t} \int\!\!\int_S \mathbf{B} \cdot d\mathbf{S}$
$\nabla \cdot \mathbf{D} = \rho$	$\int\!\!\int_S \mathbf{D} \cdot d\mathbf{S} = \int\!\!\int\!\!\int_V \rho \, dV$
$\nabla \times \mathbf{H} = \mathbf{J} + \dfrac{\partial \mathbf{D}}{\partial t}$	$\oint_C \mathbf{H} \cdot d\mathbf{l} = \dfrac{\partial}{\partial t} \int\!\!\int_S \mathbf{D} \cdot d\mathbf{S} + \int\!\!\int_S \mathbf{J} \cdot d\mathbf{S}$
$\nabla \cdot \mathbf{B} = 0$	$\int\!\!\int_S \mathbf{B} \cdot d\mathbf{S} = 0$

[6]The textbook by Plonsey and Collin listed in the Annotated Bibliography for this chapter discusses Maxwell's equations in detail.

In the integral equations, $d\mathbf{l}$ is traditionally used to indicate the vector direction along the curve C that is the boundary of the area over which the surface integrals are computed. Those integrals represent the total flux of the field over the area.

In general, each vector in Maxwell's equations is a function of space and time. The fundamental *electromagnetic fields* are the electric field strength $\mathbf{E}$ in volts per meter and the magnetic flux density $\mathbf{B}$ measured in webers per square meter, or teslas in SI units. The electric flux density $\mathbf{D}$ measured in coulombs per square meter and the magnetic field strength $\mathbf{H}$ measured in amperes per meter are typically used to describe the fields inside of a material.

In this chapter, Maxwell's equations are investigated for electrostatic and magnetostatic fields. Chapter 13 presents time-varying electromagnetic fields and the wave equation that derives from Maxwell's equations.

CONSTITUTIVE RELATIONS

The equations that describe the sources and the environment in which the electromagnetic fields occur are called the *constitutive relations*. These additional equations are necessary to solve Maxwell's equations for specific cases.

In a vacuum, the relationships between the fields are

$$\mathbf{D} = \epsilon_0 \mathbf{E},$$
$$\mathbf{H} = \frac{\mathbf{B}}{\mu_0},$$

where $\epsilon_0 = (1/36\pi) \times 10^{-9}$ farad/meter is the permittivity and the constant $\mu_0 = 4\pi \times 10^{-7}$ henry/meter is the permeability. In free space, these values are constant, and the result

$$c = (\mu_0 \epsilon_0)^{-1/2} = 3 \times 10^8 \text{ meters/second}$$

is the speed of light in vacuum. In general, the permittivity ϵ and permeability μ can be a function of many parameters including position, time, or the frequency of an applied sinusoidal wave.

Maxwell's equations can be simplified and rewritten for convenience in many cases. For example, taking the divergence of the equation for $\mathbf{H}$ and recognizing that $\nabla \cdot (\nabla \times \mathbf{H}) = 0$ yields the continuity equation

$$\nabla \cdot \mathbf{J} = -\frac{\partial}{\partial t} \nabla \cdot \mathbf{D} = -\frac{\partial \rho}{\partial t}.$$

The divergence theorem provides the integral relationship

$$\int\int_S \mathbf{J} \cdot d\mathbf{S} = -\frac{\partial}{\partial t} \int\int\int_V \rho \, dV.$$

ELECTROSTATIC
CASE
Static fields have no time variation, so Maxwell's equations predict

$$\nabla \times \mathbf{E} = 0,$$
$$\nabla \cdot \mathbf{D} = \epsilon_0 \nabla \cdot \mathbf{E} = \rho,$$
$$\nabla \times \mathbf{H} = \nabla \times \frac{\mathbf{B}}{\mu_0} = \mathbf{J},$$
$$\nabla \cdot \mathbf{B} = 0$$

for these fields in vacuum. Notice that the electric and magnetic fields are independent for static fields.

The E Field The electrostatic equations lead to a number of important conclusions. For example, since the curl of **E** is zero, this irrotational field can be represented as the gradient of a scalar, so that

$$\mathbf{E}(x, y, z) = -\nabla \Phi(x, y, z). \tag{12.81}$$

Using the fact that **E** is derived from a scalar potential, we can form a differential equation for the potential or calculate the potential from a line integral. Restricting the discussion to free space by writing $\mathbf{D} = \epsilon_0 \mathbf{E}$ and combining the equation for divergence of **D** with Equation 12.81 yields

$$\nabla^2 \Phi(x, y, z) = -\frac{\rho(x, y, z)}{\epsilon_0},$$

which is known as *Poisson's equation*.

The line integral for **E** is formed by noting that if $d\mathbf{l} = dx\,\mathbf{i} + dy\,\mathbf{j} + dz\,\mathbf{k}$, then

$$\mathbf{E} \cdot d\mathbf{l} = -\left(\frac{\partial \Phi}{\partial x}\,dx + \frac{\partial \Phi}{\partial y}\,dy + \frac{\partial \Phi}{\partial z}\,dz\right) = -d\Phi.$$

The electrostatic potential difference V_{ab} between two points in space is thus given by the line integral of **E** as

$$V_{ab} = \Phi_a - \Phi_b = -\int_b^a \mathbf{E} \cdot d\mathbf{l}. \tag{12.82}$$

Kirchhoff's voltage law states that the sum of the voltage around a closed loop is zero. This law can be expressed in terms of a line integral as

$$\oint_C \mathbf{E} \cdot d\mathbf{l} = 0$$

by letting $a = b$ in Equation 12.82 to form a closed curve. Can you determine Kirchhoff's current law in vector form from Maxwell's equations in the absence of time varying or magnetic fields? The law states that the sum of all currents flowing out of a junction is equal to zero.

Ampere's Circuital Law   Ampere's circuital law relates the magnetic field to the current that caused the field. In free space, the law can be derived by writing

$$\int\int_S (\nabla \times \mathbf{B}) \cdot d\mathbf{S} = \mu_0 \int\int_S \mathbf{J} \cdot d\mathbf{S} = \oint_C \mathbf{B} \cdot d\mathbf{l},$$

where the surface S is bounded by the closed contour C. Then, applying Stokes' theorem to the equation $\nabla \times \mathbf{B} = \mu_0 \mathbf{J}$ leads to Ampere's circuital law

$$\oint_C \mathbf{B} \cdot d\mathbf{l} = \mu_0 I \quad \text{amperes.} \qquad (12.83)$$

In this equation $I = \int\int_S \mathbf{J} \cdot d\mathbf{S}$ is the current flowing through an area bounded by the curve C. Ampere's law is convenient for solving for the field $\mathbf{B}$ in problems with appropriate symmetry.

As you can show in Problem 12.28, the magnetostatic field due to an infinitely long wire carrying a current of I amperes is

$$\mathbf{B} = \mathbf{a}_\phi \frac{\mu_0 I}{2\pi r},$$

where r is the radial distance from the wire. The field becomes infinite as r approaches zero since the wire has no diameter in this ideal representation. Problem 12.28 also treats the more practical case of a coaxial cable in which the inner conductor has a nonzero radius.

The* B *Field   Maxwell's equation

$$\nabla \cdot \mathbf{B} = 0$$

indicates that $\mathbf{B}$ is a *solenoidal* field, and thus $\mathbf{B}$ can be derived as the curl of a vector field. Using the divergence theorem, another conclusion is that the magnetic flux passing through any closed surface is zero. This means that the flux lines of $\mathbf{B}$ are always continuous and form closed loops.

The fact that the divergence of $\mathbf{B}$ is zero is often cited to prove that magnetic monopoles do not exist.[7] All magnets have both a north and south pole and the $\mathbf{B}$ field is continuous through the magnet. There is no change in this equation even if the fields are time varying.

REINFORCEMENT EXERCISES AND EXPLORATION PROBLEMS

In these problems, do the computations by hand unless otherwise indicated, and then check those that yield numerical or symbolic results with MATLAB.

---

[7]Every so often a researcher claims to have found one. None of the claims has been validated.

REINFORCEMENT EXERCISES

P12.1. Integrals   Evaluate the following integrals:

TABLE 12.6 *Table of Integrals*

Type	Integral
Power	$\displaystyle\int u^n\, du$
Log	$\displaystyle\int \frac{du}{u}$
Exponential	$\displaystyle\int e^{au}\, du$
Trigonometric	$\displaystyle\int \cos u\, du$
	$\displaystyle\int \sin u\, du$
Algebraic	$\displaystyle\int \frac{du}{\sqrt{a^2 - u^2}}$
	$\displaystyle\int \frac{du}{a^2 + u^2}$

P12.2. Regions of integration   Integrate

$$\int\int_S e^{x^2}\, dS,$$

where S is the triangle with vertices at $(0,0), (1,0)$, and $(1,1)$, as shown in Figure 12.6. Select the regions of integration as follows:

    a. Integrate over the x-limits first.

    b. Integrate over the y-limits first.

Can you show that the integral in Part a is the same as that in Part b?
Hint: Expand the integrand in a Taylor series.

P12.3. Double integrals   Using double integrals, solve the following problems:

    a. Find the area of the region bounded by the straight line $y = x$ and the curve $y = x^2$.

    b. Find the volume under the surface

$$x^2 + y^2 + z = 9, \quad x \geq 0,\, y \geq 0,$$

and between the planes $y = 1$ and $x = 2$.

P12.4. Double integration Integrate the function $z = x^2 + y^2$ over the circle $x^2 + y^2 = 1$.

P12.5. Trigonometric substitution   Integrate the function

$$\int_0^{\pi/2} (\sin^2 \theta \cos^2 \theta + \frac{1}{3} \cos^4 \theta) \, d\theta$$

as described in Example 12.3.

P12.6. Volume integral Compute the volume of a sphere of radius a in rectangular and spherical coordinates.

P12.7. Volume of a region Evaluate the integral

$$\int \int \int \frac{1}{x^2 + y^2 + z^2} \, dx dy dz$$

in the region between concentric spheres centered at the origin such that $3 \leq r \leq 6$.

P12.8. Mass and centroid Let $\rho(x, y)$ be the density of a thin plate (in units of mass per unit area) in the xy-plane. Assuming that the density is a continuous function, the mass m can be computed as

$$m = \int \int \rho(x, y) \, dA,$$

and the *centroid* or center of mass has the coordinates

$$\bar{x} = \frac{1}{m} \int \int x \rho(x, y) \, dA,$$

$$\bar{y} = \frac{1}{m} \int \int y \rho(x, y) \, dA.$$

Calculate the mass and centroid of a quarter-circle disk of radius a in the first quadrant with density

$$\rho(x, y) = k \sqrt{x^2 + y^2}.$$

P12.9. Jacobians Find the Jacobian for the following transformations:

    a. Translation $x = u + a$, $y = v + b$;

    b. Rotation $x = u \cos \theta - v \sin \theta$, $y = u \sin \theta + v \cos \theta$;

    c. Expansion $x = au$; $y = bv$.

Explain the geometric significance of the results.

P12.10. Jacobian in spherical coordinates   Given the transformation

$$x = r \cos \theta \sin \phi, \qquad y = r \sin \theta \sin \phi, \qquad z = r \cos \phi,$$

find the Jacobian

$$J(r, \theta, \phi) = \frac{\partial(x, y, z)}{\partial(r, \theta, \phi)}.$$

P12.11. Polar coordinates Integrate $f(x, y) = x^2$ over the region enclosed by the curve

$$r = (1 - \cos \theta).$$

P12.12. Tangent and normal vectors   Suppose $\mathbf{F}(t)$ is a nonzero differentiable function of constant length. Prove that $\mathbf{F}(t)$ is orthogonal to the derivative vector $\mathbf{F}'(t)$. Then, determine the unit tangent vector $\mathbf{T}$ and unit normal vector $\mathbf{N}$ to the graph of a smooth function $\mathbf{R}(t)$. Describe the tangent vector in terms of the motion of a particle along the curve defined by $\mathbf{R}(t)$.

P12.13. Space curves   Using the result of Problem 12.12, find the unit tangent vector and a unit normal vector for the space curve

$$x = 3\cos t, \quad y = 3\sin t, \quad z = 4t.$$

P12.14. 2D Fourier transform   Compute the 2D Fourier transform of the function

$$f(x, y) = \begin{cases} 1, & -1 < x < 1,\, -1 < y < 1, \\ 0, & \text{otherwise.} \end{cases}$$

P12.15. Probability distribution   Let the distribution

$$f(x) = \frac{1}{180} e^{-x/180}, \quad 0 \le x \le \infty,$$

describe the hours of service of electronic components. What is the probability that a component will operate correctly for

    a. Less than 90 hours?

    b. More than 360 hours?

P12.16. Trapezoidal integration   The trapezoidal integration formula is simple and can be used to obtain a reasonable estimate of the value of a definite integral in many cases.

    a. Derive the trapezoidal integration formula of Equation 12.46.

    b. Evaluate the (error) function

$$\mathrm{erf}(x) = \frac{2}{\sqrt{\pi}} \int_0^x e^{-\zeta^2}\, d\zeta$$

    for $x = 1$ using five points with the trapezoidal rule.

P12.17. Special functions   Compare the result of Problem 12.16 to that obtained by the MATLAB command **erf**.

P12.18. Parametric form of line integral   Evaluate the line integral

$$\int_C x e^{yz}\, dz$$

over the curve $x = t$, $y = t$, $z = -t$ for $1 \le t \le 2$.

P12.19. Line integral   Evaluate the line integral of Example 12.10,

$$I = \int_C x^2 y\, dx - (x + y)\, dy,$$

where C is the straight line between $(0, 0)$ and $(2, 1)$. Write the curve in terms of parametric equations and apply Equation 12.55.

Reinforcement Exercises and Exploration Problems

P12.20. Vector line integral   The integral

$$\int_C f(x, y, z)\, d\mathbf{r}$$

is the vector sum of $f\, d\mathbf{r}$ along C when a scalar function $f(x, y, z)$ exists in the region of space where the curve is defined. Let a scalar function be

$$f(x, y, z) = x^3 y + 2y$$

and compute the line integral between the points $(1, 1)$ and $(2, 4)$ along the paths:

a. $y = x^2$;

b. C, a straight line between the points.

P12.21. Surface integral   Given the scalar function $f = x^2 z$, evaluate the surface interval of f over the right circular cylinder $x^2 + y^2$ of height h.

P12.22. Work   Determine the work in carrying a rock of mass m up an ideal mountain of height h with shape

$$y = -x^2 + h, \qquad -h^{1/2} \le x \le h^{1/2}.$$

P12.23. Domains   Which of the following domains are simply connected?

a. The whole xy plane;

b. The annular region between two circles;

c. The interior of a circle minus the center;

d. The exterior of a sphere;

e. The space between two concentric spheres.

P12.24. Area by Green's theorem   Compute the area of an ellipse using Green's theorem in the plane. The major axis has length a and the minor axis has length b.

P12.25. Conservative force   Let a particle move along the curve

$$y = 1 + x^2$$

subject to an attractive force of magnitude $F = kr$ directed toward the origin. Determine the work in moving the particle along the curve from $(0, 1)$ to $(1, 2)$.

P12.26. Flux   Compute the flux due to the field $\mathbf{F} = v_0\, \mathbf{k}$ over the surface of a hemisphere of radius a centered at the origin

$$z = \sqrt{a^2 - x^2 - y^2}.$$

Discuss the meaning of the result if the field represents the velocity field of a fluid that is flowing in the z direction.

P12.27. Gauss' flux theorem   Let S be a closed surface and let $\mathbf{R}$ denote the position vector of a point measured from the origin $\mathbf{0} = (0, 0, 0)$. Prove that

$$\int\int_S \frac{\mathbf{N} \cdot \mathbf{R}}{R^3}\, dS$$

is equal to the following values:

a. Zero if $\mathbf{0}$ lies outside S;

b. 4π if $\mathbf{0}$ lies inside S.

Thus, the total flux of an inverse-square vector field due to a point source at the origin can be calculated by integrating over the surface enclosing the source.

P12.28. Ampere's law   Find the magnetostatic field for the following cases:

a. Compute the **B** field as a function of distance r measured perpendicular to an infinitely long wire carrying a direct current I.

b. Compute the **B** fields inside and outside a coaxial cable if the radius of the solid inner conductor is a and the cylindrical shell has inner radius b and thickness $c - b$.

P12.29. Lorentz force   The Lorentz force law states that the force exerted on a charged particle moving with velocity **v** is

$$\mathbf{F} = q(\mathbf{E} + \mathbf{v} \times \mathbf{B}).$$

Find the velocity and position of a particle of charge q, mass m, initial position $\mathbf{r}_0$, and initial velocity $\mathbf{v}_0$ which enters a uniform field $\mathbf{E} = E_0\mathbf{k}$;

P12.30. MATLAB arclength   You are to buy cable for a single-span suspension bridge that has a cable hanging in the form of a parabola $y = kx^2$. Write a MATLAB program to compute the length of the cable for the bridge with the characteristics:

a. The distance between the supports is 1 mile.

b. The height of the cable (relative to its lowest point) is 1250 feet.

Hint: Try a simple test case to check the integration.

P12.31. MATLAB symbolic arc length   Write a MATLAB program to compute the arc length of a function. If the integral cannot be solved symbolically, use a numerical approximation. Test the program with the functions

a. $y = x^2$;

b. $y = \sin x$ for $0 \le x \le \pi$.

P12.32. MATLAB normal distribution   The normal probability distribution is defined as

$$p(x) = \frac{1}{\sqrt{2\pi\sigma^2}} e^{-(x-\mu)^2/2\sigma^2},$$

where μ is the mean value and σ is called the standard deviation. Write a MATLAB function that computes the probability in the interval $[a, b]$. What is the probability that a value lies within one standard deviation from the origin if $\mu = 0$ and $\sigma = 1$?

EXPLORATION PROBLEMS

P12.33. Integral transformations   Find the area of the ellipse

$$\frac{x^2}{a^2} + \frac{y^2}{b^2} = 1$$

by transforming a double integral to integrate over a circle.

P12.34. Parameterized line integrals Let a path C be the helix defined as

$$c(t) = (\cos t, \sin t, t), \qquad 0 \le t \le 2\pi.$$

Evaluate the line integral over C for the functions:

    a. $\mathbf{F} = x\mathbf{i} + y\mathbf{j} + z\mathbf{k}$;

    b. $f(x, y, z) = x^2 + y^2 + z^2$.

P12.35. Representation of line integrals If the function $\mathbf{F}$ is defined as

$$\mathbf{F} = F_1(x, y, z)\,\mathbf{i} + F_2(x, y, z)\,\mathbf{j} + F_3(x, y, z)\,\mathbf{k},$$

show that

$$\int_C \mathbf{F} \cdot d\mathbf{R} = \int_C F_1\,dx + F_2\,dy + F_3\,dz.$$

P12.36. Independence of path Assume that $\phi(x, y)$ and its first partial derivatives are continuous for all (x, y) in a region of the plane and that $\mathbf{F} = \nabla\phi$. Show that

$$\int_C \mathbf{F} \cdot d\mathbf{R}$$

is independent of path in this region.

Hint: Suppose that C is a smooth curve parameterized by $x = x(t)$ and $y = y(t)$ and use the chain rule.

P12.37. Exact differential Let $\mathbf{F}$ be the vector field

$$\mathbf{F} = F_1(x, y, z)\,\mathbf{i} + F_2(x, y, z)\,\mathbf{j} + F_3(x, y, z)\,\mathbf{k}$$

with continuous partial derivatives. Suppose that

$$F_1\,dx + F_2\,dy + F_3\,dz = d\phi$$

is an exact differential. Show that $\nabla \times \mathbf{F} = 0$.

Hint: The mixed partial derivatives are equal.

P12.38. Frictional force Assume the particle in Problem 12.25 is subject to a frictional force along the path. If the coefficient of friction is μ, compute the work done moving the particle from $(0, 1)$ to $(1, 2)$, as before. Describe the physical meaning of the terms in the expression for the work.

Hint: The frictional force is normal to the path and presses the particle against the path.

P12.39. Work Determine the work done in moving a particle from $(0, 4, 0)$ to $(0, 4, 3)$ under the influence of an inverse square force field

$$\mathbf{F}(x, y, z) = \frac{\alpha\mathbf{R}}{R^3}.$$

P12.40. Report on electrostatics Write a report on an interesting application of electrostatic fields. A few topics are

    a. Van de Graaff generator;

    b. Triboelectricity;

    c. Electrostatic separation;

    d. Xerography.

These and other topics are discussed in the textbook *Applied Electromagnetism* by Liang Chi Shen and Jin Au Kong, published by PWS Engineering, Boston, MA, 1987.

P12.41. Relaxation time Assume a conducting body has a permittivity ϵ and a conductivity σ. Using $\mathbf{J} = \sigma\mathbf{E}$ and the divergence relationships, determine how charge density will change with time if an initial charge density $\rho_0(x, y, z)$ at $t = 0$ is placed within a conduction body. Compute the *relaxation time*, which is the elapsed time required for the initial charge distribution to decay to $1/e$ of its initial value.

P12.42. MATLAB integration Compare the numerical results with the exact solution when integrating

$$\int_0^\pi \sin x \, dx$$

using MATLAB's trapezoidal integration routine and the quadrature routines.

P12.43. MATLAB integration Using MATLAB's routines for integration, integrate the functions

    a. $\sqrt{x}$

    b. $\log x$

over the interval $0 \le 0 \le 1$. Compute the error in the result.

P12.44. MATLAB 2D FFT Plot the Fourier transform of Problem 12.14.

ANNOTATED BIBLIOGRAPHY

1. Apostol, Tom M., *Mathematical Analysis*, Addison-Wesley, Reading, MA, 1974. *A rigorous treatment of calculus, including multivariate calculus. Proofs are given for many of the theorems pertaining to multiple integrals and line integrals. For example, the proof of the change of variables formula for multiple integrals is presented in this textbook.*

2. Brigham, E. Oran, *The Fast Fourier Transform and Its Applications*, Prentice Hall, Englewood Cliffs, NJ, 1988. *A very readable treatment of Fourier techniques, including the 2D FFT. There is also an extensive bibliography.*

3. Dahlquist, Germund, and A. Björck, *Numerical Methods*, Prentice Hall, Inc., Englewood Cliffs, NJ, 1974. *A useful reference that covers important techniques in numerical computing.*

4. Forsythe, George E., M. A. Malcolm, and C. B. Moler, *Computer Methods for Mathematical Computations*, Prentice Hall, Englewood Cliffs, NJ,1977. *A classic text that describes the MATLAB integration algorithms in detail.*

5. Kaplan, Wilfred, *Advanced Calculus, Fourth edition*, Addison-Wesley, Reading, MA, 1991. *An excellent treatment of advanced calculus.*

6. Kellogg, Oliver Dimon, *Foundations of Potential Theory*, Dover Publications, New York, 1953. *Although this small text was first published in 1929, it still remains one of the best introductions to potential functions as used in physical applications. Newtonian mechanics and electrostatics are particularly well covered, as are certain advanced applications.*

7. Plonsey, Robert and R. E. Collin, *Principles and Application of Electromagnetic Fields*, McGraw-Hill, Inc.,New York, 1961. *An excellent treatment of introductory electromagnetic theory.*

ANSWERS

P12.1. Integrals

TABLE 12.7 *Short Table of Integrals*

Type	Integral
Power	$\int u^n\, du = \dfrac{u^{n+1}}{n+1} + C \quad n \neq -1$
Log	$\int \dfrac{du}{u} = \ln|u| + C$
Exponential	$\int e^{au}\, du = \dfrac{e^{au}}{a} + C$
Trigonometric	$\int \cos u\, du = \sin u + C$
	$\int \sin u\, du = -\cos u + C$
Algebraic	$\int \dfrac{du}{\sqrt{a^2 - u^2}} = \sin^{-1}\dfrac{u}{a} + C$
	$\int \dfrac{du}{a^2 + u^2} = \dfrac{1}{a}\tan^{-1}\dfrac{u}{a} + C$

Many other integrals can be reduced to one of the forms shown in the table by substitutions or integration by parts.

P12.2. Regions of integration Doing the x integration first yields

$$\int_0^1 \int_y^1 e^{x^2} \, dx \, dy.$$

This is not promising, but a series expansion could be used to compute the integral. Trying the other order of integration yields the result

$$\int_0^1 \int_0^x e^{x^2} \, dy \, dx = \int_0^1 x e^{x^2} \, dx = \frac{1}{2}[e-1].$$

P12.3. Double integrals

a. The area of the region bounded by the straight line $y = x$ and the curve $y = x^2$ can be computed as

$$\int_0^1 \int_{x^2}^x dy \, dx = \int_0^1 (x - x^2) \, dx = \frac{1}{6};$$

b. The volume under the surface $x^2 + y^2 + z = 9$ and between the regions defined by $0 \le x \le 2$ and $0 \le y \le 1$ is

$$\int_0^2 \int_0^1 (9 - x^2 - y^2) \, dy \, dx = 44/3.$$

Integrating in the opposite order yields the same result.

P12.4. Double integration Integrating the function $z = x^2 + y^2$ over the circle defined as $y = \pm\sqrt{1 - x^2}$ leads to the integral

$$\int_{-1}^1 \int_{-\sqrt{1-x^2}}^{\sqrt{1-x^2}} (x^2 + y^2) \, dy \, dx = \frac{\pi}{2}.$$

P12.5. Trigonometric substitution To evaluate

$$\int_0^{\pi/2} \sin^2 \theta \cos^2 \theta \, d\theta$$

substitute

$$\sin^2 \theta = \frac{1 - \cos 2\theta}{2} \quad \text{and} \quad \cos^2 \theta = \frac{1 + \cos 2\theta}{2},$$

which reduces the integral to the form

$$\frac{1}{4} \int_0^{\pi/2} d\theta - \frac{1}{4} \int_0^{\pi/2} \cos^2 2\theta \, d\theta.$$

Substituting $\cos^2 2\theta = (1 + \cos 4\theta)/2$ in the second integral and integrating yields $\pi/16$. The second integral,

$$\int_0^{\pi/2} \frac{1}{3} \cos^4 \theta \, d\theta,$$

can be evaluated by expanding $\cos^4 \theta$ using the double angle formula for $\cos^2 \theta$. The result is again $\pi/16$, and the original integral has the value $\pi/16 + \pi/16 = \pi/8$.

Answers

P12.6. Volume integral To determine the volume of the sphere $x^2 + y^2 + z^2 \leq a^2$, the regions of integration in rectangular coordinates could be

$$-a \leq x \leq a, \qquad -\sqrt{a^2 - x^2} \leq y \leq \sqrt{a^2 - x^2}$$

and

$$-\sqrt{a^2 - x^2 - y^2} \leq z \leq \sqrt{a^2 - x^2 - y^2}.$$

Integrating, $\int \int \int dz\,dy\,dx = 4\pi a^3/3$.

In spherical coordinates $dV = r^2 \sin \phi\, dr\, d\theta\, d\phi$, so

$$\int \int \int dz\,dy\,dx = \int_0^\pi \int_0^{2\pi} \int_0^a r^2 \sin \phi\, dr\, d\theta\, d\phi = \frac{4}{3}\pi a^3.$$

P12.7. Mass of a region In spherical coordinates, the integral becomes

$$\int_0^{2\pi} \int_0^{2\pi} \int_3^6 \frac{1}{r^2} r^2 \sin \phi\, dr\, d\phi\, d\theta = 4\pi \int_3^6 dr = 12\pi.$$

The integrals over the angles are 4π for this symmetric region.

P12.8. Mass and centroid In polar coordinates $\rho(r, \theta) = kr$. The mass of the quarter circle is

$$m = \int_0^{\pi/2} \int_0^a kr^2\, dr\, d\theta = \frac{k\pi a^3}{6}.$$

Because of symmetry, $\bar{x} = \bar{y}$ along the line $y = x$ so that

$$\bar{x} = \bar{y} = \frac{1}{m} \int \int y\rho\, dA = \frac{1}{m} \int_0^{\pi/2} \int_0^a kr^3 \sin \theta\, dr\, d\theta = \frac{3a}{2\pi}.$$

P12.9. Jacobians For the transformations in the plane, the Jacobians are

    a. Translation $J = 1$.

    b. Rotation $J = 1$.

    c. Expansion $J = ab$.

Geometrically, the translation and rotation do not change the area of a region being transformed. The transformation in Part c changes the area of a transformed region if $ab \neq 1$.

P12.10. Jacobian in spherical coordinates Writing x, y, z in spherical coordinates as

$$x = r \cos \theta \sin \phi, \qquad y = r \sin \theta \sin \phi, \qquad z = r \cos \phi$$

leads to the Jacobian

$$\frac{\partial(x, y, z)}{\partial(r, \theta, \phi)} = \begin{vmatrix} \sin \phi \cos \theta & -r \sin \phi \sin \theta & r \cos \phi \cos \theta \\ \sin \phi \sin \theta & r \sin \phi \cos \theta & r \cos \phi \sin \theta \\ \cos \phi & 0 & -r \sin \phi \end{vmatrix}$$

$$= -r^2 \sin \phi$$

P12.11. Polar coordinates In polar coordinates, the integrand is $g(r, \theta) = r^2 \cos^2 \theta$ and the integral becomes

$$\int_0^{2\pi} \int_0^{1-\cos\theta} r^3 \cos^2 \theta \, dr \, d\theta.$$

Integrating with respect to r leads to the integral

$$\frac{1}{4} \int_0^{2\pi} (1 - \cos\theta)^4 \cos^2 \theta \, d\theta = \frac{49\pi}{32}.$$

P12.12. Tangent and normal vectors Suppose $\mathbf{F}(t)$ is a nonzero differentiable function of constant length. Since $||\mathbf{F}(t)||$ is constant, so is the square $||\mathbf{F}(t)||^2 = \mathbf{F}(t) \cdot \mathbf{F}(t)$. The derivative is zero, so that

$$0 = \frac{d}{dt}[\mathbf{F}(t) \cdot \mathbf{F}(t)] = \mathbf{F}'(t) \cdot \mathbf{F}(t) + \mathbf{F}(t) \cdot \mathbf{F}'(t) = 2\mathbf{F}(t) \cdot \mathbf{F}'(t).$$

Thus, $\mathbf{F}(t) \cdot \mathbf{F}'(t) = 0$ and $\mathbf{F}'(t)$ is orthogonal to $\mathbf{F}(t)$. Since $\mathbf{R}'(t)$ is a tangent vector to the graph of $\mathbf{R}(t)$ at the point corresponding to $\mathbf{R}(t)$, a unit tangent vector is

$$\mathbf{T}(t) = \frac{\mathbf{R}'(t)}{||\mathbf{R}'(t)||}.$$

The tangent vector is not zero since the curve defined by $\mathbf{R}(t)$ is assumed to be smooth. The unit normal vector will be perpendicular to the tangent vector at each point on the curve defined by $\mathbf{R}(t)$. There are an infinite number of such vectors but we choose the vector $\mathbf{N}$ that is defined as

$$\mathbf{N} = \frac{\mathbf{T}'(t)}{||\mathbf{T}'(t)||}.$$

Since $\mathbf{T}(t)$ is a vector with length 1 for all t, it follows that $\mathbf{T}(t)$ is orthogonal to $\mathbf{T}'(t)$ for all t. Thus, $\mathbf{N}(t)$ is perpendicular to $\mathbf{T}(t)$ at every point. The tangent vector indicates the direction of the motion of the time t. Writing $\mathbf{v} = v(t)\mathbf{T}(t)$ for the velocity shows that $||\mathbf{T}'(t)||$ represents the speed at that instant.

P12.13. Space curves Letting $t = z/4$, the curve is seen to be a circular helix with the radius $x^2 + y^2 = 9$. The position vector is

$$\mathbf{R}(t) = 3\cos t\,\mathbf{i} + 3\sin t\,\mathbf{j} + 4t\,\mathbf{j}.$$

Therefore, the tangent vector is

$$\mathbf{T}(t) = \frac{\mathbf{R}'(t)}{||\mathbf{R}'(t)||} = -\frac{3}{5}\sin t\,\mathbf{i} + \frac{3}{5}\cos t\,\mathbf{j} + \frac{4}{5}\mathbf{k}.$$

The normal vector is

$$\mathbf{N} = \frac{\mathbf{T}'(t)}{||\mathbf{T}'(t)||} = -\cos t\,\mathbf{i} - \sin t\,\mathbf{j}.$$

P12.14. 2D Fourier transform For the function $f(x, y) = 1$ only in the region of the xy-plane such that $-1 < x < 1$, $-1 < y < 1$, the Fourier transform is

$$F(u, v) = \int_{-1}^1 e^{-i2\pi ux}\, dx \int_{-1}^1 e^{-i2\pi vy}\, dy = \frac{\sin(2\pi u)\sin(2\pi v)}{\pi^2 uv}.$$

P12.15. Probability distribution   The probability distribution is

$$F(x) = \int_0^x \frac{1}{180} e^{-x/180} \, dx = 1 - e^{-x/180}.$$

Thus,

    a. $P(x < 90) = F(90) = 1 - e^{-.5} = 0.393;$

    b. $P(x > 360) = 1 - F(360) = 0.135.$

P12.16. Trapezoidal integration   Using MATLAB **trapz** with five points between 0 and 1, the trapezoidal approximation to the integral is 0.83836777744120. This is correct to about two decimal places.

P12.17. Special functions   The result from the MATLAB command **erf** is 0.84270079294971.

P12.18. Parametric form of line integral

$$\int_C x e^{yz} \, dz = \int_1^2 t e^{-t^2} \frac{dz}{dt} \, dt = \frac{1}{2}(e^{-4} - e^{-1}).$$

P12.19. Line integral   By making the definitions $\mathbf{F} = x^2 y \, \mathbf{i} - (x+y) \, \mathbf{j}$ and $\mathbf{R} = 2t \, \mathbf{i} + t \, \mathbf{j}$ for $0 \le t \le 1$, $d\mathbf{R}/dt = 2\mathbf{i} + \mathbf{j}$ and $\mathbf{F} \cdot d\mathbf{R}/dt = 8t^3 - 3t$. Integrating leads to the result

$$\int_0^1 (8t^3 - 3t) \, dt = 1/2.$$

P12.20. Vector Line Integral   For the integral

$$\int_C (x^3 y + 2y) \, d\mathbf{r}$$

between the points $(1,1)$ and $(2,4)$:

    a. For $y = x^2$, $dy = 2x \, dx$ so that

$$\int_{1,1}^{2,4} (x^3 y + 2y)(dx \, \mathbf{i} + 2x \, dx \, \mathbf{j}) = \frac{91}{6} \mathbf{i} + \frac{359}{7} \mathbf{j};$$

    b. For $y = 3x - 2$ on a straight line between the points, $dy = 3 \, dx$ and

$$\int_{1,1}^{2,4} (x^3 y + 2y)(dx \, \mathbf{i} + dy \, \mathbf{j}) = \frac{161}{10} \mathbf{i} + \frac{483}{10} \mathbf{j};$$

P12.21. Surface Integral   The surface integral must be evaluated over three surfaces. On the curved surface, let $dS_1 = a \, d\theta dz$, $x = a \cos\theta$, and $z = z$ in polar coordinates, so that

$$\int\int_{S_1} x^2 z \, dS_1 = \int_0^h \int_0^{2\pi} (a \cos\theta)^2 \, z(a d\theta dz) = \frac{\pi a^3 h^2}{2}.$$

On the base S_2, the integral is zero since $z = 0$ there. On the upper surface, S_3, $z = h$ so that

$$\int\int_{S_3} x^2 z \, dS_3 = \int_0^{2\pi} \int_0^a (r \cos\theta)^2 h \, r dr d\theta = \frac{\pi a^4 h}{4}.$$

The integral over the whole surface is thus the sum of these integrals, and the value is

$$\frac{\pi a^3 h (2h + a)}{4}.$$

P12.22. Work The force on the rock as it is carried up the mountain is $\mathbf{F} = -mg\,\mathbf{j}$, and the integral of work is

$$-\int_{(h^{1/2},0)}^{(0,h)} (mg\,\mathbf{j}) \cdot d\mathbf{r} = -mg \int_0^h dy = -mgh$$

since $d\mathbf{r} = dx\,\mathbf{i} + dy\,\mathbf{j}$. The result is the change in potential energy for the rock.

P12.23. Domains The simply connected domains are the whole xy-plane, exterior of a sphere, and the space between two concentric spheres.

P12.24. Area by Green's Theorem From Green's theorem

$$A = \frac{1}{2} \oint (x\,dy - y\,dx).$$

For the ellipse $x^2/a^2 + y^2/b^2 = 1$. Let $x = a\cos t$ and $y = b\sin t$ so that $x' = -a\sin t$ and $y' = b\cos t$, and

$$A = \frac{1}{2} \int_0^{2\pi} (xy' - yx')\,dt = \pi ab.$$

P12.25. Conservative Force The tangential force along the path is $F_t = F\cos\alpha = kr\cos\alpha$. Using geometry, the various angles could be determined and the work integral calculated. However, it is easy to show that the force $\mathbf{F} = k\mathbf{r} = k(x\,\mathbf{i} + y\,\mathbf{j})$ is conservative so the work can be evaluated from the values at the endpoints. Let $\mathbf{F} = \nabla\Phi$ so that $\Phi = kr^2/2$. Thus, the work is

$$W = \int \nabla\Phi \cdot d\mathbf{r} = \frac{k}{2}(r^2)\Big|_1^5 = 2k.$$

P12.26. Flux To determine the flux on the spherical surface, we compute the normal as

$$\mathbf{N} = \frac{\mathbf{R}}{R} = \frac{x\,\mathbf{i} + y\,\mathbf{j} + z\,\mathbf{k}}{\sqrt{x^2 + y^2 + z^2}} = \frac{1}{a}(x\,\mathbf{i} + y\,\mathbf{j} + z\,\mathbf{k}).$$

Thus,

$$\int\int_S \mathbf{F} \cdot \mathbf{N}\,dS = \int\int_S \frac{v_0}{a} z\,dS.$$

Changing to spherical coordinates with $z = a\cos\phi$ and $dS = a^2\sin\phi\,d\phi d\theta$ leads to the result

$$\int\int_S \frac{v_0}{a} z\,dS = \frac{v_0}{a} \int_0^{2\pi} \int_0^{\pi/2} (a\cos\phi)(a^2\sin\phi)\,d\phi d\theta = \pi a^2 v_0.$$

The flux represents the rate of outflow of the fluid across the spherical surface. The units are cubic meters per second in SI units. Notice that the result is the rate of inflow of an incompressible fluid across the disk $x^2 + y^2 \le a^2$ representing the base of the hemisphere. Thus, the rate of inflow across the base is equal to the rate of outflow across the hemisphere, as expected.

P12.27. Gauss' flux theorem

    a. By the divergence theorem

$$\int\int_S \frac{\mathbf{N} \cdot \mathbf{R}}{R^3}\,dS = \int\int\int_V \nabla \cdot \frac{\mathbf{R}}{R^3}\,dV = 0,$$

    since $\mathbf{0}$ is outside of V so that $R \ne 0$ everywhere in V;

b. If **0** is inside S, surround **0** with a small sphere s of radius ϵ. Then, the conditions of Part a hold for the region between the surfaces, so that

$$\iint_{S+s} \frac{\mathbf{N} \cdot \mathbf{R}}{R^3} \, dS = \iint_{S} \frac{\mathbf{N} \cdot \mathbf{R}}{R^3} \, dS + \iint_{s} \frac{\mathbf{N} \cdot \mathbf{R}}{R^3} \, dS = \iiint_{\tau} \nabla \cdot \frac{\mathbf{R}}{R^3} \, dV = 0,$$

where τ is the region bounded by S and s. Thus,

$$\iint_{S} \frac{\mathbf{N} \cdot \mathbf{R}}{R^3} \, dS = -\iint_{s} \frac{\mathbf{N} \cdot \mathbf{R}}{R^3} \, dS.$$

On s, the normal extends into the region enclosed by s so that the normal is opposite to $\mathbf{R}$. At $R = \epsilon$, $\mathbf{N} = -\mathbf{R}/\epsilon$, which yields

$$\frac{\mathbf{N} \cdot \mathbf{R}}{R^3} = -\frac{\mathbf{R} \cdot \mathbf{R}}{\epsilon^4} = -\frac{1}{\epsilon^2}.$$

Finally,

$$\iint_{S} \frac{\mathbf{N} \cdot \mathbf{R}}{R^3} \, dS = -\iint_{s} \frac{\mathbf{N} \cdot \mathbf{R}}{R^3} \, dS = \iint_{s} \frac{1}{\epsilon^2} \, dS = \frac{4\pi\epsilon^2}{\epsilon^2} = 4\pi.$$

So the statement of Gauss' flux theorem concerning a source at the origin that yields an inverse-square field shows that the surface integral is 4π since **0** lies inside S.

P12.28. Ampere's Law There are four regions to be considered. Assuming that the coaxial cable is infinite in length, the field is directed in the θ direction only, using polar coordinates to describe a cross section of the cable. Let the inner conductor carry a current $I \, \mathbf{k}$ and the outer shell carry the return current $-I \, \mathbf{k}$. Using Ampere's law, we find the results

$$B_\theta = \begin{cases} \dfrac{\mu_0 I r}{2\pi a^2}, & 0 < r < a, \\[2mm] \dfrac{\mu_0 I}{2\pi r}, & a < r < b, \\[2mm] \dfrac{\mu_0 I}{2\pi r}\left(\dfrac{c^2 - r^2}{c^2 - b^2}\right), & b < r < c, \\[2mm] 0, & r > c. \end{cases}$$

P12.29. Lorentz Force Let $\mathbf{F} = qE_0 \, \mathbf{k}$. Assume the initial position and velocity of the particle is defined as $\mathbf{r}_0 = x_0 \, \mathbf{i} + y_0 \, \mathbf{j} + z_0 \, \mathbf{k}$ and $\mathbf{v}_0 = v_{0x} \, \mathbf{i} + v_{0y} \, \mathbf{j} + v_{0z} \, \mathbf{k}$. Applying Newton's law to the force $F_z = qE_0$, $dv_z/dt = (q/m)E_0$. Integrating and using the initial condition yields

$$v_z(t) = v_{0z} + \frac{qE_0}{m} t,$$

and the total velocity is

$$\mathbf{v}(t) = v_{0x} \, \mathbf{i} + v_{0y} \, \mathbf{j} + \left(v_{0z} + \frac{qE_0}{m} t\right) \mathbf{k}.$$

The position vector is

$$\mathbf{r}(t) = (v_{0x} t + x_0) \, \mathbf{i} + (v_{0y} t + y_0) \, \mathbf{j} + \left(v_{0z} t + z_0 + \frac{qE_0}{2m} t^2\right) \mathbf{k}.$$

Comment: MATLAB answers are on the disk accompanying this text.

13

PARTIAL DIFFERENTIAL EQUATIONS

PREVIEW_____

This chapter presents a brief introduction to partial differential equations. These equations define the motion or change of state of a system whose mathematical model depends on more than one independent variable. After an introduction to partial differential equations, we will study Laplace's equation of equilibrium, the diffusion equation of heat, and the wave equation as examples. As we shall see, this chapter builds on the discussions of both ordinary differential equations in Chapters 5 and 6, and vector calculus in previous chapters.

INTRODUCTION TO PARTIAL DIFFERENTIAL EQUATIONS

Partial derivatives are useful for studying phenomena in which the dependent variable is a function of two or more independent variables. A second-order partial differential equation involving two independent variables x and y can be written as

$$A\frac{\partial^2 u}{\partial x^2} + B\frac{\partial^2 u}{\partial x \partial y} + C\frac{\partial^2 u}{\partial y^2} + D\frac{\partial u}{\partial x} + E\frac{\partial u}{\partial y} + Fu = G(x,y), \qquad \textbf{(13.1)}$$

where $u = u(x,y)$ is the unknown function, G is a known function of the variables, and the coefficients A, B, C, D, E, F may depend on x and y. Equations written in this form are classified as follows:

$$\begin{aligned}
\text{elliptic} \quad &\text{if } B^2 - 4AC < 0; \\
\text{parabolic} \quad &\text{if } B^2 - 4AC = 0; \\
\text{hyperbolic} \quad &\text{if } B^2 - 4AC > 0.
\end{aligned} \qquad \textbf{(13.2)}$$

The partial differential equations are classified in the terms used for conic sections in analytic geometry because of the characteristics of the solutions. Table 13.1 lists an important example of each class of equation that will be studied in this chapter and defines the type of phenomenon described by the solution. Problem 13.1 asks for the classification of various differential equations.

TABLE 13.1 *Example partial differential equations*

Name	Equation	Classification	Solution
Laplace's equation	$\nabla^2 \Phi = 0$	Elliptic	Equilibrium
Heat equation	$\dfrac{\partial T}{\partial t} = k\nabla^2 T$	Parabolic	Diffusion
Wave equation	$\dfrac{\partial^2 u}{\partial t^2} = a^2\nabla^2 u$	Hyperbolic	Wave propagation

Notice that the equations in the table are all of second order in the spatial variables. First-order partial differential equations are not unimportant, but we restrict the discussion in this chapter to the more important second-order equations. These equations are solved using techniques similar to those developed earlier in the book for ordinary differential equations.

In this chapter, the equations will be introduced with one or two spatial dimensions. These partial differential equations are easily generalized to three dimensions and coordinate systems other than rectangular.

Partial differential equations arise from mathematical models of the physical world. The modeling usually begins with an observation about nature, such as Newton's law of gravitation or Fourier's law of heat conduction described in Chapter 11. Then, mathematical techniques are applied so that the resulting equations are transformed into a convenient form for solution.

Laplace's equation, introduced in Chapter 11, provides an excellent example of the combination of physics and mathematics leading to a mathematical model that can solve a variety of problems. Physical systems that satisfy Laplace's equation have a great deal in common, as we shall see later when the equation is studied in detail for specific problems. As a generalization, any vector field that is both irrotational and solenoidal at a point satisfies Laplace's equation since $\mathbf{F} = \nabla\Phi$ and $\nabla \cdot \mathbf{F} = 0$.

In three-dimensional rectangular coordinates, the potential $\Phi(x, y, z)$ associated with a system described by Laplace's equation satisfies

$$\nabla \cdot \nabla\Phi = \nabla^2\Phi = \frac{\partial^2\Phi}{\partial x^2} + \frac{\partial^2\Phi}{\partial y^2} + \frac{\partial^2\Phi}{\partial z^2} = 0. \tag{13.3}$$

This equation describes an equilibrium situation since there is no time dependence of the solutions. Many partial differential equations that involve time variations reduce to Laplace's equation when all changes of the solution in time have ceased.

A solution to the heat equation

$$\frac{\partial T}{\partial t} = k\nabla^2 T \tag{13.4}$$

is the temperature distribution in a region due to heat transferred by conduction. This equation is a special case of a *diffusion equation* in which heat is flowing or diffusing into a material. The constant k is called the *thermal diffusion coefficient* or *diffusivity* of the material. An important aspect of the heat equation under discussion is that heat is neither generated nor lost in the body. As t goes to infinity, the temperature distribution reaches an equilibrium state and the heat equation is equivalent to Laplace's equation.

The wave equation in the form

$$\frac{\partial^2 u}{\partial t^2} = a^2\nabla^2 u \tag{13.5}$$

could describe the motion of a perfectly flexible string or an electromagnetic wave propagating in a lossless medium. The wave equation for the electromagnetic field is derived as a special case of Maxwell's equations.

The analytical technique used to solve the partial differential equations in this chapter is called the *separation of variables* method. Although

---

this method cannot be universally applied, it is adequate for many of the elementary applications of partial differential equations in engineering and physics.

In the separation of variables approach, we attempt to convert a partial differential equation into several ordinary differential equations. Then, a solution is sought that solves the equations and satisfies any other conditions, such as initial or boundary conditions, for the problem.

□ EXAMPLE 13.1   *Separation of Variables*

Consider the partial differential equation

$$\frac{\partial^2 u(x,y)}{\partial x^2} = \frac{\partial u(x,y)}{\partial y}. \tag{13.6}$$

There are several different methods of solution but we choose the method of separation of variables.

In the separation of variables method, it is assumed that the solution can be written as a product

$$u(x,y) = X(x)Y(y)$$

so that

$$\frac{\partial u}{\partial x} = X'Y, \quad \frac{\partial u}{\partial y} = XY', \quad \frac{\partial^2 u}{\partial x^2} = X''Y.$$

Then, substituting the product solution $u(x,y) = X(x)Y(y)$ in Equation 13.6 leads to the equation

$$X''Y = XY'.$$

Dividing both sides by XY separates the variables as

$$\frac{X''(x)}{X(x)} = \frac{Y'(y)}{Y(y)}.$$

Since the left side is only a function of x and the right side is only a function of y, the conclusion is that both sides are independent of the variables and must be constant. To simplify the algebra, it is convenient to equate each side to a constant α^2 so that possible solutions of the differential equation must satisfy

$$\frac{X''}{X} = \frac{Y'}{Y} = \alpha^2.$$

The constant α^2 is called the *separation constant*. If $\alpha^2 = \lambda^2 > 0$, the separated equations become

$$X'' - \lambda^2 X = 0, \quad \text{and} \quad Y' - \lambda^2 Y = 0,$$

with solutions determined by the techniques presented in Chapter 5. The results are

$$X(x) = c_1 \cosh \lambda x + c_2 \sinh \lambda x, \quad \text{and} \quad Y(y) = c_3 e^{\lambda^2 y}.$$

The product solution for a positive separation constant is thus

$$u(x,y) = X(x)Y(y) = A_1 e^{\lambda^2 y} \cosh \lambda x + B_1 e^{\lambda^2 y} \sinh \lambda x \tag{13.7}$$

where $A_1 = c_1 c_3$ and $B_1 = c_2 c_3$.

Using the other choices for α^2, possible solutions for Equation 13.6 are

$$
\begin{aligned}
u(x,y) &= A_2 e^{-\lambda^2 y} \cos \lambda x + B_2 e^{-\lambda^2 y} \sin \lambda x \quad \text{if } \alpha^2 = (i\lambda)^2 < 0 \text{ (13.8)} \\
u(x,y) &= A_3 x + B_3 \quad \text{if } \alpha^2 = 0. \tag{13.9}
\end{aligned}
$$

It is easy to verify that the solutions given as Equation 13.7, Equation 13.8, and Equation 13.9 do satisfy the differential equation. Since the partial differential equation is linear, the sum of the solutions is also a solution. The separation constant and the constants in the solutions are determined by the boundary conditions for $u(x,y)$, as will be shown later in specific examples. $\square$

WHAT IF? Suppose you wish to solve the equation

$$
\frac{\partial^2 u(x,y)}{\partial x^2} + \frac{\partial u(x,y)}{\partial y} = y.
$$

You should verify that the separation of variables technique will not work. Thus, you need other methods to solve nonhomogeneous partial differential equations.

Boundary and Initial Conditions   As we have discussed in connection with ordinary differential equations, it is the task of mathematical analysis to determine under what conditions a solution to a partial differential equation exists and furthermore when a specific solution is unique. We will not pursue these details here, but several references listed in the Annotated Bibliography for this chapter include such discussions.

The unique solution corresponding to a particular problem is obtained by applying additional conditions arising from the initial or boundary values for the equation. A problem is called *well posed* if it has a unique, stable solution that satisfies the partial differential equation and the additional conditions. For a boundary value problem, the solution is *stable* if small changes in the boundary conditions cause appreciable changes in the solution only close to the boundary.

Suppose the solution to a partial differential condition is of the form $u(x,y,z,t)$. The condition

$$
u(x,y,z,t_0) = \phi(x,y,z),
$$

which specifies the solution at time $t = t_0$, is known as an *initial condition*. For the heat equation and the wave equation, an initial condition must be specified to apply the equation to a particular problem. Since Laplace's equation has no time dependence, only boundary values need be specified to uniquely solve the equation and satisfy the boundary condition.

The boundary values for a second-order partial differential equation are of three general types. In the first case, the value of $u(x,y,z,t)$ is defined on a boundary to yield a boundary condition called a *Dirichlet condition*. For certain problems, a condition can be defined for the normal derivative of u on the boundary. This condition, written as $\partial u / \partial n$, is known as a *Neumann condition*. For the heat equation, the Dirichlet condition specifies the temperature on the boundary. A Neumann condition defines the heat flux across a boundary. A linear combination of a Dirichlet and a Neumann condition is also possible for some problems.

Boundary Conditions for Electrostatics   The discussion in Chapter 11 showed that Laplace's equation describes the electrostatic potential. Consider a region of space in which we desire to solve for the potential, and further suppose that the region contains one or more closed conducting surfaces. It is known that specifying the potential on the closed surfaces in the region defines a potential problem with a unique, stable solution. This is the Dirichlet problem.

Similarly, specification of the electric field everywhere on the surfaces also defines a unique problem. The field is the normal derivative of the potential and its specification defines Neumann boundary conditions for Laplace's equations. In the case of Laplace's equation, only Dirichlet *or* Neumann boundary conditions may be specified on a closed surface. In general, specifying both conditions on the same surface would lead to an inconsistent result. However, *mixed boundary values* may be specified in which the potential is prescribed on some portions of the surfaces involved and the normal (field) is specified on the remaining portions.

Numerical Solutions   For ordinary differential equations, numerical techniques such as the Runge-Kutta method of solution described in Chapter 6, could in principle, be applied to any problem. When solving partial differential equations numerically, the classification of the equation as elliptic, parabolic or hyperbolic generally determines the numerical solution method that is usually applied. Examples and problems in this chapter will demonstrate the use of MATLAB to display analytical solutions. Numerical solutions are considered in the problems and in several of the books listed in the Annotated Bibliography for this chapter.

LAPLACE'S EQUATION

━━━━━━━━━━━━━━■━━━━━━━━━━━━━

As introduced in Chapter 11, Laplace's equation

$$\nabla^2 \Phi = 0$$

describes the gravitational potential, the electrostatic potential, or the velocity potential associated with steady-state fields. When the temperature is constant in time, the heat equation reduces to Laplace's equation.

In these applications, Laplace's equation is valid in regions of space where no sources or sinks are present. Poisson's equation

$$\nabla^2 \Phi(x, y, z) = \rho(x, y, z) \qquad (13.10)$$

applies if the vector field derived from the gradient of Φ has a source or sink at the point of interest. For the gravitational field, ρ is the density of a mass distribution.

Laplace's equation is an excellent example of the unifying power of mathematics, as physical systems with entirely different characteristics can be described by the same equation. The key to understanding this generality of Laplace's equation is recognizing that the equation describes equilibrium. Stated another way, equilibrium occurs when the potential energy is minimized, and Laplace's equation states this condition for systems with potential functions.

This section presents Laplace's equation in rectangular, cylindrical, and spherical coordinates. The emphasis is on problems for which analytical solutions are possible.

RECTANGULAR COORDINATES

Laplace's equation in 3D rectangular coordinates is

$$\frac{\partial^2 \Phi(x, y, z)}{\partial x^2} + \frac{\partial^2 \Phi(x, y, z)}{\partial y^2} + \frac{\partial^2 \Phi(x, y, z)}{\partial z^2} = 0. \qquad (13.11)$$

The most common approach to solving this equation is separation of variables. Letting $\Phi(x, y, z) = X(x)Y(y)Z(z)$ and dividing Equation 13.11 by the product yields

$$\frac{X''}{X} + \frac{Y''}{Y} + \frac{Z''}{Z} = 0. \qquad (13.12)$$

Each term is a function of only one variable, so each term must be a constant for the equation to hold for all arbitrary values of x, y, and z. As described in a previous section, each term can be set equal to the square of an arbitrary number. Thus, Equation 13.12 becomes three ordinary differential equations,

$$
\begin{aligned}
\frac{d^2 X(x)}{dx^2} - k_x^2 X(x) &= 0, \\
\frac{d^2 Y(y)}{dy^2} - k_y^2 Y(y) &= 0, \\
\frac{d^2 Z(z)}{dz^2} - k_z^2 Z(z) &= 0,
\end{aligned}
\qquad (13.13)
$$

where k_x, k_y, and k_z are the separation constants. Notice that the constants need not be independent for Equation 13.12 to be valid. The requirement is

$$k_x^2 + k_y^2 + k_z^2 = 0.$$

These constants will be determined by the boundary conditions for a particular problem.

As previously discussed, Dirichlet conditions specify the potential function on the boundary of the region of interest. Neumann conditions specify the flow rate through the boundary. Physically, the normal derivative at a boundary could represent quantities such as heat flow, current flow, or the electric field. If the normal derivative is zero over a region of the boundary, that region would be considered *insulated* so that no flow could occur across the boundary there. Considering the steady-state heat equation, no heat could be added or lost through a boundary for which the normal derivative is zero.

☐ EXAMPLE 13.2   *Laplace's Equation in a Rectangle*

This example presents a standard mathematical method used for solving Laplace's equation subject to Dirichlet boundary conditions. Figure 13.1 shows the region of interest. Although the region in this example is two-dimensional, the same solution method applies to three-dimensional regions. The potential function is assumed to be constant (zero) on three sides of the rectangle but can vary as function of x on the edge where $y = b$.

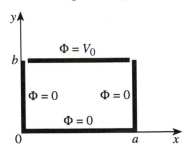

FIGURE 13.1 *Rectangular region for Laplace's equation*

Although the physical system consists of a 2D rectangular region in the figure, this region could represent a cross section of a long rectangular pipe in which the fields of interest are independent of the third dimension. Such problems arise in the study of rectangular waveguides, where the fields of interest are the electric and magnetic fields. In other physical systems that this model might represent, the potential could be temperature or voltage. In such cases, the top boundary must be insulated from the sides of the region. The textbook by Weinberger listed in the Annotated Bibliography for this chapter discusses the solution of Laplace's equation with piecewise-continuous boundary conditions. Essentially, the series solution we present here satisfies the equation and the boundary conditions except at points of discontinuity.

Laplace's equation to be solved is thus

$$\frac{\partial^2 \Phi(x,y)}{\partial x^2} + \frac{\partial^2 \Phi(x,y)}{\partial y^2} = 0 \tag{13.14}$$

for $0 < x < a$ and $0 < y < b$ subject to the conditions

$$\Phi(0, y) \;=\; 0, \qquad \text{for} \;\; 0 < y < b$$

$$\begin{aligned}
\Phi(a, y) &= 0, &&\text{for} \quad 0 < y < b \\
\Phi(x, 0) &= 0, &&\text{for} \quad 0 < x < a \\
\Phi(x, b) &= f(x), &&\text{for} \quad 0 < x < a.
\end{aligned}$$

Separating the solution to Laplace's equation as $\Phi(x, y) = X(x)Y(y)$ leads to the equations

$$\begin{aligned}
\frac{d^2 X(x)}{dx^2} - k_x^2 X(x) &= 0, \\
\frac{d^2 Y(y)}{dy^2} - k_y^2 Y(y) &= 0
\end{aligned}$$

with the requirement

$$k_x^2 + k_y^2 = 0. \tag{13.15}$$

Thus, $k_x^2 = -k_y^2$. One possibility is that $k_x = k_y = 0$. If both constants are zero, the solutions are linear in x and y and cannot satisfy the boundary conditions. Since $X(x)$ must have at least two zeros in the range $0 \le x \le a$, it appears that $k_x^2 < 0$ so that the separation constant is imaginary. Letting $k_x = ik$, $X(x)$ has the form

$$X(x) = A_1 \sin kx + B_1 \cos kx$$

and $Y(y)$ has the form

$$Y(y) = A_2 \sinh ky + B_2 \cosh ky.$$

The product solution then becomes

$$\Phi(x, y) = [A_1 \sin kx + B_1 \cos kx][A_2 \sinh ky + B_2 \cosh ky]. \tag{13.16}$$

We now apply the boundary conditions to determine the constants in Equation 13.16. The requirements that

$$\Phi(0, y) = 0 \quad \text{and} \quad \Phi(x, 0) = 0$$

imply that $B_1 = 0$ and $B_2 = 0$, respectively. This yields

$$\Phi(x, y) = A \sin kx \sinh ky$$

where A and k are constants. The boundary condition at $x = a$ requires that

$$0 = A \sin ka \sinh ky \quad 0 < y < b.$$

The potential function can be zero on the right boundary if and only if $ka = m\pi$ for m an integer. Therefore, Laplace's equation and three of the boundary conditions are satisfied by every function

$$\Phi(x, y) = A \sin\left(\frac{m\pi x}{a}\right) \sinh\left(\frac{m\pi y}{a}\right) \tag{13.17}$$

where $m = 1, 2, 3, \ldots$ is any integer.

Because Laplace's equation is a linear differential equation, the sum of the terms in Equation 13.17 satisfies the equation and yields a solution as

$$\Phi(x, y) = \sum_{m=1}^{\infty} A_m \sin\left(\frac{m\pi x}{a}\right) \sinh\left(\frac{m\pi y}{a}\right). \tag{13.18}$$

Imposing the last boundary condition at $y = b$ would require

$$f(x) = \sum_{m=1}^{\infty} A_m \sin\left(\frac{m\pi x}{a}\right) \sinh\left(\frac{m\pi b}{a}\right) \tag{13.19}$$

for $0 < x < a$. Considering the discussion in Chapter 8, this equation is seen to be the Fourier series expansion of the function $f(x)$ in the interval $[0, a]$.

As an example of the fourth boundary condition, let $f(x) = V_0$ on the top boundary, where V_0 is a constant. Equation 13.19 becomes

$$V_0 = \sum_{m=1}^{\infty} A_m \sin\left(\frac{m\pi x}{a}\right) \sinh\left(\frac{m\pi b}{a}\right). \tag{13.20}$$

Solving for the values A_m as in Problem 13.4 yields the potential as

$$\Phi(x, y) = \frac{4V_0}{\pi} \sum_{m-\text{odd}}^{\infty} \frac{\sin\left(\frac{m\pi x}{a}\right) \sinh\left(\frac{m\pi y}{a}\right)}{m \sinh\left(\frac{m\pi b}{a}\right)} \tag{13.21}$$

for $m = 1, 3, 5, \ldots$, since the sum is zero for m an even integer.

$\square$

MATLAB Solution   Although MATLAB does not have built-in functions that solve partial differential equations, MATLAB programs can be written to solve the equations numerically or to plot solutions derived analytically. In this text, numerical solutions for partial differential equations are not studied in detail. Several of the references listed in the Annotated Bibliography for this chapter treat numerical methods.

MATLAB solutions of ordinary differential equations are discussed in Chapter 5 and Chapter 6. Since the method of separation of variables involves finding solutions of ordinary differential equations, MATLAB can be helpful in solving partial differential equations by this method. Fourier series coefficients like those in Equation 13.19 can be computed with MATLAB commands, as shown in Example 13.3.

The professional version of MATLAB can be extended by the addition of a *Partial Differential Equation Toolbox*, available from The MathWorks. Applications areas for the Toolbox include structural mechanics, electromagnetics, and heat transfer.

□ EXAMPLE 13.3   ***MATLAB Plot of Laplace Equation Solution***
MATLAB can be used to help us visualize the solution to Laplace's equation for the problem presented in Example 13.2. The accompanying script uses *Symbolic Math Toolbox* commands to compute the Fourier series coefficients as defined in Equation 13.21.

As a specific case, let $\Phi(x, b) = 100$ at the top boundary of the rectangle in Figure 13.1. Then, the numerical solution is computed for the first 10 terms of the potential. Figure 13.2 shows the 3D plot of the potential.

Solution of the Laplace 2D equation

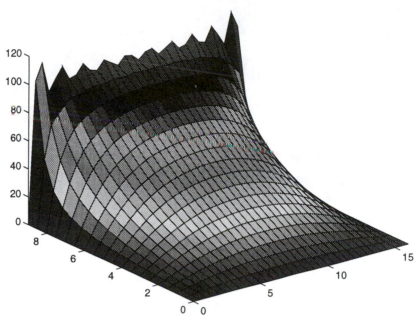

FIGURE 13.2 *Laplace's equation solution*

The solution for $\Phi(x,y)$ plotted in Figure 13.2 shows a smooth variation with x and y except near the boundary where $\Phi(x,b) = 100$. The result appears to be inaccurate near the boundary and to not match the potential. Using the Fourier series solution, values at the boundary $y = b$ exhibit the *Gibbs phenomenon* described in Chapter 8.

If a greater number of terms is taken in the series, the potential will appear smoother near the boundary, but some overshoot is always present. However, other methods of solution for Laplace's equation do not generally exhibit this behavior. Garcia's textbook listed in the Annotated Bibliography for this chapter presents methods of solution to Laplace's equation that do not involve Fourier series.

Figure 13.3 shows the equipotential contours of the potential calculated by the MATLAB program. The Fourier coefficients are computed using commands **int**, **symdiv**, and **numeric** from the *Symbolic Math Toolbox*.

At the points of discontinuity on the boundary, the series solution converges to one-half the value of the jump in potential. This is characteristic of any Fourier series solution. Away from the boundary, the series for the potential converges rapidly and 10 terms give a fairly accurate representation of the potential within the rectangular area. Several points on the curves are annotated with the value of the potential by the command **clabel**.

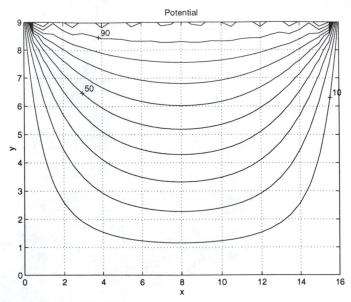

FIGURE 13.3 *Potential in 2D structure*

MATLAB Script _____

Example 13.3

```
% EX13_3.M Solve Laplace's equation in 2D with the conditions
%    bx1=f(0,y)=0; bx2=f(a,y)=0; by1=f(x,0)=0; by2=f(x,b)=100
%   The dimensions are a=16, b=9.
clear                        % Clear variables and
clf                          %   figures
%
a=16;                        % Dimensions of the region
b=9;
V0 = 100;                    % Constant potential at y=b (top)
%
% Solve for Fourier coefficients - symbolic solution
%
Nmax = 19;                   % Compute 1, ..., Nmax terms
sinhb='sinh(m*pi*b/a)';
for m=1:1:Nmax;
   fn='V0*sin(m*pi*x/a)';
   intx=int(fn,'x',0,'a'); % Symbolic integral
   am=symdiv(intx,'a/2');
   am=symdiv(am,sinhb);
   am=eval(am);
   A(m)=numeric(am);        % Convert to numbers
end
%
% Plot the results
```

```
%
dx=0.5;                         % Increments
dy=0.5;
[x,y]=meshgrid(0:dx:a,0:dy:b);
fxy=zeros(size(x));
for m=1:2:Nmax;                 % Odd components  1,3,5, ..., Nmax
   fxy=fxy + A(m)*sin(m*pi*x/a).*sinh(m*pi*y/a);
end
surf(x,y,fxy)                   % 3D view
axis([0 16 0 9 0 120])         % Set axes
title('Solution of the Laplace 2D equation')
pause
%
% Contour view
%
figure(2)                       % New figure
[Dx,Dy]=gradient(fxy,dx,dy);
v=0:10:100;                     % Define equipotentials
c=contour(x,y,fxy,v);
vc=[10 50 90];                  % Label a few equipotentials
clabel(c,vc);
hold on
xlabel('x')
ylabel('y')
title('Potential')
grid
```

□

WHAT IF?   Using the results of Example 13.3, it is possible to plot the gradient of the potential. If differences in potential represent voltage, the gradient represents the electric field. Plot the field and notice that the field lines are perpendicular to the lines of equal potential.

CYLINDRICAL
COORDINATES

Laplace's equation in cylindrical coordinates is

$$\frac{\partial^2 \Phi}{\partial r^2} + \frac{1}{r}\frac{\partial \Phi}{\partial r} + \frac{1}{r^2}\frac{\partial^2 \Phi}{\partial \theta^2} + \frac{\partial^2 \Phi}{\partial z^2} = 0, \tag{13.22}$$

where the potential is $\Phi(r, \theta, z)$. If Φ separates, so that

$$\Phi(r, \theta, z) = R(r)\Theta(\theta)Z(z),$$

Equation 13.22 can be written as

$$\frac{1}{R(r)}\frac{\partial^2 R(r)}{\partial r^2} + \frac{1}{rR(r)}\frac{\partial R(r)}{\partial r} + \frac{1}{r^2}\frac{1}{\Theta(\theta)}\frac{\partial^2 \Theta(\theta)}{\partial \theta^2} + \frac{1}{Z(z)}\frac{\partial^2 Z(z)}{\partial z^2} = 0.$$

Multiplying by r^2 leads to the relationship

$$\frac{r^2}{R(r)}\frac{\partial^2 R(r)}{\partial r^2} + \frac{r}{R(r)}\frac{\partial R(r)}{\partial r} + r^2\frac{1}{Z(z)}\frac{\partial^2 Z(z)}{\partial z^2} = -\frac{1}{\Theta(\theta)}\frac{\partial^2 \Theta(\theta)}{\partial \theta^2} = \nu^2,$$

since the term in Θ is a function of θ only and the equation can be valid for all values of r, θ, and z only if this term is constant. The separation constant is chosen as ν^2, which can be any real number.

Equation for $\Theta(\theta)$ The ordinary differential equation for Θ is

$$\frac{d^2\Theta(\theta)}{d\theta^2} + \nu^2\Theta(\theta) = 0. \tag{13.23}$$

If the separation constant is chosen so that $\nu = 0$, the solution is

$$\Theta = a + b\theta,$$

where a and b are arbitrary constants. In other cases,

$$\Theta(\theta) = c_1 e^{i\nu\theta} + c_2 e^{-i\nu\theta}. \tag{13.24}$$

As described in previous chapters, this result could be written in terms of sines and cosines if ν^2 is positive or the hyperbolic functions if $\nu^2 < 0$.

Solution for $Z(z)$ Solving for R and Z, Laplace's equation with the separated solution yields

$$\left[\frac{1}{R(r)}\frac{\partial^2 R(r)}{\partial r^2} + \frac{1}{r}\left(\frac{1}{R(r)}\frac{\partial R(r)}{\partial r}\right) - \frac{\nu^2}{r^2}\right] + \frac{1}{Z(z)}\frac{\partial^2 Z(z)}{\partial z^2} = 0.$$

Let the separation constant for the Z term be $-k_z^2$. Then, the separated potential is

$$\frac{d^2 Z(z)}{dz^2} + k_z^2 Z(z) = 0. \tag{13.25}$$

If $k_z^2 \neq 0$, the solutions for $Z(z)$ could be written in terms of sines and cosines or hyperbolic functions like the solutions for $\Theta(\theta)$ just discussed.

Solution for $R(r)$ After separating, the Θ and Z solutions to Laplace's equation and rearranging, the result for R becomes Bessel's differential equation

$$r^2\frac{d^2 R(r)}{dr^2} + r\frac{dR(r)}{dr} - \left(k_z^2 r^2 + \nu^2\right)R(r) = 0. \tag{13.26}$$

The solutions depend on the values of ν^2 and k_z^2 as summarized in Table 13.2. In the table, the primes refer to differentiation with respect to r for the function $R(r)$.

TABLE 13.2 *Solutions to Equation 13.26*

Constants	Equation	Solution
$k_z = \nu = 0$	$r^2 R'' + rR' = 0$	$a + b\log r$
$k_z = 0, \nu = n^2$	$r^2 R'' + rR' - n^2 R = 0$	$ar^n + br^{-n}$
$k_z = i\lambda$	$r^2 R'' + rR' + (\lambda^2 r^2 - n^2)R = 0$	$AJ_n(\lambda r) + BY_n(\lambda r)$

In case $k_z = \nu = 0$, there is no dependence of Φ on θ or z. When there is no z dependence and ν is an integer $n = 1, 2, \ldots$, the solution to Laplace's equation can be written

$$\Phi(r, \theta) = \frac{a_0}{2} + \sum_{n=1}^{\infty} (a_n r^n \cos n\theta + b_n r^n \sin n\theta) \qquad (13.27)$$

if we impose the condition that the solution be finite at $r = 0$. Thus, the term r^{-n} in the table is not considered. Equation 13.27 for the solution is written in the form of a Fourier series expansion since it is often possible to determine the coefficients in the solution by Fourier analysis. The coefficients in Equation 13.27 are found by the techniques described in Chapter 8.

If k_z is imaginary, the r dependent solution to Laplace's equation from Table 13.2 is

$$R(r) = AJ_n(\lambda r) + BY_n(\lambda r), \qquad (13.28)$$

where J_n is the Bessel function of the first kind and order n, while Y_n is the Bessel function of the second kind and order n. The properties of these functions are discussed in Chapter 6. If the potential is to be finite everywhere and the origin is included in the solution for $R(r)$, we conclude that $B = 0$ because $Y_n(\lambda r) \to -\infty$ as $r \to 0$.

Since Laplace's equation is linear, a superposition of the Bessel functions also solves the equation. To satisfy certain types of boundary conditions, it is convenient to express the radial part of the solution as a Fourier-Bessel series. Example 13.5 shows this approach to solving the wave equation.

SPHERICAL COORDINATES

Laplace's equation in spherical coordinates is

$$\frac{1}{r^2} \frac{\partial}{\partial r}\left(r^2 \frac{\partial u}{\partial r}\right) + \frac{1}{r^2 \sin\phi} \frac{\partial}{\partial \phi}\left(\sin\phi \frac{\partial u}{\partial \phi}\right) + \frac{1}{r^2 \sin^2 \phi} \frac{\partial^2 u}{\partial \theta^2} = 0, \qquad (13.29)$$

where the solution is $u(r, \theta, \phi)$ and ϕ is the polar angle ($0 \leq \phi \leq \pi$) measured down from the z-axis as defined in Chapter 10. The separated solution has the form

$$u(r, \theta, z) = R(r)\Theta(\theta)\Phi(\phi).$$

As a special case, we consider the boundary condition that a sphere S of radius a is kept at a fixed distribution of potential

$$u(a, \theta, \phi) = f(\phi)$$

and the potential is independent of θ so that $\partial^2 u / \partial \theta^2 = 0$. The origin of the spherical coordinate system is at the center of S and $f(\phi)$ is a known function. The problem is to find the potential at all points in space. Laplace's equation reduces to

$$\frac{\partial}{\partial r}\left(r^2 \frac{\partial u}{\partial r}\right) + \frac{1}{\sin\phi}\frac{\partial}{\partial\phi}\left(\sin\phi\frac{\partial u}{\partial\phi}\right) = 0. \qquad \text{(13.30)}$$

The solution now has the form $u(r, \phi) = R(r)\Phi(\phi)$ because the solution in space is independent of the angle θ. Problem 13.16 treats a more general case.

The solution to Equation 13.30 can be found by setting the term in r to a constant K, with the result

$$\frac{1}{R}\frac{\partial}{\partial r}\left(r^2\frac{\partial R}{\partial r}\right) = K,$$

$$\frac{1}{\sin\phi}\frac{\partial}{\partial\phi}\left(\sin\phi\frac{\partial\Phi}{\partial\phi}\right) + K\Phi = 0.$$

It is convenient to write the separation constant as $K = n(n+1)$ so that the equation for $R(r)$ becomes a *Cauchy-Euler equation*

$$r^2\frac{\partial^2 R}{\partial r^2} + 2r\frac{\partial R}{\partial r} - n(n+1)R = 0 \qquad \text{(13.31)}$$

with solutions

$$R(r) = Ar^n + Br^{-(n+1)}. \qquad \text{(13.32)}$$

Although n is an arbitrary constant because K was arbitrary, we only consider cases in which n is an integer. This assures that the solution for Φ and its derivative are continuous and finite in the interval $0 \le \phi \le \pi$.

The equation for Φ becomes

$$\frac{\partial}{\partial\phi}\left(\sin\phi\frac{\partial\Phi}{\partial\phi}\right) + [n(n+1)\sin\phi]\Phi = 0, \qquad \text{(13.33)}$$

which can be shown to be *Legendre's differential equation*, treated in Chapter 6. Making the substitution $w = \cos\phi$, $\sin^2\phi = 1 - w^2$ and

$$\frac{d}{d\phi} = \frac{d}{dw}\frac{dw}{d\phi} = -\sin\phi\frac{d}{dw}.$$

With these substitutions, Equation 13.33 becomes

$$(1 - w^2)\frac{\partial^2\Phi}{\partial w^2} - 2w\frac{\partial\Phi}{\partial w} + n(n+1)\Phi = 0. \qquad \text{(13.34)}$$

□ EXAMPLE 13.4   *Legendre Function Solution of Laplace Equation*

Consider a known distribution $f(\phi)$ defined on a sphere of radius a. This distribution could represent an electrical potential or a steady-state temperature distribution on the sphere. Thus, we assume that the solution in space is governed by Laplace's equation. From the circular symmetry of the problem, we seek a solution of the form $u(r, \theta, \phi) = R(r)\Phi(\phi)$. Furthermore, suppose we wish to find the solution at any point *within* the sphere.

The radial function $R(r)$ satisfies Equation 13.31 with solution

$$R(r) = Ar^n + \frac{B}{r^{(n+1)}}.$$

Since the solution must remain finite everywhere, the problem is divided by the spherical surface at $r = a$ into the interior problem for $r \leq a$ and the exterior problem for $r \geq a$. Inside the sphere, $B = 0$ since the origin is included. The exterior solution is presented in Problem 13.17.

Legendre's Equation 13.33 has solutions

$$\Phi(\phi) = P_n(\cos\phi) \qquad n = 0, 1, \ldots,$$

where each P_i is a Legendre polynomial, as defined in Chapter 6. The solutions to Laplace's equation then take the form

$$A_0 P_0, A_1 r P_1(\cos\phi), \ldots, A_n r^n P_n(\cos\phi).$$

To satisfy the boundary condition, we try an infinite series of the individual product solutions as

$$u(r, \phi) = \sum_{n=0}^{\infty} A_n r^n P_n(\cos\phi) \tag{13.35}$$

and form the *Fourier-Legendre series* for $f(\phi)$

$$u(a, \phi) = \sum_{n=0}^{\infty} A_n a^n P_n(\cos\phi) = f(\phi). \tag{13.36}$$

Since the Legendre polynomials are orthogonal over the interval as described in Chapter 6, the coefficients are computed as shown in Problem 13.7. These coefficients are found to be

$$A_n = \frac{2n+1}{2a^n} \int_0^\pi f(\phi) P_n(\cos\phi) \sin\phi \, d\phi. \tag{13.37}$$

Given the function $f(\phi)$, the problem is solved once these integrals are evaluated. Problem 13.8 gives a few specific examples for $f(\phi)$.

Convergence of the Fourier-Legendre series. It can be shown that if $f(\phi)$ and $f'(\phi)$ are piecewise continuous in the interval $0 \leq \phi \leq \pi$, the series of Equation 13.35 converges and is the solution to Laplace's equation within the sphere.

□

THE HEAT EQUATION

The heat equation

$$\frac{\partial T}{\partial t} = k \nabla^2 T \qquad \qquad \textbf{(13.38)}$$

describes the temperature distribution $T(x, y, z, t)$ due to heat transfer by conduction. For this ideal model, we assume that there are no heat sources in the volume of interest and the diffusivity k is constant. Specification of both boundary conditions and initial conditions is necessary to solve the heat equation.

An initial condition describes the initial temperature distribution, usually at $t = 0$. The specification could be

$$T(x, y, z, 0) = f(x, y, z),$$

where f is a function of the spatial variables only in the region of interest. Boundary conditions may take a variety of forms depending on the physical situation at the boundary. Some possible situations at the boundary include:

1. The temperature may be controlled.

2. The rate of heat flow may be controlled.

3. Heat may be transferred by convection.

The first condition is a Dirichlet condition in which T is specified around the boundary. The second is a Neumann condition specifying the gradient of the temperature at the boundaries. If the region is *insulated*, the heat flow is assumed to be zero. Mixed boundary conditions combining these two conditions are also possible. When heat is transferred by convection, the boundary condition involves the difference in temperature of the region involved and the surrounding medium. Obviously, there are many possible boundary conditions and variations involved in heat transfer. We restrict our discussion to simple cases for which the separation of variables technique can be applied.

Consider a heated rod that extends along the x-axis. If the lateral surface of the rod is insulated so that no heat can pass through it, then the temperature distribution will be a function only of x and t. If ends of the rod are held at zero temperature, the separation of variables method can be used to solve the problem. The following theorem summarizes the results.

The heated rod with zero endpoint temperatures

Given a rod of length L with ends fixed at zero temperature and initial temperature distribution $f(x)$ described by the relations

$$\frac{\partial T}{\partial t} = k\frac{\partial^2 T}{\partial x^2} \quad (0 < x < L, t > 0),$$
$$T(0, t) = T(L, t) = 0,$$
$$T(x, 0) = f(x),$$

the series solution is

$$T(x, t) = \sum_{n=1}^{\infty} b_n e^{-n^2 \pi^2 kt/L^2} \sin\left(\frac{n\pi x}{L}\right). \tag{13.39}$$

In the solution, the $\{b_n\}$ are the Fourier sine series coefficients

$$b_n = \frac{2}{L}\int_0^L f(x)\sin\left(\frac{n\pi x}{L}\right)\, dx, \quad n = 1, 2, \ldots.$$

In Problem 13.18, you are asked to prove the theorem. Problem 13.9 can be solved by a direct application of this theorem.

THE WAVE EQUATION

Wave motion can be modeled with a partial differential equation. Such motion occurs in various physical situations, including vibrating strings, vibrating membranes, and vibrating beams. For example, a drum head becomes a vibrating membrane when the drum is being played. Acoustic waves, electromagnetic waves, water waves, and waves traveling through solid media are other forms of wave motion. The wave equation to be studied is

$$\frac{\partial^2 u}{\partial t^2} = c^2 \nabla^2 u, \tag{13.40}$$

where $u(\mathbf{r}, t)$ describes the characteristics of the wave in space and in time and c^2 is a constant. The constant is written as c^2 to indicate that this constant is positive. In this section, we first introduce the wave equation in one-dimension.

WAVE EQUATION IN RECTANGULAR COORDINATES

Consider the *one-dimensional wave equation*

$$\frac{\partial^2 u}{\partial t^2} - c^2 \frac{\partial^2 u}{\partial x^2} = 0. \qquad (13.41)$$

This partial differential equation was solved by d'Alembert in 1747. By introducing the new coordinates

$$\zeta = x + ct,$$
$$\eta = x - ct$$

in Equation 13.41 and applying the chain rule, the terms in the wave equation become

$$\frac{\partial^2 u}{\partial t^2} = c^2 \left(\frac{\partial^2 u}{\partial \zeta^2} - 2 \frac{\partial^2 u}{\partial \zeta \partial \eta} + \frac{\partial^2 u}{\partial \eta^2} \right)$$

$$c^2 \frac{\partial^2 u}{\partial x^2} = c^2 \left(\frac{\partial^2 u}{\partial \zeta^2} + 2 \frac{\partial^2 u}{\partial \zeta \partial \eta} + \frac{\partial^2 u}{\partial \eta^2} \right)$$

Subtracting the terms, the wave equation becomes

$$-4c^2 \frac{\partial^2 u}{\partial \zeta \partial \eta} = 0.$$

Since $c \neq 0$, the partial derivative term is zero and a solution to the equation is

$$u(x, t) = p(\zeta) + q(\eta) = p(x + ct) + q(x - ct).$$

If time t is a parameter, the interpretation of the solution is that the first term represents a disturbance (wave) moving in the negative x direction and the second term represents a disturbance moving in the positive x direction. Both disturbances move with speed c in their respective directions.

As the solution to a specific physical problem, the solution of Equation 13.41 approximates the motion of a flexible uniform string stretched between fixed points along the x axis. If the string has a linear density ρ kilograms per meter and is stretched under a tension τ, the constant is defined by the relationship $c^2 = \tau/\rho$. Suppose that the ends of the string satisfy the endpoint conditions

$$u(0, t) = u(L, t) = 0.$$

Then, the motion of the string can be determined if the *initial position function* $u(x, 0)$ and the *initial velocity function* $\partial u(x, 0)/\partial t$ are known. The one-dimensional problem for this particular situation becomes

$$\frac{\partial^2 u}{\partial t^2} = c^2 \frac{\partial^2 u}{\partial x^2}, \qquad (0 < x < L, \ t > 0),$$
$$u(0, t) = u(L, t) = 0,$$
$$u(x, 0) = f(x), \qquad (0 < x < L),$$
$$\frac{\partial u(x, 0)}{\partial t} = g(x), \qquad (0 < x < L). \qquad (13.42)$$

It can be shown that the general solution to the wave equation with the given initial and boundary conditions is

$$u(x,t) = \frac{1}{2}[f(x+ct) + f(x-ct)] + \frac{1}{2c}\int_{x-ct}^{x+ct} g(s)\,ds. \qquad \textbf{(13.43)}$$

Although this solution to the wave equation is useful, solution methods using Fourier series apply to a wide range of partial differential equations. In keeping with the presentation of boundary value problems in Chapter 6 and partial differential equations in this chapter, we will apply the method of separation of variables to find solutions for the wave equation.

☐ EXAMPLE 13.5   *Wave Equation*

Given the one-dimensional wave equation and the boundary and initial conditions of Equation 13.42, we assume that $u(x,t) = X(x)T(t)$ so that

$$\frac{X''(x)}{X(x)} = \frac{T''(t)}{c^2 T(t)} = -\lambda^2.$$

The separation constant is chosen so that the solutions have sinusoidal variation. This is required because there is no loss term in the wave equation presented in this section and the initial energy defined by $f(x)$ or $g(x)$ is conserved. For example, initial oscillations will continue indefinitely.

Using the method of eigenvalues, the spatial solutions that satisfy the boundary conditions $X(0) = X(L) = 0$ are

$$X_n(x) = \sin \lambda_n x, \qquad n = 1, 2, \ldots,$$

where $\lambda_n^2 = (n\pi/L)^2$. The temporal solutions are then of the form

$$T_n(t) = a_n \cos \lambda_n ct + b_n \sin \lambda_n ct$$

where the constants a_n and b_n are arbitrary. Linear combinations of these solutions also satisfy the wave equation and the boundary conditions so the complete solution is

$$u(x,t) = \sum_{n=1}^{\infty} u_n(x,t) = \sum_{n=1}^{\infty} \sin \lambda_n x [a_n \cos \lambda_n ct + b_n \sin \lambda_n ct]. \qquad \textbf{(13.44)}$$

Using a Fourier series to satisfy the initial conditions yields the coefficients as

$$a_n = \frac{2}{L} \int_0^L f(x) \sin\left(\frac{n\pi x}{L}\right) dx \qquad \textbf{(13.45)}$$

$$b_n = \frac{2}{n\pi c} \int_0^L g(x) \sin\left(\frac{n\pi x}{L}\right) dx. \qquad \textbf{(13.46)}$$

This solution assumes that the series for $u(x,t)$ may be differentiated term-by-term to satisfy the condition defined by $g(x)$.

Each u_n in Equation 13.44 represents a harmonic motion having the frequency $\lambda_n c/2\pi = cn/2L$ hertz. The term for $n = 1$ is called the *fundamental mode* of oscillation.

☐

The wave equation in polar coordinates is

$$\frac{\partial^2 u}{\partial r^2} + \frac{1}{r}\frac{\partial u}{\partial r} + \frac{1}{r^2}\frac{\partial^2 u}{\partial \theta^2} = \frac{1}{c^2}\frac{\partial^2 u}{\partial t^2}. \qquad (13.47)$$

Assuming a solution in the form $u = R(r)\Theta(\theta)T(t)$, the wave equation will separate into the equations

$$T''(t) + \lambda^2 c^2 T(t) = 0, \qquad t > 0, \qquad (13.48)$$

$$\Theta'' + n^2\Theta = 0, \qquad -\pi < \theta < \pi, \qquad (13.49)$$

$$r^2 R'' + rR' + (\lambda^2 r^2 - n^2)R = 0, \qquad 0 < r < a. \qquad (13.50)$$

According to the discussion of Laplace's equation in cylindrical coordinates, the separation constant for the Θ function is chosen to be an integer so that the solution is periodic with period 2π. Also, the equation for $R(r)$ is Bessel's differential equation. Assuming that the origin is included in the problem, the solutions are $J_n(\lambda r)$.

Linear combinations of the product solutions must be formed that satisfy the boundary and initial conditions. Often the boundary conditions can be satisfied by a sum of Bessel functions. An example will be presented after Fourier-Bessel series are described.

Fourier-Bessel Series The Bessel functions of the first kind have a useful orthogonality property that allows an arbitrary function $f(r)$ defined in a circular region to be expanded in a Fourier-type series. Assume that $f(r)$ is piecewise continuous over the interval $0 \le r \le a$. Then, it can be shown that these Bessel functions are orthogonal over the interval for selected arguments.

■ THEOREM 13.2   ***Orthogonality of Bessel functions***

Let λ_{nm} for $m = 1, 2, \ldots$ be the sequence of values of λ that are the zeros of the Bessel function of order n so that $J_n(\lambda a) = 0$. Then, the Bessel functions $J_n(\lambda_{n1}a), J_n(\lambda_{n2}a), \ldots$ are orthogonal on the interval $0 \le r \le a$ with respect to the weight function r, and they have the relationship

$$\int_0^a rJ_n(\lambda_{nm}r)J_n(\lambda_{ns}r)\,dr = 0, \qquad m \ne s. \qquad (13.51)$$

The *Fourier-Bessel series* representing $f(r)$ is

$$f(r) = \sum_{m=1}^{\infty} A_m J_n(\lambda_{nm}r). \qquad (13.52)$$

When $\lambda_{ns} = \lambda_{nm}$, the value of the integral in Equation 13.51 can be computed as

$$\int_0^a rJ_n^2(\lambda_{nm}r)\,dr = \frac{a^2}{2}J_{n+1}^2(\lambda_{nm}a). \qquad (13.53)$$

The result allows the Fourier-Bessel coefficients to be computed using the orthogonality property of the Bessel functions.

Circular Membrane A common application of the wave equation is to determine the motion of circular membranes, such as drums, microphones, or pump diaphragms. We will assume that the solutions are radially symmetric and that the membrane is fixed along its boundary $r = a$. The wave equation is thus

$$\frac{\partial^2 u}{\partial r^2} + \frac{1}{r}\frac{\partial u}{\partial r} = \frac{1}{c^2}\frac{\partial^2 u}{\partial t^2}, \tag{13.54}$$

subject to the conditions

$$\begin{aligned} u(a,t) &= 0, & t \geq 0, \\ u(r,0) &= f(r), & 0 < r < a, \\ \frac{\partial u}{\partial t} &= g(r), & t = 0. \end{aligned}$$

Thus, $f(r)$ is the initial deflection of the membrane and $g(r)$ is the initial velocity.

Separating the wave equation as in Equation 13.50 without the θ dependence ($n = 0$), the radial solution that is bounded is $J_0(\lambda r)$. To satisfy the boundary condition at $r = a$, choose the argument of the Bessel function as

$$J_0(\lambda_m a) = 0,$$

where $\lambda_m = \lambda_{0m}/a$ and λ_{0m} are the zeros of J_0 for m any integer. The numbers were given in Chapter 6 and a few examples are

$$\lambda_{01} = 2.4048, \quad \lambda_{02} = 5.5201, \quad \lambda_{03} = 8.6537.$$

Thus, a complete solution to the radially symmetric wave equation is

$$u(r,t) = \sum_{m=1}^{\infty} J_0(\lambda_m r)[a_m \cos \lambda_m ct + b_m \sin \lambda_m ct] \tag{13.55}$$

The coefficients are found by matching the boundary conditions.

☐ EXAMPLE 13.6   ***Polar Wave Equation Solution***
Given the solution to the radially symmetric wave equation

$$u(r,t) = \sum_{m=1}^{\infty} J_0(\lambda_m r)[a_m \cos \lambda_m ct + b_m \sin \lambda_m ct] \tag{13.56}$$

with the conditions $u(r,0) = f(r)$ and $\dfrac{\partial u}{\partial t} = g(r)$, the coefficients are found as

$$\begin{aligned} a_n &= \int_0^a f(r)J_0(\lambda_m r)r\,dr/I_m, \\ b_n &= \int_0^a g(r)J_0(\lambda_m r)r\,dr/(\lambda_m c I_m), \end{aligned}$$

where

$$I_m = \int_0^a [J_0(\lambda_m r)]^2]r\, dr = \frac{a^2}{2} J_1^2(\lambda_m)$$

considering Theorem 13.2 and Equation 13.53.

□

TIME-VARYING ELECTROMAGNETIC FIELDS

When electromagnetic fields vary with time, the electric and magnetic fields are linked by Maxwell's equation

$$\nabla \times \mathbf{E} = -\frac{\partial \mathbf{B}}{\partial t}.$$

This equation is the differential form of *Faraday's law*, which states that the induced voltage around a closed path is equal to the negative time rate change of the magnetic flux through the closed path. The integral form of Faraday's law is thus

$$\oint_C \mathbf{E} \cdot d\mathbf{l} = \frac{\partial}{\partial t} \int \int_S \mathbf{B} \cdot d\mathbf{S}.$$

Stokes' theorem can be used to replace the line integral by a surface integral and to derive the differential relationship.

Either the differential or integral form of Maxwell's equation relating $\mathbf{E}$ and $\mathbf{B}$ states an important conclusion of electromagnetic theory: *a time-varying magnetic field is always accompanied by an electric field.*

An outstanding conclusion derived from Maxwell's equations is that the electric and magnetic fields satisfy a wave equation. At the time in the 1880s, wave propagation was predicted, but it was not realized for communication until a decade later. The wave equation for electromagnetic waves is obtained directly from Maxwell's equations that relate $\mathbf{E}$ and $\mathbf{B}$.

Considering Maxwell's equations discussed in Chapter 12, the wave equation in free space is obtained by computing

$$\nabla \times \nabla \times \mathbf{E} = -\mu_0 \frac{\partial}{\partial t} \nabla \times \mathbf{E} = \mu_0 \epsilon_0 \frac{\partial^2 \mathbf{E}}{\partial t^2} \qquad (13.57)$$

and expanding $\nabla \times \nabla \times \mathbf{E} = \nabla \nabla \cdot \mathbf{E} - \nabla^2 \mathbf{E}$. Since $\nabla \cdot \mathbf{E} = 0$ in free space, the wave equation for $\mathbf{E}$ is

$$\nabla^2 \mathbf{E} = \mu_0 \epsilon_0 \frac{\partial^2 \mathbf{E}}{\partial t^2} = \frac{1}{c^2} \frac{\partial^2 \mathbf{E}}{\partial t^2}, \qquad (13.58)$$

where $c = 3 \times 10^8$ meters per second, i.e., the speed of light in a vacuum. Once the electric field is defined in free space, the corresponding magnetic field is computed from Maxwell's equations. Problems 13.10 and 13.11 at the end of the chapter deal with the electromagnetic wave equation.

REINFORCEMENT EXERCISES AND EXPLORATION PROBLEMS

In these problems, do the computations by hand unless otherwise indicated, and then check those that yield numerical or symbolic results with MATLAB.

REINFORCEMENT EXERCISES

P13.1. Classification   Classify the partial differential equations in Table 13.1 using the categories in Equation 13.2.

P13.2. Classification of Tricomi equation   Classify the Tricomi equation

$$\frac{\partial^2 u(x,y)}{\partial x^2} + x\frac{\partial^2 u(x,y)}{\partial y^2} = 0$$

as elliptic, parabolic, or hyperbolic for various values of x.

P13.3. Derivation of a partial differential equation   A partial differential equation can be derived from a relationship between variables by differentiation. Find the partial differential equation satisfied by $z(x,y) = f(x) + g(y)$, where $f(x)$ and $g(y)$ are differentiable functions.

P13.4. Fourier series solution   Carry out the Fourier series analysis to show that the potential solution to Laplace's equation in Example 13.2 is Equation 13.21.

P13.5. Laplace's polar equation solution   Show that Equation 13.27 is the solution to Laplace's equation in polar coordinates.

P13.6. Laplace's equation in a disk   Determine the Fourier coefficients for the potential

$$f(\theta) = \begin{cases} 0, & -\pi < \theta < -\dfrac{\pi}{2}, \\ 1, & -\dfrac{\pi}{2} < \theta < \dfrac{\pi}{2}, \\ 0, & \dfrac{\pi}{2} < \theta < \pi \end{cases}$$

using the solution of Laplace's equation $\Theta(r,\theta)$ in polar coordinates given by Equation 13.27 for the potential in a disk of radius a. Also, determine the potential at the center of the disk and relate that value to the values around the edge of the disk.

P13.7. Orthogonal functions   Suppose that $g(x) = \sum_{n=0}^{\infty} a_n f_n(x)$, where the functions $f_0, f_1, \ldots$ are orthogonal on the interval $[a,b]$ with weight function $p(x)$. Show that

$$a_n = \frac{\int_a^b p(x)g(x)f_n(x)\,dx}{\int_a^b p(x)[f_n(x)]^2\,dx}.$$

P13.8. Laplace's spherical equation solution   Find the solution to Laplace's equation inside a sphere of radius a for the following distributions on the surface of the sphere:

    a. $f(\phi) = 1$.

    b. $f(\phi) = \cos\phi$.

P13.9. Heat conduction Assume that a rod of length L along the x-axis with endpoint temperatures held at zero degrees has the initial temperature distribution $f(x) = C$, where C is a constant temperature. Find the temperature distribution $T(x,t)$ if the thermal diffusivity of the material is k.

P13.10. Electromagnetic wave equation Considering the wave equation of Equation 13.58, write the equation for the x-component of $\mathbf{E}$ in rectangular coordinates if $\mathbf{E}$ is a vector field in $\mathbf{R}^3$.

P13.11. Electromagnetic fields Given that an electric field varies as

$$E_x = E_0 \sin \frac{\omega}{c}(z - ct),$$

find the corresponding $\mathbf{H}$ field.

P13.12. MATLAB solution to vibrating string Consider a string of length $L = 2$ meters with ends fixed so that $u(0,t) = u(L,t) = 0$. Determine the motion of the string if the initial deflection is

$$f(x) = \begin{cases} 0.1x, & 0 \le x \le 1, \\ 0.2 - 0.1x, & 1 < x \le 2. \end{cases}$$

Let $c = 100$ meters/second and plot the solution from $t = 0$ in increments of $\Delta t = .004$ seconds.

P13.13. MATLAB vibrating circular membrane Solve the wave equation for a vibrating circular membrane of radius a subject to the conditions

$$
\begin{aligned}
u(a, \theta, t) &= 0, \\
u(r, \theta, 0) &= f(r, \theta), \\
\frac{\partial u(r, \theta, 0)}{\partial t} &= 0
\end{aligned}
$$

in terms of the Fourier coefficients of the solution.

Plot the modes of oscillation

$$u_{mn}(r, \theta, t) = J_n\left(\frac{\lambda_{nm} r}{a}\right) \cos n\theta \cos\left(\frac{\lambda_{nm} ct}{a}\right)$$

and linear combinations of these functions. Select reasonable numerical values such as $c = 1$ meter per second and $a = 1$ meter. For example, try a combination such as

$$u(r, \theta, t) = J_1(\lambda_{21} r) \cos \theta \, \cos \lambda_{21} t + J_2(\lambda_{32} r) \cos 2\theta \, \cos \lambda_{32} t.$$

EXPLORATION PROBLEMS

P13.14. Derivation of partial differential equation Derive the partial differential equation satisfied by $z(x,y) = f(x + \alpha y) + f(x - \alpha y)$, where f is an arbitrary differentiable function.

P13.15. Second-order differential equations Compare the terms and their meaning between the partial differential equation

$$\rho \frac{\partial^2 u}{\partial t^2} + b \frac{\partial u}{\partial t} - K \nabla^2 u = F(t, x, \ldots)$$

where $u = u(t, x, y, \ldots)$ and the ordinary differential equation

$$m \frac{d^2 x(t)}{dt^2} + b \frac{dx(t)}{dt} + kx(t) = F(t).$$

Consider the coefficients as constants but describe the various types of solutions with different values of the constants.

P13.16. Laplace's equation in spherical coordinates Suppose that the term in Laplace's equation in spherical coordinates that involves θ separates as

$$\frac{\partial^2 \Theta}{\partial \theta^2} + m^2 \Theta = 0.$$

Show that Laplace's equation becomes the associated Legendre equation described in Chapter 6, and that the general solution can be written as

$$
u(r, \theta, \phi) \;=\; \sum_{n=0}^{\infty} \sum_{m=0}^{n} (a_m \cos m\phi + b_m \sin m\phi)
$$
$$
\times \left(c_n r^n + d_n r^{-(n+1)} \right) P_n^m(\cos \phi)
$$

with the assumption that terms containing Legendre functions of the second kind are ignored to avoid the singularity at $\phi = 0, \pi$.

P13.17. Laplace's solution outside a sphere Show that the solution for Laplace's equation outside the sphere in Example 13.4 can be written as

$$u(r, \phi) = \sum_{n=0}^{\infty} \frac{B_n}{r^{n+1}} P_n(\cos \phi),$$

where

$$B_n = \frac{2n+1}{2} R^{n+1} \int_0^{\pi} f(\phi) P_n(\cos \phi) \sin \phi \, d\phi.$$

P13.18. Heat conduction in a rod Prove Theorem 13.1 by assuming that the solution to the one-dimensional heat equation has the form $T(x, t) = X(x)F(t)$ and applying the separation of variables technique.

P13.19. Convergence of the series solution Show that the solution to the heat equation

$$T(x, t) = \sum_{n=1}^{\infty} b_n e^{-n^2 \pi^2 kt/L^2} \sin\left(\frac{n\pi x}{L}\right),$$

given as Equation 13.39, converges if the coefficients are bounded; that is $|b_n| < M$ for $n = 1, 2, \ldots$, where M is a constant.

Hint: Apply the Weierstrass M-test and the ratio test, as described in Chapter 6.

P13.20. One-dimensional wave equation solution Consider the wave equation with the conditions

$$
f(x) = \begin{cases} 2\dfrac{x}{L}, & 0 \le x \le \dfrac{L}{2}, \\[2mm] \left(2 - 2\dfrac{x}{L}\right), & \dfrac{L}{2} \le x \le L, \end{cases}
$$

and $\partial u(x, 0)/\partial t = 0$. Solve the problem using the Fourier series approach and then show that the result can be written as

$$u(x, t) = \frac{1}{2}[f^*(x - ct) + f^*(x + ct)],$$

where f^* is the odd periodic extension of $f(x)$.

Hint: Apply the equality $\sin \alpha \cos \beta = [\sin(\alpha - \beta) + \sin(\alpha + \beta)]/2$ to the Fourier series solution.

P13.21. Can we hear the shape of a drum? We know that vibrating structures can be analyzed by computing the characteristic frequencies. Can we infer the shape of a vibrating object by knowing its characteristic frequencies?

Hint: This question is discussed in the article " You can't hear the shape of a drum," in *American Scientist*, Volume 84, No.1, pp 46-55 January-February, 1996. An article by Mark Kac in *American Mathematical Monthly* 73 in 1966 asked the question "Can one hear the shape of a drum?"

P13.22. Finite differences and elements Two numerical methods for solving partial differential equations are the *finite difference method* and the *finite element method*. The finite difference solution requires that a differential equation with specified initial and boundary conditions be replaced by a difference equation. Historically, the finite element method was applied to large structures for which it was not possible to define a differential equation describing the entire structure.

Compare these methods in general terms and then describe how they differ in the following aspects:

    a. Treatment of boundary conditions.

    b. Use of physical insight in formulating a problem involving the method.

ANNOTATED BIBLIOGRAPHY

1. Carslaw, H. S., and J. C. Jaeger, *Conduction of Heat in Solids*, Oxford University Press, London, 1959. *A classic textbook that presents an extensive coverage of heat conduction problems.*

2. Garcia, Alejandro L., *Numerical Methods for Physics*, Prentice Hall, Englewood Cliffs, NJ, 1994. *The textbook presents MATLAB and FORTRAN solutions of partial differential equations.*

3. Powers, David L., *Boundary Value Problems*, Harcourt Brace College Publishers, Fort Worth, TX, 1987. *An introductory treatment of boundary value problems and partial differential equations.*

4. Weinberger, H. F., *A First Course in Partial Differential Equations*, Blaisdell Publishing Company, New York, 1965. *A good and fairly complete introduction to partial differential equations.*

5. Wilson, Howard B., and L. H. Turcotte, *Advanced Mathematics and Mechanics Applications using MATLAB*, CRC Press, Boca Raton, Fl., 1994. *The textbook has MATLAB solutions for partial differential equations. The finite element and finite difference methods are compared in an example.*

ANSWERS

P13.1. Classification For the equations in the table, $B = 0$ so that we have the following classifications for the partial differential equations in two independent variables:

    a. Laplace's equation, $A = C = 1$, elliptic;

    b. heat equation (x, t), $A = 0$ and $C = 1$, parabolic;

    c. wave equation (x, t), $A = 1$ and $C = -1$, hyperbolic;

using Equation 13.2.

P13.2. Classification of Tricomi equation The classifications of the Tricomi equation are

$$\frac{\partial^2 u}{\partial x^2} + x \frac{\partial^2 u}{\partial y^2} = 0 \quad \begin{cases} \text{elliptic} & \text{if } x > 0 \\ \text{parabolic} & \text{if } x = 0 \\ \text{hyperbolic} & \text{if } x < 0. \end{cases}$$

P13.3. Derivation of a partial differential equation To derive the partial differential equation satisfied by $z(x, y) = f(x) + g(y)$, differentiate with respect to x or y. Thus, $\partial z / \partial y = g'(y)$ so that $\partial^2 z / \partial x \partial y = 0$.

P13.4. Fourier series solution Using Equation 13.20 in Example 13.2

$$V_0 = \sum_{m=1}^{\infty} A_m \sin\left(\frac{m\pi x}{a}\right) \sinh\left(\frac{m\pi b}{a}\right),$$

multiply both sides by $\sin(n\pi x/a)$, integrate and use the orthogonality of the sinusoids. The coefficients are

$$A_m = \frac{V_0 \int_0^a \sin\left(\frac{m\pi x}{a}\right) dx}{a \sinh\left(\frac{m\pi b}{a}\right)} \quad m = 1, 2, 3, \ldots.$$

The integral has the value $-a[\cos m\pi - 1]/m\pi$ with the value $2a/n\pi$ for m odd and zero if m is even. Thus,

$$A_m = \frac{2V_0[1 - \cos m\pi]}{m\pi \sinh\left(\frac{m\pi b}{a}\right)} \quad m = 1, 3, 5 \ldots.$$

Substituting the coefficients in the equation for the potential, the result is Equation 13.21.

P13.5. Laplace's polar equation solution By direct substitution of Equation 13.27 into Laplace's equation in polar coordinates we have

$$\frac{\partial^2 \Phi}{\partial r^2} + \frac{1}{r}\frac{\partial \Phi}{\partial r} + \frac{1}{r^2}\frac{\partial^2 \Phi}{\partial \theta^2} = \sum_{1}^{\infty}(a_n r^n \cos n\theta + b_n r^n \sin n\theta)[n(n-1) + n - n^2] = 0.$$

P13.6. Laplace's equation in a disk Solving for the coefficients, the potential on the disk is

$$\Theta(r, \theta) = \frac{1}{2} + \sum_{n=1}^{\infty} \frac{2 \sin(n\pi/2)}{n\pi^2} \frac{r^n}{a^n} \cos n\theta.$$

The value at the center of the disk is

$$\Theta(0, \theta) = \frac{1}{2\pi} \int_{-\pi}^{\pi} f(\theta) \, d\theta,$$

which is the average value of the potential values around the edge of the disk.

P13.7. Orthogonal functions Given the expression

$$g(x) = \sum_{n=0}^{\infty} a_n f_n(x),$$

multiply both sides by $p(x)f_m(x)$ and integrate to yield

$$\int_a^b p(x)f_m(x)g(x)\,dx = \sum_{n=0}^{\infty} a_n \int_a^b p(x)f_m(x)f_n(x)\,dx. \qquad (13.59)$$

Since the functions $f_0, f_1, \ldots$ are orthogonal on the interval $[a, b]$ with weight function $p(x)$,

$$\int_a^b p(x)f_m(x)f_n(x)\,dx = 0, \quad \text{if} \quad m \neq n.$$

Thus, assuming $\int_a^b p(x)[f_n(x)]^2\,dx$ is not zero, only one term remains in Equation 13.59. Solving for a_n yields the nth coefficient, as stated in the problem.

P13.8. Laplace's spherical equation solution The solutions to Laplace's equation inside the sphere of radius a are the following:

    a. If $f(\phi) = 1$, $u = 1$.

    b. If $f(\phi) = \cos\phi = P_1$, $u(r, \phi) = r\cos\phi$ since only $A_1 = 1$ is nonzero in the Fourier-Legendre series.

P13.9. Heat conduction A rod of length L along the x-axis with endpoint temperatures held at zero degrees has the initial temperature distribution $f(x) = C$, the temperature distribution is

$$T(x, t) = \frac{4C}{L} \sum_{n=1}^{\infty} \frac{1}{2n-1} \sin[(2n-1)x] e^{-(2n-1)\pi^2 kt/L^2}.$$

P13.10. Electromagnetic wave equation Assuming that $\mathbf{E} = (E_x, E_y, E_z)$, the wave equation for E_x is

$$\nabla^2 E_x - \frac{1}{c^2}\frac{\partial^2 E_x}{\partial t^2} = \frac{\partial^2 E_x}{\partial x^2} + \frac{\partial^2 E_x}{\partial y^2} + \frac{\partial^2 E_x}{\partial z^2} - \frac{1}{c^2}\frac{\partial^2 E_x}{\partial t^2} = 0.$$

P13.11. Electromagnetic fields If the electric field varies as

$$E_x = E_0 \sin \frac{\omega}{c}(z - ct),$$

the corresponding $\mathbf{H}$ field is found from the equation $\nabla \times \mathbf{E} = -\mu_0 \partial\mathbf{H}/\partial t$. Taking the curl of $\mathbf{E}$ leads to the result

$$-\mu_0 \frac{\partial \mathbf{H}}{\partial t} = \frac{\partial E_x}{\partial z}\mathbf{j} = \frac{\omega E_0}{c}\cos\frac{\omega}{c}(z - ct)\,\mathbf{j}.$$

Integrating with respect to time shows that

$$H_y = \frac{E_0}{\mu_0 c} \sin \frac{\omega}{c}(z - ct).$$

Comment: MATLAB programs are included on the disk accompanying this textbook.

                    Chapter 13 ■ PARTIAL DIFFERENTIAL EQUATIONS

INDEX

MATLAB Commands

abs   64, 421
all   24
angle   421, 425
any   24
axis   38, 443
besselj   325
bessely   325
beta   557
bode   401
break   27, 305
cart2pol   485
cart2sph   485
charpoly   166
clabel   443, 615
clc   14
clear   9, 14
clf   17
clock   28
computer   7
cond   129
contour   443, 504
contour3   443
cross   63
cylinder   485
dbclear   28
dbcont   28
dblquad   558
dbstop   28
dbtype   28
del2   499
demo   5
det   110, 123, 127
determ   110, 114, 124
diag   102–103, 334
diary   8
diff   466, 468, 499
dir   8
disp   10, 14
dot   63
dsolve   196, 207, 216, 225,
    230–232, 314, 326
echo off   28
echo on   28
eig   166, 169

eigensys   166
ellipj   557
ellipke   557
else   26
erf   557, 593
etime   28
eval   10, 34, 37
exp   5, 39
expint   557
expm   173
eye   102–103, 334
ezplot   17, 196, 207, 443
factor   153
feval   297
fft   418, 421, 557
fft2   557
fftshift   421, 434
figure   443
fill   443
fill3   443
find   24, 476
fmin   476
fmins   476, 478
for   26, 28
format   48
format compact   6
format long   39, 49
fourier   389, 421, 557
fplot   17, 443
fprintf   10
funfun   231
funm   173
funtool   557
fzero   476
gamma   325, 557
ginput   19, 434
global   254
gradient   499, 504
grid   19, 443
gtext   207, 216
help   3, 22
hold   17
if   26, 273
ifft   421, 435
input   12
int   81, 173, 389, 557–558, 615